DEPOSITION AND GROWTH: LIMITS FOR MICROELECTRONICS

AMERICAN INSTITUTE OF PHYSICS
CONFERENCE PROCEEDINGS NO. **167**
NEW YORK 1988

AMERICAN VACUUM SOCIETY SERIES 4

SERIES EDITOR: GERALD LUCOVSKY
NORTH CAROLINA STATE UNIVERSITY

DEPOSITION AND GROWTH: LIMITS FOR MICROELECTRONICS

ANAHEIM, CA 1987

EDITOR: G.W. RUBLOFF
I B M—THOMAS J. WATSON
RESEARCH CENTER

L.C. Catalog Card No. 88-71432
ISBN 0-88318-367-6
DOE CONF-8711124

Printed in the United States of America.

Pstack SCIMON

CONTENTS

CHAPTER IV – INTERCONNECTIONS
LOW TEMPERATURE METALS AND INSULATORS

CHAPTER V – ALTERNATIVE GROWTH TECHNIQUES AT THE FOREFRONT I

CHAPTER VI – ALTERNATIVE GROWTH TECHNIQUES AT THE FOREFRONT II

PREFACE

A Topical Conference on **Deposition and Growth: Limits for Microelectronics** was held at the Anaheim Hilton Hotel in Anaheim, California on November 3–5, 1987, in conjunction with the American Vacuum Society's 34th National Symposium. It was co-sponsored by the Electronic Materials and Processing Division and the Thin Films Division of the AVS. The presence of a Topical Conference concurrent with the National Symposium continues the practice begun in 1985 with the Topical Conference on Frontiers in Electronic Materials and Processing (interrupted in 1986 due to the international conferences which ran concurrently with the National Symposium).

The presentation of a specific Topical Conference at the National Symposium provides the opportunity for a more thorough discussion of particular sets of issues. This opportunity is especially useful to the AVS and National Symposium audience because of the importance of electronic materials, processing, thin films and interfaces, and underlying surface chemistry and physics as a base for broad technological applications, most notably in microelectronics. Indeed, the title of this Symposium as "Limits for Microelectronics" conveys two important facts:

1. The further advancement of microelectronics technology is highly dependent on our understanding of, and ability to control, specific thin film materials and processes which are needed to fabricate devices and structures of ever higher performance and sophistication.
2. Advances in process control and understanding, as made by the AVS community, can contribute greatly to the progress of the technology.

This year the Program Committee decided to concentrate on a particular subset of the issues crucial to Electronic Materials and Processing and Thin Films Division interests—namely, Deposition and Growth. The motivation for focusing on a well-defined subset of issues was to exploit the specific strengths of the AVS and National Symposium for the benefit of the attendees. In particular, the Program Committee wanted to bring together insights spanning the range from the most fundamental surface physics and chemistry to the microelectronics technology applications of the materials and processes. Both evolutionary and revolution/alternative deposition and growth processes were highlighted. Finally, the Committee wished to present these issues within the broader context of process integration, whereby application of improved deposition and growth processes produce real value to microelectronics technology only when they can be successfully integrated into a complex *sequence* of processes needed for a particular device, circuit, or packaging technologies. This larger context raises key questions, such as thermal budgets of individual process steps, the sequence in which different processes are used, and the significant challenge which process integration represents for revolutionary (as opposed to evolutionary) materials and processes.

The Topical Conference program consisted of six half-day sessions, each of which is represented in this volume as a chapter (in order of appearance at the Topical Conference). Sessions were fomulated within specific areas of deposition and growth by including invited talks by internationally recognized experts in these fields, interspersed with contributed papers on current research appropriate to the Topical Conference theme.

The major themes are outlined in Chapter I: Overview, including discussions of the key role of thin film materials and processing in microelectronics, the importance of process integration, and the need for optimized thermal budgeting in process sequences, with examples in both Si and III-V technology. Chapter II: Semiconductor Growth focuses on epitaxial growth by CVD and MBE and on the fundamental surface science underlying these processes. Chapter III: Insulator Growth discusses both thermal and beam-enhanced insulator deposition and growth, together with insulator characterization and fundamental structure and chemistry. In Chapter IV: Interconnections: Low Temperature Metals and Insulators, low temperature deposition and growth of metals and insulators are discussed with particular emphasis on interconnection technology, where low thermal budgets are especially important so as not to deteriorate the active device structures already built. Finally, Chapters V and VI: Alternative Growth Techniques at the Forefront describe alternative (or revolutionary) approaches for achieving improved material characteristics through new processes.

The format of this Topical Conference was similar to that of the AVS National Symposium except for the manner in which the manuscripts are now being published. This volume continues the AVS series of topical conferences (started in 1985) and is being published through the American Institute of Physics as a sub-series of AIP Conference Proceedings. The manuscripts were reviewed in accordance with standards for the Journal of Vacuum Science and Technology and published in camera-ready format.

This Topical Conference is the product of the efforts of a considerable number of people. Its theme, approach, and content were developed mainly by the 1987 EMPD Program Committee: G. W. Rubloff (Chairman), R. Powell, J. C. Bean, T. M. Mayer, G. Margaritondo, S. D. Allen, and P. E. Luscher. Input also came from the 1987 Thin Films Division Chairman, R. Messier, and from the members of the 1987 EMPD Executive Committee: G. W. Rubloff, L. L. Kazmerski, P. Holloway, R. Powell, C. B. Duke, C. R. Aita, J. C. Bean, J. S. Harris, G. McGuire, D. J. Ehrlich, R. Ludeke, P. Petroff, and J. H. Weaver. The sessions were chaired by G. W. Rubloff, M. L. Green, J. Batey, P. Luscher, S. D. Allen, and T. M. Mayer, who deserve thanks for their coordination of the presentations, questions, and discussions.

For the success of the Topical Conference and these Proceedings, the largest piece of gratitude goes to the speakers and authors who contributed the technical content. On behalf of the Electronic Materials and Processing Division and the Thin Films Division, we wish to express our appreciation to the AVS Board of Directors for approving and supporting this Topical Conference, particularly AVS President Paul Holloway, Topical Conference Committee Chairman Fred Dylla, and 1987 National Symposium Chairman Bill Rogers and Co-Chairman Neal Shinn.

Finally, let me express my personal gratitude to: Gerry Lucovsky and Becky Gates of the Journal of Vacuum Science and Technology, for their enormous contribution in managing the review process for the manuscripts and assembling them for this Proceedings volume; Len Brillson, whose experience and success in publishing the 1985 Topical Conference Proceedings served as a valuable guide; and the EMPD 1987 Program Committee, whose stimulating ideas toward a theme and approach and whose cooperation in developing a detailed program made a successful contribution to the AVS community.

Gary W. Rubloff, Proceedings Editor

IBM Research Division
T. J. Watson Research Center
Yorktown Heights, NY

November 5, 1987

CHAPTER I
OVERVIEW

MATERIALS AND PROCESSING SCIENCE: LIMITS FOR MICROELECTRONICS

R. Rosenberg
IBM T.J. Watson Research Center, Yorktown Heights, N.Y., 10598

Abstract

The theme of this talk will be to illustrate examples of technologies that will drive materials and processing sciences to the limit and to describe some of the research being pursued to understand materials interactions which are pervasive to projected structure fabrication. It is to be expected that the future will see a progression to nanostructures where scaling laws will be tested and quantum transport will become more in evidence, to low temperature operation for tighter control and improved performance, to complex vertical profiles where 3D stacking and superlattices will produce denser packing and device flexibility, to faster communication links with optoelectronics, and to compatible packaging technologies. New low temperature processing techniques, such as epitaxy of silicon, PECVD of dielectrics, low temperature high pressure oxidation, silicon-germanium heterostructures, etc., must be combined with shallow metallurgies, new lithographic technologies, maskless patterning, rapid thermal processing (RTP) to produce needed profile control, reduce process incompatibilities and develop new device geometries. Materials interactions are of special consequence for chip substrates and illustrations of work in metal-ceramic and metal-polymer adhesion will be offered.

PROCESS INTEGRATION, INTERACTIONS & INTERDEPENDENCIES

A.K. Sinha
AT&T Bell Laboratories, Allentown, PA 18103

Abstract

A primary objective of this talk is to highlight certain recent trends in process integration that have been driven by application-specific VLSI products. Illustrations will be provided from our recently developed 0.9/0.75μm device technology[1], which incorporates several low-temperature, high rate operations designed to conserve thermal budget, preserve scaled doping profiles and ensure topological integrity of the structure. Examples include high-pressure steam oxidation of Si, rapid thermal processing of $TiSi_2$, low-temperature flow of a ternary glass, plasma-enhanced CVD of conformal oxide, and selective CVD of tungsten. The needs for improved process uniformity and rapid processing intervals now favor flexible single-wafer processors over older batch reactors. Integration of multiple single-wafer systems enables streamlined processing of critical segments of the process log with resulting quality improvements. Moreover, integration of a complex set of processes involves interactions among a multiplicity of variables. These interactions are best identified using Taguchi's orthogonal arrays, an example of which will be shown for the selective CVD tungsten process.

1. M.-L. Chen, C.-W. Leung, W.T. Cochran, R. Harney, A. Maury, & H.P.W. Hey, IEDM Technical Digest, p. 256 (1986).

LIMITED REACTION PROCESSING; FLEXIBLE THERMAL BUDGETING

J.F. Gibbons, S. Reynolds, C. Gronet, D. Vook, C. King and W. Opyd
Stanford University, Stanford, CA 94305

Abstract

Future VLSI devices will require process technology capable of providing abrupt thin layers of high quality semiconductors and dielectrics. By combining rapid thermal processing with chemical vapor deposition, many of these process requirements can be fulfilled. This new technology, Limited Reaction Processing, has been used to grow epitaxial layers of Si, SiGe, and III-V materials. Layer thickness control and interface abruptness are equivalent to MBE grown material.

With shrinking vertical device dimensions, thermal exposure of the substrate becomes critical if the required doping profiles are to be maintained. Conventional CVD technology subjects wafers to unnecessary thermal exposure. Limited Reaction Processing eliminates long ramp times and high temperature purge cycles, so that the wafer need only be hot during the actual deposition. These benefits derive from the reduced thermal mass of our single wafer process.

GALLIUM ARSENIDE INTEGRATED CIRCUIT TECHNOLOGIES

Richard Y. Koyama
TriQuint Semiconductor
Beaverton, Oregon, 97076

ABSTRACT

Advanced electronic technologies based on III-V compounds have been the subject of research for more than twenty years. The last five years have seen an explosive growth in the number of participants and in the development of this technology. Significant progress has been made in areas relating to crystal growth, micrometer and sub-micrometer lithography, dielectric deposition and etching, ion implantation and anneal, metals deposition and definition, etc. Many of these technologies are extensions of silicon technologies; others are unique to GaAs wafer fabrication. Significant progress has also been made in the understanding of the physics of FET-based GaAs devices. All of these advances have led to the commercial availability of GaAs IC technology ranging from SSI to LSI devices, as well as research demonstrations of circuits containing more than 10^5 devices. The commonly available depletion-mode GaAs MESFET IC technology is utilized in a variety of circuit topologies to fabricate digital, microwave, and analog devices ranging from a few active devices, to hundreds of devices. Although less common, enhancement/depletion-mode MESFET IC processes are also available; these are used in high speed, low power, digital LSI systems with thousands of gates. Advanced heterostructure technologies are also making rapid progress in research; these epitaxy-based devices promise even higher performance.

I. INTRODUCTION

"Gallium arsenide is the material of the future . . . . and will ALWAYS be the material of the future!"[1] This insightful statement is still partly true, but has become significantly less true recently. The commercial availability TODAY of small, to large scale gallium arsenide integrated circuits (GaAs ICs) announces that "the future is now!" The only reason that the statement is still partly true is due to the emerging realm of III-V optoelectronic integrated circuits, and III-V heterostructure integrated circuits; in this case, "the future" is still a few years out, but there is no question that it will happen.

The objective of this paper is to briefly review GaAs (and other III-V) IC fabrication technologies as they are practiced today. The prevalent commercial IC technologies are based on GaAs metal semiconductor field effect transistors (MESFETs); the primary emphasis of the paper will be on this technology. The MESFET provides the basis for circuits ranging from a few transistors for microwave amplifiers, to large scale integrated (LSI) circuits containing thousands of gates for GHz clock rate digital systems. In addition, other important III-V technologies, particularly epitaxial based technologies, will also be discussed. These epitaxial technologies in research today promise even higher performance for systems of the future.

II. GaAs SEMI-INSULATING SUBSTRATE WAFERS

No discussion of GaAs IC technology would be complete without at least a mention of the starting substrate. Most MESFET based GaAs ICs today are fabricated by ion implantation into semi-insulating (SI) GaAs wafers grown by the liquid encapsulated Czochralski (LEC) technique. The advent of commercial availability of this material in the 1980-81 time frame has had a major impact on the acceleration of the development of the technology.

The technology of GaAs crystal growth significantly trails the almost elegant technologies which have allowed the growth of near-perfect silicon crystals with diameters greater than 200 mm. In GaAs growth, impurities introduced by the starting materials and the growth environment, and native defects introduced by the growth procedure itself, limit the ultimate purity and perfection of the crystal. However, GaAs wafers are readily available today at the 3 in. diameter size, and 100 mm wafers are being qualified by IC manufacturers.

No one would argue with the fact that the development and availability of high quality GaAs wafers, and the development of GaAs IC technology have been synergistic. For today's GaAs IC technology, the starting substrates appear to be "adequate." Numerous GaAs IC manufacturers purchase SI GaAs wafers from equally numerous GaAs wafer vendors. Although the wafers can be qualified as "adequate," the technology of GaAs crystal growth will continue to evolve. On the other hand, GaAs IC fabrication technologies will also continue to evolve. If one had to determine the bottleneck in the ability to manufacture high quality, high yield GaAs ICs in today's technology, IC fabrication technologies appear to be a greater limiting factor than the availability and quality of starting SI GaAs wafers. As production volumes for GaAs ICs increase, especially with LSI devices on 100 mm and larger wafers, it is likely that issues relating to defect density, wafer flatness, reproducibility, etc. will be revisited. However, it is safe to say that today's level in both crystal growth and IC fabrication are far from the maturity of silicon technology, and both have much opportunity for improvement.

III. GaAs MESFET INTEGRATED CIRCUIT TECHNOLOGY

An early demonstration[2] of a GaAs MESFET IC was published in 1974. It was based on Schottky MESFETs fabricated on semi-insulating gallium arsenide substrates. Since then, and particularly in the early part of the 1980 decade, a large number of significant IC demonstrations have been made. As high quality undoped substrate material became available, rapid progress was also made in process development. These technological advances have spawned the foundation of a new branch of the semiconductor industry: GaAs ICs!

III.A. GaAs MESFETs and GaAs MESFET CIRCUITS

The depletion-mode (D-mode) GaAs MESFET is a fundamental building block for the implementation of GaAs IC technology. Depending on the culture and the wisdom of the process engineer, it can be fabricated in any number of ways: generically, these include planar,[3] gate recessed,[4] self-aligned,[5] etc. Again, depending on the culture and wisdom of the circuit designer, these FETs can be grouped and interconnected in innumerable ways to form circuit building blocks. In the microwave field, design has evolved around functional components such as amplifiers, switches, filters, mixers, phase shifters, etc.; components with performance from 2 to over 40 GHz have been realized.[6] In the digital world, design has evolved around specific circuit topologies that resulted in circuit families with different performance features: buffered FET logic (BFL),[2] Schottky diode FET logic (SDFL),[7] capacitor-diode FET logic (CDFL),[8] etc. These FETs and logic families make possible digital GaAs IC components ranging from SSI chips, to high clock rate (> 3 GHz) MSI devices (12 bit MUX/DMUX).[9]

More recently, the enhancement-mode (E-mode) GaAs MESFET has also made its commercial appearance.[10] In combination with the D-mode MESFET, it is possible to build devices of much higher circuit complexity. E/D (enhancement/depletion) technology, characterized by low power dissipation, has allowed the fabrication of very high performance LSI circuits (16 K RAM with more than 10^5 circuit elements).[11] In addition, the ability to integrate LSI digital logic capabilities with high performance microwave and analog devices has led to the fabrication and production of unique signal processing capabilities useful for many system functions (8 bit DAC).[12]

III.B. THE FABRICATION OF GaAs MESFETs

A fairly generic sequence for the fabrication of the basic E-mode and D-mode MESFET is illustrated in Fig. 1. Patterned photoresist is used to mask the E-mode, D-mode, and the n^+ contact implants (Figs. 1a, 1b, and 1c). Silicon ions accelerated in the range of 80 to 200 kev are used for creating the donor levels in the channel and contact regions of the FET; the implant dose will typically range from $3x10^{12}$ to $1x10^{13}$ cm^{-2}. The GaAs wafer is typically encapsulated in some dielectric (silicon oxide, silicon nitride, etc.) to minimize surface dissociation during the implant anneal; traditional furnace anneals up to 900 °C, for times up to 30 minutes, as well as rapid thermal anneal techniques are used. Following implant activation, ohmic contacts to the source and drain regions of the FET are defined (Fig. 1d). A thermally alloyed gold-germanium-nickel metallurgy is commonly used for this purpose. Gate metallization is next (Fig. 1e); in this case, titanium is the Schottky metal, palladium or platinum is used as the barrier metal, and gold is used as the high conductivity medium. At this point, the basic E-mode and D-mode FETs are complete. Subsequent process steps provide metal layers for the first level (Fig. 1f) and second level (Fig. 1g) of interconnects to complete fabrication of the IC. Typically, a dielec-

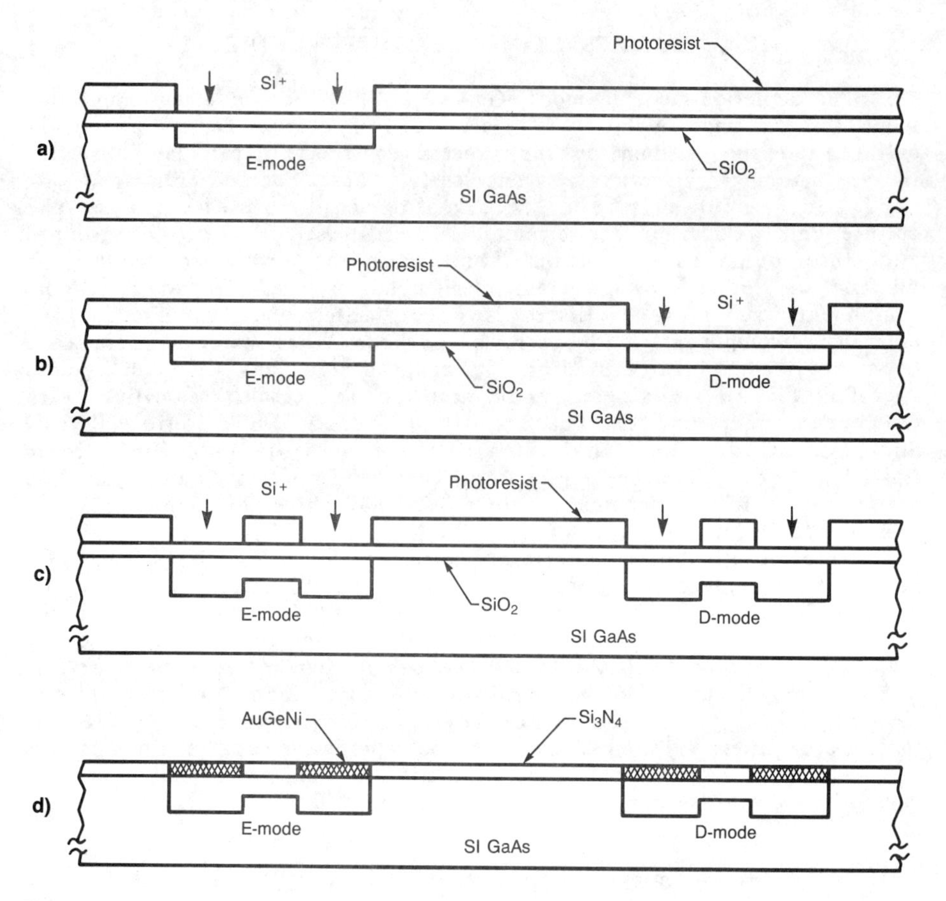

Fig. 1. Fabrication sequence for E- and D-mode GaAs MESFETs. Starting with a semi-insulating substrate, the wafer is subjected to: a) E-mode, b) D-mode, and c) n^+ implants. Then the metals are deposited: d) ohmic contacts, e) gate metal, f) first level interconnect, and finally, g) second level interconnect.

tric is used to isolate the first and second interconnect metals. In the example shown, the second level of interconnect has been executed with "airbridge" metal,[4] wherein the isolation is provided by air (see Fig. 2); the airbridge metal significantly reduces wiring capacitance to allow higher speed operation.[10] This basic E/D MESFET process takes approximately twelve mask levels and is capable of fabricating high performance LSI circuits.

In order to produce microwave or analog circuits, other passive circuit elements such as resistors, capacitors, and inductors may be required. Thin-film materials in the form of metal alloys, through cermets are deposited on the GaAs wafers to form moderately low (25 Ohm/sq.) to very high (megohms/sq.) resistors. These resistors are used for precision analog applications, and high resistance loads for memory cells, respectively. Capacitors are also fabricated as metal-insulator-metal structures formed between two available layers of metal, and the dielectric. Both the resistor and capacitor require

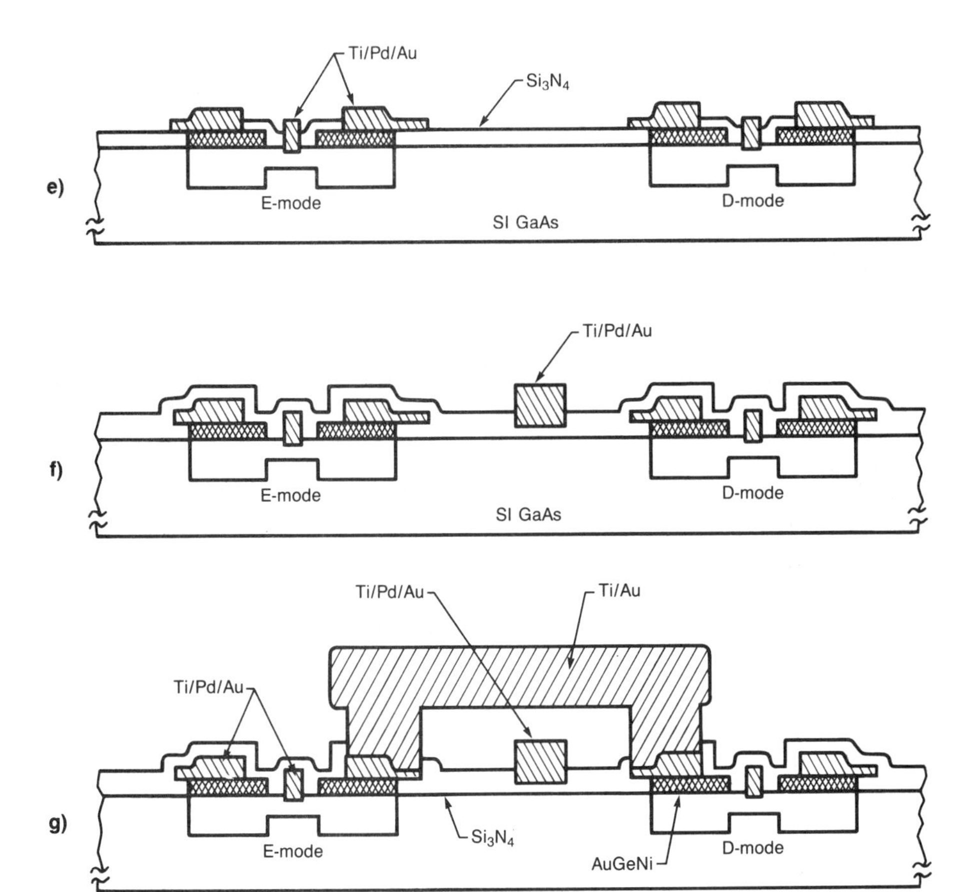

Fig. 1. (Continued)

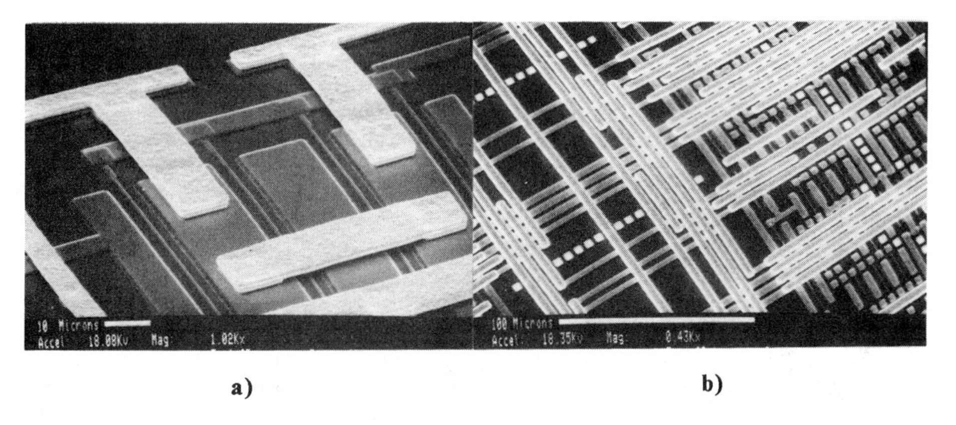

a) b)

Fig. 2. SEM photographs showing details of first and second level interconnect metals a) a close up showing short airbridge interconnects and 0.5 um gate metal fingers; b) a distant view of a gate array chip showing airbridges in the wiring channels (3.0 um wires and 2.0 um spaces).

additional mask layers for definition. The inductor is readily available in either the first or second level of interconnect, and does not require a separate mask layer.

III.C. CONSIDERATION OF DEPOSITION AND ETCHING PROCESSES

A necessary part of any semiconductor device fabrication process is the routine deposition and removal of dielectric insulators and metals; these films can be deposited and removed as continuous films, or selectively. Because GaAs lacks a counterpart to SiO_2 in the silicon technology, all insulating films are produced by deposition of dielectrics such as silicon oxide and silicon nitride; these are commonly deposited by CVD and plasma techniques. Metal films are deposited by e-beam evaporation or sputtering techniques.

From a philosophical point of view, a major consideration of the wafer process engineer is to first establish, and then to maintain a "good" interface between the GaAs surface and whatever is attached to it. The first time this happens is in the wafer preparation steps prior to the lithographic definition of the implant patterns. Typically, the wafer surface is cleaned to achieve good adhesion to the dielectric film which is deposited. This dielectric film (e.g., silicon oxide shown in Fig. 1a) now protects the GaAs surface from further abuse during processing. It is well known that even photoresist processing or de-ionized water,[13] can etch the GaAs surface. For this reason, the lithography which defines the implants, metals, etc. is ideally done on the surface of the dielectric, rather than on the semiconductor itself.

The presence of the dielectric is not only a blessing, but could also be a detriment. The very process of depositing the dielectric could damage the interface; this is particularly true of plasma/sputter deposition techniques where energetic particles can bombard the semiconductor surface prior to, and during the deposition. If this damage is excessive, or uncontrolled, it could have significant effect on the device performance. Although GaAs MESFETs are considered as "bulk" devices, most of the action takes place within the first 2000 A° or less, of the surface. Quite often, it is possible to remove much of such damage by a low temperature anneal; but generally it is preferable to avoid inducing such damage in the first place.

Another major consideration of this first dielectric layer is its ability to survive the high temperature (800 °C) implant anneal. Its major function is to prevent gross decomposition of the surface through dissociation. However, it is true that neither silicon oxide, nor silicon nitride are absolutely impervious to the diffusion of gallium and/or arsenic at high temperatures.[14] In fact, it is necessary that some surface diffusion or re-arrangement of atomic species is allowed in order to properly site and activate the implanted ions.[15] It is also possible to implant anneal GaAs wafers without an encapsulating dielectric: 1) the "proximity cap" utilizes a second wafer which is placed in face-to-face contact with the wafer to be annealed;[16] 2) an ambient which has a high vapor pressure of arsenic[17] is also suitable, and is called the "capless" anneal. Both of these techniques are more difficult to achieve on a consistent basis than the dielectric encapsulation technique.

Finally, the mismatch of the thermal expansion coefficients of GaAs and any dielectric (or metal) which is used, is a fundamental limitation. Such a mismatch leads to mechanical stress at the interface for any temperature other than the deposition temperature of the dielectric (typically 200 to 450 °C). At implant anneal temperatures, this stress often leads to failure of the film. Even at room temperature, because GaAs is piezoelectric, stress at the GaAs/metal and GaAs/dielectric interfaces can induce charges and effect surface depletion;[18] obviously, this can directly effect the device performance.

Metals deposition and definition techniques in GaAs IC technology are both similar and dissimilar to those used in silicon technology. Similarities would include deposition techniques such as e-beam evaporation and sputtering. Common metals used in GaAs IC fabrication include titanium, gold, palladium, platinum, germanium, and nickel; some refractory metals such as tungsten, molybdenum, tantalum, and their nitrides and silicides are also used. In addition, alloys such as nickel-chromium and the cermets are used as thin-film resistor material. Although etch-back and ion-milling are used for metals definition, lift-off is more commonly used in GaAs IC fabrication, especially for fine line definition.

Although direct lift-off utilizing only the properties of exposed and developed photoresist is possible and commonly used, "dielectrically-aided" lift-off is also used, especially for structures which must be defined through the protective dielectric. In these cases, the dielectrics are chosen more for their etching properties, rather than their dielectric characteristics; however, they must still retain their fundamental properties such as their optical constants, and their electrical and mechanical strength. The proper choice of composition and thickness of the dielectric allows a predictable etch profile of the lift-off structure. Such a proper choice alleviates residual metal fragments called "flags" or "wings," which are common to improper lift-off techniques. Obviously, it is important for the patterned photoresist to survive the dielectric etching environment; even if etch selectivity (dielectric versus photoresist) is not a problem, excessive particle bombardment or heating of the photoresist can destroy its properties, and prevent its subsequent removal. Lift-off is routinely used for ohmic metallization, and to define MESFET gate fingers as short as 0.5 micrometers (um) (see Fig. 2) utilizing conventional contact or 10:1 stepper lithography. Interconnect metallizations are defined by lift-off, etch back (ion-mill), and electroplating techniques.

A potential major limitation posed by today's lithographic and etching capabilities is that of "via" definition when the second interconnect metal must contact the first. A specific example (see Fig. 2b) is a 5.0 um metal pitch with 2.0x2.0 um vias (3.0 um metal width and 2.0 um space).[10] A 1.0x1.0 um via and a 1.0 um space, could reduce the pitch to 3.0 um (using a 2.0 um metal width) and thereby significantly decrease the wiring area. However, it is very difficult to achieve high yield on the lithographic definition and dielectric etching of a 1.0x1.0 um via with today's technology. Unless other circuit, device or processing approaches prevail, this problem is likely to remain a limitation.

Although advanced lithographic techniques such as far-uv, x-ray, and e-beam will benefit all semiconductor technologies of the future, most GaAs ICs are manufactured using commercially available optical lithography. For the commonly available 1.0 um MESFETs, as well as for the advanced 0.5 um MESFETs, either contact or 10:1 stepper machines are used. Similar to the needs of the silicon industry, stepper technology is important for the patterning of large volume devices; it is particularly important in order to obtain high yield of large area LSI devices on 3 in. and 100 mm GaAs wafers.

III.D. CONTROL OF THE PINCH-OFF VOLTAGE OF E/D MESFETs AND FACTORS WHICH INFLUENCE THAT CONTROL

A primary endeavor in the fabrication of the GaAs MESFET is to predictably set the pinch-off voltage (V_p) of the multitude of E- and D-mode FETs within a given IC chip in order to produce a functional device; this must be done from chip-to-chip across the wafer, from wafer-to-wafer down the boule, from boule-to-boule from a given vendor, from vendor-to-vendor (domestically, as well as across both the Atlantic and the Pacific), AND from day-to-day! This could be VERY DIFFICULT, but it can be done!

The process as described in Fig. 1 will result in E- and D-mode FETs with specific pinch-off voltages. In this case, fundamental factors which will affect the pinch-off voltages are the donor concentration (implant dose), the channel thickness (implant energy), dielectric thickness, and various substrate properties. A major substrate related factor is the degree to which the implanted ions are activated to form donors in the GaAs energy band structure. This factor will be strongly influenced by the anneal encapsulant which is used, the impurities and native defects which are present, and the degree of compensation of the semi-insulating substrate. V_p will also be affected by the quality of the Schottky interface of the gate. This style of fabrication places great faith in the process engineer's ability to control all of the relevant process parameters, and in the quality control engineer's ability to adequately screen the necessary properties of the starting SI GaAs substrates.

Another method which allows more explicit control of the pinch-off voltage is the gate recess process. Gate recess is executed after installation of the ohmic metallization (Fig. 1d), prior to the deposition of gate metallization (Fig. 1e). Typically, after removing the passivation dielectric from the gate region, the GaAs is etched by wet chemical means, or by dry plasma techniques[19] to the appropriate depth. The appropriate depth is achieved by monitoring the saturation current on a FET-like process monitor test structure. The uniformity across a given wafer due to such a recess technique is a major concern.[4, 19] Yet, it has been demonstrated in a production environment, and it is possible to simultaneously recesses both the E-mode and the D-mode devices of large area LSI circuits.[10, 20]

Discussed above are some of the "macroscopic" factors that affect the pinch-off voltage of the MESFETs. With respect to the fabrication process described in Fig. 1, there are other subtle, but known factors which can also influence the resulting V_p. These include: 1) enhanced diffusion of implanted ions,[15] defects, or impurities due to implantation damage or the mere presence of impurities or defects; 2) anisotropic etch rates on the two <011> axes of the (100) GaAs surface[21] dictate that all gates be appropriately aligned, especially for gate recess processes; 3) implant channeling due to the changing incidence angle[22] of the scanned ion implant beam can cause significant variations in the channel properties across the wafer; 4) other stress related effects; and, 5) there are also numerous observed, but unexplained results!

III.E. ADVANCED MESFET STRUCTURES

It is fair to say that the process sequence described above does not build the highest performance FET structure. There are many demonstrations of research processes, and some manufacturing processes which attain higher performance. Typically, the price that one pays for achieving higher performance is a more complex process and/or device structure, and/or processes which stretch the limits of the materials being used.

A rather "routine" method for achieving higher performance is to decrease the gate length of the FET. Although most commercial devices are produced with a nominal 1.0 um gate length, some manufacturers also produce devices with 0.5 um gate lengths. This is especially true for microwave devices where high frequency capability, and low noise figure are important. The use of electron beam exposure of resist for definition of sub-micron length gates is relatively common in research environments. Some have fabricated ICs with FET gates as short as 0.2 um.[23] Naturally, in such devices, it is also necessary to scale other dimensions of the FET structure in order to avoid short channel and other parasitic effects. Although 0.5 um structures are manufacturable using today's technologies, the manufacturability of 0.1 to 0.25 um features using e-beam and/or x-ray lithographic techniques in the near-term is an open question.

The self-alignment technique to achieve tighter dimensional control in processing is another method to fabricate higher performance devices. The primary benefit of this technique is to reduce or minimize the source resistance of the E-mode FET which has a normally closed channel. Figure 3 compares the recessed and self-aligned gate structures. In the self-aligned case, the surface depletion in the gate-source and gate-drain regions is minimized by increasing the channel doping concentration adjacent to the gate region; this is achieved by utilizing the gate metal itself, or a dummy gate feature, as the n^+ implantation mask in order to preserve the lightly doped channel region below the gate. A technique which utilizes the gate metal as the mask,[5,24] requires a refractory gate metallization scheme in order for the gate to survive the implantation anneal step. Other elegant, but complex schemes utilizing dummy gates have also been demonstrated.[25] Whereas some of the dummy gate self-alignment schemes can allow for gate recess adjustment of the pinch-off voltage, the refractory gate schemes cannot.

As sub-micron gate, self-aligned gate, and other advanced process-

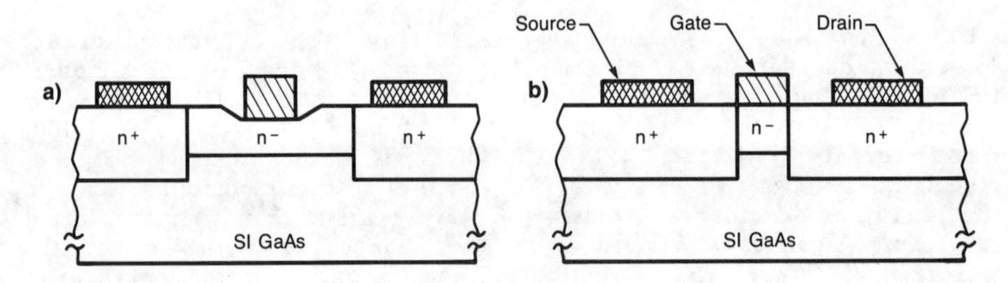

Fig. 3. Comparison of a) a recessed gate GaAs MESFET, and b) a self-aligned gate GaAs MESFET. The performance advantage gained by the self-aligned structure is due to the decreased source resistance achieved by the self-aligned n^+ implant.

es become routinely accepted (i.e., "manufacturable"), parasitic effects which heretofor were not problems (i.e., "previously ignored"), now become dominating factors, and begin to degrade the ultimate performance of the FET. One problem mentioned earlier is the short channel effect,[25] which is characterized by excessive sub-threshold current. Another rather insidious problem is that of "side/back-gating" wherein negative bias potentials on metallizations will induce potentials in the substrate and affect channel conduction in nearby devices;[26] this could be quite serious for tightly packed digital circuits. Another potential problem for tightly packed circuits is one of isolation. Although a "natural" benefit of the undoped LEC substrate is its inherent isolation, this problem may arise due to either substrate or process deficiencies; if necessary, some processes include an explicit damage implant to enhance the isolation of devices.[27] Some of these problems have been addressed recently by the use of channel confinement layers produced by the implantation of p-type regions.[28,29] These problems become greatly magnified as higher performance and larger circuits dictate even tighter device geometries.

IV. GaAs JFET INTEGRATED CIRCUIT TECHNOLOGY

In spite of the overwhelming dominance of GaAs MESFETs as the basis for III-V ICs, one other technology based on GaAs is the junction field effect transistor (JFET) technology. Whereas a Schottky barrier is used to modulate the conducting channel in a MESFET structure, a GaAs homojunction is used to control the channel in a JFET structure. In addition, in spite of the relatively low mobility of holes in GaAs (u_p~200 versus u_n~5000 $cm^2/V*s$), JFET technology has capitalized on the use of complementary[30] devices in their circuitry. This has allowed the fabrication of very low power random access memory (RAM) with good radiation characteristics.[30]

Fundamentally, the fabrication of JFET devices parallels that of the MESFET. It is also based on ion implantation into semi-insulating undoped GaAs wafers; in addition to silicon for the n-type implants, it also requires p-type implants such as berylium or magnesium. Because of the inherent buried channel structure of the JFET (see Fig. 4), device variations caused by dielectric thickness and surface

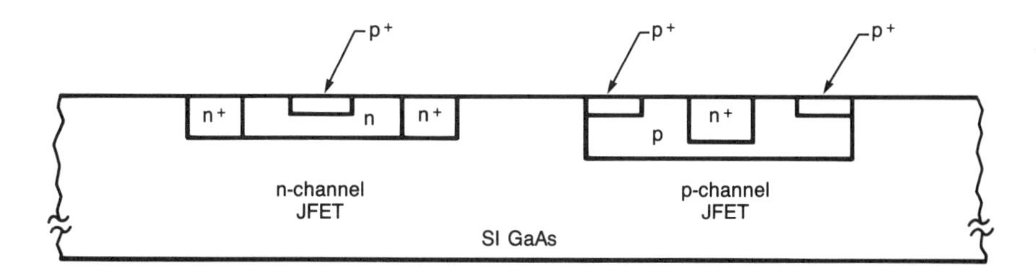

Fig. 4. Schematic diagram of the implanted cross-section of the complementary JFET device (reproduced with permission).

depletion are minimized. However, like the MESFET process, the JFET process suffers similar process control limitations.

V. HETEROSTRUCTURE TECHNOLOGIES

III-V heterostructure devices are the subject of a great deal of research (and development) activity today. These devices can generically be classified into two families: 1) FET-based devices; and 2) bipolar junction-based devices. They are all fabricated on epitaxially grown active semiconductor layers, and require some processing techniques which are different from GaAs MESFET fabrication.

The FET-based family include MESFET-like and metal-insulator-semiconductor(MIS)FET-like devices. The MESFET-like members have had the most investigation, and have demonstrated significant microwave and digital applications. These devices include the SDHT (selectively doped heterostructure transistor), the HEMT (high electron mobility transistor), the TEGFET (two-dimensional electron gas FET), and the MODFET (modulation doped FET); fundamentally, these names refer to the same structure, and are based on doped AlGaAs (SDHT is the favored name in this paper). These devices have demonstrated very high performance E/D LSI circuits including a 4K RAM.[31] The members of the MISFET branch have a similar structure, but are based on undoped AlGaAs as the insulator for the gate structure; some have been able to demonstrate complementary devices.[32]

There are two members of the heterojunction bipolar family: 1) the "collector down"[33] heterojunction emitter coupled logic (HECL); and 2) the "emitter down" heterojunction integrated injection logic (HI^2L).[34] A major advantage of the heterojunction bipolar approach is the fact that the threshold or turn-on voltage is no longer a process parameter, but a material parameter dependent on the semiconductor band gap energy. HI^2L offers a moderately simple epitaxial structure and circuit approach; it has demonstrated its capability to fabricate LSI devices.[34] The HECL technique appears to offer the highest performance due to its high current drive capability. Application to high speed/resolution analog-to-digital conversion appears to be a primary goal.[35] However, it does require a fairly complex epitaxial structure, along with a very complex fabrication procedure.

V.A. HETEROSTRUCTURE EPITAXIAL MATERIALS

All of these heterostructure devices are based on the use of III-V epitaxial materials, primarily combinations of GaAs and AlGaAs layers. Molecular beam epitaxy (MBE) has been the process of choice for fabricating the starting material for these structures. This technique has successfully demonstrated every device mentioned above, albeit in research environments. More recently, metal-organic chemical vapor deposition (MOCVD) has been making strides as a possible "high through-put" technique for providing commercial volumes of epitaxial wafers. This technique has made demonstrations for fabricating some of the devices mentioned above.

The major issues regarding the use of epitaxial material for the fabrication of heterostructure devices are the quality and commercial availability. MBE and MOCVD appear to share the same quality problems, but the severity of the different problems differ in each case. Surface morphology, defect density, thickness/doping uniformity, and reproducibility are qualities about which every epitaxial crystal grower is concerned. In order for heterostructure technology to be viable, it will be necessary for high quality 3 in. (100 mm preferred) diameter wafers to be available at "reasonable" cost. At this time it is possible to purchase epitaxial materials for these devices from either MBE or MOCVD wafer vendors.

A very interesting development which has made significant progress during the last eighteen months is the epitaxial growth of III-V compounds on other than III-V substrates. Of particular interest is the growth of GaAs on silicon substrates; naturally, this would eventually include the growth of heterostructure epitaxial layers on silicon as well. The potential availability of high quality GaAs on 100 mm (or larger) silicon substrates could have a significant impact on the large volume manufacturing of today's MESFET GaAs LSICs. Assuming that they would be regularly available and "interchangeable" with bulk undoped LEC substrates, the major advantage of such substrates would be large diameter, high mechanical strength, good thermal conductivity, and hopefully, low cost. Of course, they also offer the real possibility of co-integrating III-V electronics (and opto-electronic circuits) with silicon integrated circuits. The capability of fabricating GaAs MESFET LSI circuits on silicon has been demonstrated with a 1K RAM.[36] With respect to heterostructure devices, since the epitaxial layer is an accepted necessity, there would be no foreseeable penalty by using a silicon substrate. However, it may be appropriate to reserve judgment on this application until it has been "commercially" demonstrated. A major technical hurdle which must be overcome is the elimination of the large number of defects which are initiated by the lattice mismatch at the interface. Nevertheless, this is a very exciting development with the potential of high return.

V.B. III-V HETEROSTRUCTURE FABRICATION

Shown in Fig. 5 is a schematic representation of a simple SDHT device structure. In this case, the epitaxial layers consist of (starting at the SI GaAs substrate): 1) high purity undoped GaAs, 2) doped

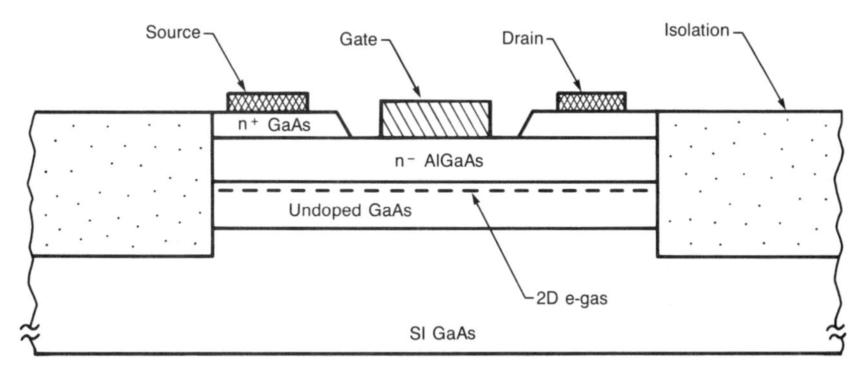

Fig. 5. Schematic diagram of a simple SDHT structure. The two-dimensional electron gas resides on the GaAs side of the lower GaAs/AlGaAs interface.

AlGaAs, and 3) a cap layer of doped GaAs. The discontinuity of the bandgap energies between layers 1 and 2, allow the formation of a narrow potential well on the GaAs side of the interface. Electrons from the doped AlGaAs layer are collected in this potential well to form a very thin conducting sheet (in fact, a two-dimensional electron gas). Because these electrons are constrained to move in a region which is free of ionized donors or impurities, the electron mobility can reach very high values, particularly at low temperatures.

As grown, the active layer is continuous across the entire surface of the wafer; therefore, one major process step is to provide for isolation of the circuit elements. For discrete devices or small integrated circuits, this can be done by removing unused regions of the field by etching. More typically, this is done by damaging the field region by implantation to re-establish semi-insulating material. The doped GaAs cap layer is used primarily to ease the ability to make a good ohmic contact to the source and drain regions of the MESFET-like structure; it must be removed from the channel region. The gate metal is applied to the AlGaAs layer to benefit from the higher barrier height, which allows for a higher voltage swing in the circuit. If the thickness and doping concentration of the AlGaAs layer is chosen properly, the pinch-off voltage can be pre-determined. Some adjustment can be made if necessary or desirable, by recessing the gate into the AlGaAs. The removal of the doped GaAs layer above the channel is done by dry etching, utilizing the differential etch stop provided by the AlGaAs layer. Because the conducting layer is very thin, it is important to avoid any damage to the semiconductor surface during any etching procedure prior to the deposition of the gate metal. Any excessive damage could strongly affect the conductivity of the electron gas layer.

Shown in Fig. 6 is a schematic diagram of the HECL collector down heterojunction bipolar transistor.[33] This complex active layer structure is fabricated by a combination of MBE grown epitaxial layers and ion implanted regions. A major process concern of this structure is to activate the implanted p-type region without damaging the properties of the grown epitaxial layers; typically, rapid thermal anneal techniques are used. In spite of the structural complexity and the great processing difficulties of this technology, significant

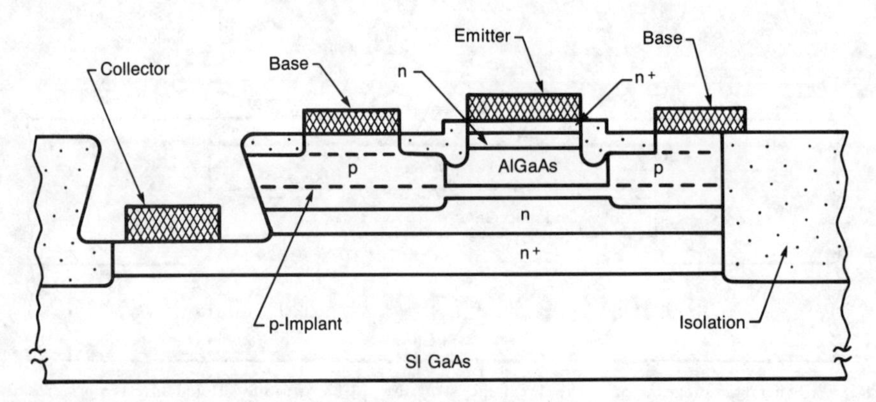

Fig. 6. Schematic diagram of the collector down HBT (reproduced with permission).

progress has been made, and impressive demonstrations of circuits have been achieved.[35]

VII. CONCLUDING REMARKS

It is impossible for this brief review to do justice to the significant research and development progress in III-V IC technologies. With the exception of only brief mention, the important areas of optoelectronic ICs and co-integration of GaAs and Si technologies has been neglected. In addition, more advanced technologies utilizing "bandgap engineered" materials and "resonant tunneling"[37] devices have not been discussed. However, it is clear that III-V IC technologies, particularly GaAs MESFET technology, have made tremendous strides in the last few years. There is no question that the material technologies and the process technologies must continue development to achieve the maturity of today's silicon based IC technologies. Nevertheless, depletion mode and enhancement/depletion mode GaAs MESFET circuits are commercially available today, and devices are being installed in commercial communication, instrumentation, and computing systems. Similarly, the large volume insertion of GaAs digital and microwave ICs into military applications will occur in the near future. The heterostructure technologies represent the leading edge of high performance electronic and optoelectronic devices, and will make their commercial appearance in the next few years.

ACKNOWLEDGMENTS

The author gratefully acknowledges the encouragement and support provided by TriQuint Semiconductor, and the members of TriQuint who have contributed to the preparation of this paper. In addition, the author's acknowledgments are also due to his many colleagues in the GaAs IC field. Those few citations which have been made are only a sampling of the significant body of work representing this technology, and should not be construed as complete.

REFERENCES

[1] The author was unable to find, and regrets his inability to cite, the proper ORIGINAL source of this quotation.

[2] R.L. van Tuyl and C.A. Liechti, IEEE J. Sol.-St. Circ. SC-9, 269 (1974).

[3] B.M. Welch, Y.D. Shen, R. Zucca, R.C. Eden, and S.I. Long, IEEE Trans. Elect. Dev. ED-27(6), 1116 (1980).

[4] A. Rode, A. McCamant, G. McCormack, and B. Vetanen, 1982 IEEE Int'l. Elect. Dev. Mtg. Tech. Dig., pg. 162.

[5] N. Yokoyama, T. Mimura, M. Fukuta, and H. Ishikawa, 1981 IEEE Int'l. Sol.-St. Circ. Conf. Tech. Dig., pg. 218.

[6] R.A. Pucel, ed., Monolithic Microwave Integrated Circuits (IEEE Press, New York, 1985).

[7] R.C. Eden, B. Welch, R. Zucca, and S.I. Long, IEEE J. Sol.-St. Circ. SC-14, 221 (1979).

[8] R.C. Eden, 1984 IEEE GaAs IC Symp. Tech. Dig., pg. 11.

[9] M.A. McDonald and G. McCormack, 1986 IEEE GaAs IC Symp. Tech. Dig., pg. 229.

[10] W.H. Davenport, 1986 IEEE GaAs IC Symp. Tech. Dig., pg. 19.

[11] Y. Ishii, M. Ino, M. Idda, M. Hirayama, and M. Ohmori, 1984 IEEE GaAs IC Symp. Tech. Dig., pg. 121.

[12] F.G. Weiss, 1986 IEEE GaAs IC Symp. Tech. Dig., pg. 217.

[13] C.E. Weitzel and T.H. Miers, IEEE Elect. Dev. Lett. EDL-2, 35 (1981).

[14] M. Helix, K. Vaidyanathan, and B. Streetman, IEEE J. Sol.-St. Circ. SC-13, 426 (1978).

[15] There is a large body of literature which discusses the interaction between GaAs, impurities, defects, encapsulants, implanted ions, gases, etc. (1978 to 1984), which provides a wealth of knowledge on these effects.

[16] R.P. Mandal and W.R. Scoble, in GaAs and Related Compounds 1978, C.M. Wolfe, ed. (The Inst. of Physics, Bristol, 1979), pg. 462.

[17] R.C. Clarke and G.W. Eldridge, IEEE Trans. Elect. Dev. ED-31, 1077 (1984).

[18] P.M. Asbeck, C.P. Lee, and M.F. Chang, IEEE Trans. Elect. Dev. ED-31, 1377 (1984).

[19] F.J. Ryan, M.F. Chang, R.P. Vahrenkamp, D.A. Williams, W.P. Fleming, and C.G. Kirkpatrick, 1985 IEEE GaAs IC Symp. Tech. Dig., pg. 45.

[20] A. Rode, T. Flegal, and G. LaRue, 1983 IEEE GaAs IC Symp. Tech. Dig., pg. 178.

[21] Y. Tarui, Y. Komiya, and Y. Harada, J. Electrochem. Soc. 118, 118 (1971).

[22] J. Kasahara, H. Sakurai, T. Suzuki, M. Arai, and N. Watanabe, 1985 IEEE GaAs IC Symp. Tech. Dig., pg. 37.

[23] J.F. Jensen, L.G. Salmon, D.S. Deakin, and M.J. Delaney, 1986 Int'l. Elect. Dev. Mtg. Tech Dig., pg. 476.

[24] K. Ueno, T. Furutsuka, H. Toyoshima, and H. Ishikawa, 1981 IEEE Int'l. Sol.-St. Circ. Conf. Tech. Dig., pg. 218.

[25] K. Yamasaki, K. Asai, T. Mizutani, and K. Kurumada, Elect. Lett. 18, 119 (1982).

[26] C.P. Lee, S.J. Lee, and B.M Welch, IEEE Elect. Dev. Lett. EDL-3, 97 (1982).

[27] D.C. D'Avanzo, IEEE Trans. Elect. Dev. ED-29, 1051 (1982).

[28] K. Yamasaki, N. Kato, and M. Hirayama, IEEE Trans. Elect. Dev. ED-32, 2420 (1985).

[29] P. Canfield, D. Allstot, J. Medinger, L. Forbes, A. McCamant, W. Vetanen, B. Odekirk, E. Finchem, and R. Gleason, 1987 IEEE GaAs IC Symp. Tech. Dig., pg. 163.

[30] J.K. Notthoff, R.B. Krien, J.S. Stephens, G.L. Troeger, C.H. Vogelsang, and C.H. Hyun, 1987 IEEE GaAs IC Symp. Tech. Dig., pg. 185.

[31] S. Notomi, Y. Awano, M. Kosugi, T. Nagata, K. Kosemura, M. Ono, N. Kobayashi, H. Ishiwari, K. Odani, T. Mimura, and M. Abe, 1987 IEEE GaAs IC Symp. Tech. Dig., pg. 177.

[32] K. Matsumoto, M. Ogura, T. Wada, T. Yao, Y. Hayashi, N. Hashizume, M. Kato, N. Fukuhara, H. Hirashima, and T. Miyashita, IEEE Elect. Dev. Letters EDL-7, 182 (1986).

[33] P.M. Asbeck, D.L. Miller, R.J. Anderson, R.N. Deming, L.D. Hou, C.A. Liechti, and F.H. Eisen, 1984 IEEE Int'l. Sol.-St. Circ. Conf. Tech. Dig., pg. 50.

[34] H.T. Yuan, W.V. McLevige, and H.D. Shih, VLSI Electronics: Microstructure Science, Vol. 11, ed. by N.G. Einspruch and W.R. Wisseman (Academic Press, Orlando, FL, 1985), pg. 173.

[35] K.C. Wang, P.M. Asbeck, M.F. Chang, G.J. Sullivan, and D.L. Miller, 1987 IEEE GaAs IC Symp. Tech. Dig., pg. 83.

[36] H. Shichijo, J.W. Lee, W.V. McLevige, and A. Taddiken, 1986 IEEE Int'l. Elect. Dev. Mtg. Tech. Dig., pg. 748.

[37] S. Sen, F. Capasso, and A. Cho, 1987 IEEE GaAs IC Symp. Tech. Dig., pg. 61.

CHAPTER II
SEMICONDUCTOR GROWTH

NON-EQUILIBRIUM PROCESSES IN LOW TEMPERATURE SILICON EPITAXY

Bernard S. Meyerson
IBM T. J. Watson Research Center, Yorktown Heights, N. Y., 10598

Abstract:

The use of both chemical kinetic and chemical equilibrium constraints in the implementation of low temperature epitaxial silicon is demonstrated. Optimizing operating conditions, we have been able to achieve epitaxial silicon layers of excellent crystallographic and electronic quality 550C. Through operation at these greatly reduced temperatures, highly non-equilibrium materials are prepared, where active boron concentrations in excess of $1x10^{20}cm^{-3}$ are obtained in as-deposited 550C materials.

Introduction:

For the field of electronic materials to advance, there must be an increased effort to employ fundamental chemical and physical data in the design of methodologies for film deposition. Most rapid progress occurs when one is able to do predictive rather than exploratory processing science, where an end goal is achieved by the incorporation of fundamental chemical and physical constraints as a guide, rather than relying upon a matrix of experiments to predict the correct set of operating parameters. We have endeavored over the past several years to employ[1-5] this predictive methodology, appealing to the most basic aspects of physical chemistry to achieve a desired result. In the following text, I will describe a method which exploits fundamental data from chemical equilibrium and chemical kinetics studies to obtain silicon epilayers at temperatures and boron doping levels thought to be generally inaccessible.

Interface Optimization:

Workers in the area of surface reactions have investigated[7,8] the equilibrium conditions for the formation of silicon dioxide on the silicon [100] surface in the presence of a number of potential oxidants. Of interest for the work reported on here, equilibrium conditions have been determined for the cases of both water vapor[7] and oxygen[8]. If one takes as a given that the silicon surface must be maintained oxide-free prior to the deposition of an epitaxial layer, such early equilibrium studies serve as a quantitative guide to system design. An example is shown in figure 1 for H_2O forming SiO_2 on a Si [100] surface, versus its etching the surface through the formation of the volatile product SiO. This supplies a quantitative set of data for the tolerable limits on water vapor content in order to maintain an oxide-free silicon surface, required to successfully prepare epitaxial silicon layers at reduced temperatures. As an example, to maintain an oxide free silicon surface at 700C, one can allow at maximum a water vapor partial pressure of 10^{-8} Torr. Similar data is also available for the case of oxygen as a contaminant. Although maintaining an atomically bare silicon surface is required to achieve high quality epitaxial layers, additional constraints exist.

It is desirable, for reasons of uniformity and throughput, that one employ a hot wall(isothermal) system for such a process. This is particularly true at lower temperatures and pressures where silicon growth rates are more likely to be thermally activated, requiring that a highly uniform temperature be maintained to obtain uniform deposition rates. However, in a hot wall reactor, the gaseous ambient is heated to the temperature of the environment, and a common argument against this technique is that the hot gaseous source would undergo significant homogeneous (gas phase) pyrolysis. This issue becomes particularly difficult when performing depositions on multiple close packed wafers, in that the reactive species(eg.SiH_2) that can be produced by the homogeneous pyrolysis of sources such as silane(SiH_4) lead to large variations[1] in film thickness across the reactor. This phenomena can, however, be eliminated through the use of greatly reduced total deposition pressure. It has been clearly demonstrated in RRKM theory that the rate of homogeneous decomposition for a given molecule becomes linear in total system pressure at the low pressure limit. Meyerson and Jasinski[9] have recently shown that when one makes the RRKM corrections to the previously employed rates for the homogeneous pyrolysis of silane, the predicted degree of silane pyrolysis across a wide temperature range ($T<800C$) is essentially zero at adequately reduced pressure($P\cong 1$ micron). The implication of this is that silane can be employed in a hot walled epitaxial reactor, at low pressures, without experiencing uniformity problems attributed to gas phase chemistry. This data was employed to design and test the Ultra High Vacuum/Chemical Vapor Deposition (UHV/CVD) apparatus employed to deposit the layers described in the text below, and is a simple case where the use of fundamental physical-chemistry data allows the predictive design of appropriate apparatus.

Film Growth:

The detailed operation of the UHV/CVD system employed here has been reported[5] elsewhere. In summary, it is a multi-wafer system employing an isothermal hot wall environment for the pyrolysis of the gaseous source onto wafers held coaxially in a furnace tube (fig. 2). Wafers were subjected to a standard RCA clean, HF dipped, and mounted in the load lock of the UHV/CVD apparatus. After transfer in the UHV section, films were deposited from CCD grade silane(SiH_4) containing up to 1 percent diborane(B_2H_6) as the dopant source. Deposition was conducted at a constant 550C, and no high temperature pre-processing or directed energy enhancement was employed. Layers were deposited on both p(10^{17}B/cm^3) and n(10^{17}As/cm^3) type substrates. Substrate dopant levels were chosen to assure that for p layers on n-type substrates resultant junctions would occur within the epilayer material.

Results:

The degree of control and abruptness of profiles that may be obtained by this method is demonstrated by the SIMS dopant profile shown in figure 3. Each plateau of dopant content shown represents an increase of 10 percent in the dopant content of the input gas stream. The abruptness of the dopant steps here is limited by the resolution of the SIMS instrument employed to obtain the data, and the transitions between neighboring dopant levels are likely more abrupt than indicated in the plot. A more remarkable result is shown in figure 4, where spreading resistance was employed to obtain the active carrier level in the most heavily boron doped layers prepared. A flat profile in excess of 10^{20} carriers/cm^3 is found for this 550C prepared sample. Samples described above were also examined to determine their crystallographic perfection by cross sectional TEM and x-ray topography, and were defect-free to the resolution limits of these techniques. To test for the existence of electrically active defects, several wafers were patterned into both large area and small area diodes, 10^6 and 400 square microns respectively. To preserve the validity of the result, all thermal cycles in forming the diodes were limited(eg.160 min. wet oxidation at 800C, 14 min. at 880C to interdiffuse the junctions). All such devices showed essentially ideal behavior, yielding junction ideality factors in the range 1-1.05, and leakage currents of 10^{-17}amperes per square micron of junction area at 5 volts reverse bias.

Discussion:

It is straightforward to demonstrate the non-equilibrium nature of our doping results above by the data shown in figure 5. The shaded region below the curve defines the limits of boron solid solubility in silicon, determined by Vick and Whittle[10]. At 550C, the equilibrium solid solubility would likely fall well below $1x10^{19}B/cm^3$, yet the data we obtain for active dopant incorporation is in excess of an order of magnitude above this. This highly non-equilibrium state is a result unique to this method of film growth. It is commonly observed in MBE[11] that one is limited to the mid 10^{18} range of B dopant concentration, and attaining this level of activation at 550C has not been reported. It must also be noted that although the material is in a non-equilibrium state as deposited, it will remain stable. This is a consequence of kinetic limitations upon the silicons ability to return to equilibrium. In order to achieve an equilibrium condition, the excess boron dopant in the material would need to precipitate out of solid solution as either B pairs or larger B clusters. This requires that the dopant diffuse, a thermally activated kinetic phenomena. There are two stages at which dopant might aglomerate. On the growth surface, dopant might undergo accelerated diffusion, leading to deactivation. Secondly, dopant could undergo bulk diffusion once incorporated in the growing layer. Bulk diffusion can be eliminated as a concern owing to the extremely low temperatures employed, where the diffusivity of boron in Si is essentially zero on the time scale of typical depositions($1<t<10$hrs). Although no data exists for the mobility of boron on the Si surface during deposition, it is apparent from the result that whatever the kinetics of diffusion are they are slow compared to the rate boron is substituted into the Si lattice. This result is reasonable as a good deal of evidence[12] to date argues for the presence of a stable layer of hydrogen passivated molecular fragments adsorbed on the growth interface. Surface diffusion would likely be inhibited by such a high level of occupancy, although this is speculative pending more direct examination of the phenomena.

Although many variety of structural data have been obtained to date implying that this material is of high quality, the most sensitive probe for hidden defects is obtained from the electrical characteristics of devices built within the layers. From the results described above, we have demonstrated that these materials are indeed of a high level of perfection, and the full function of the 1,000,000 square micron diodes demonstrates a low level of defects present. All of these results point to a need to expand upon some of the basic tenets of epitaxy. On one hand, we find that the mobility of surface species in this case is inadequate to allow for boron segregation, yet there is adequate mobility to result in atomically smooth and highly perfect layers. It is unlikely that this process can be ascribed to a form of step flow/kink adsorption model described in the BCF theory of high temperature epitaxy. Given the high degree of surface occupancy, a model allowing for some form of preferential direct adsorption on atomically correct sites may have to be invoked to explain the epitaxial process observed here. Several efforts are underway to study the surface adsorption and reaction dynamics of silicon precursors such as silane, but until this data is available, a more precise definition of the growth mechanism will have to be put off.

Conclusions:

We have demonstrated a method by which highly perfect silicon layers can be prepared at temperatures as low as 550C. Such layers are found to be highly perfect crystallographically, and p/n junctions fabricated in the in-situ doped layers demonstrate ideal rectification behavior. Boron doping levels in excess of 10^{20} B/cm^3 have been shown to be electrically active, violating solid solubility limits for boron in silicon by in excess of an order of magnitude. This demonstrates the highly non-equilibrium nature of the UHV/CVD process.

References:

1) B. S. Meyerson and W. L. Olbricht, J. Electrochem. Soc. 131, 2361(1984).

2) B. S. Meyerson and M. L. Yu, J. Electrochem. Soc. 131, 2366(1984).

3) Ming L. Yu, J. Vitkavage, and B. S. Meyerson, J. Appl. Phys. 59, 4032(1986).

4) J. F. Morar, B. S. Meyerson, U. O. Karlsson, F. J. Himpsel, F. R. McFeely, D. Rieger, A. Talebibrahimi, and J. A. Yarmoff Appl. Phys. Let. 50, 463(1987).

5) B. S. Meyerson, Appl. Phys. Let. 48, 797(1986).

6) B. S. Meyerson, E. Ganin, D. A. Smith, and T. N. Nguyen, J. Electrochem. Soc. 133 1232(1986).

7) F. W. Smith and G. Ghiddini, J. Electrochem. Soc. 129, 1300 (1982).

8) G. Ghidini and F. W. Smith, J. Electrochem. Soc. 131, 2924(1984).

9) B. S. Meyerson and J. M. Jasinski, J. Appl. Phys. 61, 785(1987).

10) G. L. Vic and K. M. Whittle, J. Electrochem. Soc. 116, 1142(1969).

11) K. L. Wang, Sol. St. Tech. 28, 137(1985).

12) A. M. Beers and J. Bloem, Appl. Phys. Let. 41, 153(1982).

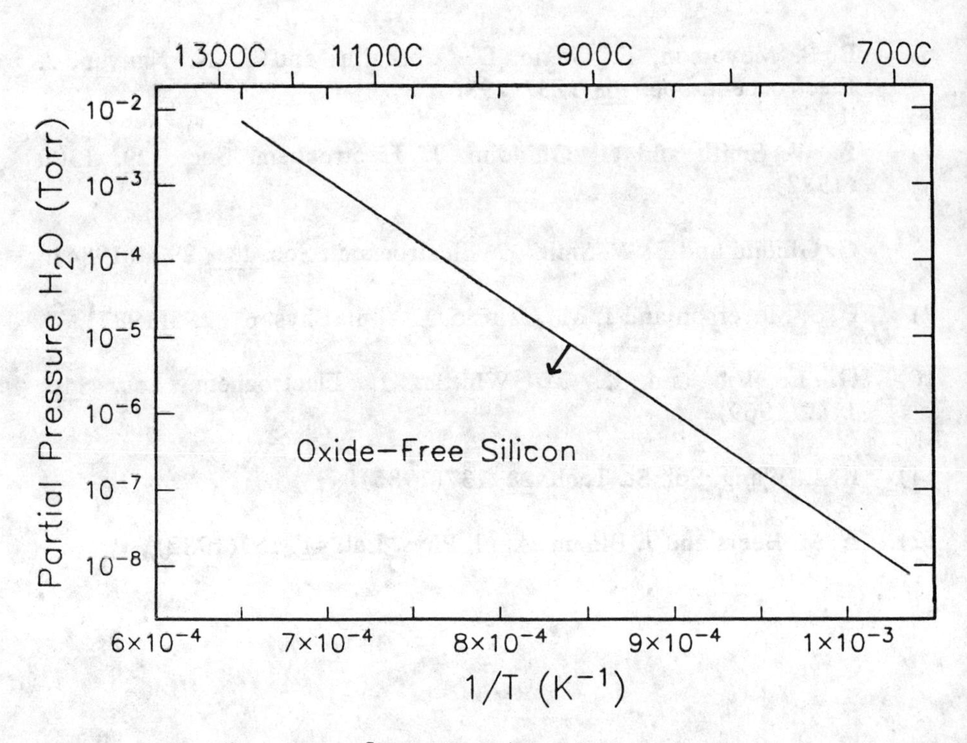

Figure 1. Equilibrium data[7] for the $Si/H_2O/SiO_2$ system.

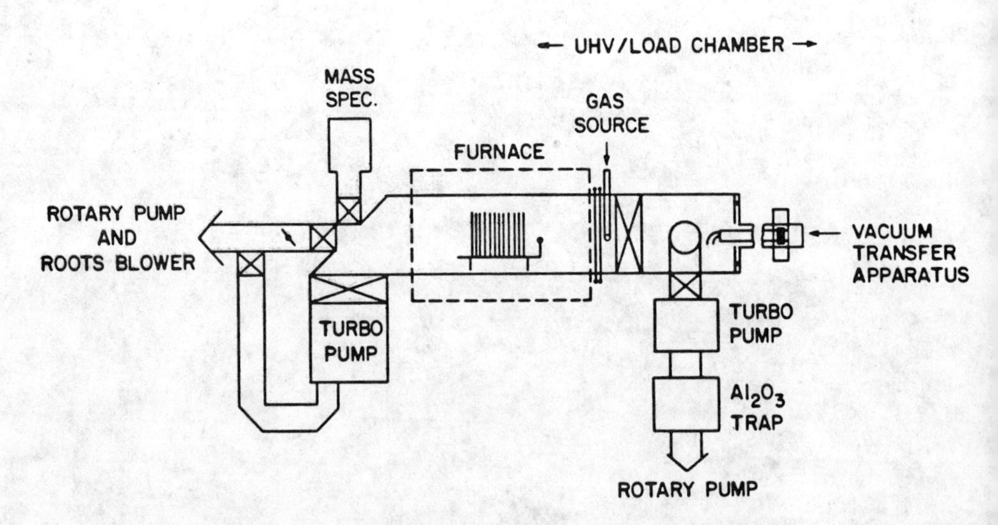

Figure 2. The UHV/CVD system.

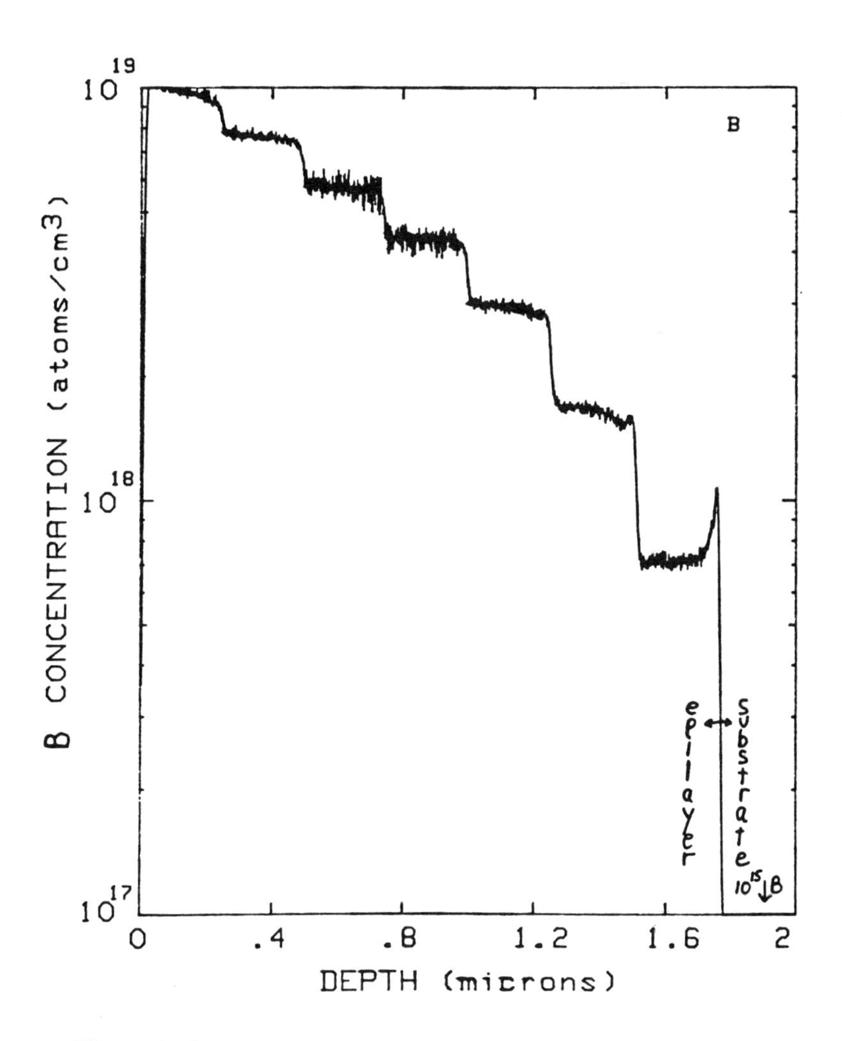

Figure 3. SIMS boron profile of 550C in-situ doped epilayer.

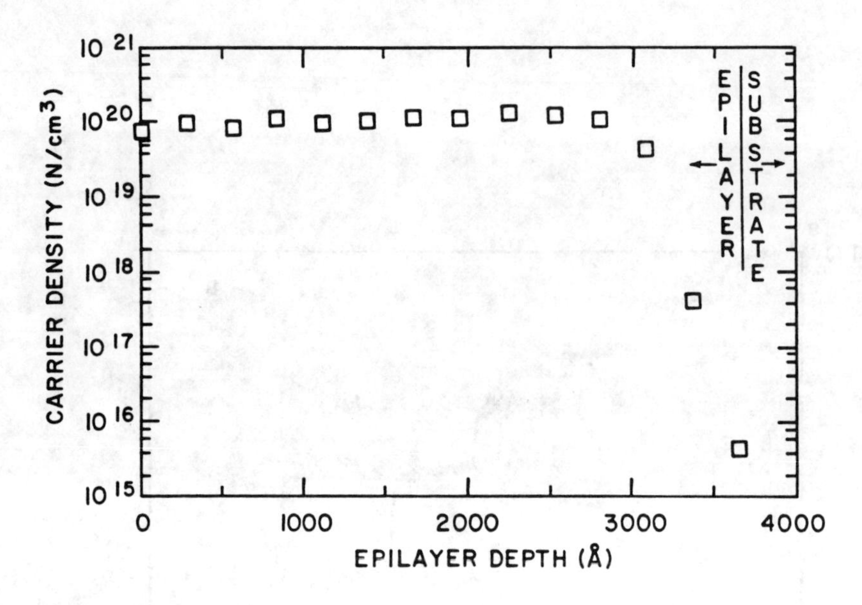

Figure 4. Spreading resistance profile of a 550C boron doped epilayer.

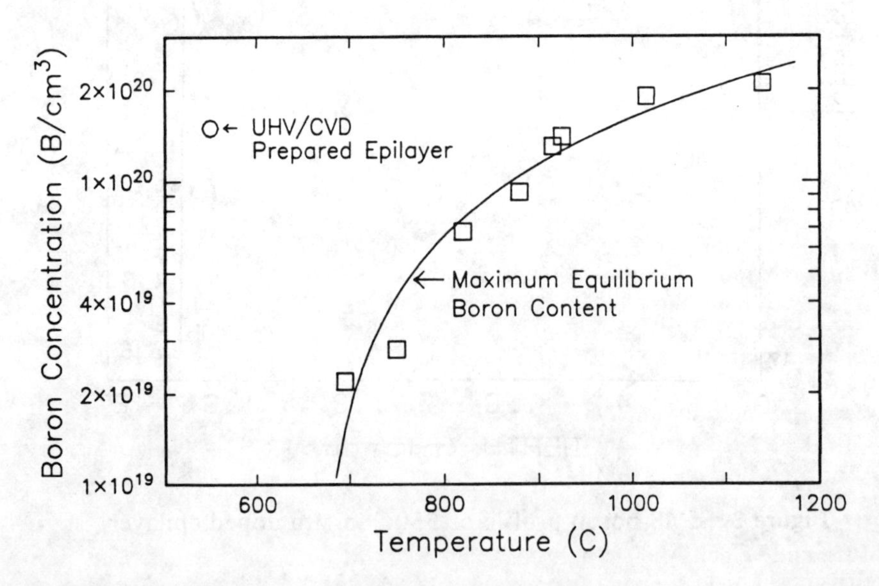

Figure 5. Equilibrium solid solubility data[10] for boron in Si as compared with the results obtained here.

LASER SPECTROSCOPY AND GAS-PHASE CHEMISTRY IN CVD

Pauline Ho, Michael E. Coltrin, and William G. Breiland
Sandia National Laboratories, Div. 1126
Albuquerque, NM 87185-5800

ABSTRACT

A coordinated program of experimental and theoretical research into the fundamental mechanisms of CVD is described. The experimental part of the program involves the use of laser spectroscopic techniques for *in situ* measurements during CVD. The theoretical part of the program consists of a computer model that contains a detailed description of the coupled fluid mechanics and gas-phase chemical kinetics for silane CVD.

The fundamental chemistry and physics of CVD can be divided into gas-phase and surface processes. Gas phase processes include chemistry and fluid mechanics, both of which are well-established fields. Gas-phase processes are better understood than surface processes, so our research to date has emphasized the events that occur in the gas phase during deposition. The silane system was chosen primarily because it is relatively simple and can serve as a model for systems with more complex chemistries.

The theoretical part of the program [1,2] consists of a computer model that contains a detailed description of the coupled fluid mechanics and gas-phase chemical kinetics for silane CVD. The model uses finite-difference methods to numerically solve the boundary layer equations of fluid flow and chemical rate expressions that describe the deposition of silicon from silane. It is a steady-state, two-dimensional model. Initially, a mechanism of 120 reversible, elementary reactions was used to describe the thermal decomposition of silane and the subsequent reactions of intermediate species. Thermochemical and kinetic data for these reactions were obtained from the literature whenever possible. Otherwise thermochemical data were obtained from *ab initio* electronic structure calculations [3,4], and kinetic parameters estimated using Benson's techniques [5]. A sensitivity analysis was used to reduce this mechanism to a set of 27 forward-and-backward reactions. The initial step in the silane decomposition mechanism is production of SiH_2. Formation of SiH_3 by loss of H from silane is slow, so silicon species containing an odd number of hydrogens are not important in silane CVD [1,2] although they are included in the model for completeness. Surface reactions are included in the model in the form of reactive sticking coefficients (RSCs). The RSC for silane on a hot silicon surface is relatively low (10^{-4} to 10^{-6}) and was derived from literature data (see Ref. 2 for details). In the absence of any experimental information, the RSC for intermediate species such as SiH_2, Si_2H_2, Si_2, etc., were set to one. The model predicts gas-phase temperature and velocity fields, and density fields for the 17 chemical species in the kinetics mechanism, deposition rates and deposition uniformity as a function of control variables such as susceptor temperature, flow velocity, total pressure, reactant gas composition, carrier gas, and cell dimension.

Silicon deposition rates predicted by the model as a function of susceptor temperature for atmospheric-pressure CVD (APCVD) compare favorably with experimental deposition rates in H_2 taken from the literature[6]. In addition to reproducing the experimental data, the model also predicts several well-known features of silane CVD: the transition between high-temperature behavior (weak temperature dependence) and low-temperature behavior (strong temperature dependence) and the higher deposition rates in He carrier gas than in H_2. The carrier gas dependence results from the fact that H_2 is a product of the silane decomposition reaction and many of the other reactions. Gas-phase reactions convert relatively unreactive silane into species such as SiH_2 and Si_2H_4, which are more reactive with the silicon surface than is silane. An atmosphere of H_2 suppresses formation of these intermediates, leading to lower deposition rates at a given temperature.

Although the correct prediction of deposition rates is a necessary test of a model for CVD, more detailed measurements are needed to to thoroughly test our complex, "first-principles" model. The experimental part of the program [7,8] provides such data through the use of laser spectroscopic techniques for *in situ* measurements during CVD. Pulsed-laser Raman spectroscopy and laser-excited fluorescence spectroscopy were used to measure spatial profiles of gas-phase temperatures and species densities. Gas temperatures were obtained from the rotational Raman spectrum of H_2 in a He carrier gas[7]. Number densities of silane were obtained from the vibrational Raman spectrum[7]. Laser-excited fluorescence was used to obtain relative density profiles of Si_2 [7] and Si atoms[8].

Profiles of gas temperatures as a function of height above the heated susceptor determined by Raman spectroscopy compared favorably with with the model predictions for three susceptor temperatures, which indicates that the 2-D boundary layer model provides a sufficiently accurate description of the temperature field (and fluid flow) in the experimental reactor to yield realistic chemical predictions. Experimental silane density profiles also compared favorably with predictions of the model, showing a substantial decrease in the silane density between 2 cm and 0 cm. The model indicates that ideal-gas expansion, gas-phase decomposition of the silane, and thermal diffusion (Soret effect) all contribute to the shape of the silane density profiles. The relative importance of these factors vary with deposition conditions[2].

Chemical species formed as intermediates in gas-phase reactions have qualitatively different profiles than silane as a function of height above the susceptor. The densities of the intermediate species Si_2 and Si atoms are low both at the susceptor and ~1 cm above the susceptor, reaching a maximum a few mm above it. The fact that these profiles have maxima in the gas phase confirms that these species are formed via gas-phase reactions rather than surface reactions. The shapes of the experimental and theoretical profiles for the Si atom densities agree well. However, the model predicts stronger temperature (Si atoms) and hydrogen (Si_2 and Si atoms) dependences than are observed experimentally, which represents a significant breakdown in the model. A kinetic analysis showed that the disagreement between the model and experiment could not be resolved by varying the rate constants for reactions in the model[8]. It may, however, be due to the omission from the model of a mechanism for gas-phase nucleation of particulates, a well known occurance under some conditions in silane CVD.

The results of the theoretical modeling and the laser diagnostic measurements show the importance of gas-phase chemistry in silane CVD. However, as the chemical environment in a CVD reactor changes with operating conditions, so do the relative importance of gas-phase chemistry, surface chemistry and fluid mechanics in determining film deposition rates. When chemical reactions are very fast, mass transfer (diffusion) rates determine deposition rates. When gas-phase reactions are very slow, gas-phase or surface-reaction rates can limit deposition rates. Despite this variability, many features of silane CVD can be understood in terms of gas-phase chemical reaction kinetics and fluid mechanics. The theoretical framework exists for treating these aspects of CVD in detail. Treating the surface reactions in comparable detail, however, is not feasible at this time.

This work was performed at Sandia National Laboratories, supported by the U.S. Department of Energy, Office of Basic Energy Sciences under contract No.DE-AC04-76DP00789.

REFERENCES

1. M. E. Coltrin, R. J. Kee, and J. A. Miller, *J. Electrochem. Soc.* **131**, 425 (1984).
2. M. E. Coltrin, R. J. Kee, and J. A. Miller, *J. Electrochem. Soc.* **133**, 1206 (1986).
3. P. Ho, M. E. Coltrin, J. S. Binkley and C. F. Melius, *J. Phys. Chem.* **89**, 4647 (1985).
4. P. Ho, M. E. Coltrin, J. S. Binkley and C. F. Melius, *J. Phys. Chem.* **90**, 3399 (1986).
5. S. W. Benson, *Thermochemical Kinetics*, Wiley and Sons, New York, 1976.
6. C. H. J. van den Brekel, PhD. Thesis, University of Nijmegen, The Netherlands (1978).
7. W. G. Breiland, M. E. Coltrin, and P. Ho, *J. Appl. Phys.* **59**, 3267 (1986).
8. W. G. Breiland, P. Ho, and M. E. Coltrin, *J. Appl. Phys.* **60**, 1505 (1986).

REACTIVE STICKING COEFFICIENTS OF SILANE ON SILICON

Richard J. Buss, Pauline Ho, William G. Breiland and Michael E. Coltrin
Sandia National Laboratories, Albuquerque, NM 87185-5800

ABSTRACT

Reactive sticking coefficients (RSCs) were measured for silane and disilane on polycrystalline silicon for a wide range of temperature and flux (pressure) conditions. The data were obtained from deposition rate measurements using molecular beam scattering and a very low pressure cold wall reactor. The RSCs have non-Arrhenius temperature dependences and decrease with increasing flux at low (710°C) temperatures. A simple model involving dissociative adsorption of silane is consistent with these results. The results are compared with previous studies of the $SiH_4/Si(s)$ reaction.

INTRODUCTION

Despite the importance of silicon chemical vapor deposition (CVD) in the manufacture of thin films for the microelectronics industry, the fundamental mechanisms of this processing technology are not well understood. Recent work combining computer modeling[1,2] and laser diagnostics[3,4] has shown that many aspects of silane CVD can be described by gas-phase chemical reactions and fluid mechanics. Under some conditions, however, the gas-phase decomposition of silane is relatively slow and the reaction of silane on the substrate surface can be a significant component of the process. A complete understanding of the silane CVD system and, in particular, the relative importance of gas-phase and surface reactions thus requires better knowledge of the reaction of silane with silicon surfaces.

Silane reacts with a hot silicon surface via a combination of physi- or chemisorption, surface reaction(s), surface migration, and molecular hydrogen desorption. Although this system has been studied previously[5–10], conflicting results have been obtained. In the absence of fundamental knowledge, it is useful to study silane-silicon surface chemistry first in terms of the overall reaction: SiH_4 + silicon surface $\longrightarrow$ deposited silicon + H_2. A convenient measure of the rate of this reaction is the reactive sticking coefficient (RSC), which we define to be the number of silicon atoms deposited per unit surface area per second divided by the number of silane molecules colliding with the surface per unit area per second (silane flux). Thus, the RSC is the probability that a silane molecule will be incorporated into the silicon solid upon collision with the surface.

In this paper we report measurements of the RSC for silane and disilane obtained from growth rates on polycrystalline silicon. Gas-phase decomposition was eliminated by using either molecular beam sources or a very low pressure deposition cell with high-velocity gas flows. For silane and disilane, we find that the temperature dependence of the RSC is not described by a single activation energy, and that the RSC is a function of incident molecular flux at low temperatures.

EXPERIMENT

The experimental apparatus used to measure RSCs is described in detail elsewhere.[11,12] Briefly, the apparatus consisted of a glass tube (5.6 cm diameter by 30 cm long) coupled to a stainless steel chamber pumped by a large oil-diffusion pump (Varian VHS–400). Two configurations were used in these experiments. In the first arrangement, the heated silicon sample was mounted in the main chamber and the glass tube served as the source for a molecular beam of silane. In the second configuration, the sample was mounted in the glass tube, which served as a reaction cell in which the silane flowed over the sample at a high linear velocity. These two configurations were used in order to cover a wide range of incident gas fluxes (pressures). The molecular beam scattering configuration provided silane fluxes at the silicon sample of 8×10^{16} to 1×10^{18} molecules cm^{-2} s^{-1} (effective pressures of 2×10^{-4} to 3×10^{-3} Torr). The high-velocity gas cell configuration produced silane fluxes of 9×10^{17} to 1×10^{20} molecules cm^{-2} s^{-1} (effective pressures of 2.5×10^{-3} to 0.28 Torr). The experimental conditions in both configurations ensured that homogeneous decomposition of the silane did not contribute significantly to the deposited solid. For the disilane RSC experiments, fluxes between 3.5×10^{16} and 3.7×10^{17} molecules cm^{-2} s^{-1} were used in the beam experiments, and fluxes between 3.1×10^{17} and 6.2×10^{18} were used in the flow cell configuration.

In the first configuration, an effusive molecular beam of room-temperature silane or disilane was scattered from the hot silicon sample. The pressures in the beam and sample chamber were such that the molecules in the beam each had only a single encounter with the sample surface. The beam orifice was 1.04 cm in diameter, and the sample was placed 3.47 cm from the front of the hole in the beam chamber. The sample chamber had a base pressure of 10^{-7} Torr, which rose to 10^{-5} Torr at the highest beam flux.

In order to obtain silane fluxes higher than could be achieved in the molecular beam configuration, the heated sample was placed inside the tube that had been used as the molecular beam source in the experiments discussed above. The surface of the substrate was parallel to the gas flow. The gas flow rate was controlled with a mass flow controller and the pressure in the tube was monitored with a capacitance manometer. A silane flow rate of 10 sccm gave a pressure of ~0.015 Torr. The linear flow rate of the gas was ~2 m s^{-1}, independent of volume flow rate.

The samples used in these studies were 1×2 cm pieces of silicon wafers doped to a resistivity of 0.2–0.4 ohm-cm. The wafers had 6500Å of undoped polycrystalline silicon on top of 1000Å of oxide. This structure provided a silicon surface for the sticking coefficient measurements but allowed the deposition rate to be determined by masking, etching to the oxide, then profiling.

The silicon samples were mounted in a holder with electrodes of tantalum foil and were resistively heated using a current-limited source. The sample temperature was measured with an optical pyrometer. The thickness of the deposited layer was determined by masking the sample, etching to the oxide layer with a HF/HNO_3 solution, and using a Dektak profilometer to measure the step height. A density of 2.33 g cm^{-3} was used to calculate the amount of silicon deposited. The RSC for silane was computed by dividing the deposition rate (in units of Si atoms cm^{-2} s^{-1}) by the incident silane flux. The disilane RSC was computed by dividing the deposition rate by twice the disilane flux. The morphologies of some of the deposited films were examined by SEM.

RESULTS

Our measured values for the RSC of silane and disilane on silicon range from 10^{-2} to 10^{-5}, depending on experimental conditions. The disilane RSCs are systematically higher than the silane RSCs for corresponding temperature and flux conditions.

Figure 1 shows an Arrhenius-type plot of the dependence of our measured silane RSC on the silicon surface temperature. These data were obtained using the molecular beam configuration at a silane flux of 5.9×10^{17} molecules cm^{-2} s^{-1} (effective pressure 1.6×10^{-3} Torr). The Arrhenius plot is distinctly nonlinear. The RSC is strongly temperature dependent at lower temperatures, but undergoes a transition to a much weaker temperature dependence at about 800°C. The slopes at the low and high temperatures correspond to activation energies of 55–60 and 0–5 kcal mol^{-1}, respectively. The curve in Fig. 1 is from a simple model for silane reacting with a silicon surface, which will be discussed below.

Figure 2 shows the dependence of the RSC on incident silane flux (F) at substrate temperatures of 710°C and 1040°C. At 710°C, the RSC decreases with increasing silane flux; the dependence is roughly $F^{-1/2}$. The data obtained from the molecular beam and high-velocity cell configurations overlap at a flux of $\sim10^{18}$ molecules cm^{-2} s^{-1} (2.8×10^{-3} Torr). At 1040°C, the RSC is relatively independent of flux. The curve in Fig. 2 is again from a simple model for silane reacting with a silicon surface, which will be discussed below.

No measurable effect of H_2 on the silane RSC was observed. The RSC was identical, within experimental error, for pure silane and a mixture of silane with a ten-fold excess of hydrogen or helium in the flow-cell configuration (2.6×10^{18} cm^{-2} s^{-1}, 710°C). In the molecular beam configuration, a one-to-one mixture of silane and hydrogen (5.9×10^{17} cm^{-2} s^{-1}, 710°C) yielded the same RSC as pure silane depositions.

Figure 3 shows an Arrhenius-type plot of the dependence of our measured disilane RSC on the silicon surface temperature. These data were obtained using the molecular beam configuration at a disilane flux of 8.7×10^{16} molecules cm^{-2} s^{-1} (effective pressure 3.4×10^{-4} Torr). The disilane data in Fig. 3 should not be directly compared with the silane data in Fig. 1 because of the difference in flux. As in the case of silane, the Arrhenius plot is distinctly nonlinear and changes from a strong to a weak temperature dependence as the temperature is increased above 800°C. The slopes at the low and high temperatures correspond to activation energies of roughly 35 and 0 kcal mol^{-1}, respectively. The temperature dependence is qualitatively similar to that observed for silane although the apparent activation energies are smaller.

Figure 4 shows the flux dependence of the disilane RSC at a silicon temperature of 710°C. Although the data are limited, the disilane RSC appears to decrease with increasing flux, with the effect being more pronounced at higher flux, above 4×10^{17} molecules cm^{-2} s^{-1} (1.55×10^{-3} Torr). This is qualitatively similar to the silane RSC data at 710°C shown in Fig. 2.

DISCUSSION

Our experimental results indicate that the decomposition of silane or disilane to give silicon atoms on a hot surface is not a simple process. The curvature of the Arrhenius plot shows that, in this system, extrapolation from measurements taken over a small temperature range can give misleading predictions outside that range. The flux dependence of the RSC at lower temperatures and higher flux introduces an additional complication. Coupled with the flux-independent RSC that would be observed from gas phase contributions to the deposition rate, it is not surprising that much of the literature on silicon CVD appears contradictory.

We have constructed a simple model to describe the interaction between silane and a hot silicon surface. The purpose of this model is to provide a physically reasonable explanation for our data. It is clear that our experimental data are not sufficient to develop a detailed description of the reaction of silane/disilane with hot polycrystalline silicon, so we have included a minimum number of reactions in the models. We expect that other models, invoking different physical mechanisms, can also be found which are consistent with our data.

Our model consists of a three-reaction system in which silane dissociatively adsorbs on the silicon surface, occupying two sites. Neighboring surface SiH_2 groups can associatively desorb silane, or decompose, releasing hydrogen. The mechanism is:

$$SiH_4 + * - * \rightarrow 2SiH_2* \tag{1}$$

$$2SiH_2* \rightarrow SiH_4 + * - * \tag{2}$$

$$SiH_2* \rightarrow * + H_2, \tag{3}$$

where $*$ represents a site on the silicon surface, and no distinction is made between Si$*$ and $*$. Adsorption of molecular hydrogen is not included because we observed no strong effects from added hydrogen, and H_2 has been found to interact only weakly with clean Si surfaces.[13] We also do not include the desorption of silicon atoms[5,7] because the vapor pressure of silicon is low even at our highest temperatures.[14] The reaction kinetics are affected by the mobility of the surface species,[15] and we assume that surface SiH_2 groups are mobile. A more detailed description and discussion of this model, including the relationships between the RSC, surface coverage and the kinetic rate parameters, will be published elsewhere.[12]

This model was also extended to disilane surface decomposition. In this case, the model includes dissociative adsorption on silicon forming three surface SiH_2 groups,

$$Si_2H_6 + * - * - * \rightarrow 3SiH_2*, \tag{4}$$

along with desorption of SiH_4 (reaction 2) and reaction to release H_2 (reaction 3). Although we have formally treated the dissociative adsorption of silane and disilane as third and fourth order reactions in our model, these processes undoubtedly consist of a number of lower-order elementary reactions.

The curves in Figs. 1–4 result from fitting the kinetic parameters in the model to our silane and disilane data. This results in the following rate constants:

$$k_1 = 9.33 \times 10^{-29} \exp(-2000/RT)\ \mathrm{cm^5 s^{-1}}$$

$$k_2 = 2.15 \times 10^{-7} \exp(-25000/RT)\ \mathrm{cm^2 s^{-1}}$$

$$k_3 = 3.95 \times 10^{14} \exp(-59000/RT)\ \mathrm{s^{-1}}$$

$$k_4 = 8.03 \times 10^{-43} \exp(-2000/RT)\ \mathrm{cm^7 s^{-1}},$$

where RT is in cal mol^{-1}. These values for the kinetic parameters were chosen to provide a good fit to the silane data and an acceptable fit to the disilane data with the same values for A_2, E_2, A_3 and E_3. Also shown in Figs. 3 and 4 are the best fits obtained using only the disilane data with the following parameters:

$$k_2 = 2.15 \times 10^{-7} \exp(-25000/RT)\ \mathrm{cm^2 s^{-1}}$$

$$k_3 = 5.34 \times 10^{13} \exp(-54000/RT)\ \mathrm{s^{-1}}.$$

$$k_4 = 9.70 \times 10^{-43} \exp(-2500/RT)\ \mathrm{cm^7 s^{-1}}$$

The numerical methods used to fit the model to our data generally did not yield a unique optimization of rate constants. A range of parameters produced comparable fits to the experimental observations. The RSCs are determined by the relative values of the rate constants, so some changes in an activation energy could be compensated by changes in the corresponding pre-exponential factor without significantly affecting the fits. The scatter in the experimental data adds to the uncertainty in the fits.

Our measured values for the RSC of silane and disilane on silicon are significantly less than one. This general result agrees with previous studies of the $SiH_4/Si(s)$ reaction. Even at the highest temperatures, less than one percent of the incident silane reacts.

We also found that disilane RSCs are roughly 10 times higher than silane RSCs. This is consistent with Gates' [16] recent observation that disilane undergoes dissociative adsorption substantially faster than silane on silicon (111)-(7×7) single crystals (at low temperatures), and indicates that the trend of higher disilane reactivity also holds under the steady-state, high-temperature conditions used for deposition.

A major result from our measurements is that the RSC is a complicated function of both surface temperature and gas flux. This indicates that comparisons between sets of experimental data must carefully account for differences in incident flux and temperatures. In comparing our results with previous work, we use the model to guide extrapolations of our observations to different conditions.

As mentioned above, there are some previous studies of the $SiH_4/Si(s)$ reaction. Figure 5 shows the RSCs we obtained for temperatures above ~ 750°C (from Fig. 1) along with RSCs obtained from the data of Farrow,[5] Farnaam and Olander,[6] Henderson and Helm,[7] and Joyce, et al.[8–10] Measurements of H_2 production (dashed lines) and Si deposition (solid lines) are both shown. Figure 5 shows that a wide range of RSC values have been reported for the same temperatures. In view of the complex nature of the RSC, differences in the details of these experiments probably lead to the disparity seen in this figure. At high temperatures we observe a weak temperature dependence which agrees better with the ~10 kcal mol^{-1} obtained by Farrow[5] and Henderson and Helm[7] for Si deposition than the ~18 kcal mol^{-1} they obtained for H_2 production. However, our values for the silane RSCs are 4–10 times higher than their Si deposition values and are, in fact, close to their values derived for H_2 production. The value of Joyce, et al.[8–10] is considerably lower, and Farnaam and Olander's[6] values are substantially higher than our observations.

According to the model, the RSC should not vary significantly with flux for temperatures above 800°C and silane fluxes below 10^{19} cm^{-2} s^{-1} (0.04 Torr). Our data in Fig. 5 can therefore be compared directly with the literature results, and our flux dependence does not explain the large differences among the sets of data. Reaction rates proportional to silane flux were reported by Henderson and Helm for both silane decomposition (monitored by total pressure changes in a static fill) and Si deposition, by Farrow for silane decomposition, H_2 appearance, and Si deposition, and by Farnaam and Olander for H_2 production. Our flux dependence is consistent with results of these three studies.

Joyce and coworkers[9,10] observed that H_2 production from silane decomposition varied as silane beam intensity (flux) to the 0.48±0.05 power but the deposition of Si(s) varied as flux to the 1.8±0.2 power. These observations differ significantly from those of other workers. However, Joyce, et al. examined the initial stages of deposition, which could be expected to differ from the steady-state deposition process emphasized in the other works. They also reported long (minutes to hours) induction times for the initial growth of silicon. Our data did not exhibit such induction times.

At the highest temperatures, the RSCs approach a limiting value considerably less than one (0.004). Our interpretation of this low value is that the deposition rate is limited by the adsorption step. One explanation for this limitation is the existence of a potential energy barrier to adsorption that could be surmounted by translational or vibrational energy of the silane molecules rather than thermal energy from the surface. This class of surface-reaction dynamics has been implicated[17–19] in the reaction of CH_4 on metal surfaces. Ceyer, et al.[17] reported a $\geq$17 kcal mol^{-1} barrier to CH_4 adsorption on Ni which may arise from the need to distort the C–H bonds in order to allow the carbon to approach the nickel. In that system, the barrier can be overcome by internal or translational energy of the incident CH_4. We calculate that a silane adsorption barrier of 4 kcal mol^{-1} would account for the low sticking probability of silane. That is, only 0.004 of room temperature silane molecules have internal or translational energy above 4 kcal mol^{-1}. If this explanation for the low RSC of silane is valid, then heating the silane should dramatically increase the reaction rate. In the flow tube geometry used in these experiments, there may be some heating of the silane which is not present in the molecular beam configuration. The flux dependence of the RSC shows no marked discontinuity in making the transition between the two experimental geometries, suggesting that the heating (of unknown magnitude) is not significant. Further experiments, such as measurements of the RSC as a function of silane kinetic energy (for example with a supersonic molecular beam) are needed to evaluate the importance of such a barrier to adsorption in this system.

CONCLUSIONS

We have investigated the reaction of room-temperature silane and disilane on a hot polycrystalline silicon surface using both a collision-free molecular beam and a very low pressure CVD cell. Reactive sticking coefficients were obtained from deposition rate data over a wide range of temperatures and silane (disilane) fluxes. The RSCs are substantially less than one, ranging from 6×10^{-5} to 4×10^{-2}. For silane we observed curved Arrhenius plots with slopes decreasing from ~60 kcal mol^{-1} at low temperatures to ~2 kcal mol^{-1} at higher temperatures. The RSCs are independent of flux (pressure) at 1040°C, but vary as flux to the $\sim -1/2$ power at 710°C. A model comprised of a dissociative adsorption mechanism with competing associative desorption and reaction was found to give reasonable agreement. For disilane, we observed RSCs that were roughly ten times higher than those for silane. We also observed a curved Arrhenius plot and a flux dependence at 710°C for disilane.

Combined with earlier studies, this work indicates that the silane RSC is not only characterized by a non-linear Arrhenius temperature dependence, but is also a function of incident silane flux and possibly the temperature of the incident silane. It also depends on crystallographic orientation,[20] and may be weakly dependent on the co-incident hydrogen flux.[21,22] This behavior indicates that, although our RSC is a convenient parameter for describing the $SiH_4/Si(s)$ reaction, this reaction is not well represented by a simple "silane-in, silicon-out" picture and the RSC thus reflects the complexity of the underlying elementary steps. This complex surface-chemistry behavior, combined with the fact that silane decomposes in the gas phase at higher pressures, may help to explain why the silane CVD literature contains many apparently conflicting results.

ACKNOWLEDGEMENTS

We thank Dr. John W. Medernach for providing us with samples, Dr. Eileen Duesler for her assistance with SEMs, Dr. J. Randy Creighton, Dr. Bruce D. Kay and Prof. Sylvia T. Ceyer for helpful discussions. We also acknowledge the technical assistance of Pamela Ward, Holly Welch and Michael P. Youngman. This work was performed at Sandia National Laboratories, supported by the U. S. Department of Energy under Contract No. DE-AC0476DP00789 for the Office of Basic Energy Sciences.

REFERENCES

1. M.E. Coltrin, R.J. Kee, and J.A. Miller, J. Electrochem. Soc. **131**, 425 (1984).
2. M.E. Coltrin, R.J. Kee, and J.A. Miller, J. Electrochem. Soc. **133**, 1206 (1986).
3. W.G. Breiland, M.E. Coltrin, and P. Ho, J. Appl. Phys. **59**, 3267 (1986).
4. W.G. Breiland, P. Ho, and M.E. Coltrin, J. Appl. Phys. **60**, 1505 (1986).
5. R.F.C. Farrow, J. Electrochem. Soc. **121**, 899 (1974).
6. M.K. Farnaam and D.R. Olander, Surface Sci. **145**, 390 (1984).
7. R.C. Henderson and R.F. Helm, Surf. Sci. **30**, 310 (1972).
8. B.A. Joyce and R.R. Bradley, Phil. Mag. **14**, 289 (1966).
9. B.A. Joyce, R.R. Bradley, and G.R. Booker, Phil. Mag. **15**, 1167 (1967).
10. B.A. Joyce, R.R. Bradley, and B.E. Watts, Phil. Mag. **19**, 403 (1968), and references therein.
11. R.J. Buss, J. Appl. Phys. **59**, 2977 (1986).
12. R. J. Buss, P. Ho, W. G. Breiland, and M. E. Coltrin, submitted to J. Appl. Phys.
13. G. Schulze and M. Henzler, Surf. Sci. **124**, 336 (1983).
14. J.F. O'Hanlon, "A User's Guide to Vacuum Technology", John Wiley and Sons, New York, 1980, pg. 372.
15. F.C. Tompkins, "Chemisorption of Gases on Metals", Academic Press, 1978.
16. S.M. Gates, in press.
17. M.B. Lee, Q.Y. Yang, S.L. Tang, and S.T. Ceyer, J. Chem. Phys. **85**, 1693 (1986).
18. C. T. Rettner, H. E. Pfnür, and D. J. Auerbach, Phys. Rev. Lett. **54**, 2716 (1985).
19. C. T. Rettner, H. E. Pfnür, and D. J. Auerbach, J. Chem. Phys. **84**, 4163 (1986).
20. T.J. Donahue and R. Reif, J. Electrochem. Soc. **133**, 1691 (1986).

21. W.A.P. Claassen, J. Bloem, W.G.J.N. Valkenburg, and C.H.J. van den Brekel, J. Cryst. Growth, **57**, 259 (1982).
22. J. Holleman and T. Aarnink, Proceedings of the Eighth International Conference on CVD, 1981, pg. 307.

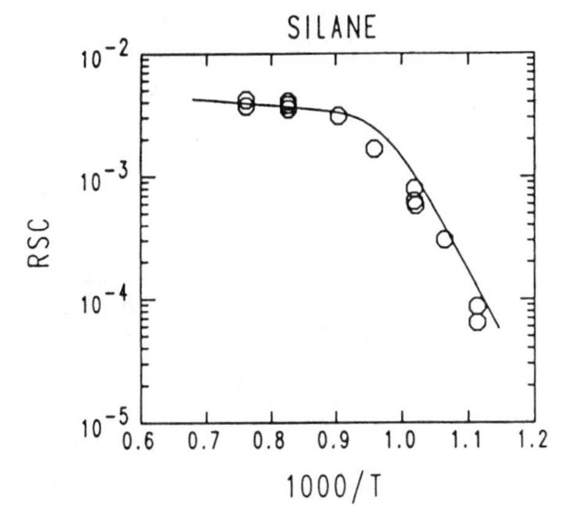

Fig. 1. Arrhenius-type plot of the silane RSC at a flux of 5.9×10^{17} molecules cm^{-2} s^{-1}. Curve shows best fit obtained by fitting the model to the silane and disilane data.

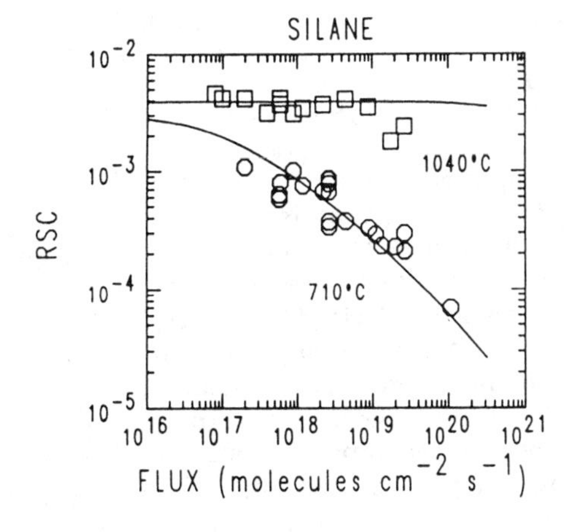

Fig. 2. Silane RSC as a function of flux at ○ 710°C and □ 1040°C. Curve shows best fit obtained by fitting the model to the silane and disilane data.

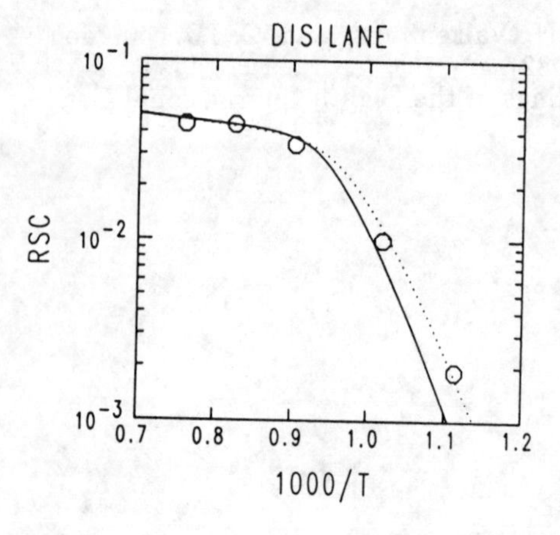

Fig. 3. Arrhenius-type plot of the disilane RSC at a flux of 8.7×10^{16} molecules cm^{-2} s^{-1}. Solid curve shows best fit obtained by fitting the model to the silane and disilane data. Dotted curve shows best fit obtained by fitting the model to the disilane data only.

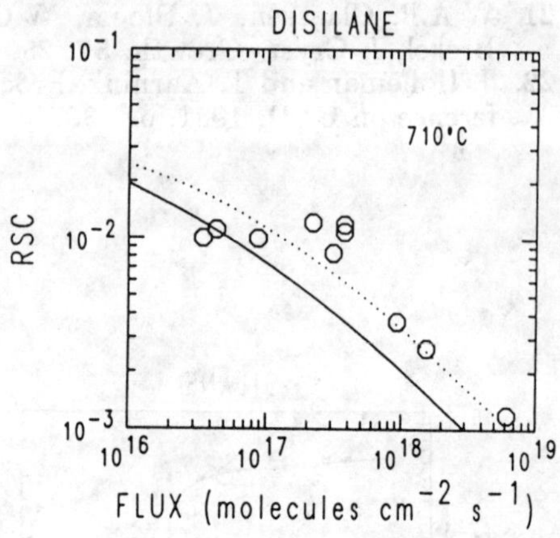

Fig. 4. Disilane RSC as a function of flux at 710°C. Solid curve shows best fit obtained by fitting the model to the silane and disilane data. Dotted curve shows best fit obtained by fitting the model to the disilane data only.

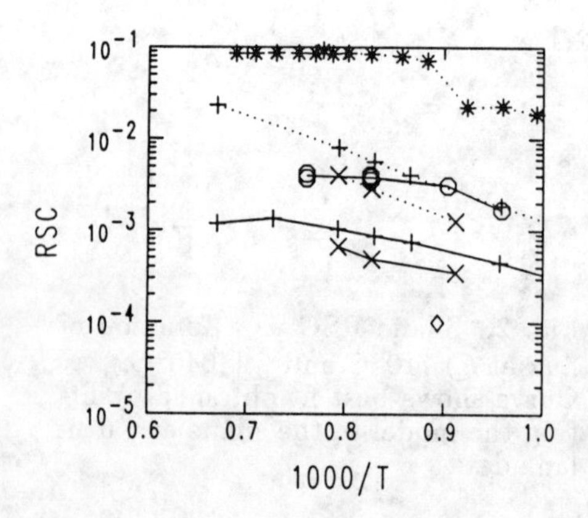

Fig. 5. Comparison of silane RSCs between this work and literature values. Chords connect points in each data set. Solid lines indicate RSCs obtained from silicon deposition. Dotted lines indicate RSCs obtained from H_2 production.
○ data from Fig. 1, silane flux 5.9×10^{17} cm^{-2} s^{-1}.
* apparent reaction probabilities from Farnaam and Olander.[6]
+ silane decomposition efficiencies, silicon deposition rates from Farrow,[5] silane flux 3.6×10^{19} (0.1 Torr).
× RSCs calculated from deposition rate data from Henderson and Helm,[7] silane flux 3.6×10^{19} (0.1 Torr).
◇ silane decomposition efficiency from Joyce and Bradley,[8] silane flux of 7×10^{16}.

FUNDAMENTAL CHEMISTRY OF SILICON CVD

S. M. GATES, B. A. SCOTT and R. D. ESTES
I.B.M. T.J. Watson Research Center, Yorktown Heights, N.Y. 10598.

ABSTRACT

Model CVD reactor studies and UHV surface adsorption kinetic measurements are a powerful combination for investigation of the chemical mechanisms active in thermal silicon CVD from silane. We use the model reactor to separate two regimes of pressure and temparature in which SiH_4 heterogeneous decomposition or homogeneous pyrolysis chemistry dominate the observed silicon film growth kinetics. Residence time of SiH_4 in the reactor hot zone and total pressure are essential quantities distinguishing the two regimes. Growth rates are controlled by surface SiH_4 adsorption kinetics in the heterogeneous regime. The regime we call the homogeneous regime is dominated by adsorption kinetics of higher silanes, Si_nH_{2n+2}. UHV adsorption kinetic measurements comparing SiH_4, Si_2H_6, and Si_3H_8 chemisorption on clean, well defined single crystal surfaces are useful in understanding the two regimes. The UHV studies also demonstrate the necessity of considering the competitive adsorption of SiH_4 with the higher silanes in film growth rate measurements because of homogeneous reactions forming higher silanes from SiH_4 under certain reactor conditions, and because of trace disilane impurities commonly present in commercially available SiH_4.

I. INTRODUCTION

Silicon hydride homogeneous and heterogeneous chemistries are combined in the complicated mechanisms of CVD silicon film growth from silane. Although many modeling studies have been carried out which presume the dominance of one or the other to explain various results {1,2,3}, there have been no definitive experiments to clearly delineate when homogeneous or heterogeneous chemistry, or a mixture of both, are controlling the observed growth kinetics in the CVD process. The present paper illustrates how a better understanding of silane CVD can be obtained by combining two very different approaches: process studies carried out in a CVD reactor, and measurements of fundamental reactant adsorption on clean silicon surfaces perfomed in a UHV system. In the former, we begin with pressure and temperature conditions used in previous silane pyrolysis studies, which are fairly well understood {4,5}. We demonstrate that careful control of CVD variables can lead to a separation of kinetic regimes where either silane gas phase chemistry, or its surface reactions, dominate the process. Interpretation of the reactor studies is facilitated by UHV studies of the surface chemistry of SiH_4 and its higher higher hydrides. Reactive sticking coefficients and activation energies for chemisorption have been measured using H_2 temperature-programmed desorption (TPD) for SiH_4, Si_2H_6 and Si_3H_8. The detailed results of these investigations, obtained under very different sets of experimental conditions, have been presented in part elsewhere {6,7,8}. Here we summarize the results and combine them in an effort to provide the guidelines for a more realistic and comprehensive model for the entire silicon CVD process.

II. EXPERIMENTAL

CVD studies were conducted in a vertical reactor using SiH_4/Ar mixtures and depositing on quartz substrates {6}. Experiments were carried out at several temperatures below the standard CVD range for reasons that will become apparent subsequently. Gas residence times and pressures were varied by adjusting flow rates and the system pumping speed. Film thickness was measured after each run using a step profiler. UHV adsorption studies were performed on both the Si(111)-(7X7) and the Si(100)-(2X1) surfaces, as characterized in situ by LEED. The procedure for measuring the reactive sticking coefficients of these molecules is described in detail elsewhere {8}, but is briefly indicated here. After exposure of the surface to one of the silanes using a calibrated effusive doser, the resulting mass spectrometric hydrogen TPD area is measured and then calibrated to a standard H_2 TPD area obtained by chemisorbing a well defined coverage of H-atoms onto a clean Si surface {9}. The number of adsorbed silane molecules is calculated from the number of surface hydrogen atoms assuming that no H_2 or SiH_x species desorb during the initial exposure, and the method is therefore restricted to below approximately 300 °C surface temperature.

III. RESULTS AND INTERPRETATION

A. CVD Reactor Results

The CVD reactor experiments described below were carried out at relatively low temperatures (400-550°C) and low total pressures (1-100 Torr), associated with previous SiH_4 pyrolysis work so as to overlap with these studies and the information they provide about gas phase silane chemistry {4,5}. It should be pointed out that the deposited silicon films are amorphous under the conditions of our CVD experiments. Therefore growth is occurring on an amorphous surface, and the films may contain 1-5% hydrogen at the lowest temperatures used in this study.

Figure 1 summarizes much of the relevant data obtained under the above range of experimental conditions. Provided that the gas residence time in the reactor hot zone (τ) and total pressure (P_{Tot}) are relatively low, silicon film growth rates (R_G) follow a simple, Langmuir-Hinshelwood-type, dependence on SiH_4 partial pressure (P_{SiH_4}). As shown in the bottom of Fig. 1 for the lowest temperatures of this study, 418°C and 443°C, the deposition rate saturates for $P_{SiH_4} > 1$ Torr. In this growth regime, termed the heterogeneous regime, R_G is dominated by the kinetics of surface SiH_4 decomposition. As is discussed in detail elsewhere {6,7}, the rate determining step is most likely hydrogen desorption from the growth interface. It is important to note, however, that relatively low total pressures and short gas residence times are required for the observation of the heterogeneous regime, and that this restriction becomes more confining with increasing temperature. For example, while $\tau < 20$ sec and $P_{Tot} < 6$ Torr are required to observe the heterogeneous regime at 418°C, we find that at 443 °C it is necessary to maintain $\tau < 6$ sec and $P_{Tot} < 5$ Torr to observe growth due exclusively to the surface decomposition of monosilane.

If the gas residence time and total pressure is not reduced further, a completely new growth regime is encountered on further increasing the temperature. Instead of saturating, R_G is observed to curve upward with increasing P_{SiH_4} at 464 °C. The effect is dramatically illustrated in the middle of Fig.1 at 480°C ($\tau = 5$ sec, $P_{Tot} = 15$ Torr versus $\tau = 0.5$ sec, $P_{Tot} = 1.5$ Torr). The superlinear increase in film growth rate now defines the homogeneous regime, where gas phase reactions contribute new channels to film growth. Higher silanes, produced in the gas, decompose on the film surface at rates that exceed those of SiH_4 surface decomposition. In fact, if the silane partial pressure is raised too high, gas phase nucleation of silicon will occur. The nucleation process involves oligomerization of higher silanes to even larger silicon clusters. Higher silane formation, oligomerization, and cluster growth may all be accompanied by dehydrogenation processes yielding hydrogen deficient silicon species. A mixture of higher silanes and dehydrogenated silicon oligomers are probably the silicon species which lead to film growth in the homogeneous regime described above.

From the data at the top of Fig. 1, it is clear that we can return to the silane heterogeneous channel by appreciably lowering residence time and/or pressure (480°C: τ= 0.5 sec, P_{Tot}= 1.5 Torr). Conditions corresponding to either regime can be found at any temperature {7}. For example, in the data at the top of Fig. 1 the heterogeneous regime has been maintained at 505°C and 533°C by using a very low τ and total pressure.

The homogeneous regime has never been explicitly extracted from CVD experiments, with the exception of HOMOCVD {10}. This is because conventional CVD experiments are carried out above 550°C, where the surface reaction rates become large, resulting in mass transport effects dominating R_G. Careful use of the interplay between τ, P_{Tot} and P_{SiH_4} allows us to obtain an activation energy for film growth in the heterogeneous regime of 37.65 kcal/mole. Previous HOMOCVD experiments find 54 kcal/mole for growth in the homogeneous regime {10}. The results summarized in Fig. 1 clearly show the importance of knowing the nature of the CVD growth regime, and determining when more than a single channel is operative, before applying mechanistic models to the process. For the case of silicon deposition from SiH_4, we can now proceed to further understand the CVD process by examining the fundamental surface reactions of the molecules involved.

B. UHV Adsorption Results

Our UHV studies have focussed specifically on the initial chemisorption step of the silicon film growth mechanism. Reconstructed single crystal surfaces of well defined structure have been used, in contrast to the amorphous surfaces used in the reactor studies. Detailed mechanistic studies on the single crystal surfaces provide a starting point for understanding the complicated surface processes active in the CVD reactor. Reactive sticking coefficients (S^R) and activation energies for chemisorption have been compared for SiH_4, Si_2H_6 and Si_3H_8 on the Si(111)-(7X7) surface {8}. SiH_4 chemisorption on Si(100)-(2X1) is presently under investigation {11}.

Room temperature exposure of the Si(111)-(7X7) surface to calibrated doses of Si_2H_6 and Si_3H_8 revealed S^R = 0.47 ± 0.1 for both molecules on the clean surface {8}. Negative activation energies were measured for di- and trisilane on the clean surface indicating chemisorption via molecular precursor states. Room temperature chemisorption of these gases produces hydrogenated silicon surfaces of unknown structure but well defined H atom coverage. Trisilane was shown to be significantly more reactive than disilane on these hydrogenated surfaces containing 1 monolayer of hydrogen {8}.

Silane is orders of magnitude less reactive than the higher silanes, exhibiting $S^R \approx 10^{-3}$ on the bare Si(111)-(7X7) surface at room temperature. Taking into account a trace disilane impurity (≈ 0.4%) S^R for silane is less than the apparent 10^{-3} value {8}.

Silane chemisorption on the Si(100)-(2X1) surface is now under investigation using purified silane with total impurity concentration < 1 part per 10^4 {11}. Preliminary results indicate that S^R is less than 10^{-5} at room temperature on the bare surface. Experiments using 2 x 10^{19} SiH_4 cm.$^{-2}$ exposure and many surface temperatures in the range 50 - 250 °C have revealed an activation energy of 1.5 ± 0.5 Kcal/mole (data not shown) for the chemisorption step. Figure 2 illustrates the adsorption kinetics observed for silane, where number of molecules adsorbed is plotted versus exposure for 50 and 250 °C surface temperatures. Considering only the 10^{21} SiH_4 $cm.^{-2}$ exposure experiment, at 250 °C , the sticking coefficient of silane on a surface containing 1/2 monolayer of H atoms is roughly 10^{-7}. We have extrapolated this UHV data at 250 °C to the pressures and temperatures of the reactor results in Figure 1. The predicted growth rate is roughly one order of magnitude too large at 418 °C and roughly one order of magnitude too small at 533 °C. The hydrogen atom surface coverage during growth must change dramatically between these two temperatures, and may account for the errors in predicted growth rate. The effects of H atom coverage and of surface structure on the rate of silane chemisorption are still under investigation {11}.

IV. CONCLUSIONS

Two conclusions may be drawn from the comparative UHV adsorption studies that are relevent to the homogeneous and heterogeneous CVD mechanisms defined by the reactor studies. First, on a hydrogenated silicon surface an order of reactivity ranking with number of silicon atoms, 3 > 2 > 1 , is observed for polysilanes, Si_nH_{2n+2}, and silane. Given the known homogeneous pyrolysis chemistry of SiH_4 {4,5}, in which di- and trisilane are observed reaction products, the observed rank order allows us to attribute film growth in the regime we have called the homogeneous regime to adsorption of higher silanes and silicon hydride clusters {6}. Second, the difference in reactivity between silane and disilane causes quantitative silane heterogeneous chemistry studies to be affected by disilane impurities at concentrations in the parts per thousand range.

The activation energy measured in our reactor for film growth strictly in the heterogeneous regime from 418 to 533 °C is about 38 Kcal/mole. This is close to the activation energy for decomposition of surface dihydride species, $SiH_2(a)$, which is 43 Kcal/mole as recently measured by Gupta and co-workers {12}, and indicates that H_2 desorption is the rate limiting step in film growth in this temperature range when P_{Tot} and τ are maintained sufficiently low. Further reactor and UHV studies of the type described here are continuing in an effort to understand the gas phase and surface chemistry aspects of silicon CVD.

V. ACKNOWLEDGEMENT

The authors thank Dr. J. Jasinski for helpful discussions regarding silane purity, and Dr. D. Beach for the purification.

REFERENCES

{1} The extensive literature on modelling of CVD has been reviewed in {2} and {3}.
{2} J. M. Jasinski, B. S. Meyerson and B. A. Scott, Ann. Rev. Phys. Chem. 38, 109 (1987).
{3} D. W. Hess, K. F. Jensen and T. J. Anderson, Rev. Chem. Eng. 3, 97 (1985), and J. Vac. Sci. Tech. A (in press).
{4} J. H. Purnell and R. Walsh, Proc. Royal Soc. A, 293, 543 (1966).
{5} J. M. Jasinski and R. D. Estes, Chem. Phys. Lett., 11, 495 (1985).
{6} B. A. Scott, R. D. Estes and D. B. Beach, Pure and Appl. Chem., in press.
{7} B. A. Scott, R. D. Estes and D. B. Beach, to be published.
{8} S. M. Gates, Surface Science, in press.
S. M. Gates, B. A. Scott, D. B. Beach, R. Imbihl and J. E. Demuth, J. Vac. Sci. Tech. A 5 (2), 628 (1987).
{9} R. J. Culbertson, L. C. Feldman, P. J. Silverman and R. Haight, J. Vac. Sci. Tech. 20, 450 (1982).
{10} B. A. Scott, W. L. Olbricht, B. A. Meyerson, J. A. Reimer, and D. J. Wolford, J. Vac. Sci. Tech. A, 2(2), 450 (1984).
{11} C. M. Greenlief, S. M. Gates, R. R. Kunz, and D. B. Beach, work in progress.
{12} P. Gupta, V. L. Colvin, and S. M. George, Phys. Rev. B, submitted.

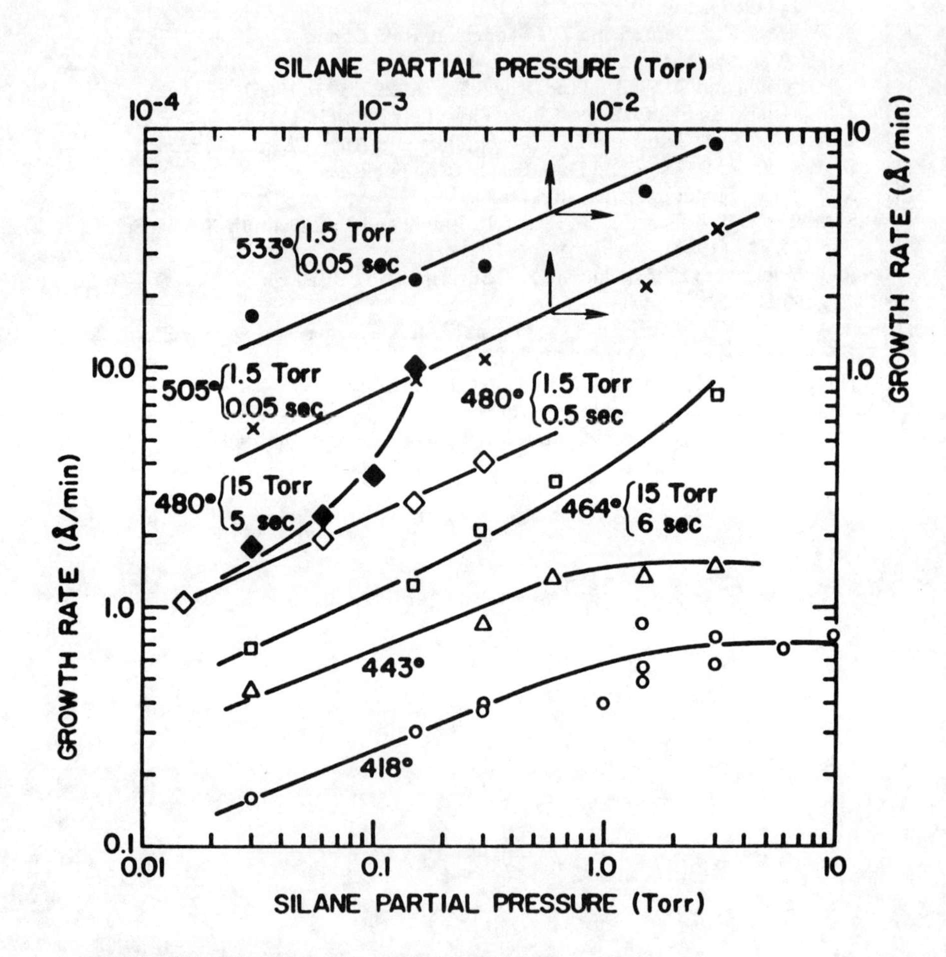

Figure 1: CVD reactor results:
Silicon film deposition rate vs. silane partial pressure at various temperatures, total pressures, and gas residence times.

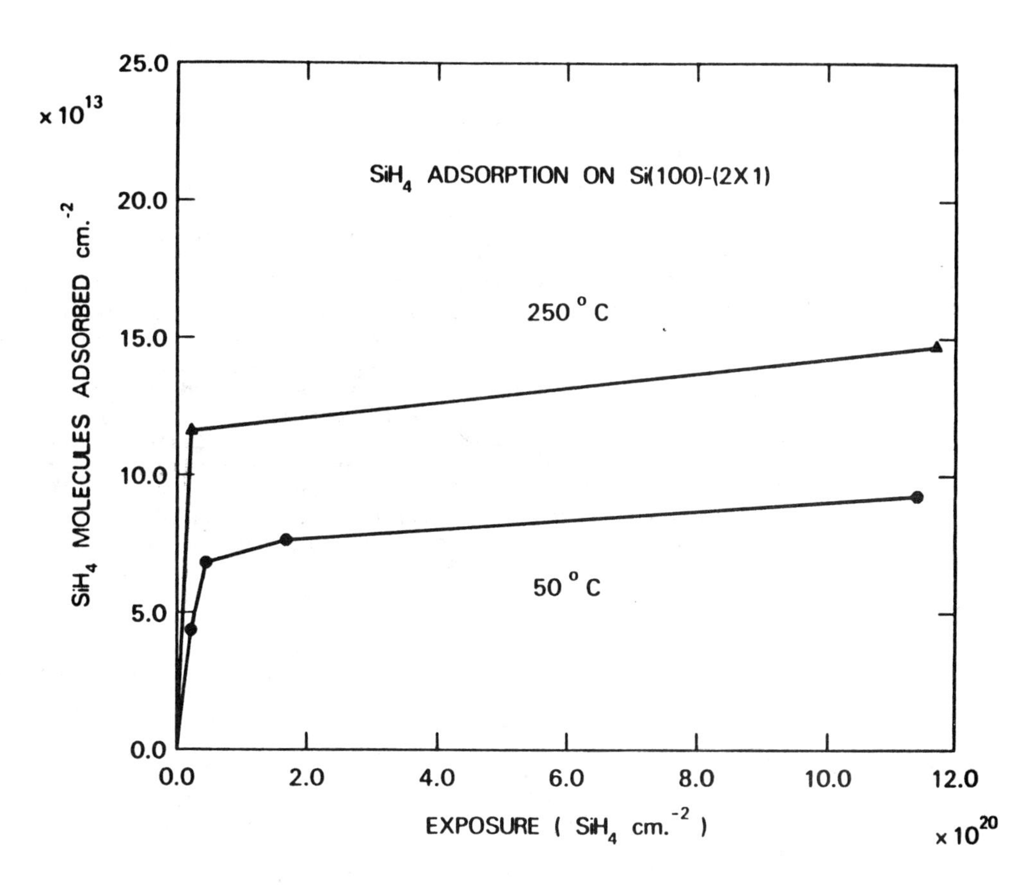

Figure 2. UHV adsorption results:
SiH_4 molecules adsorbed cm^{-2} versus molecules exposure cm^{-2} on the Si(100)-(2X1) surface for surface temperatures of 50 and 250 °C. The molecules adsorbed are measured by H_2TPD calibrated to an internal standard.

Desorption Kinetics of Hydrogen from Silicon Surfaces Using Transmission FTIR

P. Gupta, V.L. Colvin, J.L. Brand and S.M George
Dept. of Chemistry
Stanford University,
Stanford, Calif. 94305

Abstract

The desorption kinetics of hydrogen from crystalline silicon surfaces were measured using transmission Fourier Transform Infrared (FTIR) Spectroscopy. The FTIR desorption measurements were performed in an UHV chamber using porous silicon samples. The desorption kinetics for H_2 evolving from both the monohydride and dihydride species were monitored using the SiH stretch at 2110 cm^{-1} and the SiH_2 scissors mode at 910 cm^{-1}, respectively. Annealing studies revealed that the SiH_2 species desorbed between 640K-700K. Likewise, the SiH species desorbed from the silicon surface between 720K-800K. Isothermal studies displayed second-order hydrogen desorption kinetics for both the monohydride and dihydride surface species. The desorption activation barriers were 65 kcal/mol (2.82 eV) and 43 kcal/mol (1.86 eV) for the monohydride and dihydride species, respectively. Upper limits of 84.6 kcal/mol (3.67 eV) and 73.6 kcal/mol (3.19 eV) for the Si-H chemical bond energies of the SiH and SiH_2 surface species are determined from these desorption activation barriers.

I. Introduction

Hydrogen is a simple prototypical adsorbate and is ubiquitous on silicon surfaces during silicon processing. Consequently, hydrogen chemisorbed on silicon surfaces is of great fundamental and technological interest. The importance of hydrogen during silicon processing is linked to the fact that hydrogen passivates silicon surfaces by tying up silicon dangling bonds. On a single-crystal silicon surface, the reduction of dangling bonds by hydrogen chemisorption dramatically affects the surface reactivity. For example, adsorbates such as SiH_4 (1,2), PH_3 (3), NH_3 (4), H_2O (5) and C_2H_4 (6) chemisorb dissociatively and form silicon-hydrogen bonds. The kinetics of these reactions are thought to be self-poisoned as the product hydrogen ties up silicon dangling bonds. Consequently, these reactions may be rate-limited by the desorption of hydrogen.

Temperature programmed desorption (TPD) mass spectrometric studies have determined an activation barrier of 59 kcal/mol (2.54 eV) for the desorption of H_2 from Si(111) 7 × 7 (7). Activation energies for the desorption of hydrogen from amorphous silicon have also been measured by thermomanometric analysis (8), infrared (9) and mass spectrometry (10). Contradictory results have been obtained because hydrogen can exist in a variety of states (11). Likewise, the difficulty with mass spectrometric studies is that they cannnot distinguish whether H_2 is desorbing from monohydride or dihydride species.

In contrast, FTIR spectroscopy is sensitive to chemical bonding and can distinguish between the monohydride and dihydride features. However, given typical IR cross sections of 1×10^{-18} cm^2 (12), sensitivity requirements limit transmission FTIR studies to high surface area materials. Porous silicon can be utilized to obtain high surface area crystalline silicon samples. Porous silicon was first obtained by Uhlir (13) by anodizing single-crystal silicon in dilute hydrofluoric acid. TEM micrographs have demonstrated that porous silicon contains a network of nearly parallel cylindrical pores approximately 205 Å in diameter (14). This porous network gives an extremely high surface area of approximately 205 m^2/cm^3 (15). Other material properties of porous silicon have been characterized and reported (16).

Porous silicon retains the crystallinity of the original silicon wafer under ordinary preparation conditions (17). Porous silicon also exhibits sharp and pronounced infrared absorption features that can be assigned to silicon monohydride and dihydride surface species. In addition, the close correspondence between the IR spectra of hydrogen on porous silicon surfaces and hydrogen on Si(100) 2 × 1 (18,19,20) suggests that the surfaces of porous silicon are very similar to Si(100) 2 × 1.

In this work, transmission Fourier Transform Infrared (FTIR) Spectroscopy was used to observe the temperature dependent changes in the coverages of both monohydride and dihydride surface species on porous silicon (21). Concurrently, TPD mass spectrometric studies monitored the loss of surface hydrogen as H_2 in the gas phase. These FTIR studies provided quantitative measurements of the kinetics of H_2 desorption from surface monohydride (SiH) and dihydride (SiH_2) species.

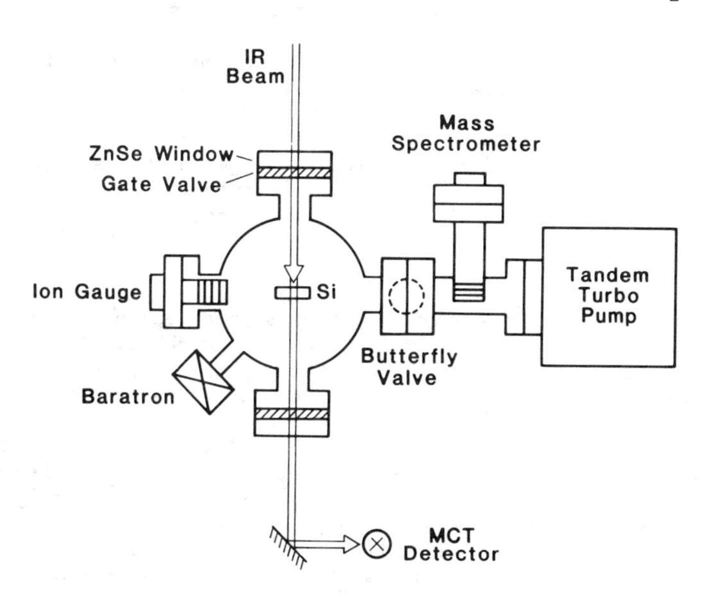

Fig. 1. Diagram of UHV chamber for in situ transmission FTIR studies.

II. Experimental

Following preparation and cleaning (21), the porous silicon samples were mounted in an UHV chamber as shown in Fig. 1. Operating pressures in this chamber were typically 2×10^{-8} Torr. These pressures were critical for consistent results because porous silicon samples can be oxidized easily at temperatures and pressures as low as 400K and 1×10^{-5} torr (22).

For the thermal annealing studies, the porous silicon samples were raised to the annealing temperature, held at the annealing temperature for approximately one second and then returned to the initial temperature. The initial temperature of the porous silicon sample was 300K and a constant heating rate of 8 K/sec was used to heat the sample. A FTIR spectrum was taken after returning to the initial temperature. This experimental sequence was performed repeatedly for annealing temperatures up to 840K.

Isothermal annealing experiments were carried out at various temperatures for both the monohydride and dihydride species. For the dihydride species, temperatures ranged from 540K to 610K. For the monohydride species, temperatures ranged from 690K to 730K. In these experiments, the sample was raised to the isothermal temperature, held at that temperature for a given time, and then returned to the initial temperature of 300K. A FTIR spectrum of the porous silicon sample was then recorded to determine the coverages of the remaining silicon monohydride or silicon dihydride species.

A new porous silicon sample prepared under identical conditions was used for each annealing run. The total concentrations of monohydride and dihydride species were the same within 5 % for each porous silicon sample. The total monohydride and dihydride concentrations were determined using the integrated areas under the absorption peaks at 2110 cm^{-1} and 910 cm^{-1}, respectively. The integrated absorbance of the monohydride species for the isothermal desorption experiments was normalized using an initial monohydride coverage which was obtained by annealing the porous sample to 640K for 10 minutes. This procedure was sufficient to remove all of the dihydride species.

III. Results

After anodization, the porous silicon samples exhibited pronounced infrared absorption features at 2087-2110 cm^{-1} (doublet), 1107 cm^{-1}, 910 cm^{-1}, 666cm^{-1} and 625 cm^{-1}. Fig. 2 shows a typical FTIR spectrum of a porous silicon sample with a 6 micron thick porous layer.

Reflectance IR studies of hydrogen on Si(100) 2 × 1 have assigned the doublet at 2087-2110 cm^{-1} to Si-H stretches (18,23,24). The mode at 910 cm^{-1} has been assigned to the scissors mode of the SiH_2 dihydride species (11,25,26). Likewise, the absorptions at 666 and 625 cm^{-1} have been assigned to the Si-H/Si-H_2 deformation modes (11,26,27). A prominent Si-Si stretch mode appears at 616 cm^{-1} in agreement with previous transmission infrared spectra of single crystal silicon (28). Due to their close proximity, the two peaks at 625 cm^{-1} and 616 cm^{-1} cannot be resolved separately.

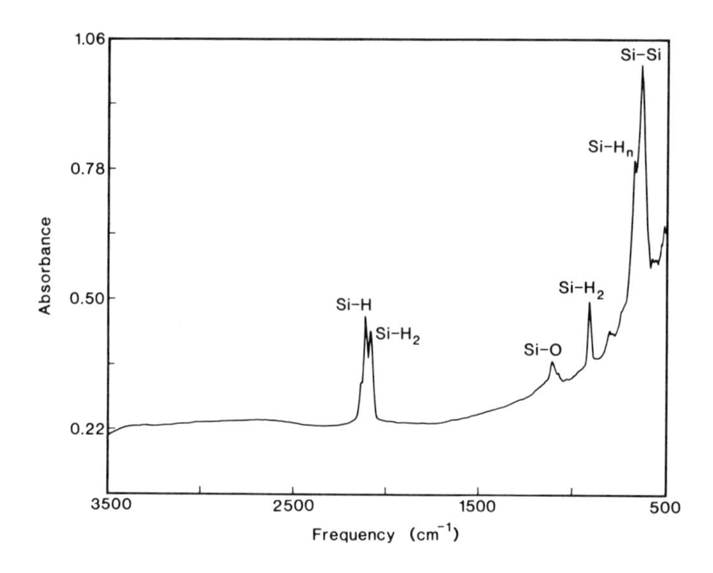

Fig. 2. Infrared spectrum of a porous silicon sample.

The absorption at 1107 cm^{-1} has been assigned to a bulk interstitial Si-O-Si asymmetric stretch (29). This feature also appears with equal intensity in unanodized silicon wafers following an HF etch. This indicates that this absorption feature is not caused by oxygen on the porous silicon surfaces.

Fig. 3 shows the integrated absorbance of the 2110 cm^{-1} and the 910 cm^{-1} absorption features plotted as a function of temperature. The integrated absorbances demonstrate that hydrogen from the silicon dihydride species at 910 cm^{-1} has desorbed from the surface by 720K. On the other hand, the monohydride species at 2110 cm^{-1} remains on the silicon surface until 820K. The integrated absorbances for the silicon monohydride and dihydride species have been weighted by a factor of two. This factor reflects the relative number densities determined by a combined infrared and mass spectrometric analysis discussed below.

The thermal annealing data shown in Fig. 3 clearly demonstrate that H_2 from monohydride and dihydride species desorbs at different temperatures. An increase in the monohydride coverage can be observed at the same point where the dihydride coverage starts to decrease. Given the second-order kinetics observed in the isothermal desorption experiments discussed below, this increase is consistent with the conversion of two dihydride species to two monohydride species plus desorbed H_2.

A complete conversion of dihydride species to monohydride species would give a larger increase in the monohydride coverage than the increase observed. However, H_2 begins to desorb from the monohydride

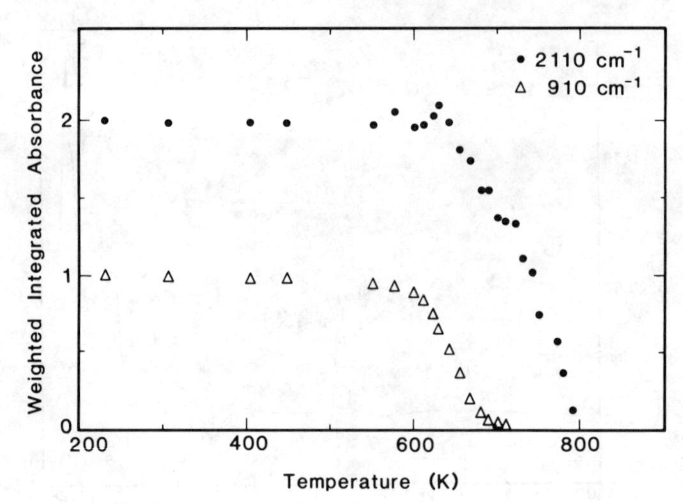

Fig. 3. Integrated absorbances for the silicon monohydride stretch at 2110 cm^{-1} and the dihydride scissors mode at 910 cm^{-1} as a function of annealing temperature.

species during the later stages of dihydride depletion. Thus the expected increase in the monohydride species from the dihydride conversion may be moderated by the concurrent desorption of H_2 from the monohydride species.

Fig. 4 displays the integrated absorbance of the 910 cm^{-1} scissors mode for the SiH_2 dihydride species as a function of time for five different temperatures. The decrease of the integrated absorbance of the 910 cm^{-1} mode reflects the isothermal desorption of H_2 from the dihydride species. Second-order rate equations of the form $d\theta/dt=k\theta^2$ were used to fit the data points. For convenience, the initial coverages θ_o (cm^{-2}) were normalized to $\theta_o'=1$ and integrated second-order fits of the form $\theta'(t)=1/(1+k't)$ were utilized. These fits are shown as solid lines in Fig 4. The rate constants k' obtained using this normalized form can be converted to absolute rate constants by $k'=\theta_o k$.

Each curve in Fig. 4 corresponds to the isothermal desorption of H_2 from silicon dihydride species at one particular temperature. These curves reflect the various temperature-dependent rate constants for H_2 desorption. The Arrhenius plot of the k' rate constants is shown in Fig. 5.

Within the temperature range of this experiment, Fig. 5 shows that the desorption follows Arrhenius behavior given by $k=v_2 \exp[-E_{des}/RT]$. The activation barrier for H_2 desorption from the silicon dihydride species was E_{des}= 43 kcal/mol (1.86 eV). Assuming an initial dihydride coverage of θ_o=2.3 × 10^{14} cm^{-2}, the preexponential for hydrogen desorption from the silicon dihydride species was 4.7 × 10^{-2} cm^2/sec.

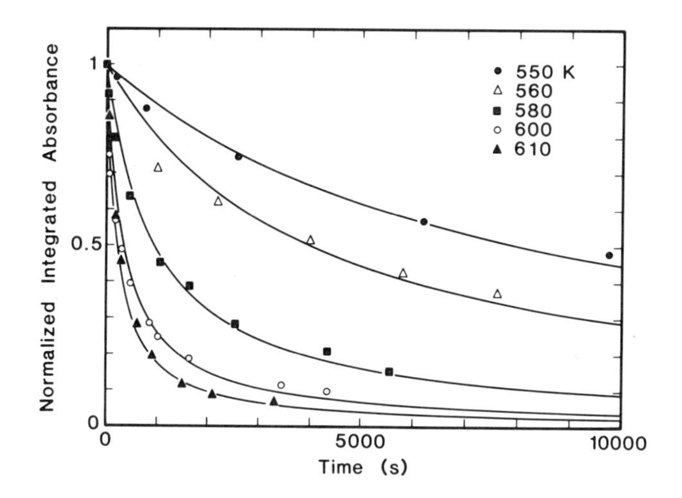

Fig. 4. Isothermal desorption of H_2 from silicon dihydride species measured by monitoring the normalized integrated absorbance of the silicon dihydride scissors mode at 910 cm^{-1} as a function of time at different temperatures. The solid lines show second-order fits of the form $\theta'(t)=1/(1+k't)$ to the data.

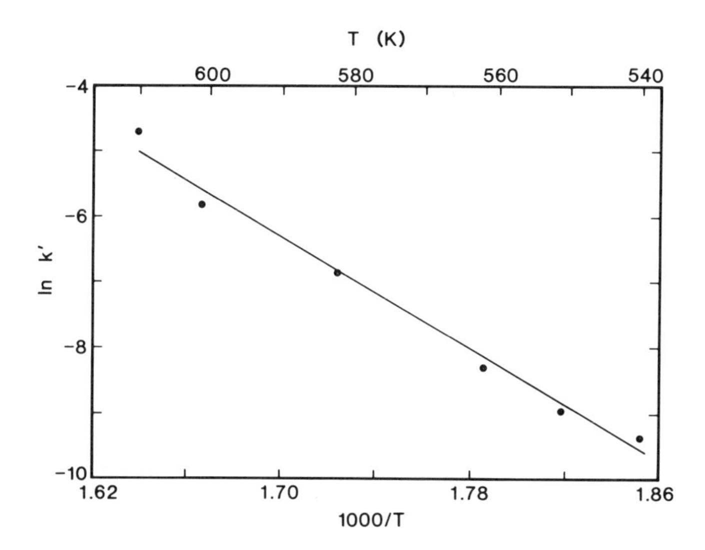

Fig. 5. Arrhenius plot of the temperature-dependent rate constants k' for H_2 desorption from the silicon dihydride species. The measured activation barrier for desorption was E_{des}=43 kcal/mol (1.86 eV).

Fig. 6 shows the isothermal desorption data for the silicon monohydride species as a function of time for five different temperatures. Like the isothermal annealing data for the dihydride species, second-order kinetics fit the data points very accurately. These fits are shown as solid lines in Fig. 6.

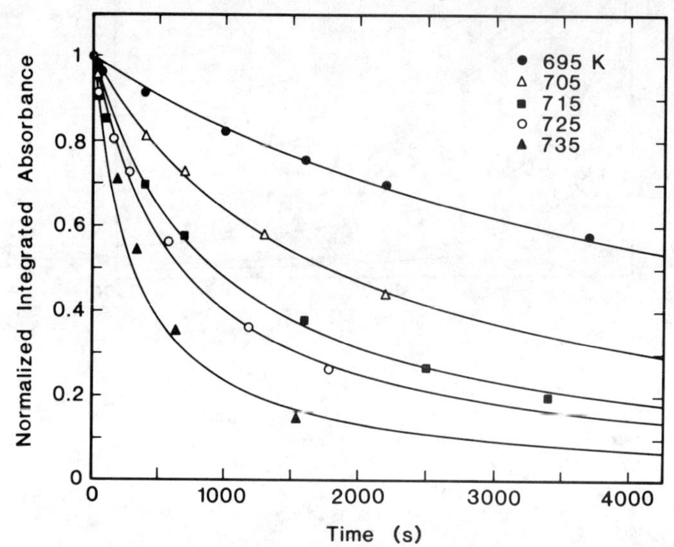

Fig. 6. Isothermal desorption of H_2 from monohydride species measured by monitoring the normalized integrated absorbance of the silicon-monohydride stretch at 2110 cm^{-1} as a function of time at different temperatures. The solid lines show second-order fits of the form $\theta'(t)=1/(1+k't)$ to the data.

Fig. 7 shows the Arrhenius plot of the temperature-dependent rate constants for the isothermal desorption of hydrogen from the monohydride species. The activation barrier for H_2 desorption from the monohydride species was E_{des}=65 kcal/mol (2.82 eV). Assuming an initial monohydride coverage of $\theta_o = 6.8 \times 10^{14}$ cm^{-2}, the preexponential for H_2 desorption from the monohydride species was 1.7×10^2 cm^2/sec. This initial monohydride coverage is equivalent to the number of silicon atoms in the uppermost layer on a Si(100) surface.

Mass spectrometric analysis has demonstrated that the ratio of the integrated desorption fluxes for H_2 evolving from monohydride and dihydride species is 3:1. The observed second-order kinetics for H_2 desorption from dihydride species argues that two dihydride species are converted to H_2 and two monohydride species during recombinatory desorption. Consequently, given the 3:1 ratio for H_2 evolving from monohydride and dihydride species, the initial ratio of the number densities for monohydride and dihydride species is 2:1.

IV. Discussion

The thermal annealing data clearly shows that the high frequency component of the doublet in the silicon-hydrogen stretching region can be assigned to a silicon monohydride species. A typical gas phase cross section for a Si-H stretch is 6.35×10^{-19} cm^2 (12). Using this cross section together with the integrated area under the silicon monohydride peak, a monohydride concentration of 1.81×10^{21} cm^{-3} was determined for a typical 2 micron thick porous layer.

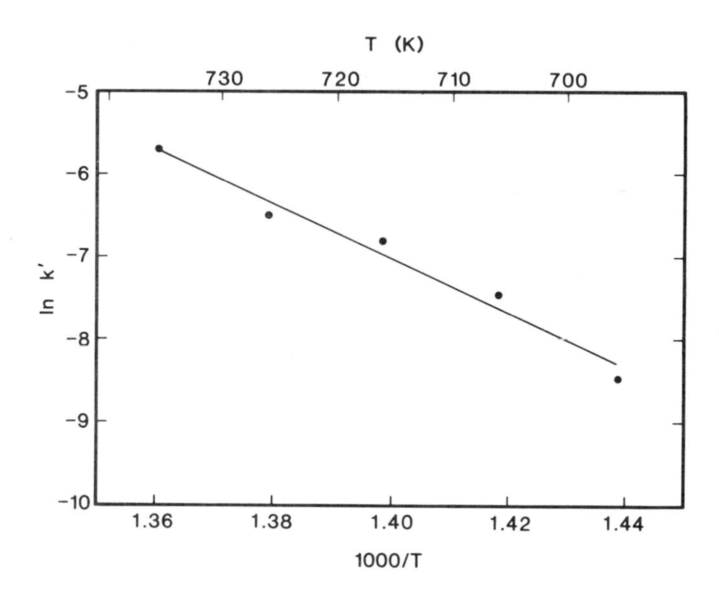

Fig. 7. Arrhenius plot of the temperature-dependent rate constants k′ for H_2 desorption from silicon monohydride species. The measured activation barrier for desorption was E_{des}=65 kcal/mol (2.82 eV).

The monohydride coverage after the thermal removal of the dihydride species was assumed to correspond to the number of silicon atoms in the uppermost layer on Si(100), i.e. a coverage of 6.8×10^{14} cm^{-2}. Consequently, a surface area of 268 m^2/cm^3 was calculated for the porous silicon. This surface area agrees well with the surface area of 205 m^2/cm^3 obtained using B.E.T. methods for porous silicon prepared under similar conditions (15).

Second-order kinetics for H_2 desorption from silicon monohydride species are expected. Two hydrogen atoms from two adjacent silicon monohydride species must recombine in order to form H_2. On the other hand, second-order desorption kinetics from the silicon dihydride species is a new observation. Previous studies had suggested that H_2 desorption from dihydride species was first-order in the silicon dihydride coverage (7,9). The rationalization for first-order kinetics was that the desorbing H_2 molecule was formed by the combination of two hydrogen atoms from the same dihydride unit.

In contrast, a theoretical simulation recently modeled the desorption of H_2 from dihydride species on Si(100) in terms of the recombinatory desorption of two hydrogen atoms from two adjacent silicon dihydride species (30). This model predicts second-order desorption kinetics in agreement with the experimental observations.

The activation barrier of 65 kcal/mol for H_2 desorption from monohydride species is in reasonably good agreement with a previously measured activation barrier of 59 kcal/mol for H_2 desorbing from Si(111) 7 × 7 (7). Recent LITD studies of hydrogen desorption from Si(111) 7 × 7 have also measured an activation barrier of 60 kcal/mol (31). Likewise, a variational transition state calculation has predicted an activation barrier of 55 kcal/mol for hydrogen desorption from an unreconstructed Si(111) surface (32).

The chemical binding energy of a hydrogen molecule is 104.2 kcal/mol (4.52 eV). If the activation barrier of desorption, E_{des}, is equal to the heat of adsorption, E_{ad}, the silicon-hydrogen chemical bond energy can be obtained using $2E(Si\text{-}H)=E_{ad}+E(H\text{-}H)$. The main assumption behind this determination of the Si-H chemical bond energy is the absence of an activation barrier for adsorption.

If an activation barrier for adsorption exists, then $E_{des} > E_{ad}$ and the silicon-hydrogen chemical bond energies will be overestimated. Consequently, upper limit values of 84.6 kcal/mol (3.67 eV) and 73.6 (3.19 eV) are obtained for the silicon-hydrogen bond energies of the SiH and SiH_2 surface species. A recent potential energy surface calculation has predicted an activation barrier of 4.2 kcal/mol (0.182 eV) for H_2 adsorption on Si(111) (32). Using this activation barrier for adsorption, chemical bond energies of 82.5 kcal/mol (3.58 eV) and 71.5 kcal/mol (3.10 eV) are obtained for the silicon-hydrogen bond energies of the SiH and SiH_2 surface species.

V. Conclusions

The kinetics of hydrogen desorption from crystalline silicon surfaces were measured using transmission Fourier Transform Infrared (FTIR) Spectroscopy. The kinetics for H_2 desorption from the monohydride and dihydride species were monitored using the Si-H stretch at 2110 cm^{-1} and SiH_2 scissor mode at 910 cm^{-1}, respectively. Annealing studies revealed that the SiH_2 species desorbed between 640K-700K, whereas the SiH species desorbed between 720K-800K. Isothermal studies revealed second-order kinetics for H_2 desorption from both monohydride and dihydride species. A desorption activation barrier of 65 kcal/mol (2.82eV) and a preexponential of 1.7×10^2 cm^2/sec were measured for H_2 desorbing from the monohydride species. A desorption activation barrier of 43 kcal/mol (1.86 eV) and a preexponential of 4.7×10^{-2} cm^2/sec were measured for H_2 desorbing from the dihydride species. These desorption activation barriers yield upper limits of 84.6 kcal/mol (3.67 eV) and 73.6 kcal/mol (3.19 eV) for the silicon-hydrogen chemical bond energies of the SiH and SiH_2 surface species.

VI. Acknowledgements

This work was supported in part by the Office of Naval Research under contracts N00014-86-K-737 and N00014-86-K-545. Some of the equipment utilized in this work was provided by the NSF-MRL Program through the

Center For Materials Research at Stanford University. We thank Prof. N.S. Lewis and his research group for providing the electrochemical instrumentation required to prepare porous silicon.

REFERENCES:

1. R. Robertson, D. Hillis and A. Gallagher, Chem. Phys. Lett. **103**, 397 (1984).
2. A.M. Beers and J. Bloem, Appl. Phys. Lett. **41**, 153 (1982).
3. M.L. Yu and B.S. Meyerson, J. Vac. Sci. Technol. A **2**, 446 (1984).
4. F. Bozso and Ph. Avouris, Phys. Rev. Lett. **57**, 1185 (1986).
5. Y. Chabal and S.B. Christmann, Phys. Rev. Lett. B **29**, 6974 (1984).
6. M.J. Bozack, W.J. Choyke, L. Muehlhoft and J.T. Yates Jr., Surf. Sci. **176**, 547 (1986).
7. G. Schulze and M. Henzler, Surf. Sci. **124**, 336 (1983).
8. J.A McMillan and E.M. Peterson, J. Appl. Phys. **50**, 5238, (1979).
9. D. Masson, L. Paquin, S. Poulin-Dandurand, E. Sacher and A. Yelon, J. Non-Cryst. Solids, **66**,93 (1984).
10. M.H. Brodsky, M.A. Frisch, W.A. Lanford and J.F. Ziegler, Appl. Phys. Lett. 30, 561 (1977).
11. M.H. Brodsky, M. Cardona, and J.J. Cuomo, Phys. Rev. B **16**, 3356 (1977).
12. L.A. Pugh and K.N. Rao, *Molecular Spectroscopy: Modern Research* (Academic Press, NY, 1976), p.206.
13. A. Uhlir, Bell Systems Tech. Journal, **35**, 333, (1956).
14. M.I.J. Beale, N.G. Chew, M.J. Uren, A.G. Cullis and J.D. Benjamin, Appl. Phys. Lett. **46**, 86 (1985).
15. G. Bomchil, R. Herino, K. Barla and J.C. Pfister, J. Electrochem. Soc. **130**, 1161 (1983).
16. R.W. Hardeman, M.I.J. Beale, D.B. Gasson, J.M. Keen, C. Pickering and D.J. Robbins, Surf. Sci. **152**, 1051 (1985).
17. I.M. Young, M.I.J Beale and J.D. Benjamin, Appl. Phys. Lett. **46**, 1133 (1985).
18. Y.J. Chabal, K. Raghavachari, Phys. Rev. Lett. **54**, 1055 (1985).
19. Y.J. Chabal, J. Vac. Sci. Technol. A **3**, 1448 (1985).
20. Y.J. Chabal, K. Raghavachari, Phys. Rev. Lett. **53**, 282 (1984).
21. P. Gupta, V.L. Colvin and S.M. George, Phys. Rev. B. (submitted).
22. P. Gupta and S.M. George, (in preparation).
23. G.E. Becker and G.W. Gobeli, J. Chem. Phys. **38**, 2942 (1963).
24. Y.J. Chabal, *Semiconductor Interfaces: Formation and properties* (Springer Verlag, NY, 1987).
25. H. Wagner, R. Butz, U. Backes and D. Bruchmann, Solid State Commun. **38**, 1155 (1981).
26. F. Stucki, J.A. Schaefer, J.R. Anderson, G.J. Lapeyre and W. Gopel, Solid State Commun. **47**,795 (1983).
27. P.B. Harwood and G.P. Thomas, Materials Research Society Symposia Proceedings, **71**, 369 (1986).
28. R.J. Collins and H.Y. Fan, Phys. Rev. **93**, 674, (1954).
29. H.J. Hrostowski and R.H. Kaiser, Phys. Rev. **107**, 966, (1957).
30. S. Ciraci and I. Batra, Surf. Sci. **178**, 80 (1986).
31. B.G Koehler, C.H. Mak, D.A. Arthur, P.A. Coon and S.M. George, (in preparation).
32. L.M. Raff, I. NoorBatcha and D.L. Thompson, J. Chem. Phys. **85**, 3081 (1986).

HETEROEPITAXY OF SEMICONDUCTOR/INSULATOR LAYERED STRUCTURES ON Si SUBSTRATES BY MOLECULAR BEAM EPITAXY

Tanemasa Asano, Hiroshi Ishiwara, and Seijiro Furukawa
Graduate School of Science and Engineering
Tokyo Institute of Technology
4259 Nagatsuda, Midoriku, Yokohama 227, Japan

ABSTRACT

This paper reviews two novel heteroepitaxial growth techniques, the thin amorphous layer predeposition method and electron beam exposure epitaxy, which have been developed for growth of high quality Si, Ge and GaAs films on fluoride/Si structures.

INTRODUCTION

As is well known, the molecular beam epitaxy(MBE) has the following two characteristics:

a) The sources of the materials are far from the substrate, and each source can be independently controlled. Moreover, the source material itself is transported without carriers. Thus a wide variety of materials can be formed.

b) Supersaturation is low so that the epitaxial growth can be obtained at low substrate temperatures.

These characteristics fit in with the requirements of fabrication technologies for future semiconductor devices, that is, precisely controlled heteroepitaxial structures. In fact, on Si substrates, heteroepitaxial growth of high quality insulators,[1-3] metal silicides,[4] Si-Ge alloys,[5] and III-V compounds[6] has been achieved.

We have been investigating heteroepitaxial growth of semiconductor/insulator layered structures on Si substrates, by employing alkaline earth fluorides such as CaF_2, SrF_2, and BaF_2 as the insulating layer[7] and Si,[8,9] Ge,[10-12] and GaAs[13,14] as the semiconductor layers. In the course of the study, we have encountered problems owing to differences in material properties between the fluorides and the semiconductors. The material properties which affect the quality of epitaxial films are as follows:

1)Crystal structure and lattice constant.

2)Thermal expansion.

3)Chemical reaction at the interface and related autodoping in epitaxial films.

4)Wettability.

These factors appear more seriously in heterostructures composed of insulator, metal, and semiconductors than in those composed of only semiconductors. Therefore, in order to grow high quality heterostructures, it is necessary to develop growth processes to suppress defects caused by these differences in material properties.

In this paper, after a brief description of growth properties of fluoride films on Si, we review novel growth techniques which have been developed for the growth of Si, Ge, or GaAs layers on fluoride/Si structures. The first one is the thin amorphous layer predeposition method which employs deposition of thin amorphous semiconductor layers on the fluoride surface prior to the growth by conventional MBE process. This method is useful to suppress interfacial reaction and to control necleation mode. The second one is the electron beam exposure epitaxy, in which an electron beam is irradiated to the fluoride surface. This method can improve the wettability of semiconductor to the fluoride surface.

GROWTH PROPERTIES OF FLUORIDE ON Si

The crystal structure of the alkaline earth fluorides is cubic fluorite structure. The lattice constants of CaF_2, SrF_2, and BaF_2 are 0.546, 0.580, and 0.620 nm, respectively. So, the lattice constants of the mixed fluorides can be continuously varied in this range.[15] Both of these properties match well to most semiconductors of interest. But the following two properties of the fluorides hamper straightforward growth of semiconductor/fluoride heterostructures; 1)the thermal expansion coefficients of the fluorides are about $20x10^{-6}$/deg which is much higher than those of semiconductors such as Si($2.5x10^{-6}$/deg) and GaAs($6x10^{-6}$/deg), and 2)the fluorides are ionic bonding material and the surface free energy of the (111) of the fluorides is particularly low.[16]

Concerning the growth of fluoride films, almost stoichiometric fluoride films can be formed by simply evaporating the fluorides in vacuum because they evaporate as molucule. CaF_2 which has the smallest lattice mismatch to Si grows epitaxially on various planes of Si substrates. Single crystal CaF_2 films can be grown on (111) and (100) oriented Si at substrate temperatures 600-800°C and 500-600°C, respectively.[7] Single crystal films of $(Ca,Sr)F_2$ mixed fluorides can be grown on CaF_2/Si structures but not directly on Si substrates.[17]

THE THIN AMORPHOUS LAYER PREDEPOSITION METHOD

Earlier attempts to form semiconductor/fluoride/Si structures have been carried out by conventional MBE growth of Si or Ge on fluoride/Si structures. When these semiconductor films were deposited onto heated fluoride surfaces, however, the following problems appeared as the substrate temperature was raised.

1) For the growth of Si films on CaF_2/Si structures, interfacial reaction takes place between deposited Si and underlying CaF_2 at substrate temperatures above 750°C.

2) For the growth of Ge films, the uniformity of Ge films becomes very poor due to island growth of Ge on fluorides.

In order to overcome these problems, the method of predepositing a thin amorphous layer has been developed.[9,11] In this method, a thin amorphous Si or Ge layer is deposited onto the fluoride surface at room temperature, prior to the deposition of Si or Ge films at elevated substrate temperatures.

Figure 1 demonstrates effects of the thin amorphous layer for

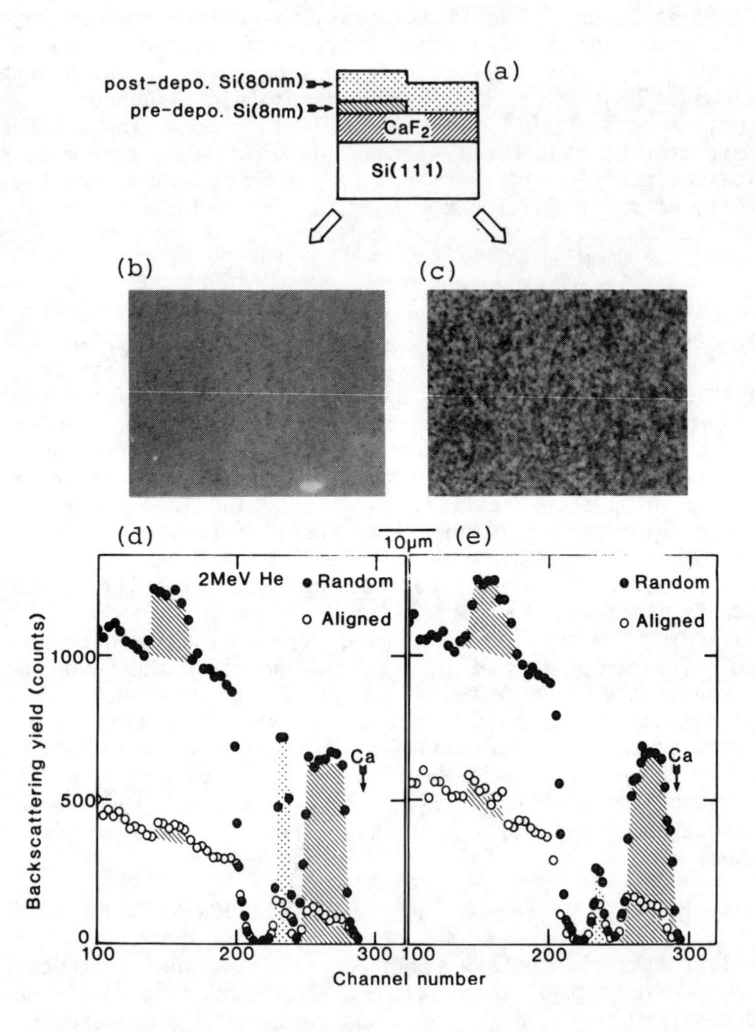

Fig. 1. (a)Schematic illustration of sample geometry. In one region(left side), an 8nm thick Si layer was deposited onto the CaF_2 surface at room temperature followed by deposition of an 80nm thick Si film at 800°C. In the other region(right side), an 80nm thick Si film was deposited directly on top of the CaF_2 film at 800°C. (b) and (c): Optical micrographs of the surfaces of the respective regions. (d) and (e): Random and aligned backscattering spectra taken from the respective regions.(Ref. 9)

preventing the interfacial reaction. The sample consists of two regions. In one region, an 8 nm thick Si layer was predeposited on CaF_2 at room temperature, followed by deposition of an 80nm thick Si films at 800°C. In the other region, an 80 nm thick Si film was directly deposited on CaF_2 at 800°C. The surface morphology of the region with the predeposited layer is rather smooth, while that of the region without the predeposited layer is very rugged. From the Rutherford backscattering spectra shown in the figure, we can see that a uniform Si film grows epitaxially in the region with the predeposited layer.

Figure 2 shows the variation of the channeling minimum yield χ_{min} near the surface of approximately 400 nm thick Si films on (111) and (100) oriented substrates with thickness of the predeposited Si layer. The χ_{min}'s of Si films on both (111) and (100) substrates are less than 8% when the predeposited Si layer is about 10 nm or less in thickness.

In order to investigate epitaxial growth mechanism in the predeposition method, the early stage of the growth of Si on the CaF_2/Si structure was analyzed by transmission electron microscopy (TEM) and ion channeling measurements. For this purpose, a sample as illustrated in Fig. 3 (a) was prepared. That is , an 8 nm thick amorphous Si layer was deposited on the whole surface of the CaF_2/Si(111) structure, and then 7 and 15 nm thick Si films were partially deposited on it at 800°C. Figures 3(b)-3(d) shows TEM pictures taken for the Si films of the respective regions, along with χ_{min} values in the ion channeling measurements. Figure 3(b) clearly shows that the predeposited Si layer forms an island structure when the sample was heated to 800°C. Although the χ_{min} value of the predeposited Si shows that the vast majority of the predeposited Si is not completely aligned epitaxially, electron diffraction analyses have shown that these islands are (111)

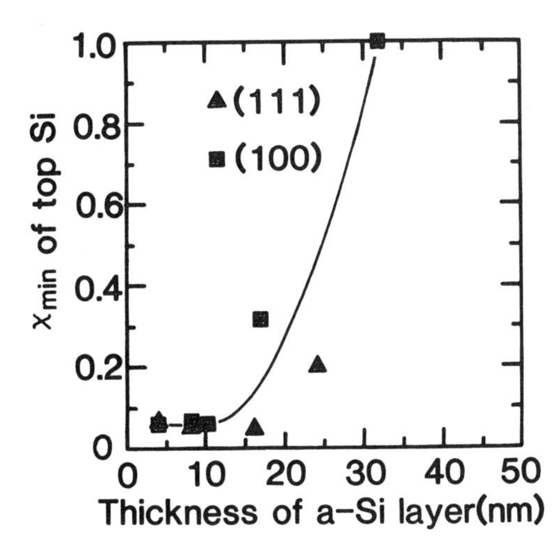

Fig. 2. Variation of near surface χ_{min}'s of 400nm thick Si films grown on CaF_2/Si structures with thickness of predeposited Si layers. (Ref. 9)

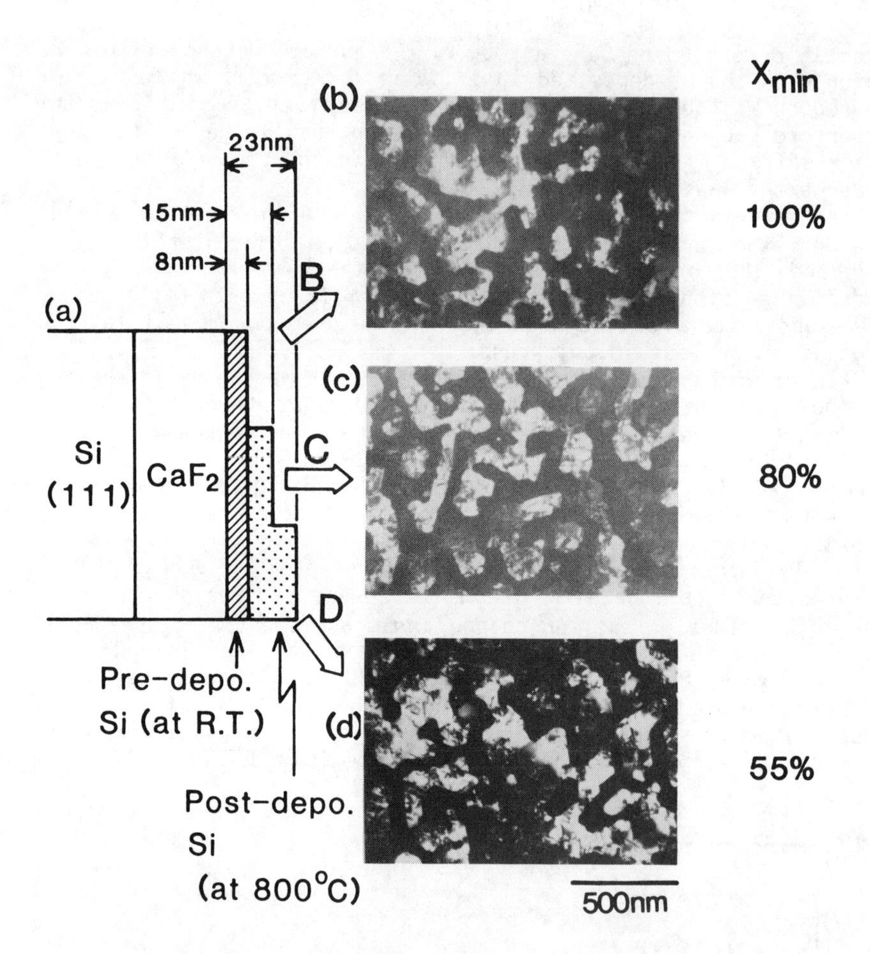

Fig. 3. (a):Schematic illustration of sample geometry. An 8nm thick Si layer was deposited onto the whole surface of the CaF_2/Si(111) structure at room temperature. After that, 7 and 15nm thick Si films were partially deposited at 800°C on top of the predeposited Si layer. (b)-(d):TEM dark field micrographs, along with χ_{min} values, taken for the Si films in the respective three regions.

oriented. When Si is deposited on the predeposited layer at 800°C, the dimension of these islands becomes larger and the epitaxial alignment become evident even in the ion channeling measuremnts. From these results, it can be said that the predeposited Si is almost aligned epitaxially in solid phase and it acts as precursor for the epitaxial growth of Si deposited at elevated temperatures.

By using the predeposition method, preparation of high quality $Si/CaF_2/Si$ heteroepitaxial structure has bocome possible. Feasibility of this structure for device application has been demonstrated by fabrication and characterization of MOSFETs[18] and test integrated circuits.[19]

The predeposition method of thin amorphous layer is also effective in improving crystalline quality and film uniformity of Ge films grown on fluoride structures.[11] Figure 4 shows the variation of the channeling minimum yield χ_{min} near the surface of approximately 300nm thick Ge films grown on CaF_2/Si structures with thickness of the predeposited Ge layers. Although use of $(Ca,Sr)F_2$ matched in lattice constant to Ge has been found to be useful for growing Ge films having better crystalline quality,[15] CaF_2 was used for these experiments in order to investigate the effects of the predeposited layer. We can see from Fig. 4 that Ge films having crystalline quality close to bulk Ge(χ_{min}=3%) is obtained when the thickness of the predeposited layer is about 1 nm. The surface of the Ge films grown on the predeposited layer was very smooth while that of the Ge films grown directly on the CaF_2 surface was very rough. It has been confirmed that interfacial reaction as observed in the Si/CaF_2 system does not occur in the Ge/CaF_2 system. These results clearly suggest that the predeposited amorphous layer is capable not only of preventing the interfacial reaction but also of changing the growth mode of films from three dimensional growth to two dimensional growth at the initial stage.

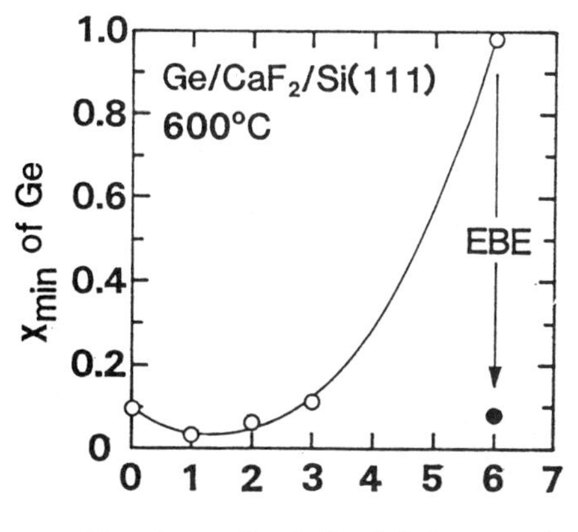

Fig. 4. Variation of near surface χ_{min}'s of 300nm thick Ge films grown on $CaF_2/Si(111)$ structures with thickness of predeposited Ge layer. The growth of the Ge films was carried out at substrate temperature of 600°C. A result of the electron beam irradiation technique was also plotted.

Methods similar to the predeposition technique have been applied successfully to the MBE growth of Si on $NiSi_2$/Si,[4] Si on SrO/Si[3] and GaAs on Si.[20] In the formation of the Si/$NiSi_2$/Si structure, it has been shown that diffusion of Ni in the top Si film can be prevented and uniform Si films having high crystalline quality can be formed by depositing a thin amorphous Si layer on the $NiSi_2$ surface. In the Si/SrO/Si system, it has also been shown that the deposition of an amorphous Si layer is useful to suppress auto-doping of Sr in the top Si film. These results coincide with the results obtained from the growth of Si or Ge on CaF_2/Si. In the formation of the GaAs/Si structures, there has been no discussion of the amorphous GaAs layer from the view point of interfacial reaction or auto-doping. But an interesting aspect of the role of the amorphous layer has been pointed out. That is, the predeposited amorphous GaAs layer has the effect to relax the lattice mismatch between GaAs and Si.[20]

THE ELECTRON BEAM EXPOSURE EPITAXY

Although the predeposition method is effective for the growth of Ge films on the fluoride/Si structures, the optimum thickness of the predeposited layer is as thin as 1 nm. TEM analyses have shown that polycrystalline nucleation takes place when the thickness of the predeposited layer is as large as 6 nm. We presume that this result is due to poor wettability of Ge to the fluoride (111) surface. Recently we have found that the optimum thickness of the predeposited layer can be drastically increased by electron beam irradiation either directly or through the predeposited layer to the fluoride surface.[12] We call this growth method electron beam exposure (EBE) epitaxy. For this growth method, an electrically scanned 3 keV electron beam was irradiated to the sample surface from the direction 3° off from the plane parallel to the sample surface. The temperature of the sample was kept at 600°C during the irradiation.

Figure 5 shows surface morphologies of the predeposited-Ge/CaF_2/Si(111) sample before and after the electron beam irradiation. The two regions were made side by side in the same substrate. The unnirradiated region shows a rugged surface, which may be due to cohesion of predeposited Ge. On the contrary, the surface becomes very smooth after irradiation of the electron beam. From reflection electron diffraction analyses, it has been found that the predeposited Ge layer becomes epitaxial after the irradiation.[12]

A result of ion channeling measurements for a Ge layer on the electron beam exposed predeposited Ge layer is also plotted in Fig. 4. Figure 6 shows TEM cross-sectional view of a Ge/CaF_2/Si(111) structure grown by using the EBE epitaxy. Although defects such as dislocations and stacking faults are present, the crystallinity is rather good. Moreover the film uniformity is excellent. In fact, the surfaces of Ge films grown by this method were featureless as observed by scanning electron microscopy.

The optimum dose range of electrons has been found to be of the

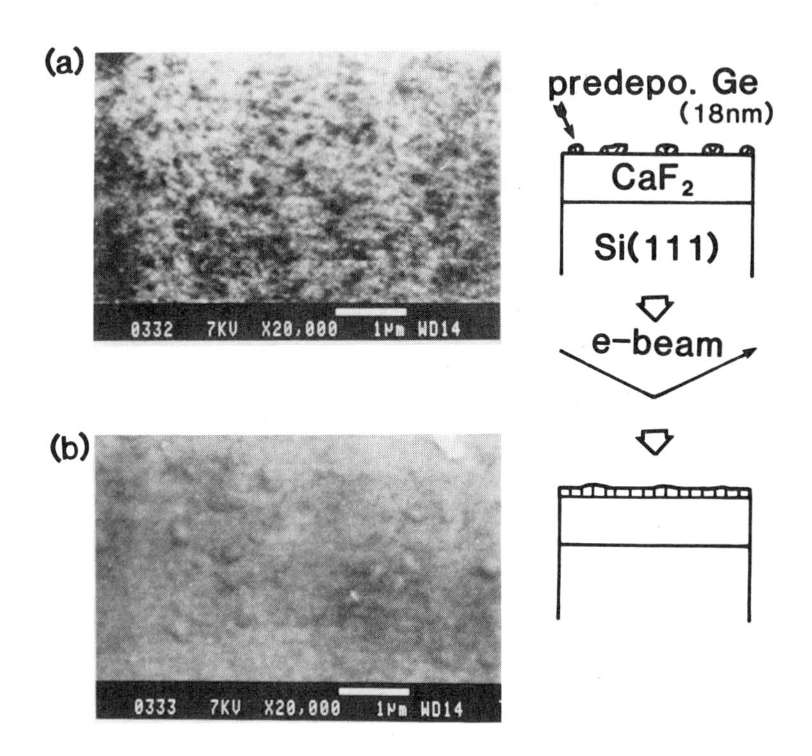

Fig. 5. Scanning electron micrographs of the surfaces of predeposited-Ge(6nm)/CaF_2/Si(111) structures before(a) and after(b) 3keV electron irradiation at 600°C. These two regions were prepared side by side on the same substrate.

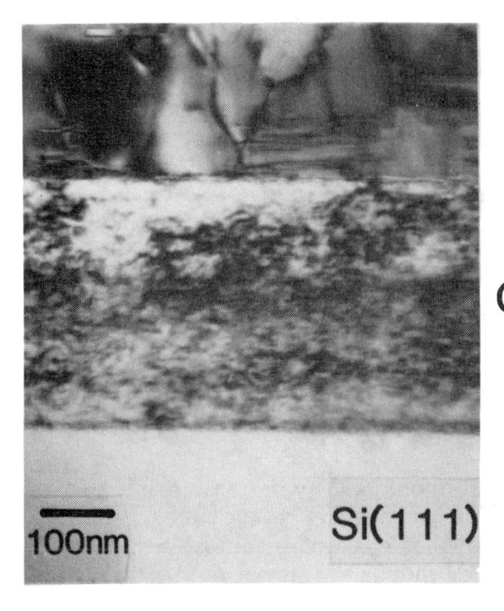

Fig. 6. Cross-sectional TEM bright field micrograph for a Ge/CaF_2/Si(111) structure. The Ge film was grown by using the electron beam exposure epitaxy technique.

order of $10^2\mu C/cm^2$. When the dose was over $10^3\mu C/cm^2$, both the crystallinity and the surface morphology became worse.

Since the electron beam irradiation is effective even when the CaF_2 surface is exposed prior to the predeposition of amorphous Ge, the electron beam irradiation is considered to decompose the CaF_2 surface so as to enhance the wettability of Ge. Actually defficiency of fluorine at electron beam irradiated CaF_2 surface has been reported.[21]

The electron beam exposure epitaxy technique has been found to be effective also for the growth of GaAs films on fluoride/Si structures. In earlier attempts to use this technique for the growth of GaAs films, electron beam irradiation was carried out through amorphous GaAs layer predeposited on the fluoride surface as in the growth of Ge films. This method was actually effective in improving the crystalline quality of GaAs films compared with conventional MBE growth. But the surface morphology was not improved. We presume that the difficulty of the use of the predeposition method in the GaAs growth arises from the fact that both Ga and As atoms can bond to the fluoride surface.

On the other hand, it has been found that electron beam irradiation to the fluoride surface under As molecular beam exposure and MBE growth onto thus irradiated fluoride surfaces is useful to grow high quality GaAs films.[22] Figure 7 shows cross-sectional TEM micrographs of $GaAs/CaF_2/Si(111)$ structures grown by the conventional MBE method and by the EBE epitaxy. The uniformity of the GaAs film grown by the EBE epitaxy is excellent while that of the GaAs film grown by the conventional method is bad having a number of facets on the surface. Furthermore the dislocation density is much lower in the GaAs film grown by the EBE epitaxy, though a small amount of microtwins appears.

Figure 8 shows dependence on the electron dose of the channeling minimum yield χ_{min} of GaAs films grown by the EBE epitaxy. The optimum dose range is several hundreds $\mu C/cm^2$. The best χ_{min} is 4.6% which is close to the value obtained from a bulk GaAs crystal (3.5%). It has also been found that the electron beam irradiation under As molecular beam impingement is effective to grow single crystal GaAs films without rotationally twined crystallites, e.g., the type A growth with respect to underlying CaF_2 films.[23] The optimum dose almost corresponds to that obtained from the growth of Ge films. This result also suggests that the electron beam modifys the fluoride surface as discussed in the above.

CONCLUSION

The growth of heteroepitaxial semiconductor/fluoride/Si structures has been reviewed. Followings summarize this review.

1) The method of predepositing thin amorphous layer is useful to grow high quality Si and Ge films on fluoride/Si structures. The predeposited layer has the ability to prevent interfacial reaction and auto-doping and to control nucleation of epitaxial

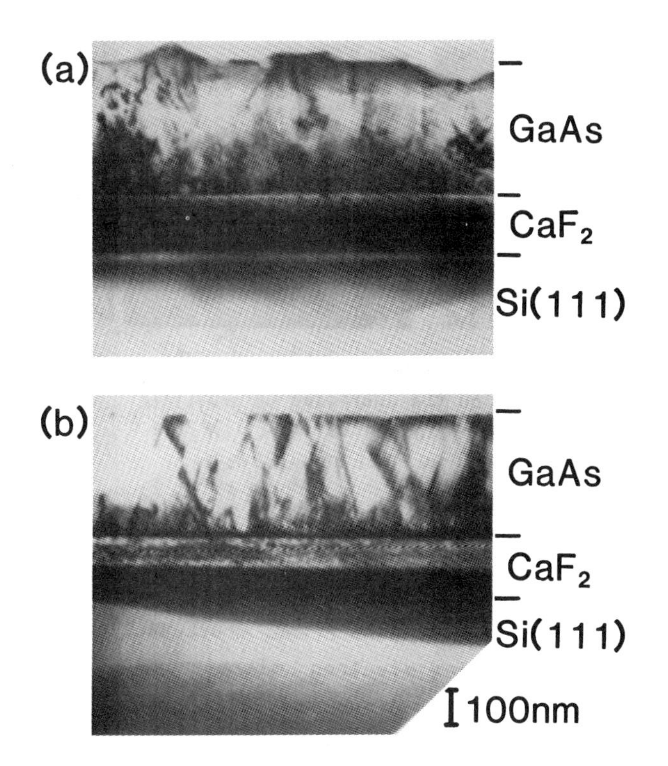

Fig. 7. Cross-sectional TEM bright field micrographs of $GaAs/CaF_2/Si(111)$ structures grown by the conventional MBE method(a) and the electron beam exposure epitaxy technique(b).(Ref. 23)

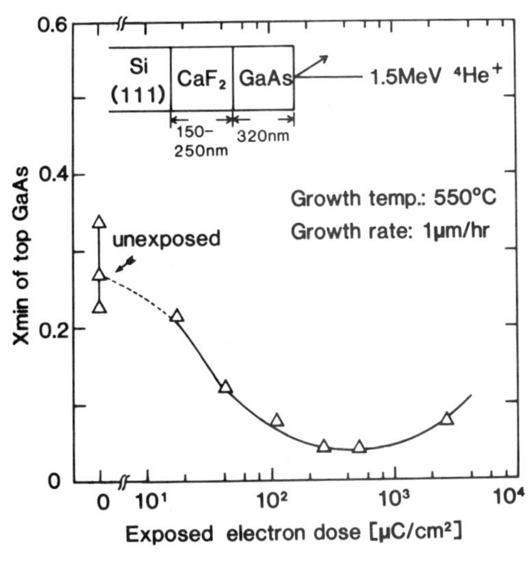

Fig. 8. Dependence on the electron dose of the χ_{min} near the surface of GaAs films grown on $CaF_2/Si(111)$ structures by the electron beam exposure epitaxy technique. The GaAs films were about 320nm in thickness.

films at the initial stage of the growth. By using this method, Si/CaF_2/Si structures having feasibility for device application can be prepared.

2) Electron beam irradiation to the fluoride surface at a moderate dose is useful to grow high quality Ge and GaAs films on fluoride/Si structures. The electron beam irradiation improves the wettability of semoconductors to the fluoride surface.

Although the MBE method has the ability to form various kinds of heterostructures, developments of such novel growth techniques as described above are necessary particularly for the growth of heterostructures composed of metals, insulators and semiconductors. These new heteroepitaxial systems have exciting prospects of application to such devices as very high speed devices and three dimensional ICs.

ACKNOWLEDGEMENTS

The authors are grateful to Dr. S. Kanemaru (presently with Electrotechnical Lab.), H. C. Lee, and Dr. K. Tsutsui for their collaborations. This work was partially supported by the 1987 Grant-in-Aid for Special Distinguished Research(No. 59060002) from the Ministry of Education, Science and Culture of Japan.

REFERENCES

1. H. Ishiwara and T. Asano: Appl. Phys. Lett. **40**, 66 (1982)
2. K. Yoneda, M. Tsuru, H. Nonaka, Y. Ishizuka and T. Nakakado: Proc. 2nd Int. Workshop on Future Electron Devices-SOI Technology and 3D Integration-, (Shuzenji, 1985) p. 75
3. Y. Kado and Y. Arita: Ext. Abs. 18th Conf. Solid State Devices and Materials (Tokyo, 1986) p. 45
4. R. T. Tung and J. M. Gibson: in *Heteroepitaxy on Si*, ed. by J. C. C. Fan and J. M. Poate (Mat. Res. Soc., Pittsburgh, 1986) p. 211
5. J. C. Bean: Proc. 1st Int. Symp. Silicon Molecular Beam Epitaxy (The Electrochem. Soc., 1985) p. 337
6. B-Y. Tsaur and G. M. Metze: Appl. Phys. Lett. **45**, 535 (1984)
7. T. Asano, H. Ishiwara and N. Kaifu: Jpn. J. Appl. Phys. **22**, 1476 (1983)
8. T. Asano and H. Ishiwara: Thin Solid Films **93**, 143 (1982)
9. T. Asano and H. Ishiwara: J. Appl. Phys. **55**, 3566 (1984)
10. T. Asano and H. Ishiwara: Jpn. J. Appl. Phys. **21**, L630 (1982)
11. S. Kanemaru, H. Ishiwara, T. Asano and S. Furukawa: Surf. Sci. **174**, 666 (1986)
12. S. Kanemaru, H. Ishiwara and S. Furukawa: J. Appl. Phys. to be published
13. T. Asano, H. Ishiwara, H. C. Lee, K. Tsutsui and S. Furukawa: Jpn. J. Appl. Phys. **25**, L139 (1986)
14. H. C.Lee, T. Asano, H. Ishiwara and S. Furukawa: Jpn. J. Appl. Phys. **25**, L595 (1986)

15. H. Ishiwara and T. Asano: Jpn. J. Appl. Phys. Suppl. **22-1**, 201 (1983)
16. L. J. Schowalter, R. W. Fathauer, R. P. Goehner, L. G. Turner, R. W. DeBlois, S. Hashimoto, J. L. Peng, W. M. Gibson and J. P. Krusius: J. Appl. Phys. **58**, 302 (1985)
17. H. Ishiwara, S. Kanemaru, T. Asano and S. Furukawa: Jpn. J. Appl. Phys. **24**, L56 (1985)
18. T. Asano, Y. Kuriyama and H. Ishiwara: Electron. Lett. **21**, 386 (1985)
19. H. Onoda, M. Sasaki, T. Katoh and N. Hirashita: in *Heteroepitaxy on Si II*, ed. by. J. C. C. Fan, J. M. Phillips and B-Y. Tsaur (Mat. Res. Soc., Pittsburgh, 1987)
20. M. Akiyama, Y. Kawarada, S. Nishi, T. Ueda and K. Kaminishi: in *Heteroepitaxy on Si*, ed. by. J. C. C. Fan and J. M. Poate (Mat. Res. Soc., Pittsuburgh, 1986) p. 53
21. C. L. Strecker, W. E. Moddeman and J. T. Grant: J. Appl. Phys. **52**, 6921 (1981)
22. H. C. Lee, H. Ishiwara, S. Kanemaru and S. Furukawa: Jpn. J. Appl. Phys. to be published
23. H. C. Lee, T. Asano, H. Ishiwara, S. Kanemaru and S. Furukawa: Ext. Abs. 19th Conf. Solid State Devices and Materials (Tokyo, 1987) p. 163

ANALYSIS OF Zn REDISTRIBUTION IN MOCVD InP LAYERS

F. R. Shepherd, C. Blaauw and C.J. Miner
Bell-Northern Research, Ottawa, Canada K1Y 4H7

Abstract

Although Zn is the source of p-type doping mostly used during MOCVD of InP, it is also known that Zn diffuses rapidly in InP at temperatures typically used for the MOCVD growth (~900K). In this paper we investigate outdiffusion of Zn from InP MOCVD layers into adjacent n-type InP spacer layers. Alternating p-type InP layers (1/2μm thick, C_{Zn}~$1x10^{18}cm^{-3}$) and n-type InP spacer layers (1/2μm, undoped or Si-doped with $10^{15}<C_{Si}<1x10^{19}cm^{-3}$) were grown by low pressure MOCVD. The distribution of Si and Zn was determined by SIMS, using implanted standards to calibrate the data. For undoped InP spacer layers, the Zn completely diffused across the spacer layers during growth (1-2 hours). When the Si concentration in the spacer layers was increased, the extent of Zn out diffusion during growth was dramatically reduced. For $C_{Si} < C_{Zn}$, Zn outdiffused into the spacer layer until C_{Zn}~C_{Si}, at which point the Zn concentration fell steeply. Conversely for $C_{Si} >> C_{Zn}$, no outdiffusion of Zn during continued growth of overlayers was detected. This behavior is similar to that found for Zn diffusion from an external source into n-type InP. In both cases (MOCVD growth, and external Zn diffusion) Zn-donor pairing can be invoked to explain the shape and general behavior of Zn. However, simple donor-acceptor pairing does not account for the low net hole concentrations found in some of these samples. Preliminary results from photoluminescence suggest that an additional Zn-defect complex may be responsible for this low electrical activity.

CHAPTER III
INSULATOR GROWTH

Thermal Oxide Growth on Silicon: Intrinsic Stress and Silicon Cleaning Effects

E. A. Irene
Department of Chemistry
University of North Carolina
Chapel Hill, North Carolina 27514

Abstract

This paper summarizes the experimental results and discusses the implications of recent research on two topics related to Si oxidation: mechanical stress effects; and the influence of impurities on the Si surface. For stress measurement, a double beam optical technique is used to measure the strain in the Si substrate due to the film stress. An intrinsic SiO_2 stress is measured which increases with decreasing oxidation temperature. Controversy exists about whether the intrinsic stress affects transport of oxidant or the interface reaction; arguments for both views are presented. A combination of in-situ ellipsometry and contact angle measurements performed on a Si surface which is immersed in various liquid media has been successfully used to determine the role of HF in Si cleaning process. A fluorocarbon film was found to replace the removed SiO_2, and the fluorocarbon renders the Si surface hydrophobic and amenable to the growth of a high quality SiO_2 film for device applications.

Introduction

The clear trend in silicon Microelectronics processing is towards lower process temperatures (1,2). The motivation is to reduce the diffusion of dopants and thereby to protect diffused junctions and eliminate unwanted interface reactions. This issue arises directly from the industrial effort to reduce device size and thereby increase the level of integration. The impact on the Si thermal oxidation process is profound and several real scientific issues emerge. Firstly, the reduction in lateral size of individual devices on a chip also requires a reduction in the gate area, A, of MOS devices. As seen in equation (1), in order to maintain the same device operating characteristics, viz., the device capacitance, C, because of the reduction of A, the oxide film thickness, L_{ox}, must also be reduced since

$$C = KA/L_{ox} \qquad (1)$$

where K is the dielectric constant for the SiO_2 film. The scientific issue here is that the contemplated integration schemes have driven the thickness range for SiO_2 films to below 20 nm. However, based on many recent studies, the presently accepted oxidation model, the Linear-Parabolic, L-P, model (3-5), is considered to be inapplicable for dry oxidations below about 30 nm SiO_2 thickness. In fact the L-P model is usually derived to contain an offset, L_o, to avert the thin film regime below several tens of nm as:

$$t-t_o = (L-L_o)/k_l + (L^2-L_o^2)/k_p \qquad (2)$$

where L and t are the SiO_2 thickness and time, L_o and t_o represent the small thickness region in L, t space which does not fit the model, viz. the offset region, k_l and k_p are the linear and parabolic rate constants, respectively (5). The initial regime, for $L< L_o$, is characterized by faster than usual oxidation kinetics. Further scientific investigation is required to determine the reasons why this regime is different from the L-P regime. Impurities are known to have a profound effect on this regime (6-8), and we will summarize some recent research in this area that is aimed at an understanding of impurity effects on the Si surface. Secondly, the required reduction in process temperatures directly affects some SiO_2 film properties (9). In particular, intrinsic film stress, σ_i, film density, ρ, Si-SiO_2 interfacial fixed charge, Q_f, and possibly Si-SiO_2 interface states, Q_{it}, all increase with decreasing oxidation temperature and all anneal to lower temperatures as depicted in Figure 1. These properties may also affect the Si oxidation mechanism. In this paper, some current research on impurity effects and mechanical stress effects on the Si oxidation mechanism is discussed. A recent review (ref. 10 and references therein) presents a more complete picture of the silicon oxidation problem. We commence with a summary of stress effects.

Stress Effects on Si Oxidation

Stress Relationships

Two components of the residual film stress, σ, are discussed: thermal stress, σ_{th}, and intrinsic stress, σ_i, and these are additive as:

$$\sigma = \sigma_{th} + \sigma_i \qquad (3)$$

The thermal stress or thermal expansion stress is attributed to the difference in thermal expansion coefficients, $\Delta\alpha$, between the SiO_2 film and the Si substrate as:

$$\Delta\alpha = \alpha(SiO_2)-\alpha(Si) \qquad (4)$$

which when multiplied by the difference between the film growth temperature, T_{ox}, and stress measurement temperature T, $\Delta T = (T_{ox}-T)$ and then multiplied by Youngs modulus, E, divided by $(1-\nu)$ where ν is Poissons ratio for the film yields an expression for the thermal stress:

$$\sigma_{th} = \Delta\alpha\cdot\Delta T\cdot[E/(1-\nu)] \qquad (5)$$

It should be clearly understood that σ_{th} cannot affect the oxidation reaction itself, since from eqn. (5) $\Delta T = 0$ at $T = T_{ox}$ and thus $\sigma_{th} = 0$ at the oxidation temperature. For SiO_2 on Si, $\alpha(SiO_2) < \alpha(Si)$, hence $\Delta\alpha$ is negative and a compressive thermal stress results upon cooling from T_{ox} to room temperature, i.e., σ_{th} is negative. The higher T_{ox}, the larger would be σ_{th} as measured at room temperature. Measurements done with oxides grown at high oxidation temperatures of 1000°C and above (11,12) have yielded stress results which are totally consistent with $\sigma = \sigma_{th}$ ($\sigma_i = 0$), viz. with a numerical value from the above expression for σ_{th} and a larger room temperature value for larger T_{ox} values. However, for oxides grown at lower oxidation temperatures, evidence for a non-zero intrinsic component was reported (13,14). In-situ stress measurements at T_{ox} (where $\sigma_{th} = 0$) have shown that σ_i is also compressive and increases with decreasing T_{ox} (i.e. oppositely to the temperature dependence of σ_{th}. The origin for this intrinsic stress was proposed to be due to the large molar volume change, ΔV, for the conversion of Si to SiO_2 which is 120% or a factor of 2.2x (14-16). The direction and order of magnitude of this stress has been confirmed (17,18).

Film Stress Techniques

Before proceeding to a summary of the stress measurements results, it would be useful to present a brief description of several stress measurement methods recently applied to SiO_2 films. Virtually all of the experimental measurements determine the strain or the deformation in the substrate, as caused by a thin uniform film on one surface of the substrate. The strain is directly proportional to stress in the elastic limit, and for the case of the substrate thickness being much larger than the film thickness, the film stress, σ_f, can be related to the radius of curvature, R, of the substrate through Stoney's formula (19):

$$\sigma_f = [E \cdot L_s^2 / (6 \cdot (1-\nu) \cdot L_f)]/R \qquad (5)$$

where the elastic constants are for the substrate which is deforming and L_s and L_f are the thicknesses for the substrate and film respectively. Thus the experimental techniques are aimed at measuring R. Three general classes of measurements are routinely found in the literature: mechanical, diffraction, and optical. These differ in the manner in which R is measured and under each general class usually several techniques have been developed. A more complete discussion of film stress measurement will be reserved for a separate review, and herein we discuss only specific reports of stress in SiO_2 films on Si substrates. The most recent reports have employed x-ray diffraction and optical techniques (13,14,17,18).

One prominent x-ray technique is based on the maintenance of the Bragg condition for diffraction from thin Si substrates (21,22). Essentially, the single crystal Si substrate is brought under Bragg conditions with respect to the incident x-ray beam, through the use of major reflections for the specific Si orientation used. While monitoring the diffracted radiation at the Bragg angle, θ_B, the sample is traversed relative to the beam while keeping the same angle to the beam. If the sample is perfectly flat, the crystal planes remain with constant angles to the x-ray beam and the Bragg conditions are maintained. If, however, the Si substrate is curved as a result of film stress, then upon traversing the sample in front of the beam, the Bragg condition will be lost and an adjustment in the crystal position will be required to reattain the Bragg condition. The amount of adjustment depends both on the substrate radius of curvature, R, which is to be measured and the distance the sample is traversed which is known. This measurement apparatus is often automated and called ABAC for automated Bragg angle camera. The advantages of this technique is the great sensitivity to stresses as low as 10^7 dyne/cm^2 and absolute reference from the lattice planes in the single crystal substrate. The disadvantage is expense and complexity, as a high intensity x-ray source, diffraction camera, and associated automation apparatus are required.

A number of optical techniques have been used on SiO_2 films. One is the so-called Newton rings technique which makes use of the optical interference pattern caused by combining the reflections from the surface of an optical flat and a curved surface in contact with the flat (23,24). Using monochromatic light, a series of rings result, Newtons rings, the number of which in a certain distance, the separation, is proportional to the curvature of the surface, R. If very uniform surfaces are obtained, this technique can be useful for stresses of about 10^9 dynes/cm^2 and upwards. Cleanliness of the surfaces in contact is a problem as is obtaining flat or uniformly curved substrates for the measurements.

Two other optical reflection techniques have been reported. One technique uses the reflection of one narrow light beam from a film covered surface (see for example ref. (25)). The argument here is that the reflected beam will be undeviated from a flat surface which is traversed during reflection. If the surface is curved, a deviation of the reflected beam will occur. As in the ABAC case, the deviation can be geometrically related to R. However, it is easy to show that a deviation due to curvature of the substrate is difficult, if not impossible, to distinguish from misalignment of a flat sample. Both situations would cause reflected beam deviation. Thus, this technique is difficult to use without unambiguous alignment procedures which, to this authors knowledge, have not been reported. The difficulty with the single beam technique is obviated with the use of two parallel reflected beams. An experimental apparatus is shown in Figure 2 (17,18). The laser light is split (at BS1) then reflected onto a silvered prism and adjusted so that the reflected beams are parallel. The parallel beams are again split (at BS2) with half going to a screen and forming two reference spots of separation X. The other is reflected from the sample surface, across the lab, in order to obtain a lever effect, and then to another plane mirror (M3) and finally to the same screen. If the sample is perfectly flat, the two sets of spots have the same X separation; but if the sample is curved, then a different separation, S, for the second set of spots is seen, viz. the reflections from the sample. R is then easily calculated from the difference between R and X and the measurement geometry. This technique has the decided advantages of being able to distinguish sample tilt from curvature, and to be absolutely calibrated. For sample tilt, the two beams would be deviated in a parallel fashion in one direction, but for curvature only the spot separation changes. To insure beam parallelism, a flat mirror is substituted for the sample and the apparatus in adjusted such that the two sets of spots are identical. For calibration of R, commercially available precision spherical mirrors with known R replace the sample thus calibrating the apparatus geometry. The results from this technique have been found to be quite reproducible and are emphasized in the discussion to follow.

SiO_2 Film Stress Results

The first studies of the SiO_2 film stress (11,12) were performed using SiO_2 films grown at temperatures of 1000°C and higher which were appropriate processing temperatures for that era. These studies used predominantly mechanical measurements of R and were concordant in that they reported residual room temperature stress values that could be completely explained based on the anticipated value for the thermal expansion stress from equation (5) above. Hence, no evidence for an intrinsic SiO_2 stress was reported. From these studies, there was no reason to include stress in any oxidation models, because at oxidation temperature σ_{th} = 0. More recent results using a double beam reflection technique in the Si oxidation environment, in-situ, revealed that a compressive intrinsic stress exists at temperatures less than 1000°C and this stress was found to increase as the oxidation temperature decreased (13). This direction of change of the intrinsic stress with oxidation temperature is opposite to that anticipated for the temperature variation of the thermal stress as obtained from equation (5). Thus, this newly measured intrinsic stress is easily discerned even in the presence of the thermal expansion stress. Of course the in-situ measurement at the oxidation temperature insures that σ_{th} = 0, hence no confusion results. More recent room temperature stress measurements using the ABAC technique (14) and the double beam reflection technique (17,18), have confirmed that an intrinsic SiO_2 film stress exists. Figure 3 shows a plot of total room temperature SiO_2 film stress measurements along with the calculated thermal expansion film stress resulting from equation (5) all as a function of the thermal oxidation temperature. The differences in the temperature variation for these two stress components is evident. Figure 4 shows σ_i which is the difference in the total and thermal stress against oxidation temperature. σ_i is seen to decrease with increasing oxidation temperature.

The origin and temperature dependence of σ_i is understood by considering the large molar volume change, ΔV, for the conversion of Si to SiO_2, and the viscoelastic nature of SiO_2 (14-16). From figure 5, we see that the as-formed SiO_2 occupies a greater volume than the Si from which it was produced, hence an expansion needs to occur into the free volume direction above the oxidizing Si surface. Using a Maxwell model for SiO_2, the rate at which the flow of SiO_2 can occur into this free direction is determined by the SiO_2 viscosity, η. At high oxidation temperatures where η is sufficiently small, the SiO_2 flows readily into the free direction with a short relaxation time relative to the time for oxidation. At lower temperatures, (below about 900°C), however, η is too large for complete relaxation, hence σ_i develops. The lower is T_{ox}, the higher is η, and the higher is σ_i. Relaxation time calculations seem to confirm this viscoelastic model for SiO_2 (14).

In terms of the relationship of σ_i to Si oxidation models, two different ideas have arisen. The first deals with the compressive σ_i in the SiO_2 film. This compression ought to reduce the diffusivity of O_2, hence the supply of O_2 to fuel the oxidation reaction. Thus a decrease in the oxidation rate should occur due to σ_i. Some confirmation for this exists from recent experiments (26) that show that when σ_i is released by long term annealing of thick SiO_2 films, the oxidation rate increases. Considering the molar volume change as the origin of σ_i, the stress distribution in the SiO_2 film should result in a large stress near the Si-SiO_2 interface with a decreasing stress towards the SiO_2 surface. This would result in a higher σ_i and a reduced oxidation rate for thinner SiO_2 films. However, it is well known that a higher oxidation rate is seen for thin films (less than 200Å), yet the shape of the thin SiO_2 oxidation data is explained by this model (27-29). Most recent σ_i measurements (20) have shown that indeed a higher stress exists for thinner films, but not as high as would be required from these diffusion models.

The second stress related oxidation model is derived from the fact that the compressive σ_i in SiO_2 gives rise to a tensile stress at the Si surface. The resulting strained Si bonds ought to yield a more reactive Si surface thereby enhancing the rate of oxidation(30). This idea seems in accord with the observation of faster oxidation rates for thinner films, where the Si-SiO_2 interface is kinetically most important, and some direct evidence for an enhanced rate of oxidation for tensile loaded Si has been presented (31). However, while there seems to be qualitative agreement with the model, quantitative scaling of the oxidation rate with σ_i is not found (32,33) and the predicted orientation dependence for the stress is not observed (18).

From Figure 6, it is seen that the intrinsic stresses for the (100), (110) and (311) Si orientations are near to each other and larger than the stress in the (111) surface. The order for oxidation rate (30,32,33) at the outset of oxidation is:

$$(110) > (111) > (311) > (100)$$

However, after several tens of nm SiO_2 growth, the order changes to:

$$(111) > (110) > (311) > (100)$$

The initial oxidation regime scales qualitatively with the number density of Si atoms in the various Si orientations, but after this regime, the change in order may be a result of the reduced compressive stress in the SiO_2, since the scaling then appears to be in the order of smaller stress higher oxidation rate (33). Since the initial oxidation rate does not scale quantitatively with the number density of Si atoms, and a crossover in rate order occurs for greater film thicknesses, a role for stress in the oxidation mechanism can be envisaged, but it should be clear that any definitive statements about the role of σ_i on Si oxidation kinetics requires further investigation.

Impurity Effects on Si Oxidation

It has long been recognized (6,7) that a variety of impurities can alter the rate of Si oxidation, the mechanism for oxidation, and in many cases the resultant interfacial $Si-SiO_2$ electronic properties. One difficulty with the detailed investigation of these effects is that the chemical analyses used to identify and quantify many of the affecting impurities is far less sensitive to the impurity than is the electronic property or oxidation rate which is altered as a result of the presence of an impurity. However, while the oxidation rate and electronic effects are sensitive, they are not specific and thus there has been great difficulty in establishing clear cause and effect relationships. A similar situation surrounds our understanding of the details of the cleaning process of semiconductor surfaces. Virtually all common semiconductors are subjected to wet and/or dry chemical processes prior to commencing the device fabrication process (34-36). This exposure is usually termed "cleaning" and the intent is to remove any impurities. However, it is well known that for many of the cases examined, some impurities are indeed removed but often impurities are merely replaced. The success of such a cleaning process relies on the innocuous nature of replacement impurities.

Some recent research in the area of impurities and impurity effects suggests that sensitive analytical techniques applied during the cleaning process, so called "in-situ" analysis should prove useful in elucidating the area of semiconductor surface cleaning and the role of impurities (7,8,37). Among the techniques explored and herein reported are in-situ ellipsometry and in-situ contact angle used in the solution cleaning environment of Si.

In-situ Solution Techniques

A. Ellipsometry. Ellipsometry is known to be sensitive to submonolayer coverage of a surface (38). The measurables in ellipsometry are the amplitude change in the light upon reflection, Ψ, and the phase change, Δ. The measurables are related to the other parameters of the reflection problem as:

$$\tan\Psi\exp(i\Delta) = f(n_A, n_F, n_S, \lambda, \varphi) \qquad (6)$$

where the n's are refractive indices (which are in general complex) for the ambient, A, film, F and substrate, S, respectively and λ is the wavelength, L_F, the film thickness and φ the angle of incidence. Usually n_A, n_S, φ, and λ are known. Thus from a single measurement of Ψ and Δ, n_F and L_F are obtained assuming that n_F is real. Figure 7 shows the fused silica sample cell used for the in situ measurements. Alignment of this cell with sample vertical in the solution was a non trivial procedure with the details in the literature (37).

B. Contact Angle. The contact angle, θ, is defined by the equilibrium of three surface tension rectors, γ_{ij}, at the solid, S, liquid, L, vapor, V, interface as γ_{SV}, γ_{SL}, γ_{LV}:

$$\cos\theta = (\gamma_{SV} - \gamma_{SL})/\gamma_{LV} \qquad (7)$$

Figure 8 shows the relationship. Usually only γ_{LV} is known for a liquid, so that since θ is measured only the difference $\gamma_{SV} - \gamma_{SL}$ is obtained. Yet the wetting behavior of solids is directly related to θ where large angles indicate hydrophobic behavior (non wetting) and small angles hydrophilic behavior (wetting). Most metals, oxides, and semiconductors are high energy solids with surface tensions, γ_{SL}'s, of from 500 to 5000 dynes/cm. Waxes and some polymers are low energy solids with surface tensions of less than 100 dynes/cm$_x$. Virtually all liquids, except liquid metals, have surface tensions less than 100 dynes/cm. From these values, it can be argued on thermodynamic grounds that virtually all liquids will wet most solids, so as to lower the surface energy of the solid and thus yield a small, hydrophilic θ. It has been shown that the critical surface tension, γ_c, obtained when $\cos\theta = 1$ and thus $\gamma_c = \gamma_{LV} = (\gamma_{SV} - \gamma_{SL})$ is specifically related to the surface structure and composition of a solid surface (39).

For the measurement of θ on semiconductor surfaces, during cleaning, and with the surface protected from the atmosphere, an inverted bubble technique (40) was adapted. The apparatus used for the experimental results and the experimental details are reported elsewhere (37). A gas bubble (N_2) can be released from the capillary and when held at the solid surface will establish the three phase equilibrium. It is important to realize that even though the bubble here is gas, the contact angle measured is the same as for the case shown in Figure 8 .

Results of In-situ Ellipsometric and Contact Angle Measurements

The first experimental result makes use of in-situ ellipsometry to follow the process of HF etching of SiO_2 on Si. The use of HF in Si technology is widespread for both the patterning and complete removal of SiO_2 from Si and the pre processing cleaning step to remove a contaminated or damaged native SiO_2 film. Furthermore, it is established that the exposure of Si to HF alters the oxidation rate of Si(7,37). The use of HF is an integral part of the successful cleaning of Si. It is anticipated that the etching process of SiO_2 in dilute HF should be able to be followed using in-situ ellipsometry, and Figure 9 shows the rather linear decrease in SiO_2 film thickness with etch time. In this experiment, it was expected that the bare Si surface would be reached. This seems not to be the case, as a minimum of about 20Å SiO_2 film thickness is reached, and it even appears as if the oxide grows slightly. However, in order to correctly interpret these results, it must be remembered that an ellipsometric model of the film covered surface is required for the analysis of the Ψ,Δ data obtained during the etching process. Up to here, we have used model comprised of three components: air - SiO_2 film - Si substrate and the Ψ,Δ data is always interpreted in terms of a calculated SiO_2 film thickness. It is seen in Figure 9 that this model works quite well down to about 20Å but below this SiO_2 thickness, the situation is not as clear. Using this simple model, the SiO_2 etching first nearly ceases and then the SiO_2 grows. Neither of these events is entirely plausible considering what we know about the virulence of the HF attack on SiO_2. If the ψ,Δ trajectory is tracked below 20Å as the SiO_2 film is removed, the situation becomes somewhat clarified and this unmodeled data is shown in Fig. 10. The solid line represents a theoretical calculation of ψ,Δ values for the situation: air - SiO_2 film - Si substrate. The zero SiO_2 film thickness, i.e. ψ,Δ for a bare Si surface is about

178°, 10.5° and as SiO_2 grows, Δ decreases and ψ increases. The etch experiment shown in Fig. 10 commenced at an SiO_2 film thickness of 85nm with a ψ,Δ of about 103°, 15.5° and proceeds towards the bare Si surface value with increasing ψ along the theoretical curve (open circles). Excellent agreement is seen along this line from 85 to about 2nm. However, near the 2nm ψ,Δ value, a deviation from the theoretical curve is seen (triangles) with Δ decreasing again and ψ increasing, but for ψ a much slower increase than if the correct model included a growing SiO_2 film. Thus, while a value of 2-3nm for SiO_2 is obtained for the ψ,Δ data in the triangle region, the data is seen to deviate substantially from the theoretical curve for SiO_2 on Si and hence from the simple model for a film on a substrate. This strongly suggests that the model is not correct and possibly that a different film is growing on the Si surface which traces out a different ψ,Δ trajectory.

In the effort to better determine the nature of the new film forming on the Si surface, in-situ contact angle measurements have proven useful(37). First it was observed that the contact angle, θ, on SiO_2 on Si was about 8° which indicates strong hydrophilic behavior of SiO_2 as anticipated for a high energy solid in contact with an essentially aqueous media. When θ is followed during HF in H_2O etching of SiO_2, an abrupt change from 8° to 78° occurs when the SiO_2 is thought to be removed (near 2nm SiO_2). The abrupt change from hydrophilic to hydrophobic behavior is a commonplace observation in HF treatment of a SiO_2 covered Si surface(42). However, based on what we know about the surface energy of Si, namely it is high, the bare Si surface should be hydrophilic as is SiO_2, and thus no abrupt change in θ is expected. We conclude that there is something else on the Si surface, something other than SiO_2 or bare Si. In order to elucidate the nature of the hydrophobic Si surface resulting from HF exposure, the critical surface tension γ_c was measured. For this purpose, a plot of measured cosθ versus γ_{LV} is obtained on the Si surface in contact with a number of liquids with various γ_{LV} and all with HF. This was done using solutions of H_2O-CH_3OH all with 1% HF. γ_{LV} varied from 72 dynes/cm for pure H_2O to 23 dynes/cm for pure HF. Figure 11 shows the plot. First it is seen that the plot is not linear. This has been shown to be the case whenever H bonding between liquid and solid can occur (40) which is expected for H_2O-CH_3OH solutions. More importantly is the extrapolated value of γ_c = 27 dynes/cm for cosθ = 1. A comparison with literature values of γ_c leads to the conclusion that the new film on Si is either a hydrocarbon and/or fluorocarbon species. Since F is present and indeed crucial, and F-C bonds are polar enough to cause H bonding between Si and liquid, we conclude that a fluorocarbon is adsorbed on the Si surface.

It is interesting that this fluorcarbon film on Si renders the surface hydrophobic thereby likely precluding much foreign and potentially degrading impurities from attaching to an otherwise high energy Si surface. In addition, this treatment leads to clean MOS devices. The effect of a final HF treatment is essentially to yield the largest oxidation rate in comparison with other accepted cleaning solutions, e.g., H_2O_2, HCl, NH_4OH (8). It is not yet known why this is the case and other effects of HF have not been determined, e.g. any long term effects of residual F after thermal oxidation.

Summary

The details of the mechanism for the reaction between Si and oxidant to form an SiO_2 film is complex with many aspects: chemical, mechanical, electrical, morphological etc.. Many reviews on silicon oxidation have treated the problem in detail. The present paper focuses on two very recent results, namely mechanical intrinsic film stress implications and impurity effects on the oxidation mechanism. For these new studies, novel techniques were used and are in themselves interesting means to study surface films. The results, while by no means have settled the major issues, have helped to gain further insight into the important problem of the mechanism for Si oxidation.

Acknowledgement

The author is indebted to G. Gould and E. Kobeda for access to their original data and for helpful discussions. This work was supported in part by the Office of Naval Research, ONR.

References

1. E.A. Irene, Semiconductor International, April 1983, p. 99.
2. E.A. Irene, Semiconductor International, June 1985, p. 92.
3. B.E. Deal and A.S. Grove, J. Appl. Phys., 36, 3770 (1965).
4. W.A. Pliskin, IBM J. Res. Dev., 10, 198 (1966).
5. E.A. Irene and Y.J. van der Meulen, J. Electrochem. Soc., 123, 1380 (1976).
6. A.G. Revesz and R.J. Evans, J. Phys. Chem. Solids, 30, 551 (1969).
7. F.N. Schwettmann, K.L. Chiang and W.A. Brown, 153rd Electrochem. Soc. Meeting, Abs. #276, May 1978.
8. G. Gould and E.A. Irene, J. Electrochem. Soc., 134, 1031 (1987).
9. E.A. Irene, Phil. Mag. B, 55, 131 (1987).
10. E.A. Irene, CRC Reviews in Solid State and Materials Science, "Models for the Oxidation of Silicon," in press 1987.
11. R.J. Jaccodine and W.A. Schlegel, J. Appl. Phys., 37, 2429 (1966).
12. M.V. Whelan, A.H. Gormans and L.M. Goossens, Appl. Phys. Lett., 10, 262 (1967).
13. E.P. EerNisse, Appl. Phys. Lett., 30, 290 (1977); 35, 8 (1979).
14. E.A. Irene, E. Tierney and J. Angillelo, J. Electrochem. Soc. 129, 2594 (1982).
15. T.Y. Tan and U. Goesele, Appl. Phys. Lett., 39, 86 (1981); 40, 616 (1982).
16. W.A. Tiller, J. Electrochem. Soc., 128, 689 (1981).
17. E. Kobeda and E.A. Irene, J. Vac. Sci. Technol. B, 4, 720 (1986).
18. E. Kobeda and E.A. Irene, J. Vac. Sci. Technol. B, 5, 15 (1987).
19. G.G. Stoney, Proc. R. Soc., London Ser. A 82, 172 (1909).
20. E. Kobeda and E.A. Irene, J. Vac. Soc. B., submitted 1987.
21. G.A. Rozgonzi and D.C. Miller, Thin Solid Films, 31, 185 (1976).
22. A. Segmuller, J. Angillelo, S.J. La Placa, J. Appl. Phys., 51, 6224 (1980).
23. A.G. Blachman, Metal. Trans., 2, 699 (1971).
24. E.A. Irene, J. Electronic Mat., 5, 287 (1976).
25. A.K. Sinha, H.J. Levinstein, and T.E. Smith, J. Appl. Phys., 49, 2423 (1985).
26. J.K. Srivastava and E.A. Irene, J. Electrochemical Soc., 132, 2815 (1985).
27. A. Fargeix, G. Ghibaudo and G. Kamarinos, J. Appl. Phys., 54, 2878 (1983); 54, 7153 (1983); 56, 589 (1984).
28. G. Camera Roda, F. Santarelli and G.C. Sarti, J. Electrochem. Soc., 132, 1909 (1985).
29. R.H. Doremus, Thin Solid Films, 122, 191 (1984).
30. E.A. Irene, H.Z. Massoud and E. Tierney, J. Electrochem. Soc., 133, 1253 (1986).

31. C.K. Huang, R.J. Jaccodine and S.R. Butler, Abs. 34, Electrochemical Society Meeting, Extend. Abstracts, Vol. 86-2, San Diego, CA, Oct. 19-24 (1986).
32. E.A. Lewis, E. Kobeda and E.A. Irene, "Proceedings of Fifth International Symposium on Silicon Materials Science and Processing," Ed. H.R. Huff, Boston, Mass., May 1986.
33. E.A. Lewis and E.A.Irene, J. Electrochem. Soc., in press 1987.
33. E.A. Irene, H.Z. Massoud and E. Tierney, J. Electrochem. Soc., 133, 1253 (1986).
34. W. Kern, Semiconductor International, p. 94, April 1984.
35. B.F. Phillips, D.C. Burkman, W.R. Schmidt and C.A. Petersen, J. Vac. Sci. Technol. A, 1, 646 (1983).
36. R.C. Henderson, J. Electrochemical Soc., 119, 772 (1972).
37. G. Gould and E.A. Irene, J. Electrochem. Soc., submitted 1987.
38. R.M.A. Azzam and N.M. Bashara, "Ellipsometry and Polarized Light," North Holland Publishing Co., New York (1977).
39. Zisman, "Contact Angle: Wetability and Adhesion," Advances in Chemistry Series Vol. 43, Ed. F.M. Fowkes, American Chem. Soc., Washington, DC (1964), Chap. 1.
40. D. McLachlan Jr., and H.M. Cox, Rev. Sci. Instrum., 46, 80 (1975).
41. E.A. Irene, Phil. Mag. B., 55, 131 (1987).

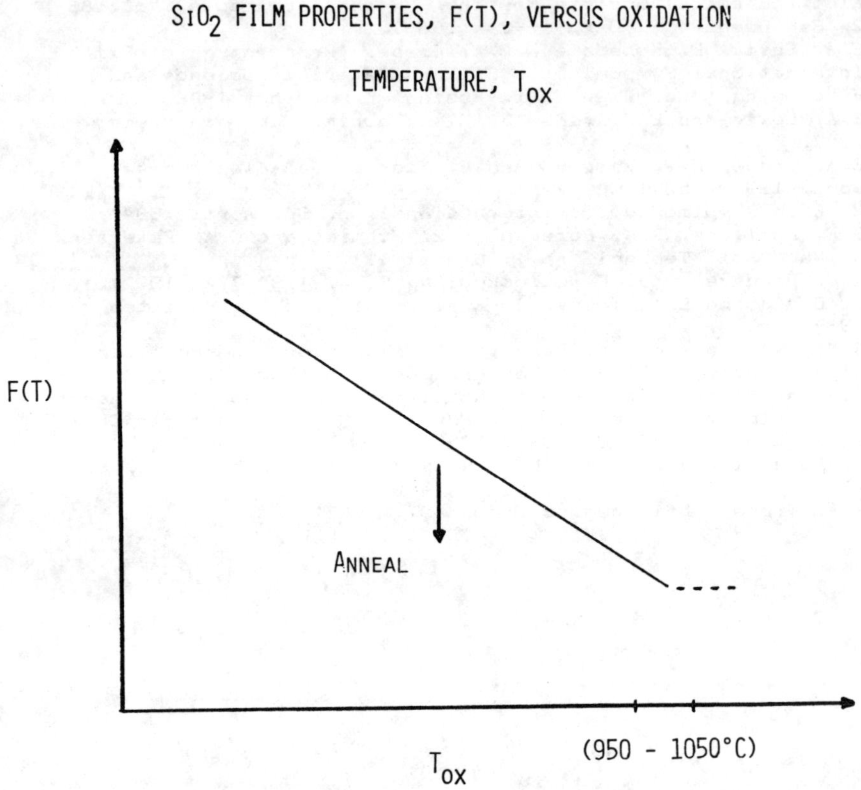

SiO_2 FILM PROPERTIES, F(T), VERSUS OXIDATION TEMPERATURE, T_{OX}

F(T)

ANNEAL

T_{OX}

(950 - 1050°C)

F(T)	
INTRINSIC STRESS	σ_i
DENSITY	ρ
REFRACTIVE INDEX	N
FIXED OXIDE CHARGE	Q_F
INTERFACE TRAPPED CHARGE	Q_{IT}

Figure 1. A schematic summary of F(T) versus oxidation temperature, T_{Ox}, where F(T) are the various oxidation temperature sensitive properties for SiO_2 such as: refractive index, density, intrinsic stress, interface fixed charge, and interface trapped charge. (after ref. (41) and with permission of Phil. Mag.).

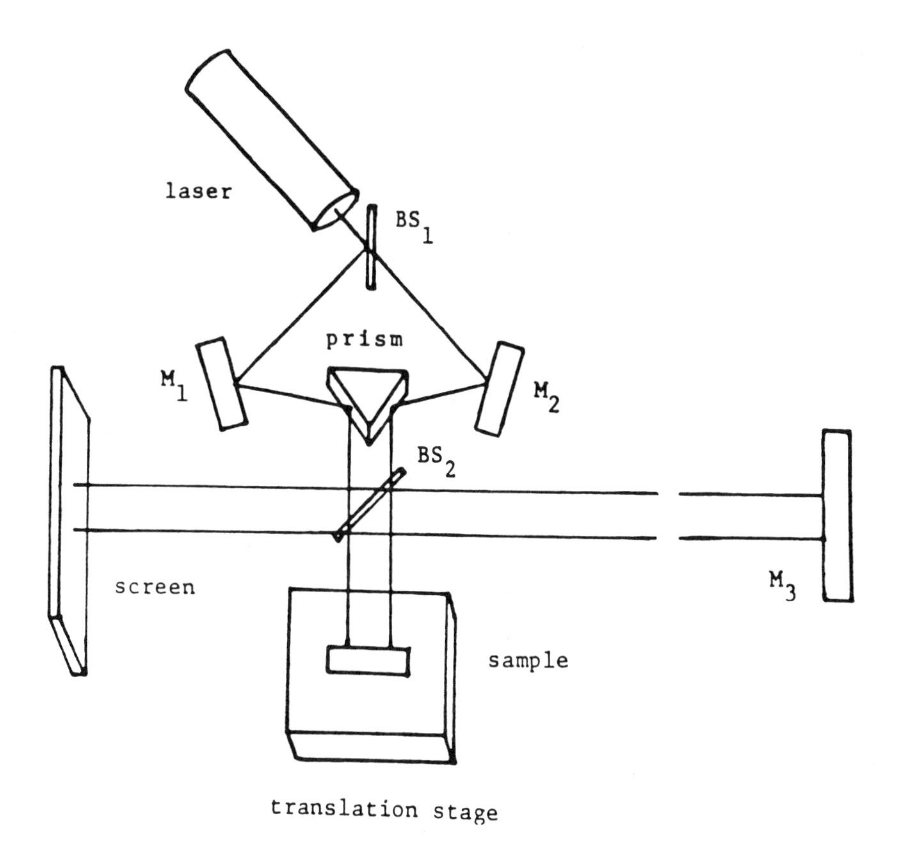

Figure 2. Parallel laser beam apparatus for the measurement of wafer curvature. BS is a beamsplitter and MS is a flat mirror. (after ref. (17) and with permission of the American Vacuum Society).

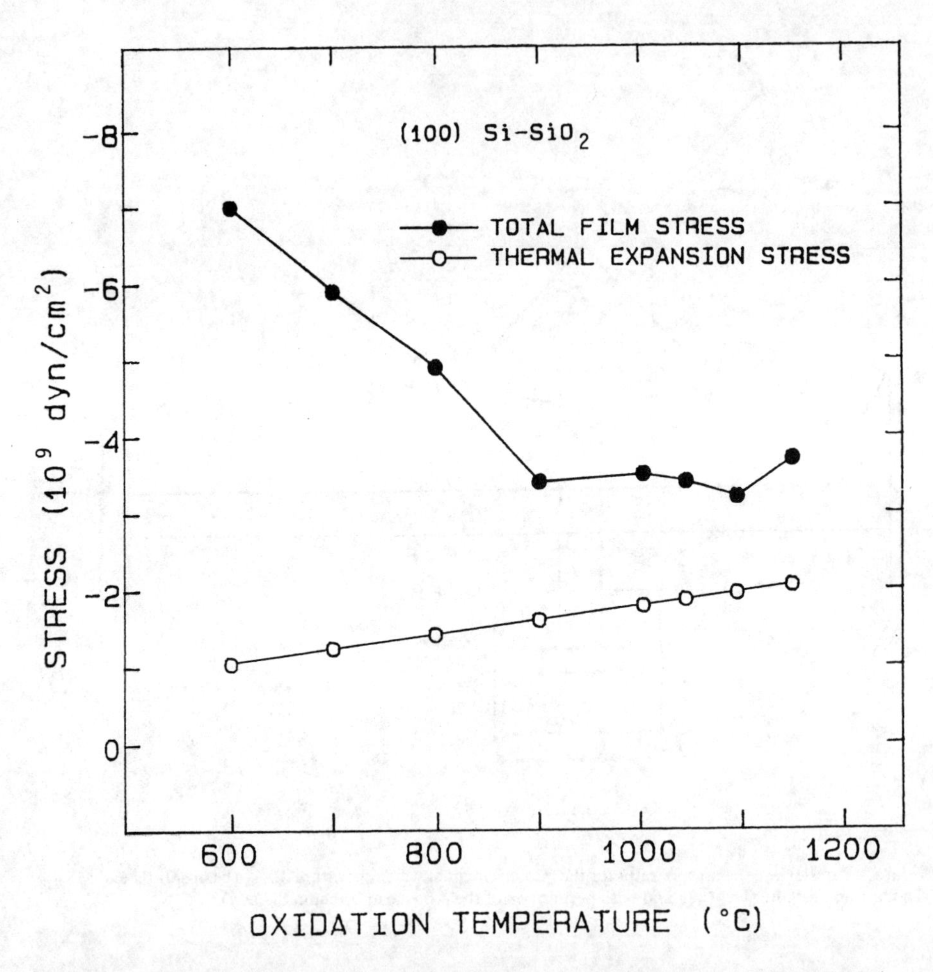

Figure 3. Total measured SiO_2 film stress and calculated thermal expansion stress for various oxidation temperatures for SiO_2 grown on Si in dry O_2. (after ref. (17) and with permission of the American Vacuum Society.)

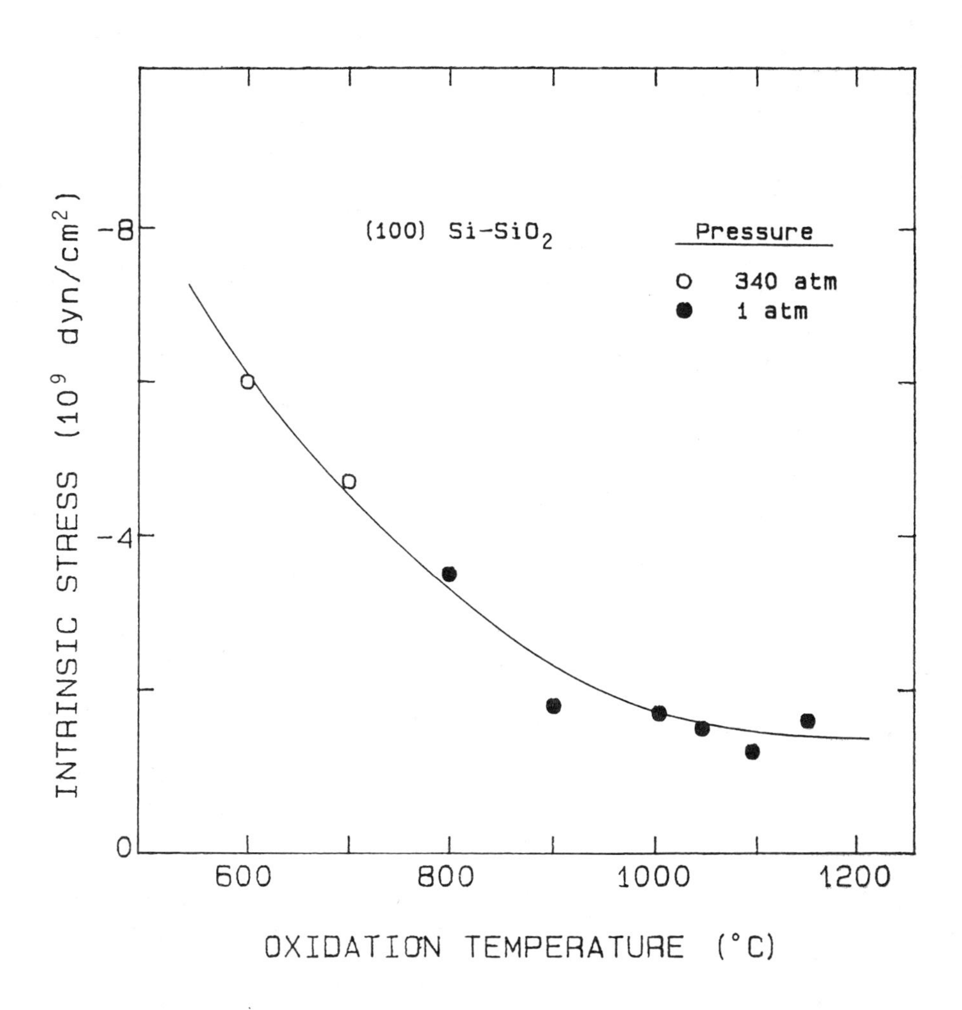

Figure 4. Intrinsic film stress versus oxidation temperatures for SiO_2 film grown on Si in dry O_2. (after ref. (17) and with permission of the American Vacuum Society.)

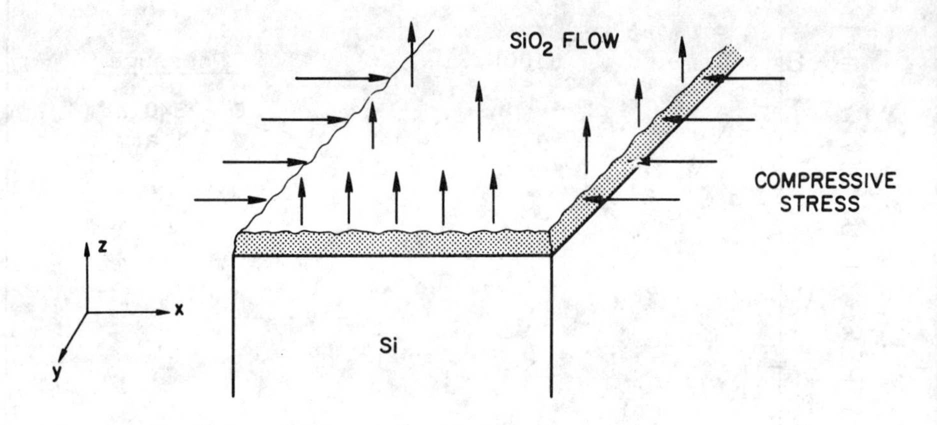

Figure 5. Pictorial representation of viscous flow in SiO_2 as a result of the molar volume change for the reaction of $Si + O_2 = SiO_2$ (after ref. (14) and with permission of the Electrochemical Society, Inc.)

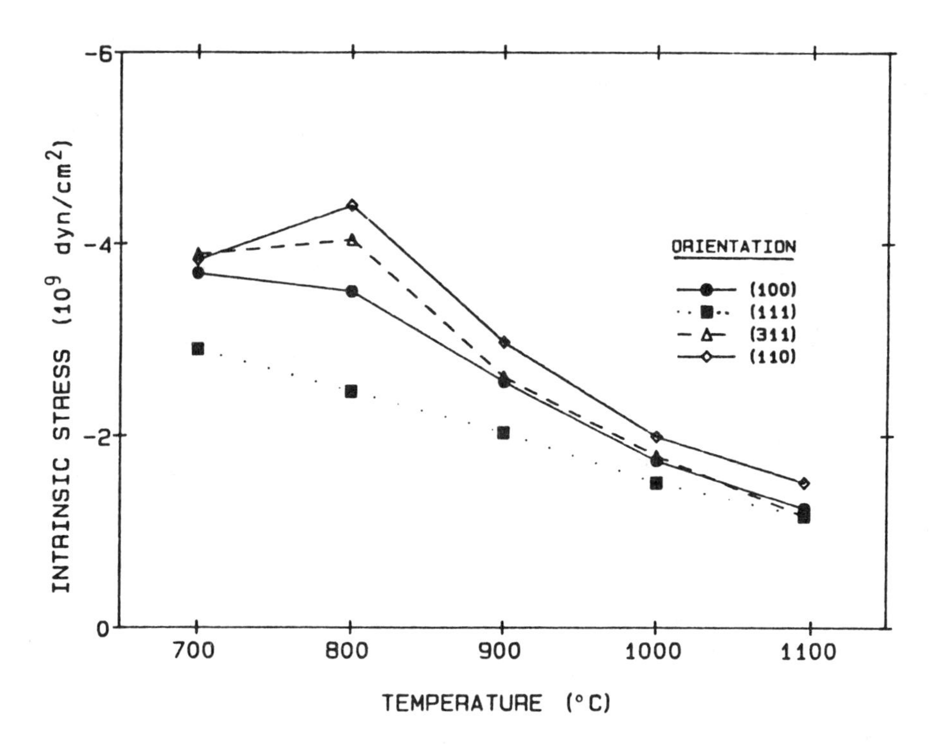

Figure 6. Intrinsic stress for a SiO_2 film on Si versus oxidation temperature in dry O_2 at one atm. for various Si orientations (after ref. (18) and with permission of the American Vacuum Society).

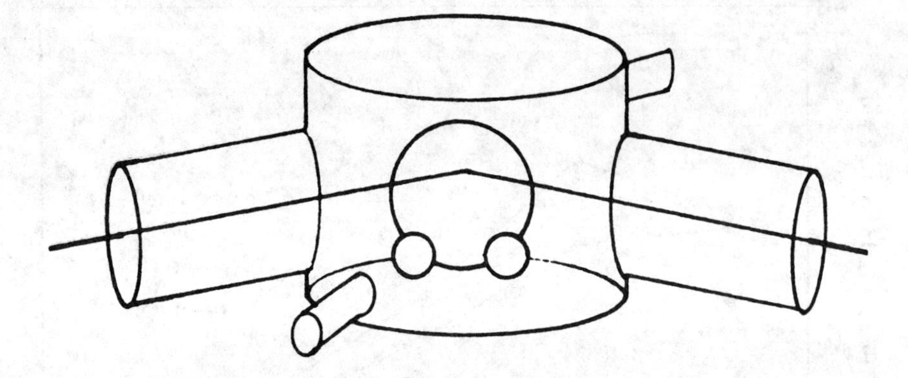

Figure 7. Fused silica sample cell used for in-situ ellipsometric measurements. (after ref. (37)).

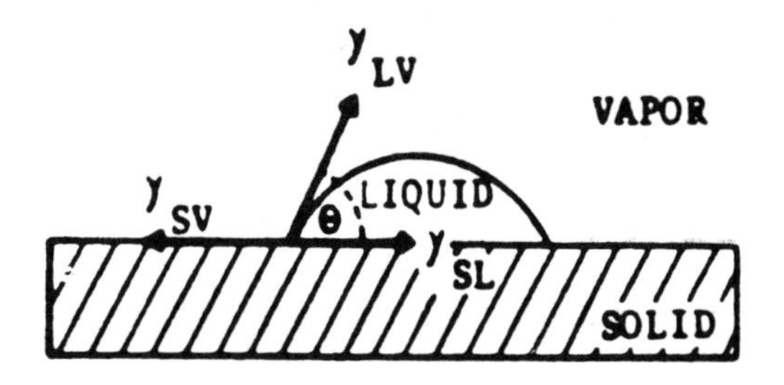

Figure 8. Representation of the important surface tension vectors, γ_{ij}, for the equilibrium between solid, S, liquid, L, and vapor, V, and with θ as the contact angle. (after ref. (37)).

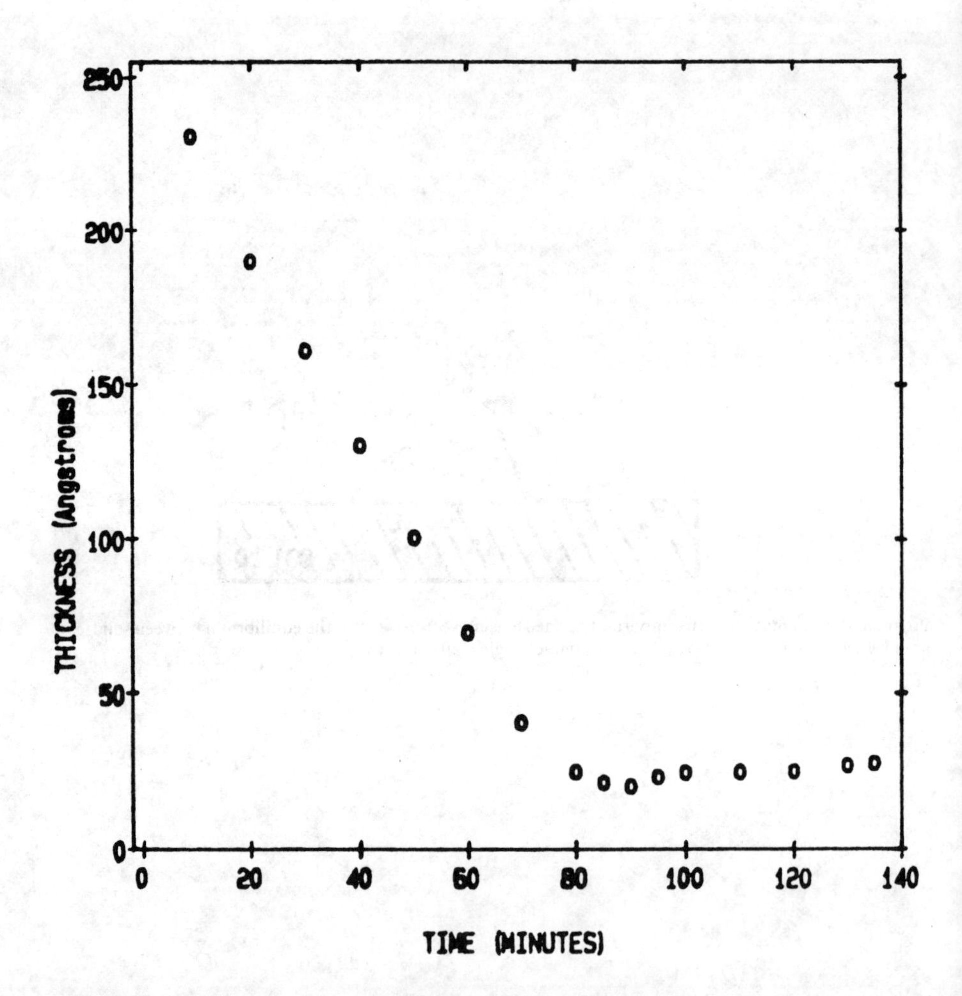

Figure 9. SiO_2 film thickness versus etch time in HF-H_2O solution from in-situ ellipsometry measurements (after ref. (37)).

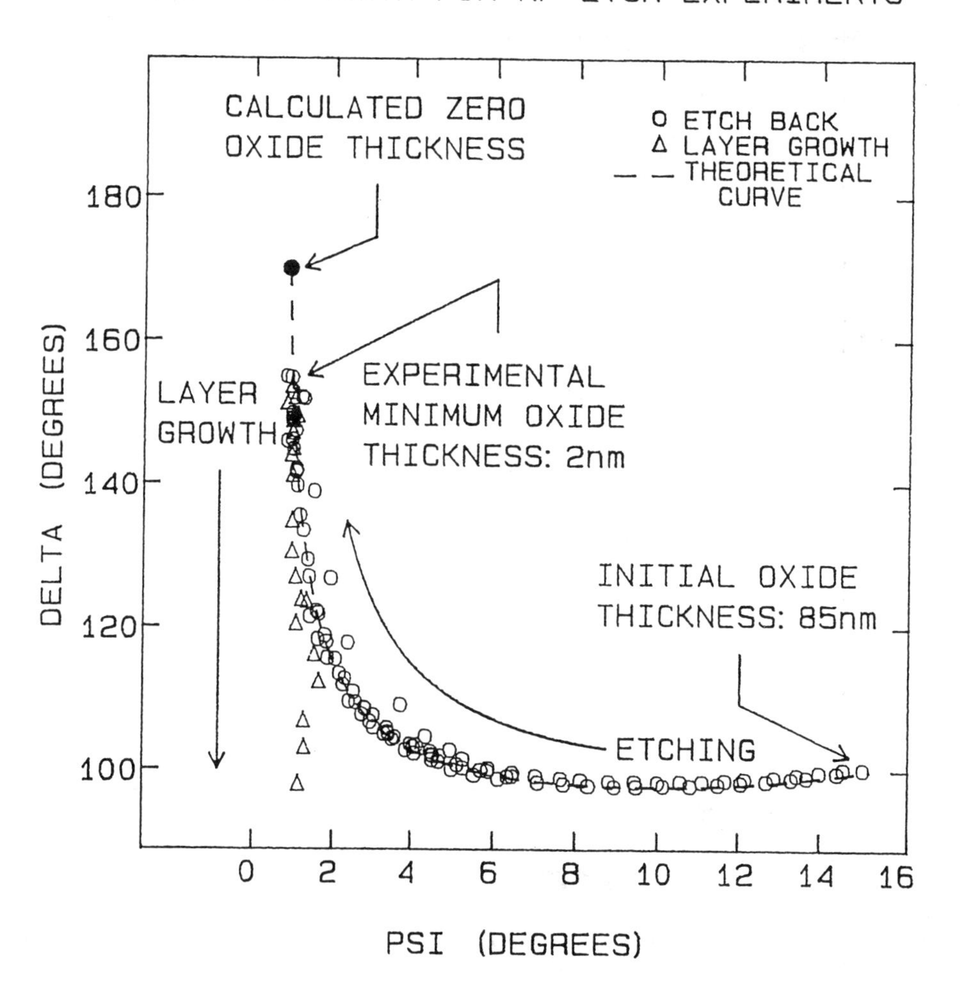

Figure 10. Ψ-vs-Δ trajectories for the etching of an 85nm SiO_2 film in HF-H_2O solution (after ref. (37)).

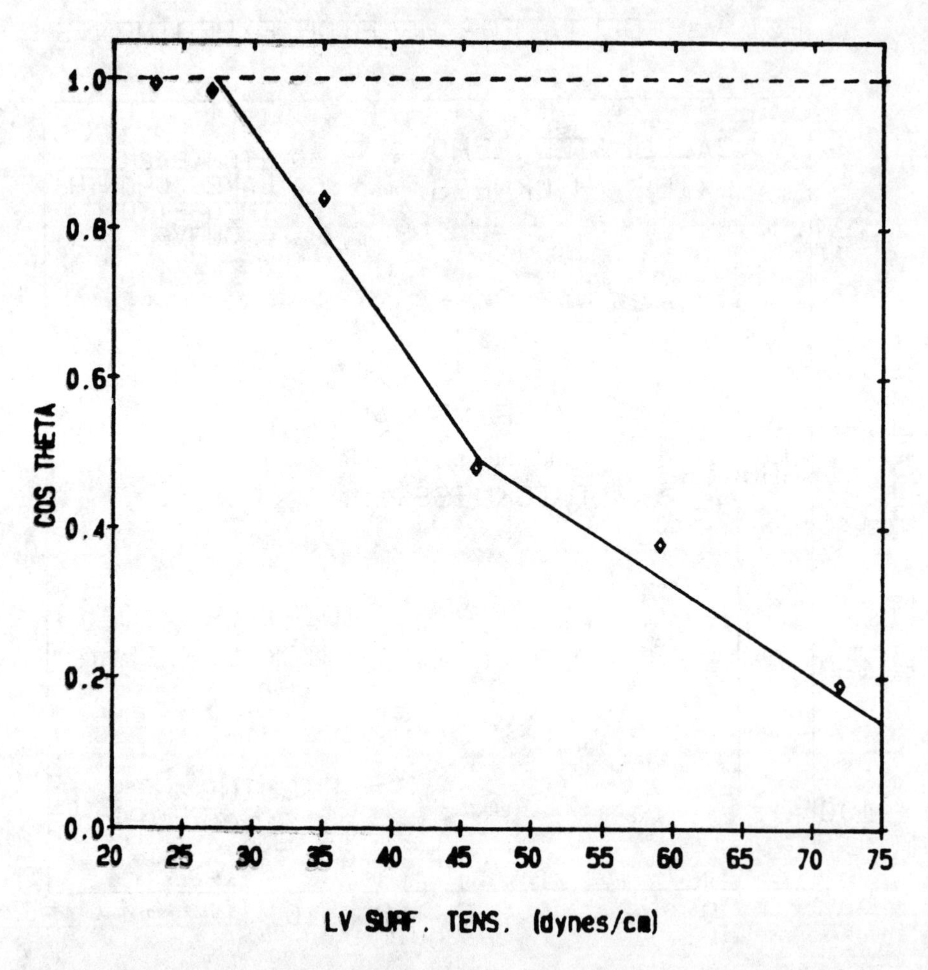

Figure 11. cos θ-vs-γ_{LV} plot of SiO_2 on Si in various CH_3OH-H_2O-HF solutions. γ_c is shown at cos $\theta = 1$ to be 27 dynes/cm. (after ref. (37)).

THE CHARACTERIZATION OF DIELECTRIC FILMS AND THE SCIENCE OF INSULATORS

Frank J. Feigl
Physics Department and Sherman Fairchild Center
Lehigh University, Bethlehem, PA 18015

ABSTRACT

Since the dielectric films incorporated within integrated circuit "chips" are susceptible to damage during processing or operation, the study of these films via measurements on Metal-Oxide-Silicon test structures has long been a technological necessity. In the present decade, however, MOS studies have become a frontier for the science of insulating solids. A number of such studies emphasizing the atomic and electronic structure of silicon dioxide films will be reviewed. The principal results are: (1) The SiO_2/Si interface region differs radically from the SiO_2 film bulk because of stress and interfacial impurity segregation. (2) Within the interface region, two distinct classes of dangling Si defects are involved in the transfer of electrons to and from the Si substrate via thermionic or tunneling processes. (3) Within the bulk, neutral defects related to water impurities can capture electrons from the SiO_2 conduction band.

INTRODUCTION

Metal-oxide-semiconductor (MOS) devices are subject to a catalogue of damage and degradation mechanisms sufficient to fill books[1-4]. These instabilities result from chip fabrication and operation. Prominent among them are the following: (1) field effect transistor (FET) channel conductance degradation and fluctuations, which result from electron scattering and trapping by defects near the oxide-silicon interface, (2) slow-trapping drifts in p-channel FETs, which are due to hole tunneling and trapping into near-interface and oxide defects, (3) hot carrier instabilities in short channel FETs, which are caused by electron or hole injection into the oxide film and their subsequent trapping at defects, (4) transient instability and total dose degradation due to ionizing radiation, which involve the creation of electron-hole pairs and the subsequent trapping of holes at interface defects, and (5) destructive dielectric wearout and breakdown, which is apparently not truly intrinsic at oxide fields below 30 MV/cm. Some of these instabilities are aggravated by VLSI/ULSI processing techniques[5]. Electron beams, ion beams, and plasmas produce ionizing radiation damage and may indirectly enhance other damage mechanisms. Hot carrier effects are aggravated because of the ever-higher electric fields resulting from device and circuit design with ever-thinner oxides and constant supply voltages.

All of the instabilities listed above are related at least in part to defects of the glass film dielectrics used for FET gate insulation, device and circuit isolation, and chip encapsulation. Many of these instabilities have been categorized within a compact classification scheme in which adverse electrical effects are described in terms of trapped and mobile charges within the MOS SiO_2 (silicon dioxide) dielectrics[1-3]. Within a modified version of this MOS oxide charge classification, the several charge components of present interest are:

Q_{otb}, bulk <u>oxide trapped charge</u>, distributed within the oxide film, but outside a region approximately 3nm wide and beginning at the atomically abrupt SiO_2/Si chemical interface. Q_{otb} is generally negative and is due to electrons trapped at defects of the oxide glass network or at impurities incorporated within that network.

Q_{oti}, also <u>oxide trapped charge</u>, but located within the 3 nm wide region adjacent to the chemical interface. This charge is generally positive and is due to holes trapped at defects or impurities.

Q_{it}, <u>interface trap charge</u>, localized spatially at the abrupt SiO_2/Si chemical interface and energetically within the silicon band gap. The distribution of interface traps within the gap (density D_{it}) can be determined experimentally. Electrically, these traps equilibrate rapidly with the silicon band carriers and for that reason are often described as "fast" interface traps.

TECHNIQUES AND METHODS FOR DEFECT CHARACTERIZATION

The physical and chemical basis for the oxide charges of figure 1 are defects within the SiO_2 glass films. There are three types of experiments used to study these defects: (1) electrical measurements on MOS devices, (2) solid state spectroscopic measurements on bulk materials, and (3) spectroscopic measurements on MOS devices.

Electrical Measurements on MOS Capacitors

The most widely used experiments are electrical measurements on MOS devices, and in particular on simple MOS capacitors. Stripped of a complex array of sample chambers, electrical and mechanical controls, and computer interfacing, these measurements are direct and simple. In the presence of oxide charge, the MOS device characteristics (capacitance C vs. V or current I vs. V) are displaced or distorted along the voltage axis, relative to a reference characteristic. The reference characteristic may be experimental or theoretical, and the oxide charge Q_{otb}, Q_{oti}, or Q_{it} is obtained directly from the data via the summary equation $Q = C_{ox}\Delta V$, where ΔV is some measure of the voltage shift and C_{ox} is the parallel plate capacitance of the oxide dielectric. These measurements have several important advantages. First, the direct use of MOS devices gives very high sensitivity relative to VLSI technology. Second, the use of simple capacitors facilitates the study of individual processing steps in isolation. Third, the measurements can be used in process monitoring via on-chip test structures.

A full program of electrical characterization of a silicon dioxide film consists of some particular research sample fabrication sequence, an oxide charge generation and measurement sequence, and correlation of these measurements with chemical or structural analyses. Data from a representative experiment within such a program is presented in figure 1a. This particular program addressed the effect of water impurities on oxide charge and mobile hydrogen species[6]. The data shown were obtained by measuring the C-V voltage shift of an MOS capacitor while a constant electron current density J flowed across the oxide film. The shift is plotted as a function of the electron fluence Jt/e, where t is the cumulative time of current flow and e is the electron charge. The measured voltage shifts are converted directly to a net oxide charge density on the right hand ordinate scale, which shows the net number of trapped electronic charges per unit area. These data constitute an oxide charging curve, and similar curves are used to represent the effects of ionizing radiation, etc., on MOS oxides.

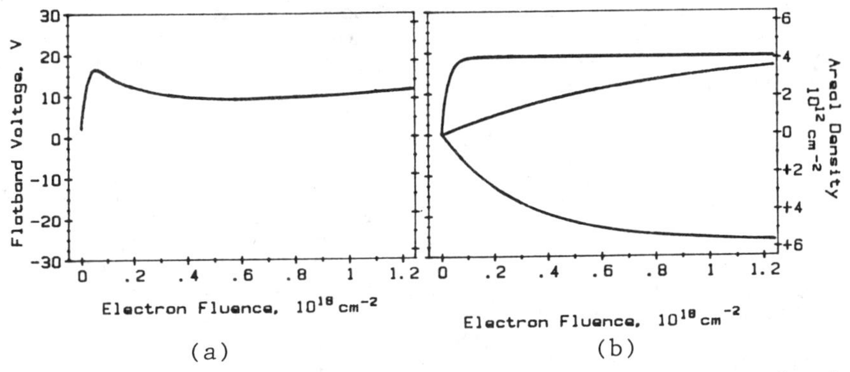

Fig. 1. An experiment on oxide charge: (a) data and (b) kinetic model. "Flatband Voltage" is a fixed point on an MOS C-V characteristic and "Areal Density" is Q/e, where oxide charge Q is related to flatband voltage shift ΔV by $Q = -C_{ox}\Delta V$.

The analysis of these data is based on a physical model for the separate components of oxide charge, Q_{otb} and Q_{oti}, and a statistical fit of first-order kinetic equations to the data. The separation of Q_{otb} and Q_{oti} is an experimental process involving comparison of voltage shifts observed in C-V measurements with those observed in I-V measurements. In the simplest case, the C-V measurement is sensitive to the total charge $Q_{otb} + Q_{oti}$, while the I-V measurement is sensitive only to the bulk charge Q_{otb}. The model is formulated as an exponential variation of the individual components with time or, equivalently, with electron current fluence. The results of the analysis are the three separate processes displayed in figure 1b. The top-most curve represents the fastest process, which is the development of negative Q_{otb} by capture of conduction band electrons at defects within the oxide film. The bottom curve represents the development of positive Q_{oti}, ultimately due to hole trapping at defects within the SiO_2/Si interface region. The remaining curve, in the middle of the diagram, represents a negative component of Q_{oti}. The sum of these three curves reproduces the data in figure 1a.

The analysis of oxide charging curves yields kinetic parameters for individual defect species. For Q_{otb}, the analysis produces the capture cross sections σ_c for electron trapping defects. A large number of studies correlating electron trapping with fabrication variations

and/or chemical analyses has produced the following summary results: Trapping in wet oxides involves a defect with $\sigma_c = 10^{-17}$ cm^2 at densities up to 10^{18} cm^{-3}. This defect has been associated with SiOH impurities via infrared stretching mode absorptions. The dominant trap in dry oxides, with $\sigma_c = 10^{-18}$ cm^2 has similarly been associated with H_2O species loosely bound within the oxide glass network. Ultradry oxide films exhibit bulk electron trapping below levels conveniently measured, indicating a concentration of electrically active defects below $2x10^{15}$ cm^{-3}.
The development of positive charge Q_{oti} at the interface has been described as anomalous, since it is produced by an electron current. None of several mechanisms proposed to explain this anomaly has proved fully satisfactory, and the generation kinetics for Q_{oti} are the object of continuing investigation.

Solid State Spectroscopy

The disadvantage of MOS electrical measurements is that they have no direct relation to the most useful theoretical models for defects, and in particular do not provide critical tests of these models. Capture cross sections, obtained from experiments of the type described in figure 1, are generally analyzed by theoretical calculations on models which specify defects as delta-function or square well or coulomb potentials[7]. On the other hand, solid state spectroscopic measurements, such as optical and infrared absorption and emission and electron paramagnetic resonance microwave spectroscopy (known by the acronym EPR), can be analyzed by theoretical calculations on molecular models indicating the type and location of atoms within the defect structure[8].

Such information is readily available only from studies of bulk SiO_2 crystals and glasses with volumes exceeding 10^{-4} cm^3 and not from MOS device dielectrics with volumes much less than 10^{-6} cm^3. Obviously, however, defects in bulk materials have no necessary connection to defects in oxide films, and hence to VLSI technology. Such a connection must be demonstrated directly on a case-by-case basis.

The most useful experiments for connecting oxide films with theoretical models are spectroscopic measurements on the films. Over the last decade very exciting advances have been made in this direction. Optical, EPR, and other spectroscopic experiments performed on oxidized silicon and on MOS capacitors have produced surprising and fundamental information on both SiO_2 film properties and SiO_2/Si interface structure. These measurements typically have low-to-moderate sensitivity in the VLSI context, but they do relate directly to both VLSI technology and appropriate solid state theory. Scientifically, these studies are among the most important and fruitful current research on dielectric materials.

THE STRUCTURE OF THE SILICON-SILICON DIOXIDE INTERFACE

As noted above in connection with figure 1, electrical measurements on MOS capacitors indicate that a region in the vicinity of the SiO_2/Si interface differs significantly from the "bulk" of the dielectric film. The critical interfacial region is 1-5 nm wide, as determined by various electrical measurements. It responds to ionizing radiation, to the passage of hole currents into the oxide, and to the passage of electron currents across the oxide predominantly (but not exclusively) by building up a positive charge Q_{oti} (that is, by trapping holes). The nature of this interface region is delineated by experiments designed to determine the structural and chemical composition of MOS oxide films. Results from two such experiments are shown in figures 2 and 3, which are models developed from three very different experiments, each of which yields a depth profile of the interface region.

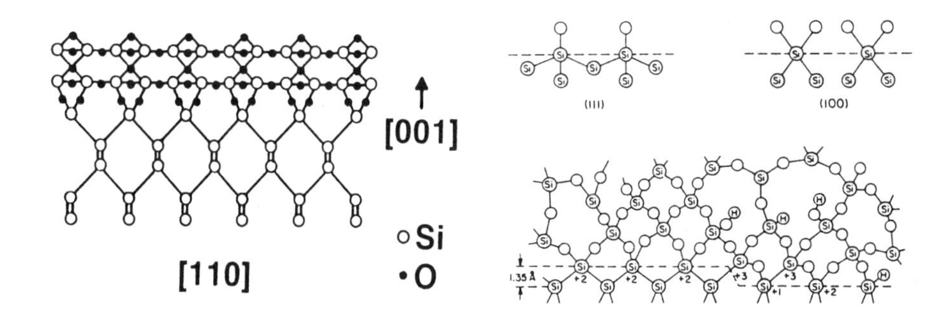

Fig. 2. Electron diffraction model of the tridymite/Si interface. Si (100) surface along [110].

Fig. 3. XPS profile model of the SiO_2/Si interface region. Si (100) surface along [110].

Figure 2 displays a crystal Si to crystal SiO_2 interface model. Such a crystal-to-crystal transition was suggested by Frank Herman as a theoretical concept a decade ago[9]. The model in figure 2, however, was constructed by Ourmazd and coworkers from their very refined electron diffraction studies on rather exotic interfaces[10]. The suggestion these workers make is that the native oxide produced on (100) Si at room temperatures consists of a strain-layer epitaxial crystalline film less than 1 nm thick and a transition to glassy SiO_2 resulting from strain relief. They further suggest that this structure persists in high temperature thermal oxidation, at least to temperatures of the order of 800°C. The particular crystal form of SiO_2 deduced from their atomic structure diffraction studies is tridymite, which is commensurate with the Si surface atom arrangement. To date, no detailed coupling has been made between the known electrical properties of the abrupt Si-SiO_2 interface and the crystal-to-crystal transition models. At present, however, theoretical investigators are examining the unattached fraction of the Si termination plane (fifty percent of the interfacial Si atoms) in this context.

Another summary description of the SiO_2/Si interfacial region is shown in figure 3. F. J. Grunthaner and P. J. Grunthaner and their coworkers have constructed this model, which is based predominantly on x-ray-induced photoelectron spectroscopy (XPS) data[11]. These data are produced by measuring the energy of electrons ejected from the SiO_2/Si interface region by 2 keV photons (x-rays) and are interpreted as the relative number of silicon atom 2p core electrons ejected, as a function of their chemical binding energy. A chemical etch-back spatial profile is constructed from such data. At the top of figure 3, in separate insets, are the dominant Si_3-Si-O transitional bonding configuration observed via XPS on (111) Si/SiO_2 interfaces and the dominant Si_2-Si-O_2 transitional bonding configuration observed on (100) Si/SiO_2 interfaces. These identifications of the anticipated bridging structures within the XPS spectra demonstrate the internal consistency of the interpretive model used. At the level of percent to tens-of-percent relative concentration, the important defect features of the atomically abrupt interface are single step ledges. These are illustrated for a (100) interface on the lower diagram of figure 3. The ledges produce irregular bridging configurations such as Si-Si-O_3 (designated "+3") and Si_3-Si-O (designated "+1"). The investigators have interpreted a binding energy shift of the Si-in-SiO_2 2p core electrons as stress-related bond strain extending approximately 3nm from the atomically abrupt interface.

Finally, hydrogen impurities occur in large concentrations, both on the atomically abrupt interface and spread throughout the interfacial strained bond region. The hydrogen impurities are pictured as SiH and SiOH structures in figure 3. The hydrogen species are not seen directly in the XPS spectra, but have been deduced from a complicated and rather tricky bond count. The volume concentration of hydrogen impurities as a function of the depth within an oxide film has been studied by C.W. Magee of the David Sarnoff Research Laboratory. Among many different films studied was a nominally dry thermal oxide. The chemical analysis technique used was secondary ion mass spectrometry (SIMS). The hydrogen concentration is the sum of all chemical species (SiH, SiOH, loosely bound H_2O, etc.) and the depth scale was generated by ion sputtering the oxide film. Within the bulk of the oxide film, the H concentration is less than one percent of the concentration of Si in SiO_2. The profiling technique has a resolution of approximately 5 nm, and the feature of most interest is the high concentration of H impurities in the immediate vicinity (less than 5 nm) of the SiO_2/Si interface. Within this region the H concentration approaches or exceeds 5% of the concentration of Si in SiO_2.

The role of these near-interface defects and impurities in the oxide charge components is currently unresolved. A very important boundary condition is the fact that the total concentrations of chemical (H impurities) and physical (step ledges and strained bonds) defects are of the order of 10^{14} cm^{-2}, while Qoti, for example, seldom exceeds 10^{13} cm^{-2}. The simplest interpretation of the accumulated evidence is that hydrogen impurities are incorporated at strained bonds within the SiO_2 glass network, forming SiOH defects in the bulk of the film and SiH defects near the Si interface. The SiOH defects have been associated with Q_{otb}, as discussed above. However, there has been no convincing association of either H impurities or strained bonds with hole traps related directly to Q_{oti}.

SPECTROSCOPIC STUDIES OF OXIDE CHARGE COMPONENTS

Selected experiments and summary results for the individual oxide charge components Q_{it}, Q_{oti}, and Q_{otb} and will now be presented. The work described represents the scientific interests of the author and the research activities of his students and professional collaborators, and is not intended as a comprehensive review.

Q_{it} and Amphoteric Interface Defects

Historically, interface trap charge Q_{it} has been the most studied of all of the oxide charge components. The interface traps responsible for Q_{it} were early-on characterized by spectroscopic measurement techniques, that is, the determination of the density of interface trap states D_{it} as a function of their energy. For fast interface traps, the range of appropriate energies is between the silicon valence band edge E_v and conduction band edge E_c, since these so-called "fast" states are in relatively direct communication with the valence band hole reservoir and conduction band electron reservoir of the silicon surface and bulk.

The spectroscopic techniques are electrical measurements of the dynamic capacitance and conductance of MOS capacitors which are analyzed within the framework of ideal static and infinite-frequency limits to yield $D_{it}(E)$. The Si band gap energy E is scanned by varying the MOS gate voltage V. Several experimental approaches to this problem are well documented in the review literature. Such studies of D_{it} have produced an abundance of data and, in isolation, a dearth of atomic-scale understanding. The fundamental problem arises from the expression for interface charge, which requires integration of the interface state density D_{it} over the band gap:

$$Q_{it} = e \int_{E_{v-Si}}^{E_{c-Si}} \{D_{it-d}(E)[1-f(E)] - D_{it-a}(E)f(E)\}dE$$

Evaluation of this simplified expression requires knowledge of the electronic character of the interface traps (whether, as indicated, they produce donor states "d" or acceptor states "a") and of their occupancy probability f (which depends on the Si doping, the oxide thickness, and V). The electrical measurements determine only the total density D_{it}, not the electronic character of the states. The latter requires either independent knowledge of the absolute sign of the interface charge or a detailed model of the interface defects.

Over the last decade the interface trap spectroscopic techniques were coupled with EPR spectroscopy on MOS structures by a wide collaboration led by E. H. Poindexter and by a collaboration at Sandia Laboratories. This work was focussed on a family of defects called P_b centers[12]. Poindexter and Caplan first demonstrated the true interfacial character of the P_b defects. Their EPR response was attributed to an unbonded electron in a singly-lobed orbital pointing along a [111] direction of the silicon crystal lattice. The orbital is fixed to a dangling Si atom which terminates the Si lattice but forms no bridge to the SiO_2 network. On (111) interfaces, the orbital lobe is normal to the interface; on other oxidized surfaces, the lobe is skew to the interface. This symmetry is unique to an interfacial defect and was the key to identification. Later workers, notably Brower, verified predicted details of this model and provided data required for quantitative theoretical studies of the P_b defects[13,14]

Poindexter and his collaborators and Lenahan and Dressendorfer[15] also demonstrated that the triply bonded silicon interface defect is an "amphoteric electron trap." This means that the energy of the P_b defect lies within the Si bandgap for both one- and two-electron occupancy of the defect. The defect is electrically neutral when the orbital is singly occupied, as described in the last paragraph. The neutral defect can emit this electron and become positive (that is, behave like a deep donor); it can also trap an additional electron and become negative (like a deep acceptor). The deep donor energy level is in the lower half of the silicon bandgap and the deep acceptor level is in the upper half. The defect contains an unpaired electron only in the neutral state. This means that the EPR signal is observed only when the interface Fermi level is positioned between the lower (donor) and upper (acceptor) defect energy levels. This was demonstrated directly in several critical experiments in which the EPR signal was monitored as the interface Fermi level was scanned from accumulation to inversion via adjustment of the MOS gate voltage V.

This two-level or amphoteric defect model represents a great advance over the pictures of interface trap charge developed by earlier workers. The model applies in detail to one particular defect, designated P_{b0}, which has been observed on (111), (110), and (100) interfaces. Figure 4 is a current working model of the (100) interface with two distinct P_b centers contributing to Q_{it}. The structure of the P_{b1} defect is still uncertain, and theorists are studying Brower's detailed P_{b0} data to determine its dependence on first neighbor and second neighbor oxygen bonding to the dangling Si atom.

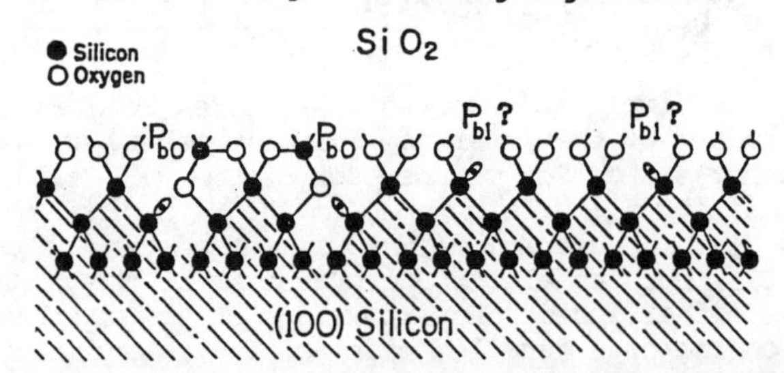

Fig. 4. Dangling neutral Si defects on the (100) interface.

The P_b defects have been shown to exist in large concentrations in MOS structures which are, from a device standpoint, poorly prepared. At present, there is conflicting evidence on their role in device quality oxides and in the development of Q_{it} in response to ionizing radiation and hot carrier production in MOS devices. S. Lyon of Princeton University has argued that interface traps produced at high-densities by a variety of means are in fact amphoteric defects.

Q_{oti} and Hole Trapping Defects

Models for the electronic structure of defects responsible for Q_{oti} have been derived from studies of gate-bias-controlled internal tunneling spectroscopy. The "anomalous positive charge" produced by electron injection was described above in connection with figure 1. This charge, once generated, can be neutralized by application of a positive gate voltage (positive bias stress) and subsequently regenerated by negative bias stress. This behavior is mediated by the SiO_2/Si interface traps responsible for Q_{it}. A mechanism proposed for positive bias stress is tunneling of electrons from the interface traps into the near-interface defects responsible for Q_{oti}. This requires defect energy levels within the SiO_2 which are energetically within or close to the Si band gap. The reverse tunneling process presumably occurs under negative bias stress.

Positive Q_{oti} produced by direct injection of holes can only be neutralized under bias stress. The mechanism proposed for this process is electron tunneling from Si valence band states into the near-interface defects responsible for Q_{oti}. Once neutralized, this "trapped hole" charge cannot by regenerated by negative bias stressing. Gate bias variations only affect the time required for neutralization, presumably because such variations alter the tunneling barrier for electrons[16,17]. The near-interface defect energy levels must be energetically degenerate with the Si valence band, and must, in particular, be displaced in energy from the Si band gap.

These models describe only the electronic energy level structure of the defects responsible for Q_{oti}. They do not address the atomic structure of these defects. Even so, these models for "anomalous positive charge" and "trapped holes" are not universally accepted. Work by other researchers, notably S. Lyon of Princeton University, indicates that the two defects are either identical or closely related structurally and may be distinguished primarily by spatial location. The relevant experiments have been conducted on a variety of oxide films and under a variety of experimental conditions, and these variations may be significant. A correlation of the several experiments is needed.

There is one observation about the atomic structure of the defects responsible for near-interface positive charge that has been made consistently (although not universally) in recent years. Several investigators have demonstrated that Q_{oti} produced by ionizing radiation and by bias stressing at high oxide electric fields correlates with a class of bulk SiO_2 defects called E' centers[18,19]. These defects have been extensively studied in both crystal quartz and bulk glass via EPR. Lenahan and his coworkers have argued recently that the particular positively charged defect responsible for Q_{oti} is a simple oxygen vacancy analogous to the E_1' defect in crystal quartz[20].

M. E. Zvanut at Lehigh is studying band-to-trap tunneling under bias stress in MOS capacitors fabricated in collaboration with J. D. Zook at the Honeywell Physical Science Center[21]. The SiO_2 films were ion beam sputtered from an SiO_2 target in high vacuum and deposited onto a variety of (100) or (111) silicon substrates at room temperature. The substrates were not acid cleaned prior to deposition;

therefore, a 0.5-1.5 nm thick native oxide layer existed between the Si surface and the sputtered SiO_2 film. The experiment pertinent to the model for Q_{oti} to be presented here is an isochronal bias stress sequence. The individual steps in this sequence consisted of applying a gate bias for 1 second and measuring the resulting shift of the CV curve. Various I-V and D_{it} measurements were made in these experiments to eliminate ambiguities in charge determinations.

Figure 5 illustrates the results of one isochronal bias stress cycle executed on a single capacitor. The capacitor was subjected to an initial one-second stress at +40 V. This was followed by a trap emptying process, a sequence of one-second stresses at systematically reduced biases to a minimum of -55 V. Subsequently, the capacitor underwent a trap filling process, a sequence of one-second stresses at systematically increased biases to a maximum of +40 V. The results shown in figure 5 were interpreted in terms of the tunneling of electrons between the silicon valence band and oxide band-gap states produced by near interface defects. A model has been developed based on two simplifying assumptions: that electron trapping defects have a single spatial location within the MOS oxide film and that these defects produce an energy level distribution within the SiO_2 band gap. The data in figure 5 can then be interpreted in terms of two distinct distributions of trapping state energies. The model, illustrated schematically in figure 6, is described in the following paragraph.

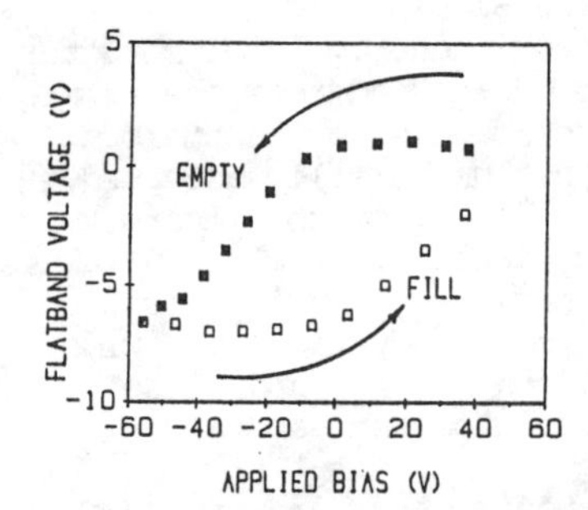

Fig. 5. Isochronal bias stress sequence. The vertical axis represents a fixed point on an MOS C-V characteristic.

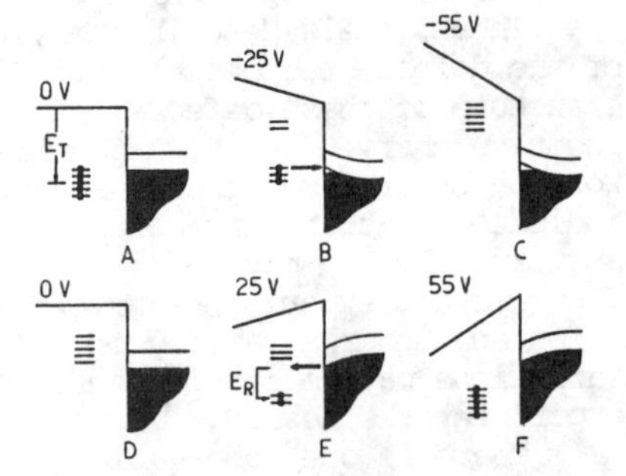

Fig. 6. Band-to-trap tunneling model for the isochronal bias stress sequence on an MOS capacitor.

When the traps are filled, as at zero bias, the entire energy level distribution must be below the Si valence band edge (figure 6a). As the applied bias is decreased to -25 V, for example, the oxide bands tilt such that the distribution is raised with respect to the Si valence band edge. Thus a portion of the distribution may be emptied (figure 6b). Finally, after sufficiently large negative biases, the energy level distribution is completely emptied and the entire distribution is be located above the Si valence band edge (figure 6c).

After emitting an electron, the energy of a trap does not remain fixed at its original level across from the SiO_2 valence band edge. Rather, the defect structure relaxes and its energy level rises by an amount E_R. Because of this, it takes a relatively large change in bias to refill the trap level distribution. This is seen at negative bias values of the fill sequence data in figure 6 and is explained schematically in figures 6d-f. As the applied bias is increased to positive values, the oxide bands and defect levels tilt such that a portion of the distribution may be filled as electrons tunnel from the Si valence band (figure 6e). As indicated, each defect relaxes by an energy E_R upon capture of an electron. Finally, at sufficiently high bias the energy level distribution is completely filled. At this point, the distribution is located at an average energy E_T below the SiO_2 conduction band edge (figure 6f).

In summary, we propose that an electron tunnels into an unrelaxed defect configuration located at an energy E_T-E_R below the SiO_2 conduction band edge. However, the electron tunnels out of the relaxed state of the defect, located at a trap depth E_T. From the data in figure 5, the unrelaxed and relaxed energy distributions may be determined. The results of figure 7 were calculated assuming a trap location 1 nm from the native SiO_2/Si interface, and we suggest that the traps are most likely at the sputtered oxide/native oxide interface. With this specific assumption, a trap depth E_T of approximately 5 eV and a relaxation energy of approximately 1 eV are obtained.

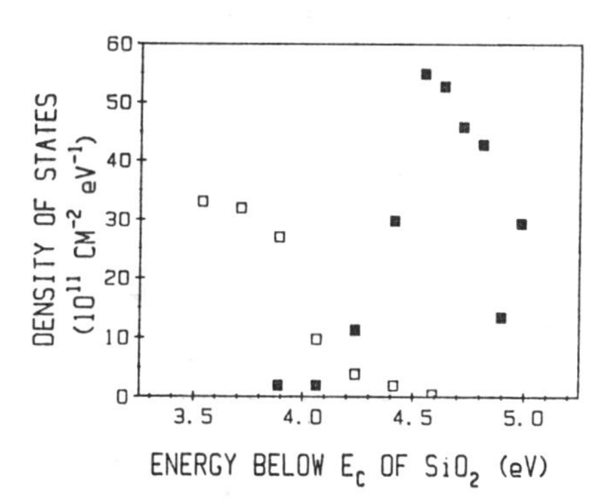

Fig. 7. Density of states diagram for unrelaxed state (open points) and relaxed state (solid points) of the near-interface hole trap.

Poindexter and Caplan have performed EPR measurements on the Honeywell sputtered oxide/native oxide films, and have found high concentrations of E' defects. Using corona charging techniques, they demonstrated that the E' EPR signal is greatly reduced by negative bias stress (as in the "empty" sequence of figure 5) and restored by positive bias stress (as in the "fill" sequence). We therefore suggest that the near-interface positive defect from which electrons tunnel into the Si valence band is an E' defect, that is, an oxygen vacancy defect. Fowler and Rudra have calculated the appropriate energies for the +1 and +2 charge states of the oxygen vacancy in crystalline quartz[22]. They obtain a value for E_T of approximately 7.5 eV and for E_R of approximately 1 eV. We are currently investigating the origin of the discrepancy between theory and experiment concerning the value of E_T.

These experiments represent the first direct measurement of the energy levels of the ubiquitous E' defects, and provide a critical test of the theoretical defect models. These experimental results, based on electrical measurements on MOS capacitors, are therefore important to the general study of silicon dioxide materials.

Q_{otb} and Electron Trapping Defects

Outside of the structurally or chemically distinct transition region adjacent to the SiO_2/Si interface, electron trapping defects must either replace or dominate hole trapping defects in SiO_2 films. As noted above in connection with figure 1, one basic electrical characteristic of an electron trapping defect is its capture cross section. Different defects dominate the bulk charge trapping behavior of wet oxide films and annealed oxide films, and "ultradry" oxide films are relatively free of bulk traps.

Additional information on these traps is obtained from internal photoemission studies. Light with sufficiently high photon energy $h\nu$ incident on the MOS oxide can excite electrons from the Si into the SiO_2 conduction band. These mobile electrons can be captured by unoccupied trapping centers within the oxide. The overall process, which is designated as photoinjection, results in a buildup of negative Q_{otb}. However, ultraviolet light with sufficiently high photon energy can also reduce negative Q_{otb}. This process, which is designated as photodepopulation, occurs when an electron trapped within the oxide film absorbs a photon and is excited into the oxide conduction band. If the now-mobile electron is removed from the oxide film, the magnitude of negative Q_{otb} is reduced. The threshhold condition for photoinjection is $h\nu > E_B$, where E_B is the SiO2/Si interfacial barrier energy, for example; that for photodepopulation is $h\nu > E_T$, where E_T is the trap depth. It is possible to control experimental conditions so that one or the other of these process is dominant.

Figure 8 displays data obtained in an experiment in which trapping centers in a wet oxide film are alternately filled by photoinjection and emptied by photodepopulation[23,24]. Photoinjection increases the magnitude of negative Q_{otb}, which is measured as an increase in a voltage $V\phi$ determined by the I-V characteristics of the MOS capacitor (a voltage shift of the type described in connection with figure 1). Conversely, photodepopulation reduces $V\phi$. The data shown consist of a sequence of photoinjection steps (all at fixed photon energy) interspersed with photodepopulation steps. Two photodepopulation steps at different photon energies are represented in the figure. Detailed analysis of the photodepopulation data yields a photoionization cross section and a photoionization threshhold energy E_T for each type of defect being studied. The cross section, shown in figure 9, is the probability that an individual trap will capture a photon.

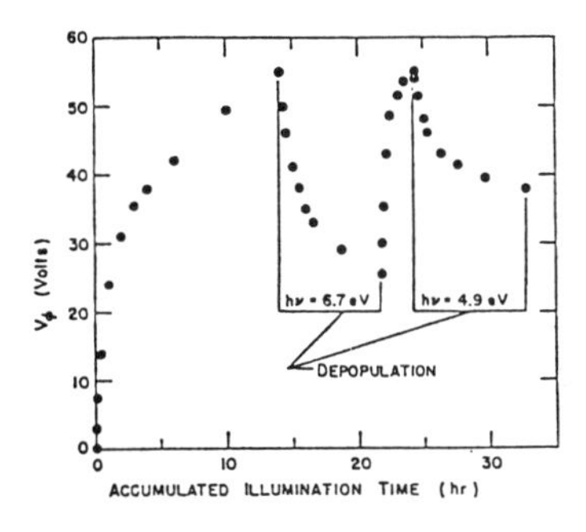

Fig. 8. Data from photoinjection-photodepopulation sequence. The vertical axis is a fixed point on an I-V sequence for photoinjection. Two photodepopulation steps are shown.

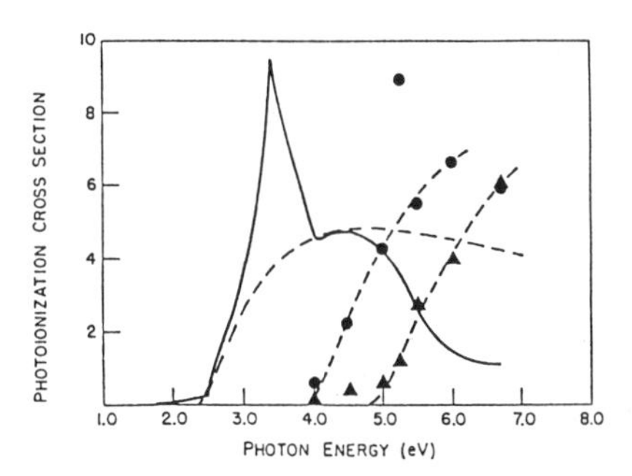

Fig. 9. Photoionization cross sections for three defects (solid line, circles, and triangles). The dash lines are a model calculation. The vertical scale is arbitrary.

Results from these experiments are presented in figure 10 in the form of a highly schematic state density diagram for SiO_2. The particular oxides examined in this study were thick thermal oxides grown in water vapor at 1300°C. Three distinct traps are illustrated, with the total density N_T of each trap plotted on the abcissa and the ground state energy level on the ordinate. The trap depth and density are tabulated for each defect. The dominant trap in this wet oxide has an electron capture cross section $\sigma_c = 10^{-17}$ cm^2 (as discussed above)

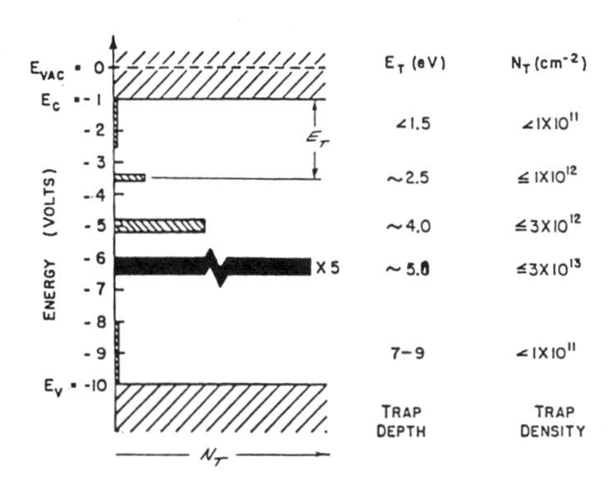

Fig. 10. Electron trapping defects in wet oxide films.

and a trap depth E_T = 5 eV. As mentioned previously, this trap has been associated with OH impurities. Other experiments at Lehigh have indicated that an additional impurity, possibly Na, is involved in this bulk silicon dioxide electron trap. The identification of a family of electron traps in MOS oxides is extremely important, since no such traps have been firmly identified in bulk SiO_2 materials. The association of negative charge with nonbridging oxygen defects (SiO defects or SiOH hydrogen species) should be investigated theoretically. Fowler and Edwards have suggested that SiO defects may be stable in a negatively charged state[25].

CHARGES IN MOS OXIDES AND DEFECTS IN SILICON DIOXIDE

A summary of the state of understanding of the defects in SiO_2 films and at the SiO_2/Si interface which are responsible for the various MOS oxide charge components can now be presented.

Q_{otb} is due to oxide bulk electron traps. These traps may involve SiOH impurities in wet or unannealed thermal oxides and probably in chemically deposited oxides. In well-annealed thermal oxides, the dominant electron trap involves H_2O loosely bound within the SiO_2 glass network. In ultradry oxides, the density of bulk electron traps is greatly reduced.

Q_{oti} is due to oxide interface hole traps. Q_{oti} due to ionizing radiation involves E' defects. These have been studied extensively in SiO_2 crystals and glasses. There is incomplete and conflicting data on the E' character of Q_{oti} due to injection of electrons or holes.

Q_{it} is due to amphoteric electron/hole traps localized on the atomically abrupt SiO_2/Si interface. The large Q_{it} in unannealed dry thermal oxides is due to P_b centers, which are triply-bonded silicon atoms. On (111) Si, these atoms terminate the silicon crystal lattice. On (100) interfaces, the precise structure of P_b centers has not been determined. There is incomplete and conflicting information on the structure of the defects responsible for Q_{it} produced by ionizing radiation or injection of electrons or holes.

This summary picture is the result of the intensive effort to relate the very sensitive techniques of MOS charge measurement to the atomic and electronic structure information available from solid state spectroscopy. This was most successful when a spectroscopic measurement was directly coupled to MOS device manipulation. One example is the voltage gated EPR used to determine the energy levels of the amphoteric interface defects responsible for Q_{it}. A second example is the photoinjection-photodepopulation techniques developed for the determination of optical trap depths and optical cross sections of the defects responsible for Q_{otb}. Finally, band-to-trap tunneling spectroscopy is being used to determine energy levels of the defects responsible for Q_{oti}.

Important and basic details of all of the oxide charge components are either uninvestigated or obscured by controversy. The only well defined defect structure is the P_b center on unannealed, unirradiated, and uninjected (111)-substrate MOS devices. However, very powerful experimental tools are now available and are being systematically refined and applied to a variety of dielectric films. These tools are being used to produce data directly relevant to serious theoretical investigations. Research on MOS insulators, developed and applied within the VLSI technological frontier, has itself become a scientific frontier[26].

ACKNOWLEDGEMENTS. The author thanks W. B. Fowler, A. H. Edwards, and M. E. Zvanut for their advice and counsel, and the Office of Naval Research Electronic and Solid State Science Program for its support.

REFERENCES

1. J. Davis, Instabilities in MOS Devices (Gordon and Breach, 1981).
2. E. H. Nicollian and J. R. Brews, MOS Physics and Technology (Wiley, New York, 1982), especially chapters 11 and 15.
3. F. J. Feigl, in vol. 6 of VLSI Electronics: Microstructure Science, Materials and Process Characterization, N. Einspruch and G. Larrabee, eds. (Academic Press, New York, 1983), p.147.
4. G. Barbottin and A. Vapaille, Instabilities in Silicon Devices, (North-Holland, Amsterdam, 1986).
5. C. M. Dozier, Solid State Technology (October, 1986), p. 105.
6. F. J. Feigl, R. O. Gale, H. Chew, C. W. Magee, and D. R. Young, Nucl. Inst. and Methods in Phys. Res. B1, 348 (1984).
7. G. Lucovsky, Solid State Commun. 3, 299 (1965).
8. A. Edwards in Defects in Glasses, F. Galeener, D. Griscom, and M. Weber, eds. (Materials Research Society, Pittsburgh, 1986), p. 3.
9. F. Herman, I. Batra, and R. Kasowski in Physics of SiO_2 and its Interfaces, S. Pantelides, ed. (Pergamon, New York, 1978), p. 333.
10. A. Ourmazd, D. Taylor, J. Rentschler, Phys. Rev. Lett. 59, 213 (1987).
11. F. Grunthaner and P. Grunthaner, Mat. Sci. Repts. 1, 65 (1986).
12. E. Poindexter and P. Caplan, Prog. Surf. Sci. 14, 201 (1983).
13. K. L. Brower, Appl. Phys. Lett. 43, 1111 (1983).
14. K. L. Brower, Z. Phys. Chem. 151, 165 (1987).
15. P. Lenahan and P. Dressendorfer, J. Appl. Phys. 55, 3495 (1984).
16. L. P. Trombetta, "Electrical Characterization of Low Temperature, High Pressure Thermal SiO_2" (Dissertation, Lehigh Univ., 1984).
17. S. Manzini and A. Modelli, in Insulating films on Semiconductors, J. Verweij and D. Wolters, eds. (North Holland, 1983), p. 112.
18. W. L. Warren and P. M. Lenahan, Appl. Phys. Lett. 49, 1296 (1986).
19. C. Marquardt and G. Sigel, IEEE Trans. Nucl. Sci. NS-22, 2234 (1974).
20. H. S. Witham and P. M. Lenahan, Appl. Phys. Lett. 51, 1007 (1987).
21. M. E. Zvanut, F. J. Feigl, W. B. Fowler, and J. K. Rudra in The Physics and Technology of Amorphous Silica (Les Arcs, France, 1987), conference proceedings to be published.
22. J. K. Rudra, "O- and Si-Vacancy and H-Related Defects in Silicon Dioxide" (Dissertation, Lehigh University, 1986).
23. D. D. Rathman, F. J. Feigl, and S. R. Butler in Insulating Films on Semiconductors 1979, G. G. Roberts and M. J. Morant, eds. (Institute of Physics, London, 1980), p. 48.
24. D. D. Rathman, F. J. Feigl, S. R. Butler, and W. B. Fowler, in The Physics of MOS Insulators, G. Lucovsky, S. T. Pantelides, and F. L. Galeener, eds. (Pergamon Press, New York, 1980), p. 142.
25. A. H. Edwards and W. B. Fowler, in Structure and Bonding in Noncrystalline Solids, G. E. Walrafen and A. G. Revesz, eds. (Plenum Press, New York, 1986), p. 139.
26. F. J. Feigl, Physics Today (October, 1986), p. 47.

Elemental And Electrical Characterization Of Thin SiO_2 Films Deposited Downstream From A Microwave Discharge.

B.Robinson, T.N.Nguyen and M.Copel

IBM T.J. Watson Research Center, Yorktown Heights, New York 10598.

ABSTRACT

A microwave induced plasma enhanced chemical vapor deposition (PECVD) technique is used to deposit gate quality thin ($\leq$ 10 nm) films of SiO_2. The deposition takes place downstream and at right angle to a N_2O microwave discharge. A mixture of 2% SiH_4, in He, is injected into the after-glow of the discharge to initiate the gas-phase precursor reactions with the resulting oxide deposited onto a Si<100> substrate held at 350C. The dielectric integrity of the SiO_2-Si system was determined by fabricating and measuring the electrical properties of metal-oxide-semiconductor (MOS) devices. High frequency and quasi-static capacitance measurements consistently gave values for the density of interface traps (D_{it}) in the low to mid 10^{10} eV^{-1} cm^{-2} range after post metallization annealing. The average measured breakdown field of the thin oxides is 8 MV/cm. Infrared (IR) spectroscopic measurements showed no SiH, SiN, SiOH or Si-H_2O complexes in the thick oxides. Secondary ion mass spectroscopy(SIMS) detected large concentrations of H, F and N in the oxide relative to gate quality thermal oxides of the same thickness. Medium energy Rutherford backscattered ion beam channeling measurements, made on both the PECVD and gate quality thermal oxides, showed no measurable difference in the interface sharpness or elemental structure.

INTRODUCTION

The potential application of PECVD oxides as gate, storage node, or isolation dielectrics in VLSI technology, as well as gate dielectrics in compound semiconductors or thin film transistor (TFT) devices has been documented[1,2]. Recently, low temperature ($\leq$ 400C) gate quality SiO_2 thin films ($\leq$ 10 nm) have been fabricated using two different PECVD techniques. This has generated considerable interest in understanding, at the fundamental level, the elemental composition and the chemical defect structure of the PECVD oxide and the oxide-semiconductor interface. An attempt to understand the PECVD oxide structure should first correlate the effects of the system configuration and the deposition parameters with the defect structure of oxide and the electronic properties of the SiO_2-Si interface. Our research has focused on understanding the system-dependent gas-phase plasma precursor chemistry as it relates to the origin of the chemical defect structure in the oxide and the electronic properties of the oxide-semiconductor interface. The two different plasma assisted deposition techniques that are currently being used to deposit gate quality SiO_2 are: in situ PECVD [3,4] and downstream PECVD [5,6].

The "in situ" and "downstream" PECVD terminology refers to the degree to which the dielectric film is exposed to potentially damaging radiation (energetic particle and optical radiation) generated by the discharge, during deposition. In the downstream PECVD technique, only the oxidant is exposed to the plasma radiation. The in situ technique exposes the SiH_4, the oxidant and the dielectric film to the energetic plasma environment which clearly must affect the initial stages of the gas-phase precursor chemistry.

In this paper we demonstrate that thin gate quality SiO_2 can be deposited in a downstream, radiation free, PECVD system using a microwave discharge source. To establish this, the electrical, chemical and elemental structure of the downstream oxide-semiconductor system is measured and compared to gate quality thermal oxides.

EXPERIMENTAL METHOD

The thin gate oxide is deposited downstream, in a quartz process chamber and at right angle to a 2.45 GHz N_2O microwave discharge. To accomplish this, atomic oxygen is generated by passing N_2O through the microwave discharge, at a pressure of 1.0 Torr. Two percent SiH_4, in He, is then injected into the after glow of the discharge. The flow rate of the N_2O is fixed at 700 sccm and the flow rate of the SiH_4/He mixture is varied from 950 sccm to 1000 sccm. The Si substrates are p-type, 2 Ω-cm and oriented in the <100> direction. A standard Huang cleaning technique[7] is used before deposition. Prior to deposition, and downstream of the discharge, the substrate is exposed to a H_2/He plasma at the deposition temperature. The substrate is maintained at 350C during the deposition process and the deposition rate is approximately 5.0 nm/min. The oxides used for the IR studies were deposited onto thin wedge shaped Si<100> wafer, polished on both sides, using the same deposition conditions as those used for the thinner gate oxides discussed above. The wedge shaped polished wafers are used to avoid internal reflections and to increase the signal to noise ratio during IR data collection. A discussion of the deposition system and the effects of the system parameters(pressure, power and substrate temperature) on the deposition rate, can be found in Ref. 5. MOS capacitors were prepared by radio frequency evaporation of 200 nm thick aluminum gate electrodes through a shadow mask. High frequency and quasi-static capacitance vs. voltage(C-V) and current vs. voltage(I-V) measurements were made on MOS devices fabricated from the PECVD oxide and a 10 nm thermally grown oxide.

High resolution ion beam channelling measurements were made on the same thermal and PECVD oxide wafers after the electrical characterization. Details of the medium energy ion scattering technique can be found in Ref. [8,9,10]. The Si substrates were aligned to a 300 KeV $^4He^+$ ion beam channelling along the $[11\bar{1}]$ direction. Exiting $^4He^+$ were detected with a commercially available electrostatic energy analyzer[8] placed to detect particles over a 25° range, centered on the [111] blocking dip. The energy resolution of the analyzer is $\Delta E/E = 4 \times 10^{-3}$, giving a depth resolution of 1 nm.

SIMS measurements were performed with O_2^+ ion beam primaries at 8 KeV and positive ion detection. The PECVD and thermal oxides were pumped down together overnight and the measurements were made contiguously to avoid instrumental drift error.

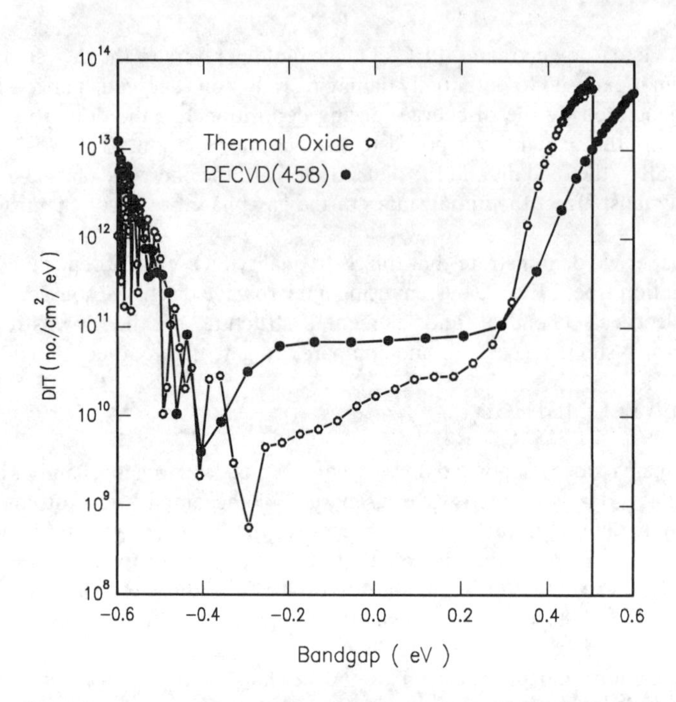

Fig. 1 The D_{it} for the thermal and the PECVD oxides are 1.5E+10. and 7.04E+10 after annealing, respectively.

RESULTS AND DISCUSSION

The density of interface traps(D_{it}) for the thermal and PECVD oxides, as a function of energy in the silicon band gap, are plotted together in Fig. 1. These data are calculated from the high frequency and quasi-static C-V measurements shown in Fig. 2 and Fig. 3. The mid gap D_{it} for the PECVD oxide is 7.04 x 10^{10} eV^{-1} cm^{-2} and this is to be compared to 1.5 x 10^{10} eV^{-1} cm^{-2} for the thermal oxide. The best PECVD oxides have a mid gap D_{it} of 3 x 10^{10} / eV cm^2 and were obtained by optimizing the deposition process. The flatband voltage as measured from the high frequency data of Fig. 3 shows a shift toward larger negative values(-0.749V) relative to the thermal data(-0.681V) which indicates that the PECVD oxide has approximately 1.5 x 10^{11}/ cm^3 more positive charge than the thermal oxide. The average breakdown field for the oxide was 8MV/cm. After the electrical measurements, the same wafers were used for SIMS and XPS measurements.

The D_{it}, the breakdown field and the trapped charge are properties of the oxide and are related to the structural and/or chemical defects generated during the deposition process. As a note, for thin films ($\leq$ 10 nm), defects can be generated during post deposition processing which are not related to the intrinsic quality of the oxide.

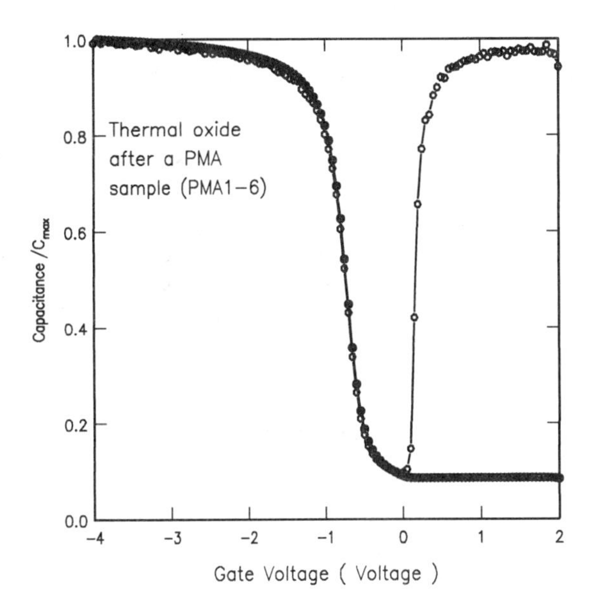

Fig. 2 Quasi-static and high frequency C-V measurements after a PMA. The oxide is 10 nm thick and was thermally grown on Si<100> in dry O_2 at 800C.

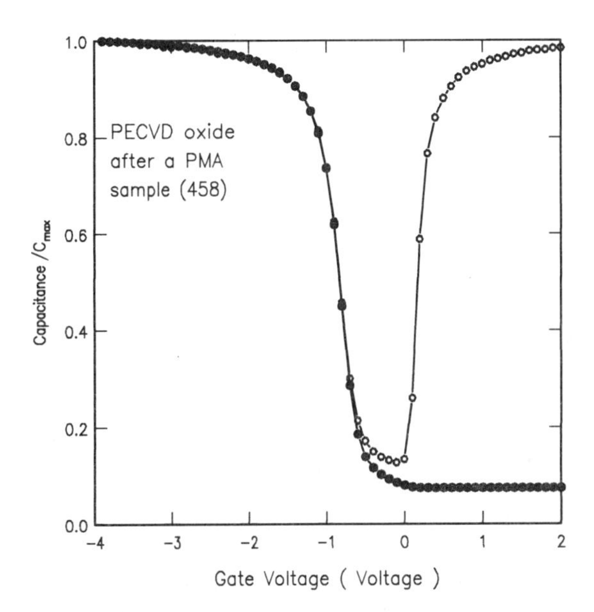

Fig. 3 Quasi-static and high frequency C-V measurements for a PECVD oxide after a PMA. The PECVD oxide is 9.3 nm thick and was deposited on Si<100> at 350C.

For example, the effects of stress and gate alloying during metal gate fabrication can negatively affect the D_{it} and the breakdown measurements. These defect artifacts can be difficult to separate from the intrinsic quality of the oxide.

Stress measurements were made on 1.0 μm, 500 nm and 100 nm thick PECVD oxides, using a wafer deflection measurement technique and gave values of 2.0 to 5.0 x 10^9 dyne/ cm^2 before, and 3.0 to 7.0 x 10^7 dyne/ cm^2 after, a 20 min. post-deposition anneal at 400C. The stress in the oxide before and after the anneal was compressive. These results demonstrate that the downstream PECVD oxides can be deposited with little or no stress after annealing. The change in the stress can also be related to shifts in IR absorption bands of the oxides. The initial location of the Si-O-Si stretch absorption band is a function of the oxide thickness, the film stress and the oxide density. On annealing the oxide at 400C, the Si-O-Si IR absorption band always shifted to higher wavenumbers regardless of the initial thickness of the oxide. We attribute this initial shift to higher wavenumbers on annealing primarily to internal stress relaxation at the Si-SiO_2 interface and not to oxide densification, at this low temperature[11,12,13].

The ion channelling spectra for the thermal and PECVD oxides are shown in Fig. 4. A single angular region of the analyzer, detecting particles exiting 28° from the surface, was used to obtain the spectra. The O/Si ratio, from the backscattered channeling data, is on average 2.00 ± 5% for both the oxides. This indicates that the oxides are stoichiometric. In calculating stoichiometries, corrections were made for the energy dependences of the analyzer transmission, the neutralization of backscattered He, and the electronic stopping power of SiO_2. The analysis of the data suggest that at this level of experimental sensitivity, the elemental composition and near interface oxide structure for the PECVD and the thermal oxides are essentially indistinguishable.

The SIMS data shown in Fig. 5 illustrates that at least an order of magnitude higher concentration of F, N, and H exist in the PECVD oxides relative to a thermal oxide of approximately the same thickness (Fig. 6). The sputtering time in Fig. 5 and Fig. 6, from 0 to approximately 15 sec. represents the elemental composition in the oxide, above 15 sec. the O_2^+ ion beam is sampling the elemental composition in the Si substrate. A silicon wafer implanted with a known concentration of F was used as a calibration standard for the SIMS analysis. Using this standard, the F concentration in the bulk PECVD oxide was calculated to be ~ 10^{18} /cm^3 and the concentration at the SiO_2-Si interface is 10^{19}/cm^3, which is below the detectable limit of the IR technique(~ 0.5 at. %, at best). This data is consistent with the observation that the IR spectrum for the 0.80 μm thick PECVD oxides show no measurable intensity for the SiH, SiN, SiOH and Si-H_2O absorption bands (Fig. 7). Although the absolute concentration of hydrogen or nitrogen in the PECVD oxide has not been measured, it is reasonable to assume that the concentrations are below 10^{19}/ cm^3 ,since 0.1 at. % is the best detection limit of the technique.

The initial XPS measurements were taken at 90 degree (normal) and 20 degree photoelectron take-off angles. The thickness of the PECVD oxide on these wafers was 4.6 nm and there was little or no observed charging during data collection.

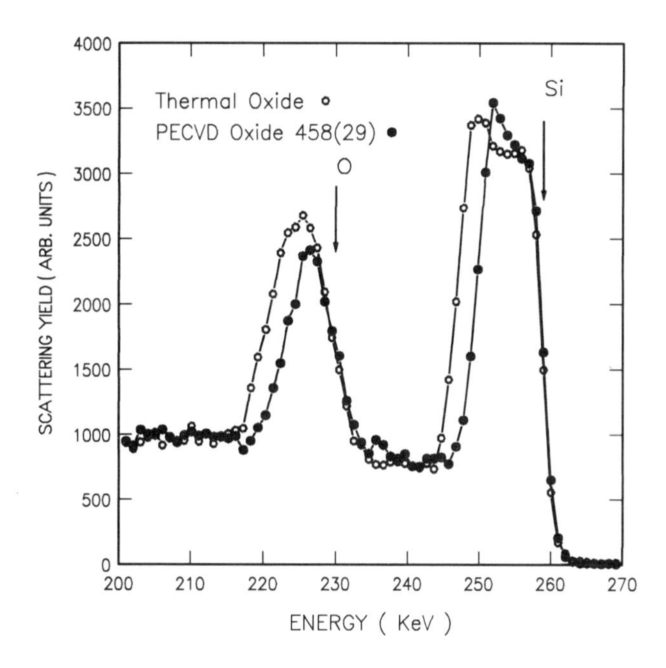

Fig. 4 Ion beam channeling spectra on high purity thermal and PECVD oxide. The thermal oxide is 10 nm thick, the PECVD oxide is 9.3 nm thick.

A core level energy shift at 399.4 eV is associated with N(1s) and is not present in the thermal oxide, see Fig. 8 and Fig. 9. This feature was only seen in the normal take-off angle spectrum of the oxide and is consistent with a silicon-nitrogen bonded complex localized to within 1 nm of the Si interface[14]. Two additional structures associated with nitrogen were identified at 400.6 eV and 402.9 eV. At this particular stage in the investigation no chemical complex has been assigned to these nitrogen core level shifts. The PECVD oxides are, as expected, predominantly SiO_2 as shown by the Si(2p) peak at 104.2 eV in Fig. 10. The elemental silicon signal at 99.3 eV originates from the underlying Si substrate. A low intensity silicon feature at 100.6 eV may be assigned to nitrogen-bonded silicon or to a SiO_2 sub-oxide. The peaks located at 99.9 eV and 104.1 eV can be associated with Si_2O (the Si^{+2} oxidation state) and SiO_2 (the Si^{+4} oxidation state), respectively. While the PECVD oxides used for the XPS measurements were 4.6 nm thick and grown on wafers without Huang cleaning , the SIMS measurements agree qualitatively with the observation of N in the oxide structure. Even with the higher concentrations of impurities in the PECVD oxides, the mid gap D_{it} is only a factor of five(in the worst case) larger than the mid gap D_{it} for the best thermal gate oxide. Preliminary experiments indicate that increasing the SiH_4/He flow rate from 950 sccm to 1000 sccm ($\pm$ 0.5 % of full scale) lowers the D_{it}. However, SIMS and channeling measurements indicate no significant change in the impurity concentration or the elemental structure of the oxide.

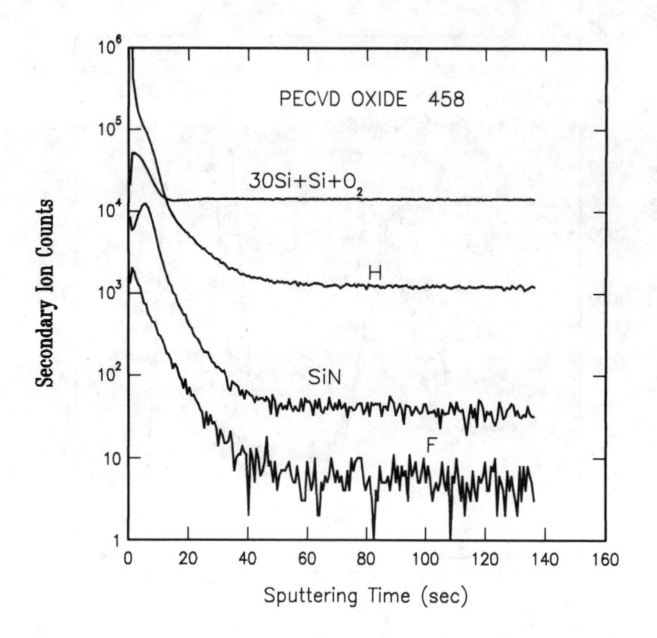

Fig. 5 SIMS analysis on a 9.3 nm thick PECVD oxide. This is the same oxide that was used for the electrical measurements discussed in the paper.

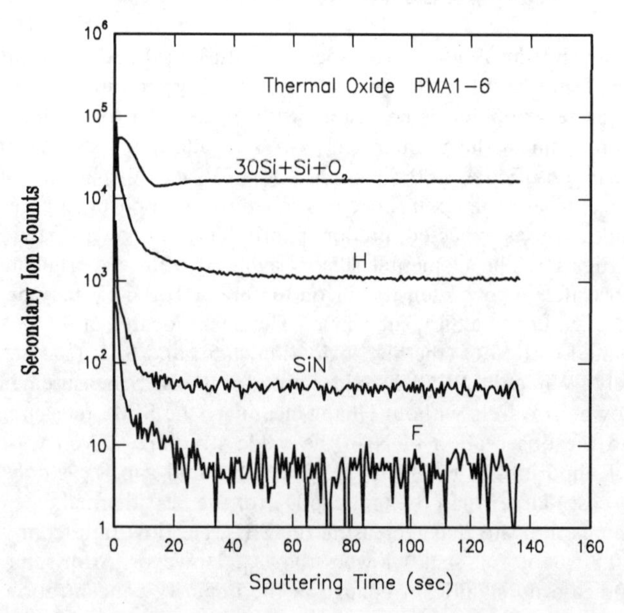

Fig. 6 SIMS analysis on a 10 nm thick thermal oxide. This is the same oxide that was used for the electrical measurements discussed in the paper.

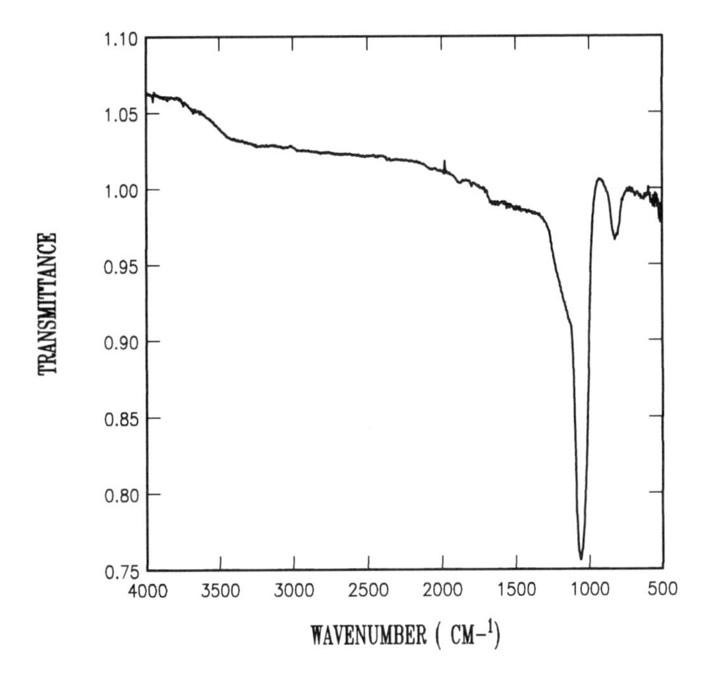

An IR spectroscopic scan of the silicon oxide film approximately 900nm thick deposited downstream at 350C and 1.0 Torr.

Although changes in the impurity concentration in the oxide may be below the detection limits of this technique, the difference in the gas flow rate may be sufficient to alter the precursor chemistry, thus changing the structure of the oxide at the atomic level (e.g., depositing Si in SiO_2 network can reduce the local stoichiometric structure to a sub-oxide[15]).

Based on the available data, a reasonable assumption as to the origin of the impurities in the oxide can be made. The F at the interface can be attributed to wafer cleaning and system related processing conditions. In preparing the wafers for deposition, the final HF dip in the Huang cleaning procedure is known to leave F at the surface. A plasma discharge of SF_6 is used to remove oxide deposits from the walls of the process chamber. The F or F_2 which remains on the walls of the quartz process chamber eventually ends up in the bulk oxide during subsequent oxide depositions. This may also explain the different F concentration at the interface and in the bulk of the oxide. We speculate that the pre-deposition exposure of the Si wafer downstream of the H_2/He plasma will clean the surface by generating atomic H from the discharge, which will abstract F from the surface of the silicon wafer when it is held at 350C.

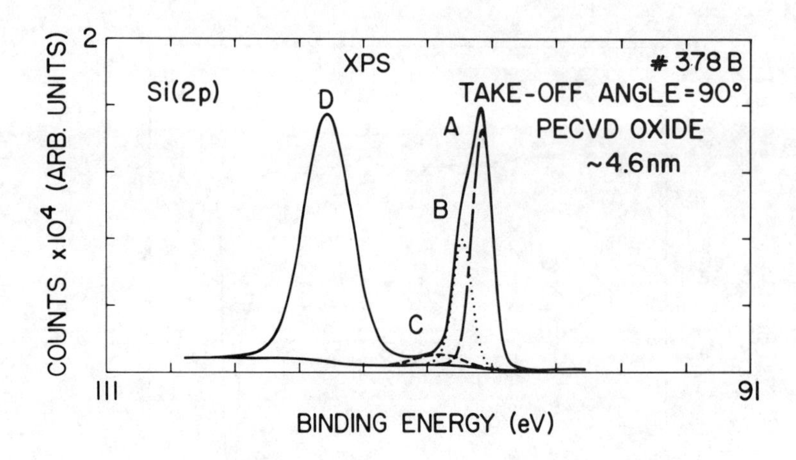

Fig. 4 Si(2p) spectrum of a 4.6 nm thick PECVD oxide at a photoelectron take-off angle of 90 degrees. The four peaks A, B, C, and D are located at 99.3 eV, 99.9 eV, 100.6 eV, and 104.1 eV, respectively.

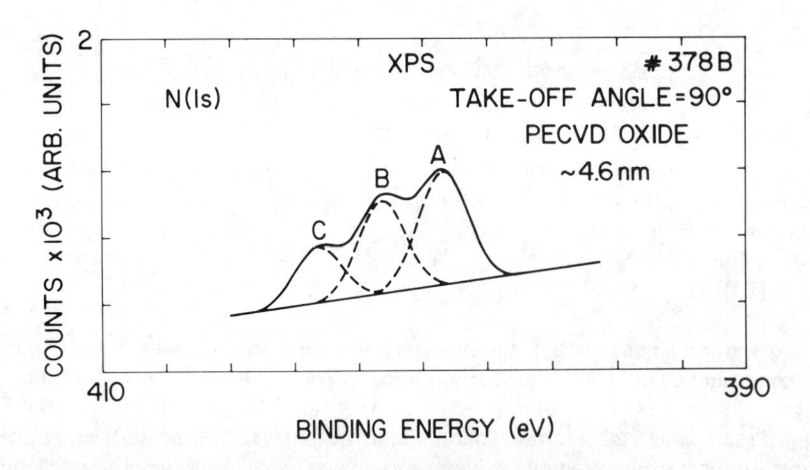

Fig. 5 N(1s) spectrum of a 4.6 nm thick PECVD oxide at a photoelectron take-off angle of 90 degrees. The three peaks A, B, and C are located at 399.4 eV, 401.3 eV, and 403.3 eV, respectively.

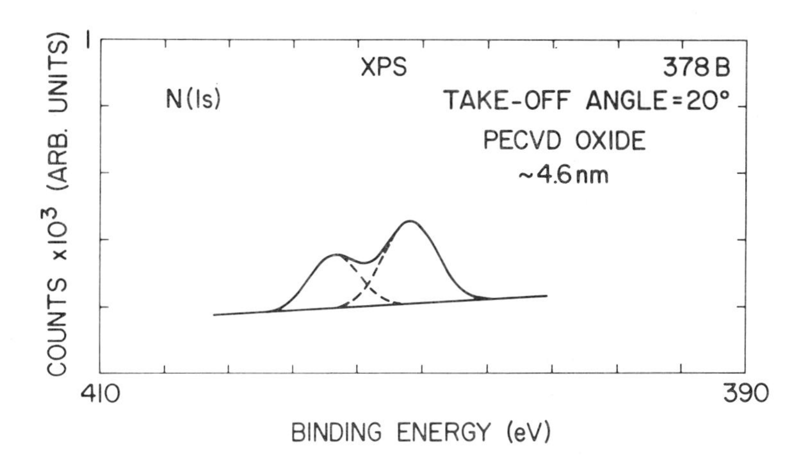

Fig. 6 N(1s) spectrum of a 4.6 nm thick PECVD oxide at a photoelectron take-off angle of 20 degrees. The structure at 399.4 eV in Fig. 5 is not present in this spectrum.

Extended exposure of the wafer to the H_2/He plasma at 350C (1 to 3 min), the lowers the measured mid gap D_{it}, however we are currently exploring the limits of this technique in reducing the D_{it} by exposing the wafer for longer periods of time. The N concentration observed in the oxide is not surprising given that the atomic oxygen needed for the PECVD reaction is generated from a N_2O microwave discharge. What is surprising is the low D_{it} in the PECVD oxides, since it is known that additional interface traps are generated on prolonged exposure to N_2 during annealing at high temperature[16]. In the context of N_2 generating interface traps, the presence of atomic N should be more effective in creating these types of defects. The hydrogen in the oxide comes from the decomposition of the SiH_4 during deposition.

If the PECVD oxides are to become part of a device structure more detailed electrical measurements will have to be made. One needs to determine the nature, the spatial location and origin of the defects in the bulk oxide, the so called E' centers, and the defects at the oxide-semiconductor interface, known as the P_b centers. We have clear evidence as to the identity and origin of the impurities in the oxide and at the interface. We are currently obtaining information on the chemical structure of the complexes in the oxide and the structure of the near-interface region of the oxide. With the information we have and are currently obtaining on the PECVD, it is of interest to determine if the PECVD oxides is a replication of the thermal oxide at the level discussed in the references[15,17,18,19].

CONCLUSIONS

It has been demonstrated, by <u>electrical</u> measurement techniques, that high quality thin films of SiO_2 can be deposited at 350C downstream from a microwave discharge. MOS electrical measurements give values for the D_{it} in the range of 3 to 7 x 10^{10} eV^{-1} cm^{-2} after annealing, which compare favorably to a D_{it} of 1.5 x 10^{10} eV^{-1} cm^{-2} for the best gate quality thermal oxide. The PECVD films exhibit little or no stress, 10^7 dyne/cm^2, after annealing at 400C for 20 min.. The high resolution backscattered channelling data on the PECVD oxides showed no significant compositional or structural changes in the depth profile as compared to a thermal oxide. Indeed, the sharpness of the SiO_2-Si interface for the PECVD oxide is strikingly similar in detail to the thermal oxide. The combined analysis of SIMS and XPS show a chemical impurity background of F, H and N which is integrated into the SiO_2 network. We have a reasonable explanation as to the origin of the impurities at the interface and in the bulk oxide.

ACKNOWLEDGEMENT

We would like to thank P.D.Hoh, J.Batey and E.Tierney for helpful discussions concerning their in situ plasma deposition process. Chris. C. Parks and G. J. Coyle for providing the SIMS and XPS data, respectively. Also, we thank S.Cohen for the MOS electrical measurements, M.Lilie for helpful suggestion in regard to data acquisition and manipulation and R.M.Tromp for assisting in the ion beam channelling experiments.

References

1. T.B.Gorczyca and B.Gorowitz, VLSI Electronics 6, N.Einspunch and G.B.Larrabee Ed. Academic Press., N.Y. Chap. 4 (1983).
2. A.C.Adams, VLSI Technology, S.M.Sze Ed., McGraw-Hill, N.Y. Chap.3 (1983).
3. J.Batey, E.Tierney and Thao N. Nguyen, IEEE Electron Device Lett., EDL 4, 148 (1987);
4. J.Batey and E.Tierney, J.Appl. Phys. 60, 3136 (1986).
5. B.Robinson, P.D.Hoh, P.Madakson, T.N.Nguyen, and S.A.Shivashankar To be published in the 1987 MRS Symposia proceedings, spring meeting, Anaheim, California.
6. B.Robinson, P.D.Hoh, P.Madakson, T.N.Nguyen and S.A.Shivashankar, ECS extended abstracts # 673 87-2 948 (1987).
7. E.A.Irene, J.Electrochem. Soc. 121 1613 (1974).
8. R. M. Tromp, H. H. Kersten, E. Granneman, F. W. Saris, R. J. Koudjis, and W. J. Kilsdonk, Nucl. Instrum. Methods B 46, 155 (1984).
9. O.Meyer, J.Gyulai and J.W.Mayer, Surface sci. 22 263 (1970).
10. J.F.Van Der Veen, Surf. Sci. Rep. 5 199 (1985).
11. M.Nakamura, Y.Moshizuki, K.Usami, Y.Itoh and T.Nozaki, Solid State Commun. 50 1079 (1984).
    Naoyuki Nagasima, J. Appl. Phys., 43,3378 (1972).
12. I.W.Boyd and J.I.B.Wilson, Appl. Phys. Lett. 50 320 (1987).
    W.A.Pliskin and H.S. Leman, J. Electrochem Soc., 112 1031 (1972).
13. J.Wong, J.Appl. Phys., 44 5629(1973).
14. C.C.Parks, B.Robinson, H.J.Leary, Jr., K.D.Childs and G.J.Coyle, Jr., presented at the 1987 MRS Fall Meeting, Boston (to be published)
15. F.J.Grunthaner, P.J.Grunthaner, R.P.Vasquez, B.F. Lewis, and J.Maserjian, J. Vac. Sci.Technol. 16 1443 (1979).
16. Hisham Z. Massoud, Technical Report No. G502-1 Stanford Electronics Lab. page 55 (1983).
17. C.R.Helms, J. Vac. Sci. Technol. A 16 608 (1979).
18. P.J.Grunthaner, M.H.Hecht and F.J.Grunthaner, J. Appl. Phys. 61 629 (1987).
19. F.J.Feigl, VLSI Electronics 6, N.Einspruch G.B.Larrabee Ed. Academic Press., N.Y. Chap.3 (1983).

LOCAL ATOMIC STRUCTURE OF THERMALLY GROWN AND RAPID THERMALLY ANNEALED SILICON DIOXIDE LAYERS

J.T. FITCH AND G. LUCOVSKY
Department of Physics, North Carolina State University
Raleigh, North Carolina 27695-8202

ABSTRACT

This paper is a study of the local atomic structure of silicon dioxide films formed by thermal oxidation in both dry oxygen and steam ambients. Films were grown over a temperature range of 800 to 1150°C. Low temperature films grown at 800°C (steam) and 850°C (dry) were subjected to rapid thermal annealing for 1, 10, and 100 sec at temperatures of 800 to 1200°C in 100°C increments. Local atomic structure was studied via infrared spectroscopy (IR). We have investigated the temperature dependence with respect to growth and annealing of the frequency, v, and half width Δv, of the bond stretching feature, the refractive index, and the film stress. Data from dry and steam oxides were compared and, though the steam oxides were generally less dense, the variations in v, Δv, n, and stress with thermal history were similar and could be readily explained in terms of systematic changes in the bond angle at the oxygen bonding site between corner connected tetrahedra.

INTRODUCTION

Since nearly all silicon based MOS integrated circuits use thermally grown SiO_2 for a gate insulator, the study of thermal oxidation of silicon is of tremendous technological and economic importance. In fact, thermally grown films of SiO_2 for electronic applications present a special opportunity to study the structure of SiO_2 because they are grown under carefully controlled conditions. We are engaged in a study of thermally grown and annealed SiO_2 films. In previous studies[1,2] thermal oxides grown in a dry O_2 ambient and annealed oxides of this type were examined and it was demonstrated that systematic correlations between (1) v and Δv, (2) n and v could be explained in terms of a model based on the angle between the corner connected SiO_4 tetrahedra at the oxygen bonding site. This paper is a continuation of that study with emphasis on a comparison between films grown in a dry versus a steam ambient. These differences will be explored by (1) comparison of infrared spectroscopy (IR) features such as stretching peak frequency and width, (2) comparison of refractive index data and correlation with IR data, and (3) a new aspect of this study, measurement of thin film stress.

EXPERIMENTAL PROCEDURE

Three different kinds of (100) silicon wafers were used for the oxide growth: P-type 30-60 ohm-cm, N-type 17-20 ohm-cm, and P-type 0.01-0.02 ohm-cm wafers which were only 3 mils thick. The 30-60 ohm-cm P-type wafers were given an HF/HNO_3 based silicon etch called CP-4A to smooth the backside of the wafers to facilitate IR. Also, the 3 mil thick wafers were scribed and broken into small strips approximately 1 by 4 cm. All of the starting material was given a Huang clean (a shorter version of the standard RCA clean procedure) and then prexodized in a diffusion furnace in an HCl containing steam ambient at 1000°C for 17 min to grow 1000 A of oxide. Oxide from the preoxidation was stripped in buffered HF. This step was done as a precaution to completely remove most residues. All of the wafers were given another Huang clean just before final oxidation. Final oxidation of the wafers was done in a dry ambient at temperatures of 850, 950, 1050, and 1150°C, and in a steam ambient at 800, 900, and 1000 °C. For the dry oxidations a flow rate of oxygen of 2.86 liters/min was used, and for the steam oxidations pyrolytic steam was used with oxygen and hydrogen flow rates of 1.0 and 2.0 liters/min, respectively. A nitrogen purge of 7.86 liters/min was used during temperature ramping. Oxidation times were intentionally adjusted to give final oxide thicknesses of approximately 1,300 A. This thickness corresponds to half an ellipsometric period, the optimum thickness for accurate determination of refractive index. The IR absorption (in the region of the three first order Si-O bands at about 1075, 800, and 460 cm^{-1}) was obtained using a Perkin Elmer PE 983 Double-Beam Grating Spectrophotometer, and the index of refraction (at 632.8 nm) and film thickness were determined by ellipsometry using a Gaertner L115A automatic ellipsometer. Stress measurements were obtained by stripping the oxide from the backsides of the 3 mil silicon strips mentioned earlier and measuring the radius of curvature of the strips induced by the oxide. Radii of curvature were calculated from the displacement of two parallel laser beams reflected off of the curved strips in a manner described by Kobeda and Irene[3]. Stress values were calculated from the usual beam bending equation

$$\sigma_F = \frac{E}{6(1 - P)} \frac{t_s^2}{t_f} \frac{1}{R} \qquad (1)$$

where σ_f is the film stress, E is the Young's modulus of silicon, P is Poisson's ratio for silicon, t_s is the thickness of the substrate, t_f is the film thickness, and R is the radius of curvature of the silicon strip. Following these measurements oxide films grown at 850°C in dry oxygen and films grown at 800°C in steam were subjected to rapid thermal annealing for a matrix of temperatures and times ranging from 800 to 1200°C and from

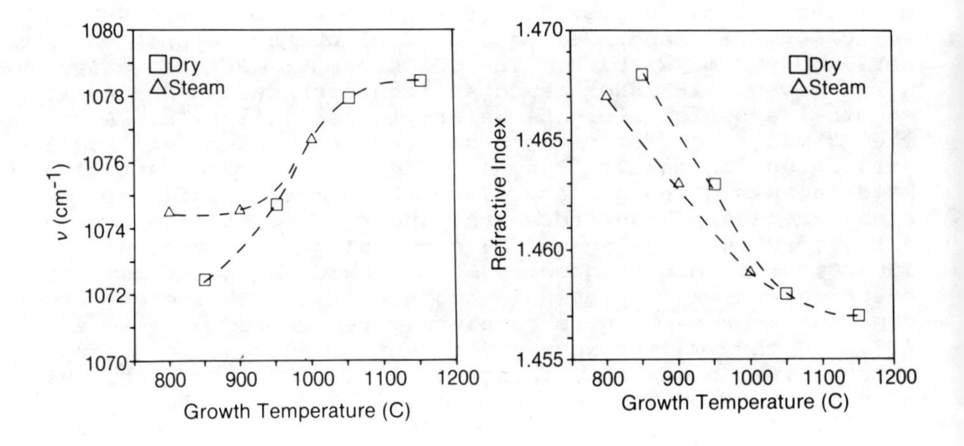

Fig. 1 v vs. growth temp. for as grown thermal oxides.

Fig. 2 Refractive index vs. temp. for as-grown thermal oxides.

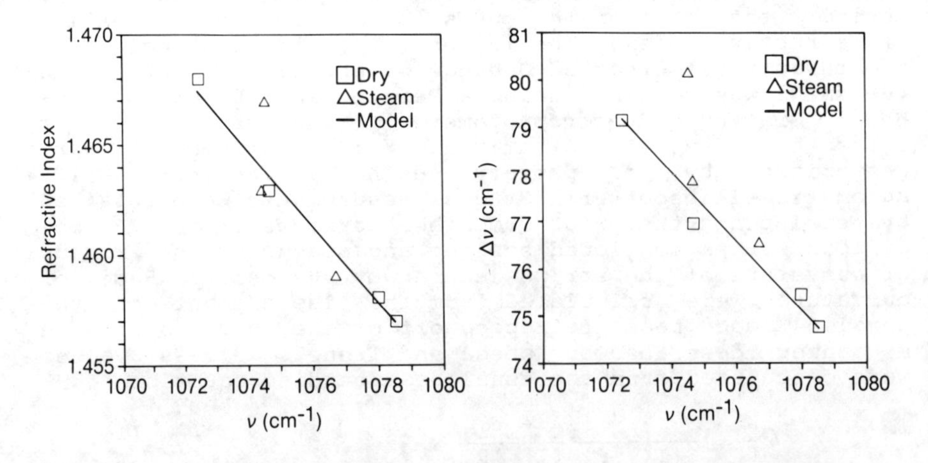

Fig. 3 n vs. v for both types of grown oxides with model fit superimposed (see text).

Fig. 4 Δv vs. v for both types of as-grown oxides. Model was calculated under constraint dθ/θ = const.

1 to 100 sec using an AG Heatpulse 210M. The ambient for this process was argon and the chamber was purged for 1 minute prior to annealing. Measurements were repeated on the annealed films.

DATA AND DATA REDUCTION

Results of the measurements discussed above are plotted in Figs. 1-9. For IR, ellipsometry, and stress measurements the plotted data points are an average of eight readings and the respective uncertainties are 0.5 cm^{-1}, 0.002, and $0.3x10^9$ $dynes/cm^2$. Figures 1 and 2 indicate the growth temperature dependence of v and n. Differences between the dry and steam oxides are readily apparent especially at low temperatures. Fig. 3 indicates a linear correlation, independent of growth temperature, between the data in Figs. 1 and 2; i.e. n vs. v. Fig. 4 suggests another linear correlation between width, Δv, and frequency, v, of the bond-stretching feature. The solid lines are from models[1,2] which will be discussed later. Figs. 5-7 illustrate the annealing temperature dependence of v, Δv, and n. Data from 1 and 10 sec anneals were not included because they represented "incomplete" stages of annealing. Trends in the steam oxide data in Figs. 1-7 are parallel to results previously reported in Ref. 2 for dry oxides. We have annealed a steam grown oxide (800 C) at 800 C in a dry ambient, as shown in Fig. 8, and have observed that the bond-stretching frequency falls to a value approximately equal to that of a dry oxide (grown at 850 C) for annealing times greater than about 10 sec. A new aspect of our work is the measurement of film stress shown in Fig. 9. Dry oxides consistently showed higher levels of compressive stress. Throughout this paper "intrinsic stress" is defined as the growth stress remaining after the stress from the mismatch in coefficients of expansion between the silicon and the oxide is subtracted. Stress measurements were also made on samples of annealed dry and steam oxides grown at 850 and 800^oC respectively. The data showed a continuous and gradual reduction in the compressive growth stress. Regarding annealed oxide data, the term "intrinsic stress" should be taken to mean remaining growth stress.

MODEL RESULTS

The systematic variations in v, Δv, and n in Figures 3 and 4 can be understood in terms of a model for the local atomic structure at the oxygen atom bonding sites. Oxygen atoms in SiO_2 are two-fold coordinated with an average bond angle in the most 'relaxed' high temperature forms of vitreous silica of about $2\theta = 145^o$[4]. The basis for the correlation between n and v was first discussed in Ref. 1 in terms of a model which assumed that differences in film properties with thermal history derived from changes in the bond angle at the Si-O-Si bonding sites.

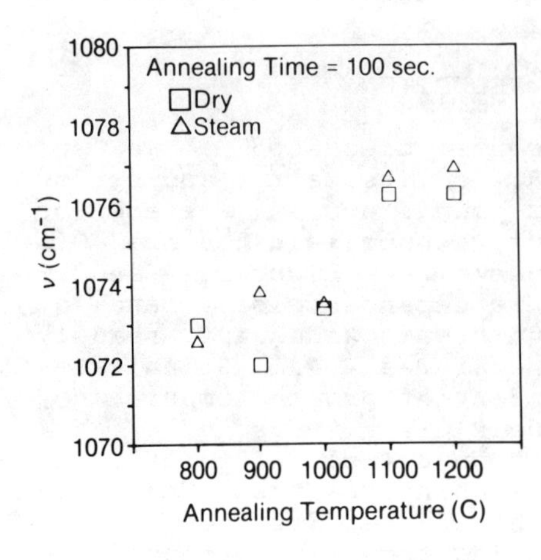

Fig. 5 ν vs. T for dry and steam oxides grown at 850°C and 800°C, respectively, and annealed.

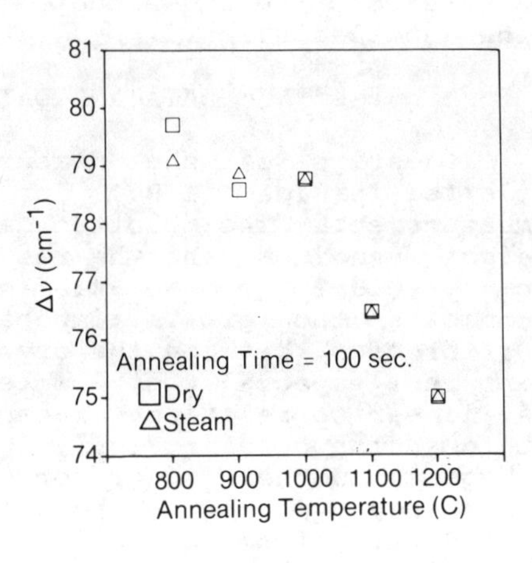

Fig. 6 Δν vs. T for dry and steam oxides grown at 850°C and 800°C, respectively, and annealed.

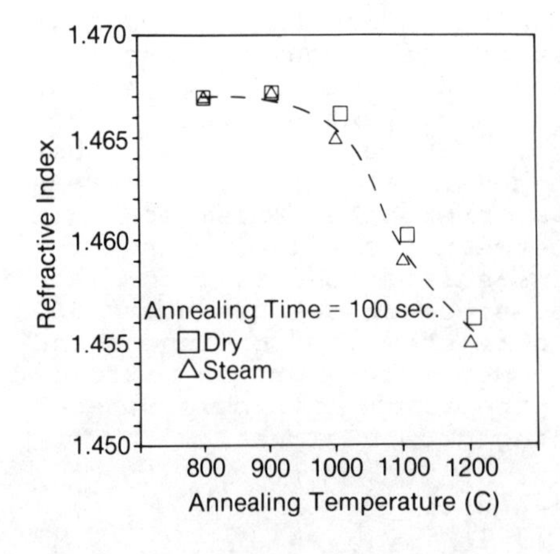

Fig. 7 Refractive index vs. T for dry and steam oxides grown at 850°C and 800°C, respectively, and annealed.

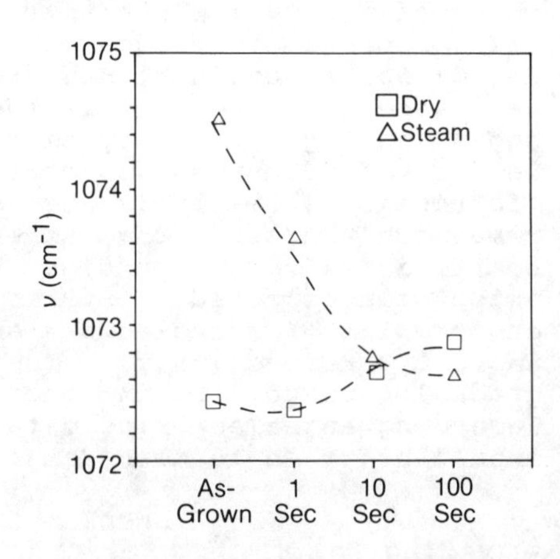

Fig. 8 ν vs. annealing time at 800°C for dry oxides grown at 850°C and steam oxides grown at 800°C.

Within the framework of this model the frequency of the IR bond stretching feature is

$$v = v_o \sin\theta; \text{ where } v_o = 1117 \text{ cm}^{-1} \tag{2}$$

The relationship is an approximation because only central forces are considered and any motion of the silicon atoms is neglected[5,6]. v_o is evaluated from the mass of the oxygen atom and the bond-stretching force constant. Experimentally, v_o was calculated by assuming that an oxide grown at 1150 C with v = 1078.5 cm-1, has a bond angle $2\theta = 150^o$. A value somewhat larger than 145° was chosen because most of the literature values are for vitreous silica solidified from a melt.

A Clausius-Mossotti model is assumed for the dielectric constant, $e = n^2$; e is related to the molar volume, V, by:

$$\frac{(e - 1)}{(e - 2)} = \frac{A}{V} \tag{3}$$

where A is a constant that includes the polarizabilities of the Si and O atoms and a normalizing density factor. It is further assumed that the molar volume is proportional to the cube of the silicon-silicon distance. An expression for d_{Si-Si} can be written in terms of the Si-O bond length r_o as

$$d_{Si-Si} = 2r_o \sin\theta \tag{4}$$

The normalization factor A in Eq. (3) can be calculated by noting that in a high temperature film with v = 1078.5 cm^{-1}, that n = 1.458. With Eqs.(2), (3), and (4) a relationship between n and v, both of which depend on $\sin\theta$, can be calculated. The solid lines in Figs. 3 and 10 derive from this model and accurately reflect the trends in the data.

The central forces model can be extended to predict a relationship between v and Δv by differentiating Eq. (2):

$$dv = v_o \cos\theta \, d\theta \tag{5}$$

Dividing Eq. (5) by Eq. (2),

$$\frac{dv}{v} = \cot\theta \, d\theta \tag{6}$$

The solid lines in Figs. 4 and 11 accurately reflect the trends in the data under the constraint that the relative angular dispersion in the bond angle θ is constant, i.e. $d\theta/\theta = 0.197$. $d\theta$ can in turn be estimated from measurements of dv. Using this type of analysis values for $2d\theta$ fall in the vicinity of 30°, which is in excellent

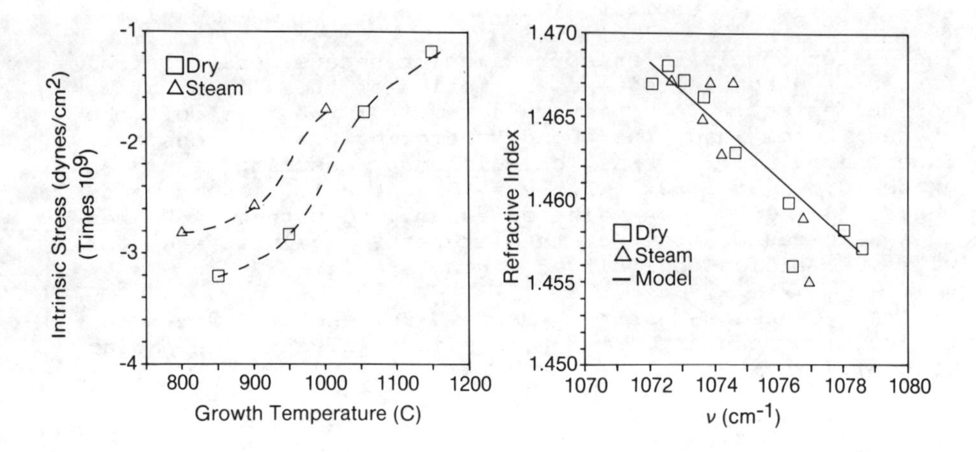

Fig. 9 Intrinsic stress vs. growth temperature.

Fig. 10 Refractive index vs. v with all data points included. Solid line is a model fit (see text).

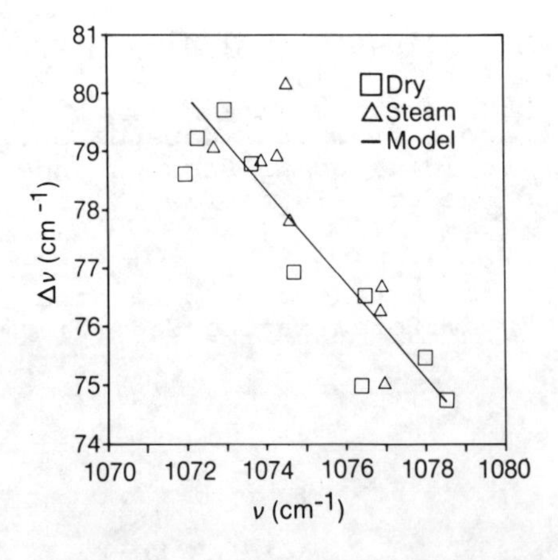

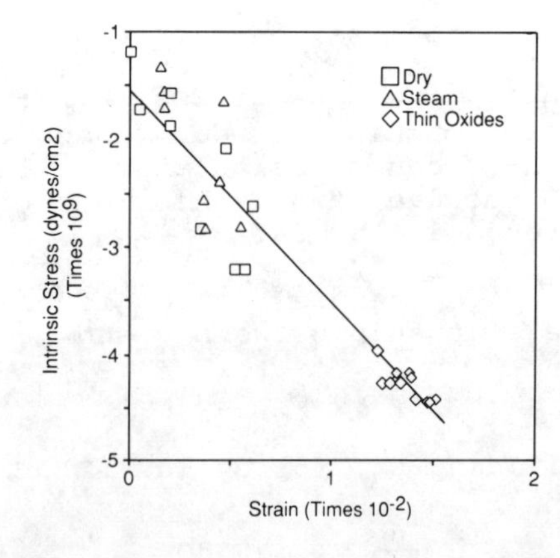

Fig. 11 Δv vs. v with both as-grown and annealed oxides included.

Fig. 12 Intrinsic stress vs. strain. Data from as-grown, annealed, and thin oxides (see text) is plotted.

agreement with values obtained from other experimental studies[5].

Film stress measurements can be correlated with the central forces model of Eq. (2) if the strain in the oxide is taken as the fractional change in the silicon-silicon distance. Assuming that the bond angle of an oxide grown at 1150°C is $2\theta_1$ and that the bond angle of a low temperature film is a smaller angle $2\theta_2$ then the strain is given by

$$\text{strain} = (\sin\theta_1 - \sin\theta_2)/\sin\theta_1 \qquad (7)$$

Using a trigonometric identity for the difference of the sine of two angles and the small angle approximation $\sin(\theta_1-\theta_2)/2 = (\theta_1-\theta_2)/2$,

$$\text{strain} = 2\cos((\theta_1+\theta_2)/2)*((\theta_1-\theta_2)/2)/\sin\theta_1 \qquad (8)$$

In Fig. 12 the intrinsic stress is plotted as a function of the strain given above. The data marked "thin oxides" were part of another study and are included for comparison. These samples were grown in a dry ambient at 700°C and had thicknesses between 100 and 600 A. The correlation in Fig. 12 is very significant because, in linear elastic solids, stress is proportional to strain by a constant E called Young's modulus. Since the oxide is under plane stress, a modified plane stress modulus, E/(1-P), is necessary, where P is Poisson's ratio. The elastic modulus can be calculated if several assumptions are made. First, use a spatially averaged stress, i.e. $(\sigma_x + \sigma_y + \sigma_z)/3 = (2-2P)\sigma_x/3$ where σ_x and σ_y are the in-plane stresses and σ_z is stress normal to the film, because in a real amorphous network the stress would be acting on a randomly oriented distribution of the Si-O-Si bond angles. Second, use a geometric factor of 1/6 in the computation of strain to be consistent with Eq. (1). Proceeding in this manner a value of the plane stress modulus of 6.1×10^{11} dynes/cm^2 was calculated. This compares favorably with the published value for E/(1-P) of 8.8×10^{11} dynes/cm^2 for SiO_2[7]. Thus, values predicted by our microscopic model of the Si-O-Si bond angle correspond closely with macroscopic material parameters.

DISCUSSION

At low growth temperatures the steam oxides have a higher stretching peak frequency v, while the dry oxides exhibit a higher refractive index. These two observations are consistent within the framework of eqns. (2), (3), and (4) because, in the case of dry oxides, a lower stretching peak frequency would indicate a smaller bond angle and hence a compacted oxide relative to the steam grown samples. At high growth temperatures the v and n values of the grown oxides converge to similar values and indicate that the network is in a relaxed state relative to low temperature grown samples. Trends of v, Δv, and n with increasing annealing temperature parallel those trends observed with increasing growth temperature. Furthermore, the stretching peak frequency, v, can be used to calculate the strain in the SiO_2 network. This calculated strain has been shown to be proportional to the measured film stress. The implication here is that the density of an oxide is affected by the remaining compressive growth stress and the IR stretching peak frequency can be used as a measure of this stress.

Although dry and steam oxides show the same trends in growth and annealing with respect to the various measurements, the data also shows significant differences. A steam oxide grown at 800°C and annealed at its growth temperature (Fig. 8) showed a high v which decreased to the value for a dry oxide grown at 850°C after only 10 sec of annealing. Since water, dissolved in SiO_2 as OH, can chemically act as a network terminator attacking the oxygen bridging atoms[8], outward diffusion of water during the anneal could explain the change in IR stretching frequency, which would be a function of the mean bond angle at the oxygen bridging atom sites. We intend to extend our present measurements to determine if the observed growth ambient related differences arise from the nature of the growth interface.

ACKNOWELDGEMENTS

This work is in part supported under ONR Contract N00014-79-C-0133. We wish to acknowledge Eddie Kobeda and Prof. E.A. Irene of the University of North Carolina at Chapel Hill for use of their experimental apparatus, for unpublished data on thin oxides, and for many helpful discussions. We also acknowledge the support of the Microelectronics Center of North Carolina for providing facilities for oxidation, annealing, and ellipsometry.

REFERENCES

1. G. Lucovsky, M.J. Mantini, J.K. Srivastava, and E.A. Irene, J. Vac. Sci. Tech. B 5, 530 (1987).
2. J.T. Fitch and G. Lucovsky, Proc. MRS, Vol. 92, Rapid Thermal Processing of Electronic Materials, p 89.
3. E. Kobeda and E.A. Irene, J. Vac. Sci. Tech. B 4, 720 (1986).
4. R.L. Mozzi and B.E. Warren, J. Appl. Cryst. 2, 164 (1969).
5. F.L. Galeener and P.N. Sen, Phys. Rev. B 17, 1928 (1978).
6. A.E. Geissberger and F.L. Galeener, Phys. Rev. B 28, 3266 (1983).
7. S. Isomae, J. Appl. Phys. 57, 216 (1985).
8. E.A. Irene, J. Electrochem. Soc. 121, 1613 (1974).

Thermal Nitridation of Si(100) Using Hydrazine and Ammonia

J. W. Rogers, Jr., D. S. Blair, and C. H. F. Peden
Sandia National Laboratories
Albuquerque, New Mexico 87185

ABSTRACT

X-ray photoelectron spectroscopy (XPS) was used to investigate the direct thermal nitridation of Si(100) using hydrazine and ammonia as nitrogen sources. The initial rate of film formation between 873 and 1173 K was markedly higher at a given temperature using hydrazine. Over this temperature range, the measured XPS N(1s) binding energy using either nitrogen source remained constant at 398.1 eV during nitridation suggesting that the nitride films were chemically similar. In addition, no oxygen was detected in the films. The increased rate of nitride formation using hydrazine is attributed to the lower energy required for dissociative adsorption relative to ammonia.

INTRODUCTION

Thin silicon-nitride films are used extensively in the semiconductor industry for their passivating and/or dielectric properties. Traditional methods for the formation of these films include chemical vapor deposition (CVD) and direct thermal nitridation (DTN) by nitric oxide [1], ammonia, or nitrogen [2]. These techniques, though widely used in industry, can result in nonideal silicon-nitride films. For example, nonuniform film growth and film impurities are two problems associated with the use of CVD techniques [3,4]. However, film formation by DTN requires elevated temperatures (1120-1375 K) [1,2,5] which can result in dopant diffusion and subsequent degradation of the desired electrical properties. Further miniaturization of semiconductor devices for VLSI places new constraints on the purity and uniformity of the silicon-nitride films. As such, attempts to better understand and improve these techniques continue [6,7].

Nitridation of silicon using different silicon substrates or nitrogen sources reportedly does not affect the properties of the nitride film. For example, DTN of Si(100), (111), and (311) using ammonia results in similar nitride films [8]. Maillot et al., [2], report that DTN using nitrogen or ammonia at 1200 K produces films that grow at different rates but have the same electronic and surface crystallographic properties. These authors further suggest that the silicon-nitride films grow in a poorly crystallized β-phase [2]. Finally, it has been reported that DTN results in stoichiometric Si_3N_4 films and that the Si-Si_3N_4 interface is abrupt [2,9].

Film growth by thermal nitridation of Si(100) with ammonia (T > 1123 K) has been reported to proceed in a layer-by-layer fashion [10]. After the first nitride layer is formed, reaction rates were found to decrease substantially. To explain these results, it was suggested [10] that diffusion of the reacting species through the nitride film becomes the rate limiting step.

Silicon-nitride film growth on silicon by DTN at room temperature is limited to the formation of one monolayer (ML) of reacted nitrogen species [11,12]. Bozso and Avouris [11] have proposed that DTN using ammonia at temperatures below 700 K becomes self-limiting as hydrogen atoms, formed by the decomposition of ammonia on the surface, bond to surface Si dangling bonds. Since these dangling bonds are likely the active sites for the dissociative chemisorption of ammonia [11], the H-Si bonds that are formed deactivate the surface for further reaction. Temperatures above 700 K are required to thermally desorb surface hydrogens from silicon and restore the initial activity of the surface [11,12].

To better understand the DTN process, we have undertaken a comparative study utilizing both hydrazine (N_2H_4) and ammonia (ND_3) as nitrogen sources. The use of hydrazine was suggested by a previous report of its application for the low temperature nitridation of Al [13]. The experiments reported here were carried out with a Si(100) sample at temperatures ranging from 673 to 1123 K in low pressures (< 10^{-6} Torr) of the nitrogen source gas.

EXPERIMENT

The experiments were performed in a Vacuum Generators (East Grinstead, England) ESCALAB 5 Surface Spectrometer. The ultra-high vacuum (UHV) analysis chamber had a base pressure of 3 X 10^{-11} Torr and was equipped with a hemispherical electron energy analyzer, residual gas analyzer for monitoring purity of gases, and Mg(Kα) X-ray source for XPS measurements. The Mg anode was operated at 240 W (12 kV, 20 mA) and the electron energy analyzer was operated to give a working resolution (including broadening due to the X-ray width) of 0.87 eV. All binding energies (BEs) were referenced to the Fermi level of gold [$Au(4f)_{7/2}$ = 84.0 eV] and were reproducible to ± 0.2 eV.

The silicon sample was a phosphorus-doped, p-type Si(100) single crystal which measured 1.5 x 1.5 x 0.15 cm. It was heated by electron bombardment and was supported between 0.1 mm W wires to minimize heat conduction away from the sample. The crystal was biased to +1200 V and electron-beam heated from the backside utilizing a W filament held near ground potential. The sample temperature was monitored using a W-5%Re/W-26%Re thermocouple which was arc welded to create a ball at the thermocouple junction. The lead wires were then threaded through a small hole sandblasted through the Si sample [14]. The hole was slightly larger than the thermocouple ball on one side and slightly smaller on the other side of the sample so as to trap the thermocouple and ensure good thermal contact. The thermocouple junction was further secured by a Ta strap, formed to provide compression between the junction and the Si substrate. All temperature measurements above 770 K were confirmed with an optical pyrometer. Sample temperatures up to 1500 K were obtainable with this mount, and were homogeneous across the face of the sample.

The nitriding experiments were performed by backfilling the chamber to the desired pressure with either ammonia or hydrazine. ND_3 (99.3 atom % D) was obtained from MSD Isotopes (Montreal, Canada), and anhydrous N_2H_4 (97+% purity) from Matheson, Coleman, & Bell (Norwood, OH). Decomposition of the hydrazine into NH_3, N_2, and H_2 over time could be detected in the mass spectra obtained after allowing the N_2H_4 to sit overnight in the glass/stainless steel ampule. This necessitated the routine purification of the hydrazine, accomplished by pumping the gaseous impurities from frozen N_2H_4 which was cooled by an ethylene glycol/dry ice bath at a temperature of 258 K.

Upon initially introducing the Si crystal into the UHV chamber, carbon and oxygen were the only impurities detected on the surface by XPS. Due to the large sampling depth of XPS (~ 35 Å), periodically a low level Ni impurity (a few percent) was detected in the subsurface region which did not affect the present results. The carbon was removed by heating the sample in oxygen at 1100 K. The oxide was then removed by a high temperature anneal at 1300 K. This cleaning procedure is known to cause some surface roughening whose effect on the nitriding rate was not assessed.

A typical experiment proceeded in the following way: (1) the crystal was heated to the desired temperature; (2) the vacuum chamber was backfilled to the desired pressure with either ND_3 or N_2H_4 for a predetermined time; (3) the gas supply was terminated and the sample was cooled to near room temperature; and (4) XPS analysis was performed. After each nitridation experiment, the surface was cleaned by heating the crystal to 1380 K. Up to one hour anneals were required to remove the thicker silicon-nitride films.

RESULTS

The Si(2p) and N(1s) XPS spectra of the Si(100) sample as a function of N_2H_4 exposure are shown in Fig. 1. The spectra illustrated are from experiments performed at a substrate temperature of 973 K. Similar spectra were obtained at different substrate temperatures and when using ND_3 as the nitridant gas. The binding energy (BE) and full width at half maximum (FWHM) of the silicon elemental peak (99.6 and 1.2 eV, respectively) remains essentially unchanged as the film thickness increases. However, as the nitride layer grows a higher BE shoulder (due to Si bonded to nitrogen in the silicon-nitride film) develops on the elemental peak. This feature continues to grow as the thickness of the film increases. The resulting silicon-nitride Si(2p) peak is well shifted from the elemental peak, with a BE of 102.2 eV and a relatively constant FWHM of 1.9 eV. The integrated areas of the XPS spectral features were determined by approximating the peaks with Gaussians. In order to successfully fit the total Si(2p) spectra it was necessary to include a third peak. This peak has a BE intermediate between that of elemental Si and Si_3N_4 (100.9 eV) and a low intensity (~10% of the total Si(2p) peak area) which remains constant as a function of film thickness. This feature is attributed to a monolayer of Si on top of the Si_3N_4 film and will be discussed in a future publication.

The XPS N(1s) peaks from the Si_3N_4 films are slightly asymmetric having a constant BE of 398.1 eV with a corresponding FWHM of ~1.8 eV. Fig. 2 is a plot of the integrated N(1s) peak area as a

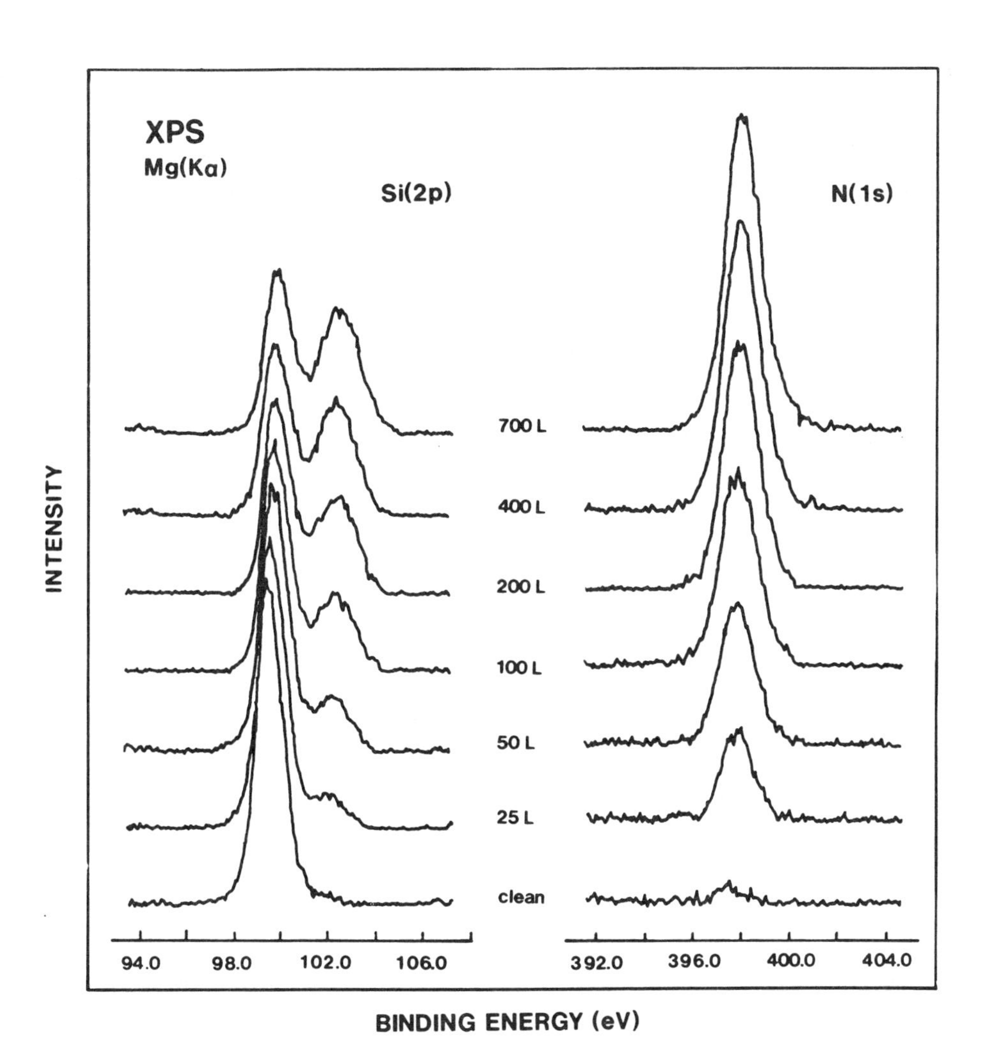

Figure 1. XPS Si(2p) and N(1s) spectra as a function of hydrazine exposure for the direct thermal nitridation of Si(100) at a substrate temperature of 973 K.

function of ND_3 and N_2H_4 exposure. The exposures are expressed in Langmuirs (1 L = 10^{-6} Torr-sec) and were performed at a N_2H_4 and ND_3 pressure of 5×10^{-8} Torr; the pressure was not corrected for the small difference in ion gauge sensitivity between N_2H_4 and ND_3. The Si(100) substrate temperature ranged from 923 to 1123 K as indicated in the figure.

The ordinate of Fig. 2 can also be expressed as a nitride film thickness. This was determined by first calculating the XPS Si(2p) nitride signal intensity that would result from an infinitely thick slab of Si_3N_4, $I^{Si(2p)}_{Si_3N_4\infty}$, by

$$I^{Si(2p)}_{Si_3N_4\infty} = I^{Si(2p)}_{Si\infty}(N^{Si}_{Si_3N_4}/N^{Si}_{Si})(\lambda^{Si}_{Si_3N_4}/\lambda^{Si}_{Si}) \quad (1)$$

where $I^{Si(2p)}_{Si\infty}$ is the experimentally obtained Si(2p) intensity from elemental silicon, N^{Si}_{Si} and $N^{Si}_{Si_3N_4}$ are the atomic densities of silicon in pure silicon and silicon nitride, respectively, and λ^{Si}_{Si} and $\lambda^{Si}_{Si_3N_4}$ are the respective electron mean-free-paths in the two materials. Using $N^{Si}_{Si} = 5.0 \times 10^{22}$ atoms/cm^3 [15], $N^{Si}_{Si_3N_4} = 4.4 \times 10^{22}$ atoms/cm^3 [16], λ^{Si}_{Si}=23 Å [17], and $\lambda^{Si}_{Si_3N_4}$ = 30 Å [2], $I^{Si(2p)}_{Si_3N_4\infty}$ was then calculated. This signal intensity was, in turn, used to calculate $I^{N(1s)}_{Si_3N_4\infty}$, the N(1s) intensity from an infinitely thick Si_3N_4 slab by

$$I^{N(1s)}_{Si_3N_4\infty} = I^{Si(2p)}_{Si_3N_4\infty}(4/3)(S^{N(1s)}/S^{Si(2p)}) \quad (2)$$

where 4/3 corrects for atomic density difference of N and Si in stoichiometric Si_3N_4. XPS sensitivity factors, $S^{N(1s)}$ and $S^{Si(2p)}$, of 0.49 and 0.36 for N(1s) and Si(2p), respectively [18], were used. Finally, nitride film thickness, d, is obtained using a standard attenuation model

$$d = -\lambda^{i}\ln(1-I^{i}/I^{i}_{Si_3N_4\infty}) \quad (3)$$

where λ^i, I^i, and $I^i_{Si_3N_4\infty}$ are the electron mean-free-path, experimentally obtained signal intensity, and the calculated (eq. 1 or 2) intensity from an infinitely thick sample of silicon nitride for the appropriate transition used (i = Si(2p) or N(1s)).

In Fig. 2, the value of d obtained from the N(1s) peak area is plotted, assuming that $\lambda^{N(1s)}_{Si_3N_4}$ = 26 Å, an average value of the mean-free-path of N(1s) electrons in silicon nitride [2]. The film thickness was also determined from the Si(2p) spectra, based on the intensity of the Si_3N_4 feature and agreed with the N(1s) derived values to within ~10%. Film thickness can also be expressed in monolayers by assuming the film has grown as β-Si_3N_4 [2] which has a reduced hexagonal unit cell with two molecules per unit cell [16], resulting in a monolayer thickness of 1.45 Å.

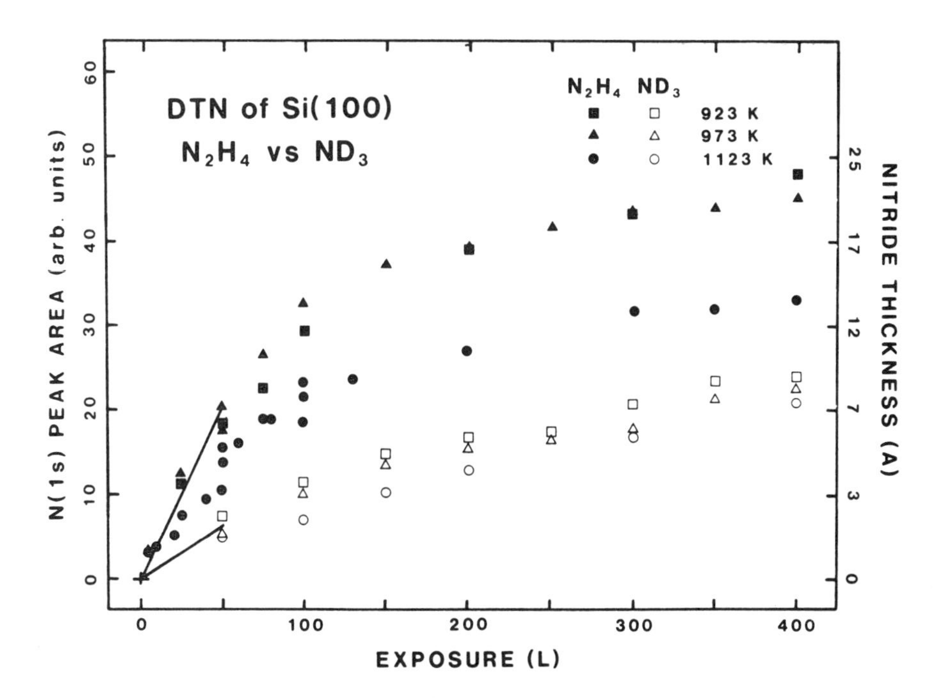

Figure 2. N(1s) peak areas as a function of hydrazine and ammonia exposure for the direct thermal nitridation of Si(100) at various temperatures.

For films grown from ND_3, Fig. 2 clearly shows that the nitride film thickness increases monotonically as a function of exposure. Nitride films of similar thicknesses are grown at substrate temperatures between 873 and 1073 K (data taken at several temperatures is not shown in Fig. 2). The films grown after 400 L exposure to ND_3 are approximately 7.5 Å, or ~5 monolayers. This result is in excellent agreement with the results of Maillot, et al. [2], but is in poor agreement with those of Glachant, et al. [9,10], whose reported ammonia exposures to obtain a given film thickness seem unrealistically high. As the silicon substrate temperature is raised above 1073 K, the thickness of the resulting Si_3N_4 film at a given exposure, decreases slightly compared to those grown at the lower temperatures.

For films grown from N_2H_4, again a monotonic increase in film thickness as a function of exposure can be seen. Similar film growth was observed at substrate temperatures of 923 and 973 K yielding a film thickness of 20-25 Å after a 400 L exposure. Nitriding at 1123 K results in a lower film thickness at comparable exposures yielding a 13 Å thick film at 400 L. Below 773 K, only one monolayer of nitride is grown with N_2H_4 or ND_3 regardless of the exposure.

Notice that films grown by exposure to N_2H_4 and ND_3 grow at markedly different rates. This is illustrated by comparing the curves for films grown from ND_3 and N_2H_4 at a substrate temperature of 923 K. After a 400 L ND_3 exposure, the film is 7.5 Å thick versus a 25 Å thick film grown at an equivalent exposure of N_2H_4. Silicon-nitride film growth shows similar behavior for both nitridation sources; a rapid initial growth characterized by the large slope below 50 L exposure, is followed by a substantial slowing of the film growth (decreasing slope) until no measurable film growth changes can be observed at the highest exposures. The N(1s) to Si(2p) peak area ratio remains constant a function of exposure for films grown from ND_3 and N_2H_4 again indicating that similar films are grown in both cases; the value of the peak area ratio confirms that the films are stoichiometric Si_3N_4. No oxygen was present in films grown by either method.

DISCUSSION

Any model of the low pressure DTN of Si using N_2H_4 or ND_3 needs to explain at least three experimental observations. These include: (1) the fact that thicker films are grown at a given temperature and exposure with N_2H_4 than with ND_3 under our experimental conditions (i.e., low pressure nitridation), (2) the decrease in film thickness for a given exposure of both N_2H_4 and ND_3 above 1073 K, and (3) the faster initial (total) rate of nitridation at a given temperature using N_2H_4 compared to ND_3. A complete model for the DTN process is beyond the scope of this paper but several points should be addressed concerning each of these observations.

Data on the pressure dependence of the nitriding process (not shown) suggests that nitridation is ultimately diffusion controlled, but under these low pressure conditions, the diffusion is mitigated by a surface reaction. Even the simplest mechanism for surface reactions preceding nitridation involves several steps including dissociative adsorption of N_2H_4 or ND_3, further dissociation to N(ads) and H(ads), desorption of H_2, and diffusion of N or Si through the growing Si_3N_4 layer. Assuming a steady-state surface concentration of nitriding species under our experimental conditions, two ways to obtain thicker films with N_2H_4 than ND_3 at a given exposure are (1) for the steady-state surface concentration of the nitriding species to be greater for N_2H_4 than for ND_3 at a given temperature (<1173 K) and pressure or (2) that there are more sites available for dissociative adsorption on these surfaces (i.e., more nucleation sites) for N_2H_4 than ND_3. These two possibilities are not mutually exclusive and could be occurring simultaneously. Further experiments at higher pressures could discriminate between these two mechanisms but these pressures cannot be tolerated in the present apparatus. Thermodynamically, one would expect the steady-state surface concentration to be higher for N_2H_4 than ND_3 under equivalent conditions due to the relative ease in breaking the N-N bond in N_2H_4 (71 kcal/mole) versus the N-H bond in ND_3 (110 kcal/mole) which is required for dissociative adsorption.

The lack of a strong temperature dependence below 1073 K for both nitridant gases is noteworthy. The substrate temperature will affect both the rate of adsorption (and thus the steady-state surface concentration) as well as diffusion of the nitriding species. Diffusion coefficients usually have an Arrhenius-type temperature dependence and, thus, one expects film thickness to increase with temperature. On the other hand, the probability for dissociative adsorption for ND_3 or N_2H_4 on the surface and, thus, the steady-state surface concentration, decreases as the temperature is raised. A decreased steady-state surface concentration for a given exposure ultimately leads to thinner films. Apparently the latter mechanism dominates under these conditions and above 1123 K this effect becomes rather dramatic for N_2H_4 as shown in Fig. 2.

The faster initial rate of nitridation with N_2H_4 is illustrated in Fig. 2 where the initial slopes of the nitride layer as a function of exposure at 923 K for N_2H_4 and ND_3 are compared. From the slopes of these curves at <50 L exposure, the initial rate of nitridation for N_2H_4 is ~3.5 times faster than for ND_3. This has important consequences when considering growing films for use as dielectric gates for microelectronics. Because film growth appears to stop after one monolayer for either N_2H_4 [12] or ND_3 [11] at substrate temperatures below 700 K due to hydrogen blocking the sites for dissociative adsorption, these results suggest that at 800 K (a temperature well above the hydrogen desorption temperature) nitridation will proceed in a facile fashion and, in addition, the initial rate using N_2H_4 will be at least 3.5 times faster than with ND_3. Since diffusion interferes with gate integrity during high-temperature processing (T>1173 K), lowering the processing temperature by three hundred degrees is a substantial improvement and should lead to a concommitant improvement in gate morphology.

CONCLUSIONS

The DTN of Si(100) using ND_3 and N_2H_4 at low pressure over the temperature range 673 to 1123 K has been investigated. Below 773 K, the hydrogen desorption temperature, only about one monolayer of nitride is formed using either gas as a nitrogen source. Between 923 and 1073 K, stoichiometric Si_3N_4 films are readily grown using N_2H_4 with the maximum film thickness approaching 25 Å at the highest exposures. Under similar conditions, Si_3N_4 films can also be grown using ND_3 but the maximum film thickness is only about 7 Å at the highest exposures. This difference in film thickness at a comparable exposure appears to be a result of a lower steady-state surface concentration for ND_3 as compared to N_2H_4 which, in turn, is probably related to the relative ease of breaking the N-N bond in N_2H_4 compared to the N-H bond in ND_3. Above 973 K, desorption of the nitriding gas is postulated to compete with dissociative adsorption leading to thinner films at a given exposure using either gas, but the effect is much greater for N_2H_4. The initial rate of film growth under comparable conditions is ~3.5 times faster with N_2H_4 as compared to ND_3 which may have important consequences for the production of Si_3N_4 gates for VLSI applications.

ACKNOWLEDGEMENT

The authors wish to thank J. R. Creighton for helpful discussions on mounting the Si crystals, for supplying the crystals, and for critically reviewing the manuscript. Partial support of this work by the U.S. Department of Energy, Office of Basic Energy Sciences, Division of Material Sciences is gratefully acknowledged. This work was performed at Sandia National Laboratories, supported by the U.S. Department of Energy under Contract No. DE-AC04-76DP00789.

REFERENCES

1. M. D. Wiggins, R. J. Baird, and P. Wymblatt, J. Vac. Sci. Technol., 18 (1981) 965.
2. C. Maillot, H. Roulet, and G. Dufour, J. Vac. Sci. Technol. B, 2 (1984) 316.
3. J.A. Wurzbach and F. J. Grunthaner, J. Electrochem. Soc. Solid-State Science and Technology, 130 (1983) 691.
4. H. H. Madden and G. C. Nelson, Appl. Surf. Sci., 11 (1982) 408.
5. J. A. Nemetz and R. E. Tressler, Solid State Technology (Feb. 1983) 79 and (Sept. 1983) 209.
6. D. V. Tsu, G. Lucovsky, and M. J. Mantini, Phys. Rev. B33 (1986) 7096.
7. P. Soukaissian, T. M. Gentle, K. P. Schuette, M. H. Bakshi, and Z. Hurzch, Appl. Phys. Lett., 51 (1987) 346.
8. R. Heckingbottom and P. R. Wood, Surf. Sci., 36 (1973) 594.
9. A. Glachant, D. Saidi, and J. F. Delord, Surf. Sci., 168 (1986) 672.
10. A. Glachant and D. Saidi, J. Vac. Sci. Technol. B, 3 (1985) 985.
11. F. Bozso and Ph. Avoures, Phys. Rev. Lett., 9 (1986) 1185.
12. C. H. F. Peden, J. W. Rogers, Jr., and D. S. Blair, in preparation.
13. H. H. Madden and D. W. Goodman, Surf. Sci., 150 (1985) 39.
14. D. S. Blair and G. L. Fowler, J. Vac. Sci. Technol., shopnote, submitted.
15. L. C. Feldman and J. W. Mayer, Fundamentals of Surface and Thin Film Analysis, (North-Holland, New York, 1986) p. 334.
16. R. W. G. Wycoff, Crystal Structure, 2nd Edition, Vol. 2, (Wiley, New York, 1964) pp. 157-158.
17. R. Flitsch and S. I. Raider, J. Vac. Sci. Technol., 12 (1975) 305.
18. C. D. Wagner, L. E. Davis, M. V. Zeller, J. A. Taylor, R. H. Raymond, and L. H. Gale, Surf. Interface Anal., 3 (1981) 211. (These values were corrected for the energy dependence of the hemispherical electron energy analyzer.)

CHAPTER IV
INTERCONNECTIONS: LOW TEMPERATURE METALS AND INSULATORS

MATERIAL AND PROCESS TRENDS IN CHIP INTERCONNECTION TECHNOLOGY

Dr. A. S. Oberai
IBM Corporation, Essex Junction, VT 05452

ABSTRACT

The demand for higher density and more function in semiconductor integrated circuits continues unabated. As semiconductor lithographic processes progress into 0.25 micrometer and below, we will enter the gigabit domain in DRAMS and see logic levels of integration rise into few millions of circuits on a chip. In this paper we will examine the effect of this progress on the chip interconnection technology.

INTRODUCTION

At the outset, a brief review of the trends[1,2,3] in the level of integration and size of DRAMS and logic chips is presented. The resulting increase in the burden on the interconnection technology[4,5] is then examined with a special emphasis on the following areas:

1) Propagation Delays - These will become comparable to gate delays and new techniques need to be developed to contain the problem. Some of these might imply more levels of interconnects, new materials for interconnects and for interlevel insulation, and perhaps low temperature operation.

2) Power - The drive for increased performance will force more attention on power dissipation and electromigration limits.

3) Input/Output (I/O) Terminals - Increased integration levels will require more signal I/O's and increased power will require more power I/O's to take care of power distribution problems.

Additionally, the impact on manufacturing technology will be examined from the point of view of three key drivers: process control, contamination control, and cycle time.

MICROCIRCUIT TRENDS

Advanced lithography continues to be the key factor in the drive for higher levels of microcircuit integration. Dynamic RAMs have been quadrupling in the level of integration every three years or so with leading edge production now at the 4 Mbit level. In the recent past, this trend seems to have accelerated even further. Figure 1 shows an approximate representation of the reduction in minimum geometries and increases in chip size as driven by DRAMs. Lithography results in a factor of two improvement in density for every generation and the remaining factor of two in the level of integration comes from increases in the chip size as well as improvements in circuit technology and device vertical structures. This last point cannot be overemphasized; lithography alone is not likely to provide all the improvements necessary to reach the goal of a gigabit chip in the 1990s.

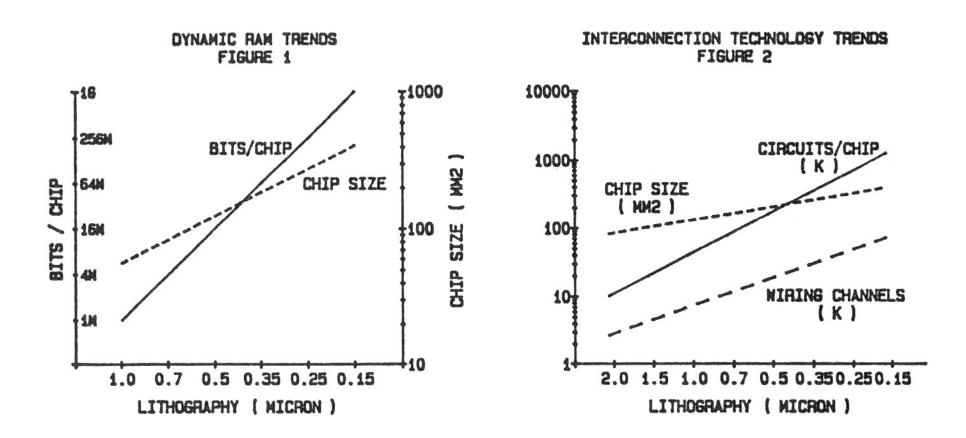

Logic follows a similar trend line (Figure 2), although later in time. Logic applications, as a general rule, do not lend themselves to the level of fine tuning and easy diagnostics available in DRAMs. Due to the application specific nature of logic parts, lithography exploitation as well as the level of integration tend to lag DRAMs. Interconnection requirements strongly dominate the logic chip size and, in turn, this chip size continues to be constrained by the limits of field size in lithography tools.

SIGNAL PROPAGATION DELAYS

At device groundrules in the range of 2 microns and above, the propagation delay on chip was relatively a small portion of the gate delay. As we shrink to smaller geometries in the range of 0.5 micron and below, it is expected that gate delays will continue to trend downwards. On the other hand, with increasing chip sizes at the smaller geometries, the propagation delays tend to dominate the design criteria for high performance chips (Figure 3). The wiring delay has shown an uptrend from 5% to 10% of the total delay at 2 micron geometries to 30% to 50% as devices size shrinks to submicron geometries. This is happening despite the fact that the number of interconnection levels have gone up from single level metal in most MOS applications to two or three levels of metal. At the same time, advanced bipolar technology has migrated from two levels of interconnect to three and four levels. A similar trend is also evident in the DRAM technology. Firstly, RC delay is on a continuing uptrend (Figure 4); secondly, for the higher performance DRAM chips, there is going to be a continuing demand for increase in the number of interconnection levels to keep the interconnection length and resistance under reasonable control (Figure 5).

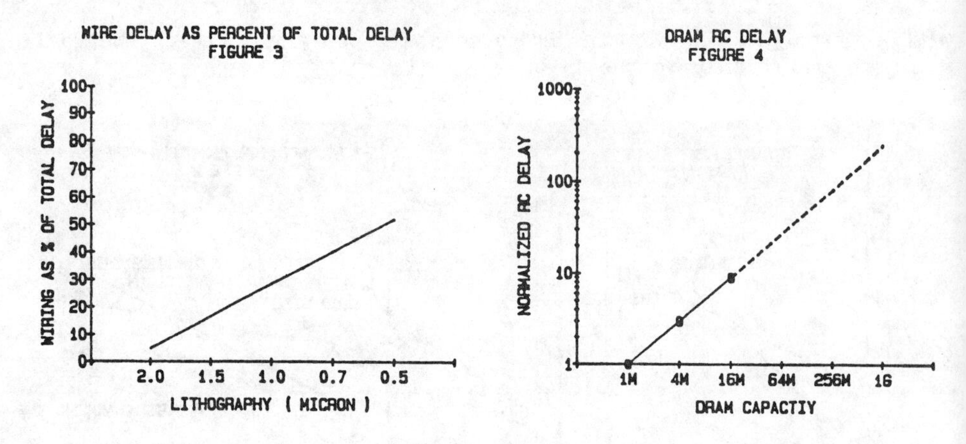

While increasing the number of interconnection levels is a potential solution to counter the trend of increasing wire propagation delay, we need to examine the material and process improvements that could reduce the RC delay. The material choices involve the use of lower resistivity interconnect materials and lower dielectric constant insulating materials. This would dictate metallurgies with resistivities around 2.5 micro-ohm-cm and insulators that are likely to be organic in nature (Figures 6 and 7). Of course, we also have

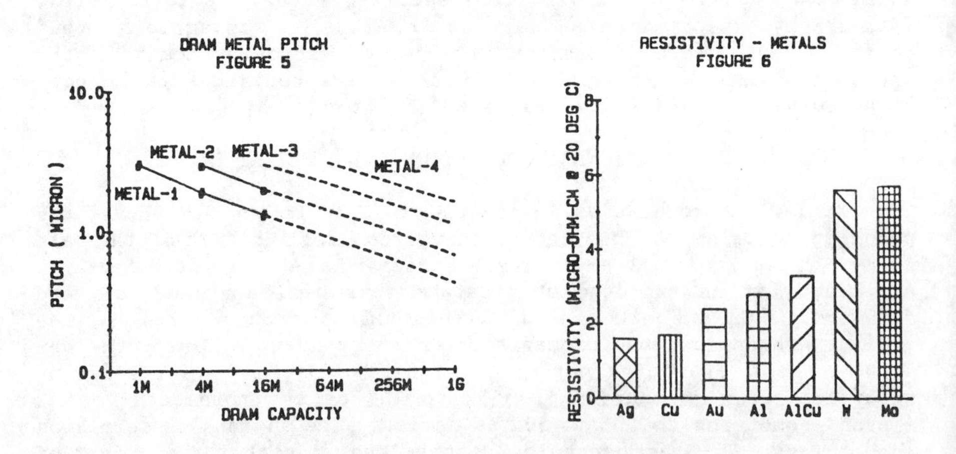

the alternative of cooling to liquid nitrogen temperatures and gaining factors of five or so improvement in resistivity (Figure 8).

The composite effect of the material choices and cooling technology on chip propagation delay is shown in Figure 9. The broken horizontal line corresponds to the lossless transmission line criterion.

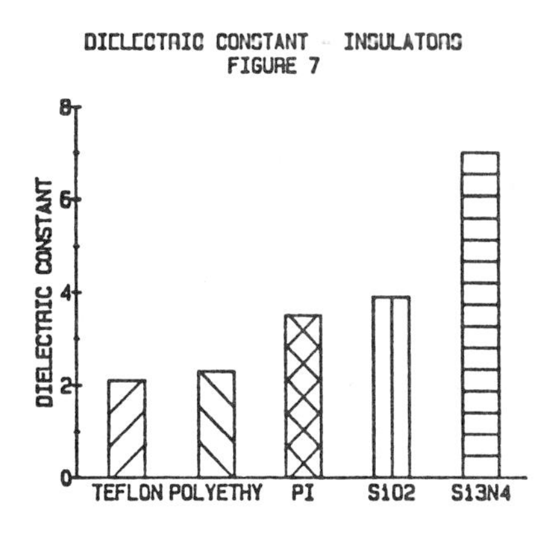

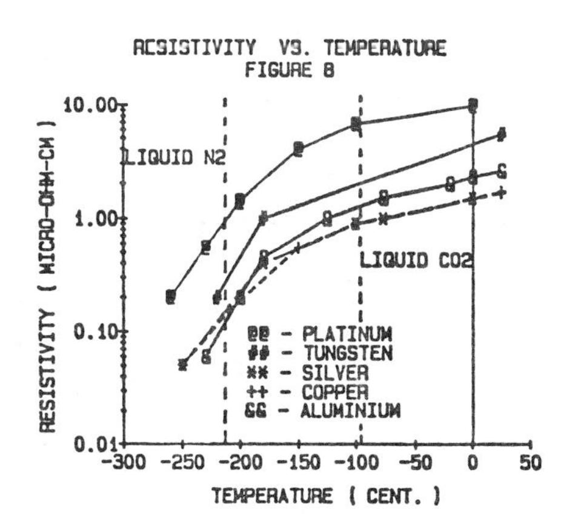

The interconnection technology chosen must position the operating point below the broken line. It is clear that the aluminum/copper and SiO2 system is beginning to run out of steam in the 0.7 - 0.5 micron geometry range. LN2 cooling of this system would extend this system to the 0.25 micron range. Alternatively, the low resistivity metallurgy/low dielectric constant insulator system coupled with LN2 cooling would be able to get down to 0.15 micron range.

POWER DISSIPATION

The increasing performance with smaller geometries dictates larger power dissipation as indicated in Figure 10. The increase

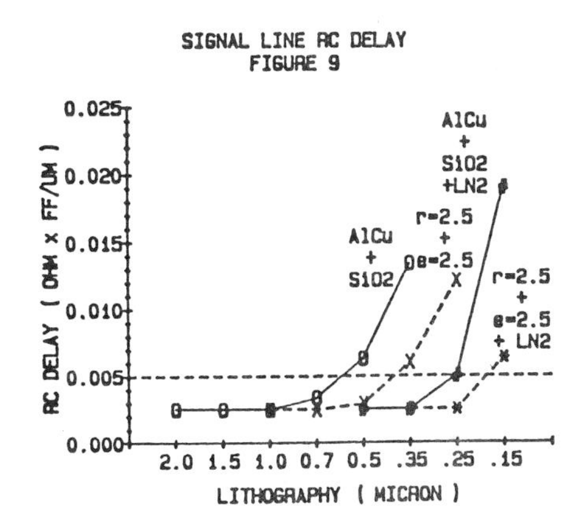

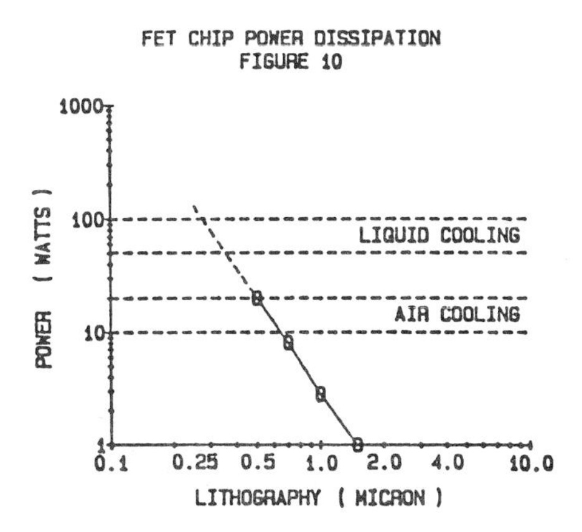

in the power is also a result of the much higher level of integration made possible by the lithography scaling. The power trends suggest that liquid cooling may become necessary in the sub-0.5 micron arena. The electromigration requirements and the interconnection material capabilities are shown in Figure 11. Apparently, this seems to be under control.

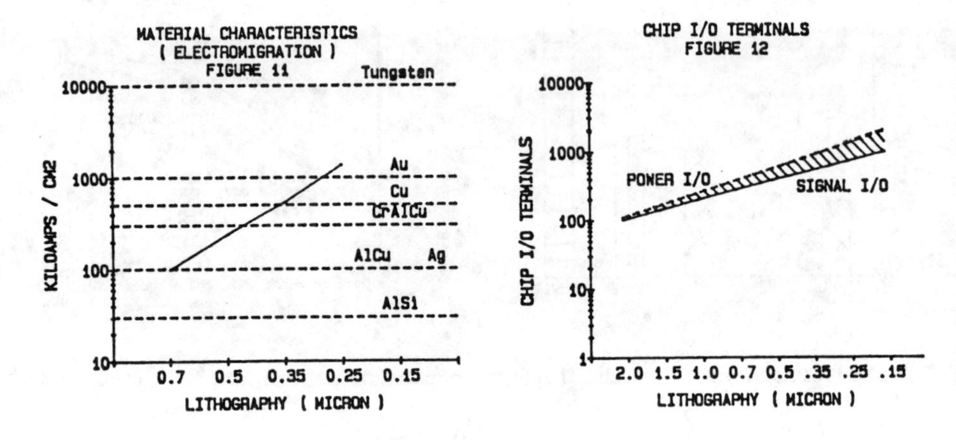

INPUT/OUTPUT TERMINALS

The increased levels of chip integration demand greater number of signal input/output (I/O) terminals as shown in Figure 12. With the expected increased power levels, more power I/Os on shorter spacings will be required to insure optimum power distribution through the chip. This has a strong impact on the chip terminals arrangement - perimeter versus area terminals. Figure 13 indicates that the future chip I/O requirements are such that area terminals will be the preferred path.

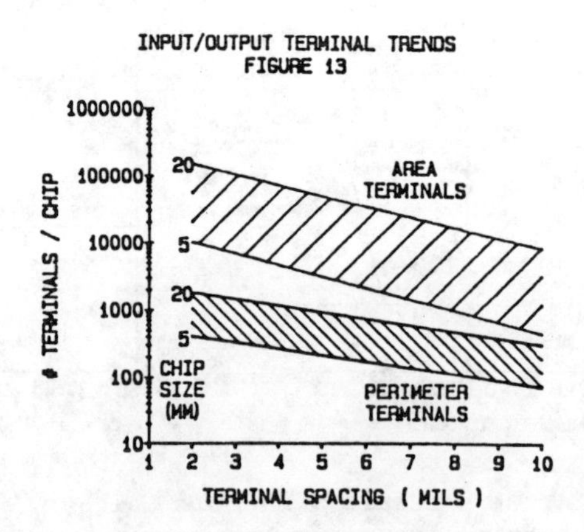

PROCESS/MATERIALS TRENDS

These trends are indicated in Table 1.

	1u ----------> 0.5u ----------> 0.25u
Interconnects	AlSi =====> High Conductivity Materials
Insulators	Inorganic/ Organic ==================> Organic
Metal Deposition	Evaporation/ Sputter =============> Sputter/CVD
Metal Etch	RIE =======================> RIE/Laser
Planarization	Resist ====================> SOG, etc.

Table 1

CRITICAL TECHNOLOGY DEMAND

All of the requirements discussed in this paper are pointing towards a highly demanding manufacturing technology environment. It is critical that the developers and manufacturers of tools, materials and processes recognize the complexity of these trends. If the present trend towards three-dimensional structures in silicon as well as the trend towards four to six levels of interconnects in the 1990s is to continue, the process complexity will increase very substantially (Figure 14). Obviously, one hopes that, through innovative techniques and materials, this trend can be slowed to a manageable level.

A very profound effect of this increase in the fabrication steps as well as level of integration complexity is to put a heavy demand

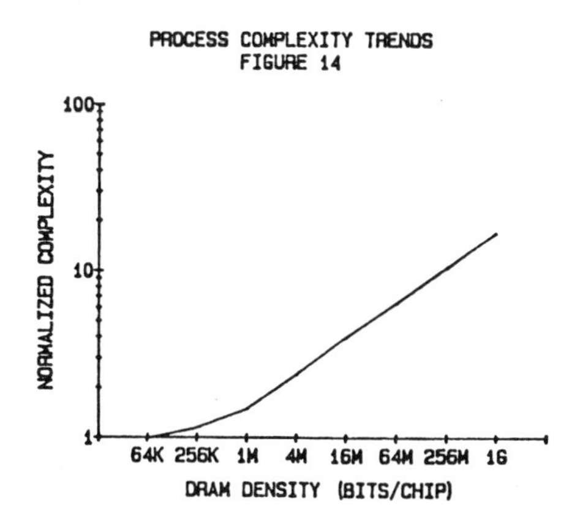

PROCESS COMPLEXITY TRENDS
FIGURE 14

on all aspects of semiconductor manufacturing (that is, materials, equipment, facilities and manufacturing systems - Figure 15). It is imperative that every element of manufacturing technology be examined from the point of view of three cardinal drivers: process control, contamination control, cycle time.

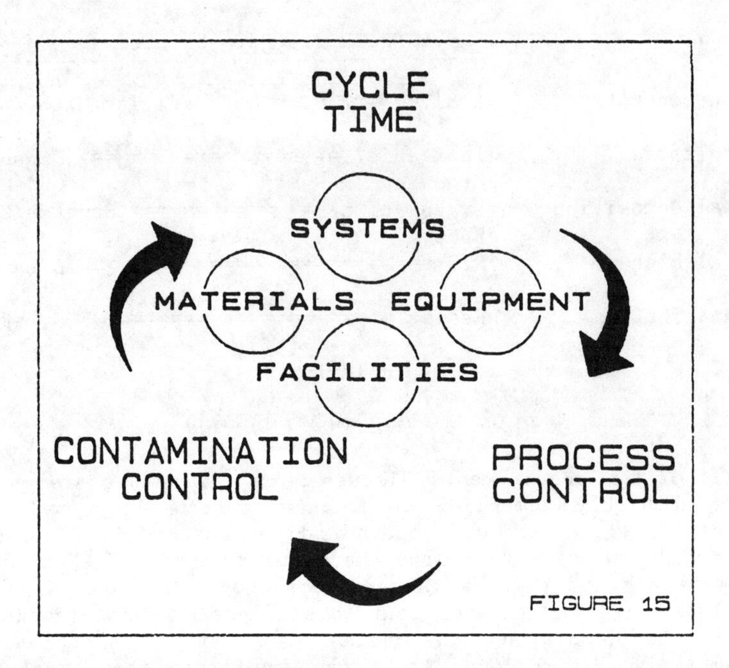

FIGURE 15

DEFECT DENSITY REQUIREMENTS

It is important to realize that the level of integration of semiconductor chips in the 1990s is more likely to be gated by the requirements of low defect levels and improved process control. Figure 16 shows a plot of defect density versus DRAM level of

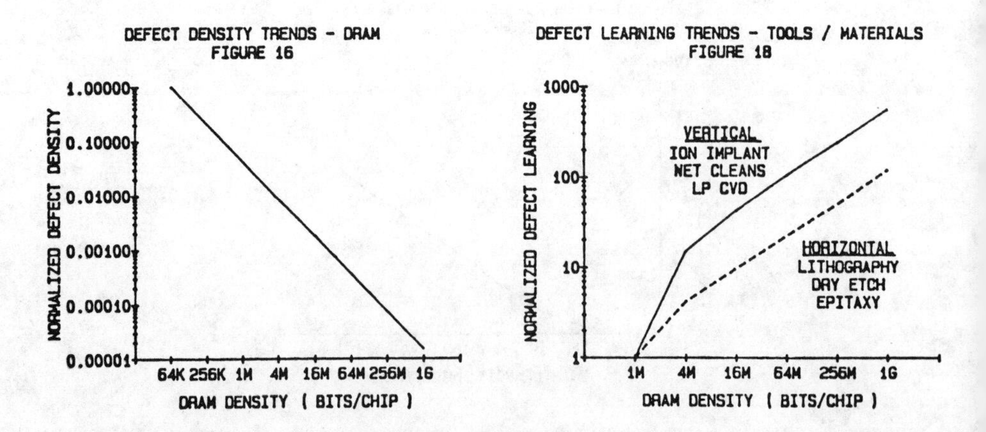

DEFECT DENSITY TRENDS - DRAM
FIGURE 16

DEFECT LEARNING TRENDS - TOOLS / MATERIALS
FIGURE 18

integration. It shows that in the next decade semiconductor equipment, processes, and materials need to be improved very substantially. A very appropriate way to attack this problem is to take a systems approach to defect control (Figure 17).

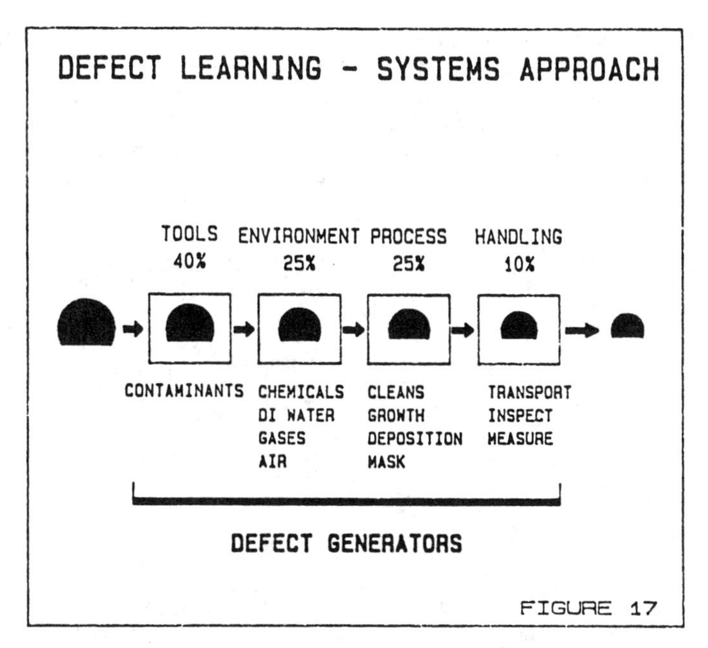

FIGURE 17

Every possible encounter that a wafer has in the factory, whether it be with equipment, materials, processes, or handling, contributes in general to a degradation of defect density. The methods by which equipment, materials, and processes are produced, transported and then utilized at the point of use in the fabrication are all critically important. For instance, it is highly unlikely that a deposition tool that was exposed to an unclean environment in production, shipping and receiving is ever likely to be clean enough to meet the needs of megabit technologies of the late 1980s and beyond.

In the 1960s and 1970s, there was much stronger focus on lithographic defects. In the late 1980s and 1990s, we need to continue to reduce lithographic defects but a much stronger emphasis needs to be placed on vertical structure defects. The thickness of films (especially gate and node oxides) has reduced dramatically and, at the same time, more and more levels of interconnects and insulators continue to be added. Therefore, a strong emphasis needs to be placed on understanding and eliminating the defect causing mechanisms. Defect learning targets (Figure 18) must be established and systematically followed for each process, chemical, gas and equipment that semiconductor manufacturing utilizes.

PROCESS CONTROL AND METROLOGY

Since the processes are getting more complex and longer, it is very important that the cycle times be controlled carefully in order to achieve reasonable rates of yield learning (Figure 19). In addition, the exceedingly lengthly and complex technology is going to demand a much better understanding of the processes, materials and tools from both the process parameter as well as contamination control point of view. So the first line of attack for the measurement technology is to help us characterize equipment and materials to a very high level of understanding. Secondly, based on this fundamental understanding, a conscientious effort is needed to provide the appropriate in situ measurement and adaptive control suitable for manufacturing.

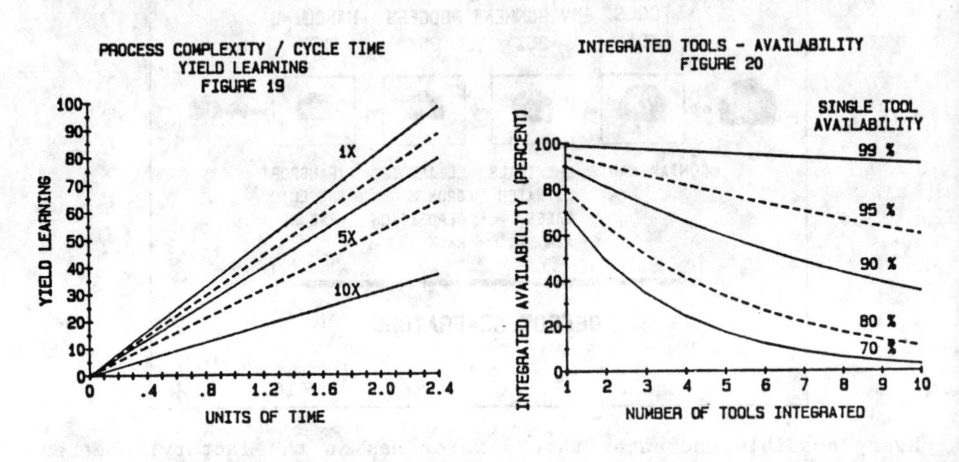

Last but not the least, it is necessary that we have appropriate detection and analysis technologies available to detect the exceedingly small defects (100 nanometers and smaller).

MULTI-STEP PROCESS EQUIPMENT

A way of achieving all three key objectives (i.e., cycle time, contamination control, process control) is to integrate process/tool steps (islands of automation). As we try to integrate more and more of the process steps together in an interconnect cell, it is appropriate to recognize that there is a trade-off between flexibility and the length of the cell. In addition, as shown in Figure 20, if the availability of each individual element of the cell is not high enough (i.e., in the 90+% range), the overall availability of the cell will be at an impractical low level, thus negating the benefits of integration. It is key, therefore, that each process, equipment, and material in the cell be well designed and characterized. As can be seen in Figure 21, the path to implementation of integrated multi-step tools requires focus on improving tool availability and process control by designing for high reliability and including features such as in situ measurements, adaptive process control, self cleaning, etc. This approach will assure a high level of availability for each individual element which is the most fundamental need for automation or integration of the interconnect cell.

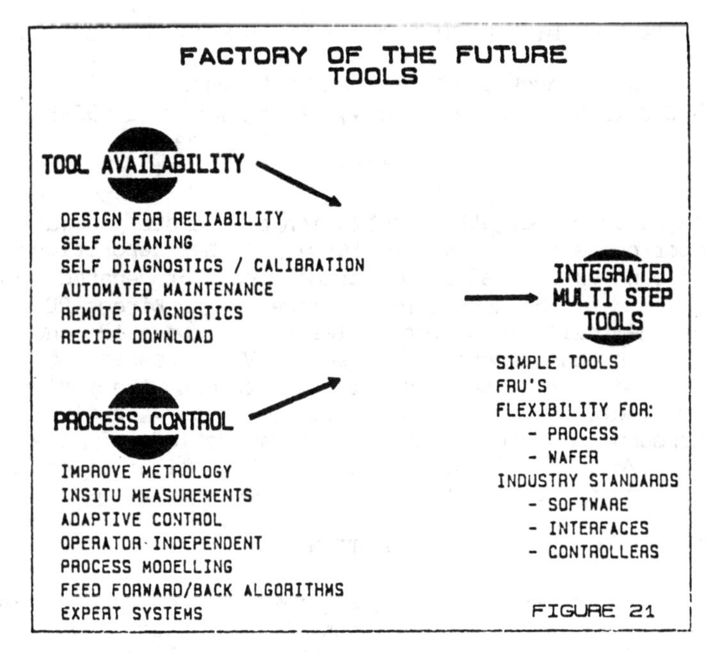

FIGURE 21

Lastly, the factory of the future is a close integration of processes, materials, equipment, and data systems. The integration of this diverse set will require increased interface standards and a close synergism between the user and equipment/material vendors.

SUMMARY

This paper has discussed the trends in future interconnection technology with the associated challenges as well as potential solutions. It is shown that low resistivity metallurgy/low dielectric constant insulator interconnection systems operating in LN2 environment will be able to contain chip propagation well into the sub-quarter micron arena. Increasing performance and integration levels will require more attention on chip cooling issues. It is noted that area terminals may be the preferred path for chip I/Os in the future.

It is also highlighted that the areas of defect control, process control, equipment and material design all need a very focussed effort from the outset. The developers of processes, materials and equipment have to build in the contamination control, process control and reliability features to assure the needs of manufacturing.

Acknowledgements

I would like to thank V. Ramakrishna, J. Canning, R. Fay, and Z. Apgar of IBM for the technical discussions, review, and their support in the preparation of this paper. I would also like to acknowledge the support of C. Dewyea of IBM in the preparation of this manuscript.

DEPOSITION OF DIELECTRICS BY REMOTE PLASMA ENHANCED CVD

G. Lucovsky, D.V. Tsu and G.N. Parsons
North Carolina State University, Raleigh, N.C. 27695-8202

ABSTRACT

This paper describes the application of remote plasma enhanced chemical vapor deposition (Remote PECVD) to the deposition of thin films of silicon oxide, silicon nitride and hydrogenated amorphous silicon. We compare the Remote PECVD process with Direct PECVD, and contrast the chemical bonding properties of silicon oxide and nitride films deposited by the Remote and Direct PECVD processes. We present an empirical model for the deposition process chemical reaction pathways that includes: (i) gas phase reactions involving either chemical consumption or selective fragmentation of the silane reactant, and/or (ii) surface reactions which determine the stoichiometry and hydrogen content of the deposited films.

INTRODUCTION

There is considerable interest in the development of low temperature (T < 500°C) processing techniques for advanced semiconductor device fabrication, both for silicon and compound semiconductor technologies. Different 'qualities' of dielectrics and semiconductor/dielectric interfaces can be used in different device applications; e.g., gate insulators for MOS or MIS structures must have lower bulk and interfacial defect densities than anti-reflection layers for optical detectors and emitters or dielectric layers for inter-level metal isolation. The most successful application of a dielectric material in a semiconductor technology is thermally grown SiO_2 for use as a gate dielectric in silcion MOS devices. Temperatures in excess of 1000°C are required for growing SiO_2 with bulk and interface defect densities that are sufficiently low for high performance FET device operation. Dielectric layers grown by 'high' temperature (>700°C) Thermal CVD, or 'low' temperature Direct PECVD (about 300 to 500°C), generally have high defect densities and are used for applications other than gate dielectrics in silicon device structures. However, in technologies using compound semiconductors (III-V's and II-VI's), the native oxides (thermally grown or formed by anodization) are not of sufficient insulating quality for device applications. Therefore, options for MOS (or MIS) and/or FET device structures include deposited oxides (and other dielectrics) where the requirements for low temperature processing generally restrict the deposition to some type of 'enhanced or assisted' CVD process [1].

We compare and contrast Remote PECVD with Thermal CVD and Direct PECVD, and then discuss the specific Remote PECVD processes that we have employed for the deposition of hydrogenated amorphous silicon, a-Si:H, silicon oxide and silicon nitride [2-5]. We emphasize the way the chemical bonding in the films varies with changes in the processing variables, highlighting the effects of He dilution of the oxygen and nitrogen reactants. We present an empirical model for the reaction pathways involved in the Remote PECVD process. Finally, we discuss some of the applications of Remote PECVD dielectrics in device structures.

CVD DEPOSITION PROCESSES

Chemical vapor deposition, CVD, is a materials synthesis process in which gas species react at a heated surface to form a solid thin film. The surface chemical reactions used in CVD include pyrolysis, reduction, oxidation, hydrolysis and disproportionation [1]. We do not use the nature of the reaction chemistry as a basis for differentiating between different CVD processes, but focus instead on energetic considerations. The 'basic' CVD process is one in which the energy necessary to promote surface chemical reactions is supplied at a heated substrate; we designate this process as Thermal CVD. For the deposition of the silicon oxide and nitride, the source of Si atoms in Thermal CVD is generally silane, SiH_4, or a silane derivative (e.g., SiH_2Cl_2, etc.), and the source of O (or N) can be O_2, N_2O, CO_2, etc. (or N_2, NH_3, etc.). Temperatures required for deposition rates of 0.X to X angstroms/second are generally about 700°C. We write a general reaction for the Thermal CVD process as:

A(gas) + B(gas) --(heat)--> C(solid) + D(gas)

where A and B are the gas reactants, C is the deposited thin film; and D is a gas 'waste' product. The deposition rate for Thermal CVD is determined by: (i) the arrival rates of A and B; (ii) the chemical reactivity of A and B at the substrate temperature, Ts; and (iii) the removal rate of D. The most important limitation on the growth rate in Thermal CVD is the reaction rate between A and B which is generally determined by Ts.

In order to reduce the Ts required for film deposition via a CVD process, an additional source of energy must be supplied to the gas reactants and/or to the surface. Under these condtions the the process becomes an 'enhanced or assisted' CVD process. This additional 'energy source' can be hot electrons, created in an RF or microwave plasma as in PECVD, or photons as in laser or UV assisted CVD. The purpose of the plasma excited hot electrons, or laser or UV photons is to generate chemically active (and generally short-lived species) that produce significant increases in the CVD reaction rates, and thereby eliminate the need for high Ts. In most plasma CVD processes the substrate is also exposed to plasma generated visible and UV radiation, and it not possible to rule out an additional assist in the deposition process chemistry from UV photons incident at the substrate.

There are two different approaches to PECVD; (i) Direct PECVD in which all of the reactant gases are subjected to RF or microwave plasma excitation, and in which the substrate is immersed in the plasma [1] and (ii) Remote PECVD in which plasma excitation is selective, and in which the substrate is outside of the plasma [2-7]. We represent the overall chemical reaction pathway in Direct PECVD by:

[A(gas) + B(gas)]* ---> C(solid) + D(gas)

where A, B, C and D have been previously defined, and where the * indicates the plasma excited gas mixture (A AND B). In a complementary way, we represent the Remote RECVD process by:

[A(gas)]* + B(gas) ---> C(solid) + D(gas),

where ONLY ONE of the gas reactants, A, is plasma excited.

Silicon oxide and nitride films deposited by Direct PECVD at a Ts between about 200 and 400°C are non-stoichiometric in the sense that they have SiSi bonds and also contain significant amounts of bonded hydrogen, about 5 to 15 at.% in the oxides as SiH and SiOH, and up to 30 at.% in the nitrides as SiH and SiNH [8]. The source of both the silicon atoms and the hydrogen atoms (except perhaps some fraction of the H atoms in the OH groups, as discussed later) is the silane (or silane derivative) and the reason for the departures from ideal oxide and nitride bonding derive from the 'way' the silane reactant is used [9]. In Direct PECVD, the silane reactant is plasma excited, along with the oxygen or nitrogen species and any rare gas diluents. It has been shown that the plasma dissociation of silane (and silane derivatives) leads to a large number of active fragments, SiH_3 and SiH radicals, SiH_2 molecules, H radicals, as well as positive ions, SiH_3^+, SiH_2^+, etc. [10]. These fragmentation species can react with the plasma generated oxygen and nitrogen species (e.g., atomic and/or molecular metastables) in a number of different ways with reactions occuring in both the gas phase (precursor production) and at the substrate (thin film generation by CVD). In either case, the resulting large number of possible parallel deposition reaction pathways leads to many different and unwanted bonding groups, e.g. SiSi, SiH, SiOH groups in SiO_2, in the deposited films. Our objective has been to reduce the number of plasma generated species in order to limit the deposition pathways to those that produce films with the desired stoichiometry (e.g., SiO_2, Si_3N_4, etc.) or alloy composition (e.g., a-Si:H) This has been accomplished by using the selective plasma excitation characteristics of Remote PECVD, but generally at the expense of a reduction in the deposition rate relative to the 'associated' Direct PECVD process, i.e., the process using the same reactants and diluents [9].

The Remote PECVD process is best described as a four step process consisting of: (i) plasma EXCITATION of the oxygen or nitrogen species, and generally a rare gas diluent, He; (ii) EXTRACTION of active species from the plasma via high gas flow rates; (iv) MIXING of these active species with silane; and (iv) film DEPOSITION via a CVD reaction on a heated substrate outside of the plasma [3,4,9]. Figure 1 of Ref. 2 indicates a schematic diagram of the deposition chamber configuration we have employed [2-5,9]. The chamber design, and the gas flow rates and deposition pressure provide the necessary conditions for a Remote PECVD process. Specifically they prevent back-streaming of the silane reactant into the plasma region, and promote gas phase reactions between plasma generated species and the silane reactant. Thses reactions generate specific precursor species by selective fragmentation of the silane by rare gas metastables [5], or by chemical reaction between the silane and active oxygen or nitrogen atomic and/or molecular species [9,11]. The plasma is established by RF excitation, typically 25 Watts at 13.56 MHz. The gas flow rate into the excitation tube is at least 100 sccm, and the associated velocity of gas flow prevents any significant diffusion of the silane reactant from the deposition chamber into the plasma glow region. The silane reactant is introduced through a showerhead dispersal ring with flow directed toward the heated substrate. The dispersal ring is about 5 cm below the entry port for the plasma generated reactants, and the heated substrate is another 5 cm below the showerhead. The gas flow rate of the silane/argon mixture (10% silane/90% Ar) is 10 sccm. The chamber is operated at a pressure of 300 mTorr. Prior to deposition the chamber is pumped to pressure of about 1 x 10^{-9} Torr and baked. This reduces (and hopefully eliminates) undesired species, mostly water vapor, from the chamber walls and fixtures. The deposition pressure of 300 mTorr ensures a high binary collision rate between the plasma extracted species and the silane reactant. We estimate that there are 10^5 binary collisions per silane molecule in transiting the region between the showerhead and the substrate. The substrate is heated by a quartz/halogen lamp (maximum Ts = 600°C). The majority of the depositions discussed in this paper have been performed in this system, and have been carried out with Ts between 200 and 550°C. In addition, we have employed two other systems with Remote PECVD capabilities. These are: (i) a multi-chamber system for MOS and MIS device fabrication [12]; and (ii) a deposition/analysis system that provides for a analysis of gas phase constituents by optical emission spectroscopy (OES), and mass spectrometry (MS) [11].

DEPOSITION OF HYDROGENATED AMORPHOUS SILICON

We discuss the deposition of a-Si:H by Remote PECVD to provide an alloy system, end-member reference point for the deposition studies of silicon dielectrics [5]. The link between the PECVD processes for a-Si:H, and the silicon dielectrics relates to the specific roles that He metastables, He*, play in the process chemistry. Thin films of electronic or photovoltaic (PV) grade a-Si:H are usually produced by the glow discharge (GD) decomposition of silane (or silane/rare gas mixtures) [13,14]; this GD process is the same as Direct PECVD. We have already noted that the plasma dissociation of silane leads at a number of different species, and that these species provide 'parallel' reaction pathways for film deposition. One 'signature' of multiple reaction pathways in a-Si:H films is in the incorporation of bonded hydrogen, in particular in the Ts dependence of the SiH, SiH_2 and $(SiH_2)_N$ bonding groups in the deposited thin films [14]. Studies using triode structures have lead to the conclusion that SiH_3 radicals in the gas phase lead to monohydride (Si-H) bonding in the deposited films, and that SiH_2 molecules in the gas phase can promote significant SiH polymerization both in the gas phase and at the film surface [14]. Films grown by direct PECVD (with no grid electrodes) show predominantly monoydride bonding at high Ts (>230°C), predominantly polyhydride bonding at low Ts (<200°C), and the presence of both types of bonding over a range of Ts from about 100 to 350°C [13].

We have grown a-Si:H films by Remote PECVD and have used the following process [5]: (i) a rare gas, either He or Ar is plasma excited (15 - 50 Watts at 13.6 MHz) and active species (He* or Ar*) are extracted from the plasma region and injected into the deposition chamber; (ii) these rare gas metastables are mixed with silane delivered via a showerhead dispersal ring; and (iii) a film is deposited on a heated substrate. The deposition mechanism involves the fragmentation of silane through collisions with the rare gas metastables. By analogy with experiments relating the break-up of small organic molecules (CH_4, CH_3OH, etc.) by rare gas metastables, we infer different fragmentation products for the mixing of Ar* and He* with the SiH_4 [15].

The films grown using He excitation are qualitatively different from films grown by Direct PECVD with respect to incorporation of bonded hydrogen. For Ts from 100 to 400°C the dominant configuration for bonded hydrogen in the Remote PECVD films is monoydride or SiH groups. By analogy with results reported for Direct PECVD [14], we propose that the dominant fragment of the He metastable initiated silane break-up is SiH_3. The amount of bonded hydrogen varies approximately linearly with Ts over this range, with about 4 at% H at Ts = 325°C and 17 at%H at Ts = 100°C. The deposition rate is approximately constant over this temperature range, nominally 0.1 Angstroms/second for an RF power input of 25 Watts, and changes linearly with RF power over the range from about 15 to 50 Watts.

We have studied the photoconductivity and defect states as functions of Ts and conclude that Remote PECVD films deposited with Ts = 230°C are 'equivalent' to the so-called 'electronic' grade material (defect densities < 1016/cm3). These films display about a two- to threefold increase in the defect state density when Ts is reduced from 230°C to approximately 150°C. In constrast, films grown by Direct PECVD display an increase in defect state density by at least a two orders of magnitude between 230 and 150°C. This increase in defect density in the Direct PECVD films has been attributed to the dominance of polyhdride bonding groups that occurs when Ts is less than about 200°C [16]. Polysilane bonding appears in the Remote PECVD films when Ts < 100°C, and these films also display a correspondingly large decrease in the photoconductivity for Ts less than 100°C.

We find results for Ar* excitation that infer a different fragmentation pattern for the silane. In addition to film deposition, that is qualitatively similar with respect to bonded hydrogen incorporation for the He* excitation initiated deposition process, we observe gas-phase generation of polysilane powders. This polymerization process is known to derive from the generation of SiH_2 groups via a sequence of reactions that first produce disilane, then trisilane etc. [17]. From our studies of a-Si:H depositions we conclude: (i) He* fragmentation of silane leads to SiH_3 groups, which in turn lead to film deposition with Si-H and Si-Si bonding groups when Ts > 100°C; whereas (ii) Ar* fragmentation of silane leads to both SiH_2 and SiH_3 groups with the possibility of gas or surface generation of polysilanes, primarily via the SiH_2 molecules.

DEPOSITION OF DIELECTRIC FILMS BY RPECVD

We describe the bonding environments in deposited silicon oxide and nitride films, emphasizing stochiometry and hydrogen incorporation. We first discuss the oxides where the overall reaction for film depositions is:

$$(O_2)^* \text{ or } (O_2/He)^* + SiH_4 \text{ ---> } SiO_2 + H$$

We observe that the deposition rate and film character are determined by a combination of Ts and He dilution. In general, decreasing Ts increases the deposition rate; this is taken as evidence for an exothermic surface CVD reaction. For He dilution up to about 98 % He, the deposition rate remains changes fairly slowly from about 1 to 4 Angstroms/second, and the bonding groups detected in the films by infrared (ir) include only SiO and SiOH (the limit for ir detection of SiH and SiOH is about 0.5 at.%). We have demonstrated in experiments with deuterated silane, SiD_4, that the OH incorporation derives mostly from impurities at concentrations less than 100 ppm in the oxygen source gas (either water vapor or methane, that is converted into water vapor by the plasma excitation) [18]. For He dilution exceeding 98 %, the deposition rate drops significantly and the bonding environemnts in the films change in two ways: (i) the films become 'non-stoichiometric' displaying SiSi bonds in addition to SiO bonds; and (ii) the films show spectroscopic features associated with SiH bonding groups (there are two SiH ir signatures, a strong bond bending absorption at about 875 cm^{-1}, and a weaker bond stretching absorption at about 2250 cm^{-1}) [19]. We have shown by SiD_4 substitution that the hydrogen atoms in SiH bonding groups derive from the silane reactant [18]. Further increases in the He dilution promote greater departures from SiO_2 stoichiometry (SiSi and SiH bond formation at the expense of SiO bonding), and also produce changes in the frequencies of the SiH absorptions that associated with changes in the nature of the atoms backed bonded to the Si atoms of the SiH groups, i.e., the substitution of Si atoms for O atoms [20]. As the limit of 100% He dilution is approached, the deposition rate cnverges smoothly to that for a-Si:H. We will show in a later section of this paper that these changes in bonding with He dilution, are correlated with changes in the plasma generation rate of atomic oxygen and metastable He [11].

We have used two different sources of nitrogen atoms for the deposition of silicon nitride films, nitrogen and ammonia [4,9]. We first discuss the nitrogen process where the deposition reaction is given by:

$(N_2)^*$ or $(N_2/He)^* + SiH_4$ ----> $Si_3N_4 + H$

In the limiting case of no He, we produce films of stoichiometric Si_3N_4, but with a very low deposition rate, approximately 0.02 Angstroms/second. As the He dilution increases to about 80% He, the deposition rate increases linearly (and modestly) to about 0.1 Angstroms/second, and the nitride stoichiometry is maintained. At higher He dilutions, the deposition rate increases more rapidly and both SiSi (by index of refraction measurements) and SiH (by ir) bonding groups become evident in the films. Again, as in the case of the oxide depositions, as the He dilutions is increased to 100%, the deposition rate and film character change 'monotonically' to what we have previously found for a-Si:H. We do not detect any ir absorptions associated with SiNH bonding groups for any level of He dilution. SiNH groups are found only in Remote PECVD 'nitride' films grown from ammonia (NH_3) or NH_3/He source gases [4,20].

The reaction chemistry for 'nitride' deposition using ammonia is:

$(NH_3)^*$ or $(NH_3/He)^* + SiH_4$ ----> 'silicon nitride' + H

where we write the thin film product as 'silicon nitride' to emphasize that it is generally not stoichiometric nitride, Si_3N_4. Over most of the range of He dilution, the reaction product is a silicon diimide/silicon nitride alloy, $(Si(NH)_2)_{(1-x)}(Si_3N_4)_x$; however under certain specific deposition conditions, we can also grow either of the end-members from the NH_3 source gas [4,9,20].

For He dilution up to about 80%, the deposited films are diimide/nitirde alloys. In this range of alloy formation, the deposition rate increases from about 0.5 to 0.8 Angstroms/second, and the source of hydrogen in SiNH groups of the diimide constituent is the ammonia reactant. This has been demonstrated by the use of deuterated ammonia, ND_3 [9,20]. As the He dilution is increased further, the deposition rate drops below that of the a-Si:H (about 0.2 Angstroms/second) and the film composition approaches Si_3N_4. For Ts = 250°C, the concentration of hydrogen in residual NH groups is about 5 at. %, and for Ts = 550°C, it is below the level of ir detection (< 1 at. %). As the He dilution is increased still further into the 90% regime, the deposition rate starts to increase, and we observe some evidence for a change over from hydrogen incorporation in SiNH to SiH groups, and for SiSi bonding groups as well.

We have correlated the changes in the various SiO, SIN, SiH, and SiNH bonding groups in the oxide an nitride films with changes in the nature of the species found in the plasma region optical emission spectroscopy (OES) and mass spectrometry (MS) [11]. This is discussed in the context of the emirical model for deposition reaction pathways that is developed in the next section of the paper.

EMPIRICAL MODEL FOR REACTION PATHWAYS IN FILM DEPOSITION

The model that we have developed for the deposition of the silicon-based dielectrics is based on two competing chemical reaction pathways: the first involves the chemical consumption of the silane reactant by active oxygen and nitrogen species and leads to the deposition of films with SiO and SiN bonding groups; and the second involves the fragmentation of the silane reactant by the rare gas metastables (used as diluents in the Remote PECVD process) and leads to the deposition of a second phase (essentially a-Si:H) which contains SiSi and SiH bonds. The experimental basis for this model derives from studies of the deposition of a-Si:H and silicon dielectric thin films by the Remote PECVD process, with particular emphasis on the effects of He dilution of the oxygen or nitrogen species on the nature of the bonding groups in the dielectrics and on the thin film deposition rate, and from studies by OES and MS of the molecular and/or atomic species generated in RF plasmas containing He and O_2, N_2 or NH_3 [11]. It should be noted that there are a number of limitations in the use of OES and MS to study plasma generated species: (i) not all of the potentially active species generated in the plasma contribute to the optical emission spectrum; and (ii) studies of plasma generated species based on MS are subject to uncertainties that are related to secondary fragmentation effects in the ionizer. In spite of these limitations, we have made significant progress in determining some of the important aspects of the deposition process reactions for the Remote PECVD deposition of silicon oxides and nitrides.

The model that we develop for the deposition of both a-Si:H and stoichiometric oxides and nitrides is additionally based on a two-step reaction mechanism. The first step involves the gas phase generation of 'silicon-based' precursor species through comsumption of the silane reactant. This process step involves either selective fragmentation of the silane reactant by rare gas metastables, or chemical consumption of the silane by active oxygen or nitrogen species. The second step involves a CVD reaction at the substrate in which the precursor species formed in the gas phase are converted to the solid film [3,9].

We condsider first the depostion process for a-Si:H films, where the initial step is the selective fragmentation in the gas phase of the silane reactant by either He* or Ar* metastables. The reaction pathways of used in our proposed model draw on the studies of numerous workers in the field of a-Si:H thin film deposition; these have been summarized in a recent review article by Matsuda and Tanaka

[14]. Consider first the a-Si:H deposition process that is based on the plasma excitation of He and the mixing of the plasma extracted He species with silane. We have confirmed the existence of He* in the plasma using OES, and the existence of extracted He* in the chamber using MS [11]. We have noted above that the mixing of He* with silane results in the deposition of an a-Si:H film in which the dominant mode of hydrogen atom incorporation for Ts between 100 and 400°C is in monohydride or SiH bonding groups. Additionaly, the a-Si:H films deposited by Remote PECVD at 230°C are 'electronic grade' with majority carrier electronic properties similar to the best PV grade GD or Direct PECVD films.

It has been established that the gas phase precursor for the deposition of electronic grade a-Si:H with hydrogen atom incorporation in monohydride bonding groups is the SiH_3 radical [14]. Based on the similarity between our films deposited at 230°C and the best GD films, we propose that the dominant precursor species formed by the fragmentation of silane by He* is also the SiH_3 radical (we plan to perform additional experiments to confirm this proposal). We further observe that for Ts between 100 and 400°C, the the amount of hydrogen retained in the films is decreases apporiximately linearly with increasing Ts. Over this range of Ts, as noted above, the only ir observable hydrogen is in monohydride groups. For films formed by Direct PECVD in this same Ts range, there are additional precursor species formed by the plasma breakup of the silane, e.g., Si, SiH, SiH_2, etc., and these promote parallel deposition reaction pathways for the incorporation of polysilane groups as well. For Ts < 100°C we observe the polysilane bonding groups in the Remote PECVD films. We believe that this derives from inefficient rejection of hydrogen (in the form of some 'SiH_n' species) at the growth surface [14]. The observation of polysilane groups in the Remote PECVD films is also accompanied by an increase in the deposition rate, hence the proposal that the rejection of hydrogen must involve an 'SiH' species. Over the entire Ts range, the deposition rate for the Remote PECVD process is proportional to the RF power into the He plasma. We believe that this dependence on RF power reflects the rate of plasma generation of He*.

The results obtained from mixing silane with plasma generated Ar* are different in as much as two reaction pathways that are evident. One pathway leads to film formation with an incorporation of bonded hydrogen that is similar to what we observe in the He* initiated Remote PECVD processs. This leads us to propose that one of the products of silane fragmentation by Ar* is the SiH_3. In addition to film formation, we also observe the gas phase generation of polysilane powders. Formation of polysilane powders are known to derive from following set of reactions that are based on the generation of SiH_2 molecules via Ar* induced fragmentation of the silane [10,17]:

$SiH_2 + SiH_4$ ----> Si_2H_6

$Si_2H_6 + SiH_2$ ----> Si_3H_8; etc.

The conditions of gas flow rate, pressure, etc. in our system favor high gas phase collision rates between the gas species injected into the chamber from the plasma discharge tube and the silane reactant delivered from the showerhead dispersal ring [9]. This accounts for the polysilane powder production. In summary then, we propose that two 'active' fragmentation products of the reaction between Ar* and silane are SiH_2 and SiH_3 and that these lead respectively to the simultaneous observation of polysilane powder generation in the gas phase and deposition in the temperature range between 100 and 400°C of thin film a-Si:H alloys with SiH bonding groups.

We have achieved deposition of stoichiometric silicon nitride (Si_3N_4) films using either N_2 or N_2/He as the source gas [4,9,20]. The deposition process involves the plasma excitation of N_2 or N_2/He, the extraction of active species from the plasma, the mixing of these species with silane (10% SiH_4/ 90% Ar), and the deposition of a thin film on a heated substrate. We have observed the following: (i) stoichiometric silicon nitride films (Si_3N_4) can be deposited from mixtures of plasma excited nitrogen and neutral silane with a deposition rate of about 0.02 Angrstroms/second; (ii) the deposition rate can be enhanced by the dilution of the N_2 source gas with He; (iii) the deposition rate is proportional to the generation rate of nitrogen atoms in the N_2/He plasma over a range of He dilution extending to about 85% He; (iv) for increasing dilution with He, the deposition rate continues to grow even though the atomic nitrogen population decreases; but (v) in this high He dilution range the deposited films display deviations from nitride stoichiometry that show up as the additional incorporation of SiSi and SiH bonding groups. In the high dilution range, the films can then be described as solid solutions of Si_3N_4 and a-Si:H with SiN, SiH and SiSi local bonding.

We first consider possible reaction pathway chemistries when only N_2 is plasma excited. There are two pathways that are distinguished by the way the N_2 reactant is consummed. Consider first a process in which the reactions between N atoms extracted from the plasma and the silane injected into the chamber occur only at the the substrate. In this case, the formation of a silicon nitride thin film, without any ir-detectable SiH or SiNH bonds requires a complete substitution of SiH bonds of the silane molecule by SiN bonds of the thin film material. The surface reaction for this type of CVD process can then be represented in the following way:

$N + SiH_4$ ----> $Si_3N_4 + H$

We believe that this type of reaction process is not very likely to occur due to: (i) the relatively high chemical reactivity between nitrogen atoms and silane, and (ii) the relatively high collision rate between nitrogen atoms and silane molecules in the gas phase in the space between the dispersal ring and the substrate.

The high binary collision rate suggests a second reaction chemistry pathway in which gas phase precursor formation occurs. We propose a precursor reaction which results in a molecular species that contains SiN bonds; this involves the insertion of N atoms into the silane molecule with the removal of one or more of the hydrogen atoms. The precursor species then undergoes a surface reaction with additional nitrogen atoms in which: (i) the remaining hydrogen atoms (in SiH bonds) on the precursor species are replaced by nitrogen atoms; and (ii) in which a continuous random network structure with only SiN bonds is formed. We are presently studying the reaction products between plasma excited nitrogen atoms and neutral silane by MS in order to determine if molecular species with SiN bonds are formed in the gas phase; this would be evidence for a two step deposition process involving gas phase precursor formation.

We now consider the changes that occur when N_2/He mixtures are plasma excited. There is a range of He dilution extending to about 80 % He in which the only potentially active species detected in the plamsa region are nitrogen atoms and where their number increaes as the He dilution increases. In this same range of diluction there is emission associated with neutral He atoms as well. Along with the spectroscopically determined increase in the nitrogen atom population, we find a proportional increase in the deposition rate for stoichiometric Si_3N_4. We propose that the deposition chemistry for this range of He dilution parallels what we have proposed above for the plasma exictation of undiluted N_2, and involves a two step reaction process. We note that the deposition rate is clearly limited by the availability of nitrogen atoms.

At higher levels of He dilution we initially observe (by OES) a mixture of nitrogen atoms and He*, and then only He* at the highest dilution levels. We propose that the generation of He*, their extraction into the deposition chamber and mixing with SiH_4 opens up a parallel or competing deposition pathway in which collisons between He* and the silane generate SiH_3 radicals. This generation of SiH_3 opens up a reaction pathway by which SiH and SiSi bonds can be 'codeposited' with SiN bonds. This proposal is consistent with what we find in the deposited films. At about the same level of He dilution that we observe both He* and nitrogen atoms by OES, we find spectroscopic evidence (ir and ellipsometry) for both SiH bonds and SiSi bonds in the deposited films, in addition to SiN bonds. In this regime of high He dilution, the deposition rate increases even though the rate at which nitrogen atoms are being produced in the N_2/He plasma is decreasing. The films produced in this regime of dilution then make a transition form nitrides with SiH and SiSi bonding 'defects' to a-Si:H with SiN bonding 'defects'.

The model for the deposition of silicon nitride films, and silicon nitride/a-Si:H alloys then involves two parallel and competing reaction pathways. Plasma generation of nitrogen atoms, and their subsequent interaction with silane through gas phase and surface CVD ractions generates either a stoichiometric silicon nitride film, or SiN bonding groups in silicon nitride/a-Si:H alloys. Plasma generation of He* opens up a deposition channel for the 'codeposition' of SiSi and SiH bonds (the bonding groups of a-Si:H). From our observations, we also conclude that interactions between SiH_3 radicals and nitrogen atoms have a smaller cross section than interactions between SiH_4 molecules and nitrogen atoms; this point will also be subjected to experimental verification in our deposition/analysis system.

We now apply our model to the deposition of silicon oxide films by Remote PECVD. These studies have been restricted to deposition chemistries in which: (i) the plasma excited species have been either oxygen (O_2) or oxygen/helium mixtures (O_2/He); and (ii) the neutral gas injected into the deposition chamber has been silane (SiH_4) diluted with Ar (10% SiH_4 in Ar). We have observed the following for films grown from plasma excited O_2 and Ar diluted SiH_4: (i) the films display only SiO and SiOH bonding groups, never any detectable SiH; (ii) SiOH bonds are evident by ir in films grown with Ts = 250°C, but are not detected in films grown at about 500°C; and (iii) over this range of Ts the deposition rate is relatively high, several angstroms/second (for 2% O_2 in He, the deposition rate decreases by a factor of about two as Ts is increased from 200°C to 400°C. We propose a specific reaction pathway model for the deposition of SiO_2 films that is based on the studies of gas phase species and thin films generated from SiH_4/NO mixtures using Direct PECVD [21]. MS studies have indicated the presence of N_2O and $H_3Si{-}O{-}SiH_3$ (disiloxane) in the gas phase, and deposited thin films display both SiH and SiO bonding. The reaction pathway proposed in Ref XX is the following:

(i) $NO + NO \text{---}> N_2O + O$;

(ii) $O + SiH_4 \text{----}> H_3SiOH$, and

(iii) $H_3SiOH + H_3SiOH \text{----}> H_3Si{-}O{-}SiH_3 + H_2$.

The disiloxane species then becomes the precursor for the SiO bonds observed in the deposited thin film. The SiH bonds can derive from either the disiloxane or more likely from SiH_3 species that are also generated in the direct plasma excitation of the silane. Note that there is no direct observation of silanols, so that this step in the reaction pathway chemistry needs verification.

By analogy with the model given above, we propose a model in with the silane molecules react with plasma generated metastable oxygen atoms to form precursor species with SiO bonding groups. We do not have sufficient data to specify whether these molecules are siloxanes, etc. or whether the process involves the formation of an intermediate species such as a silanol. In any case, we propose that any precursors formed undergo a surface CVD reaction involving the 'complete and efficient' replacement of hydrogen in any terminal SiH_3 groups with oxygen. We have additionally observed that the deposition rate is not simply proportional to any power of the atomic oxygen generation rate as determined by OES. We are presently usiing MS to determine the nature of the precursors (if any) that are formed when the active oxygen species and silane are mixed.

We now condider the changes in deposition that result from He dilution of the O_2 source gas. We find relatively small changes in the film chemistry and the deposition rate for He dilutions up to about 80 %He. In this range the only ir detectable bonding groups in the deposited thin films are SiO and SiOH (we have already noted that by usiing an SiD_4 source gas, that primary source for SiOH in the films was an impurity in the O_2 source gas, rather than the result of a reaction between oxygen species and silane). For this range of He dilution (0 to 80 %), the species we observe in the plasma by OES are oxygen atoms and oxygen molecular ions, O_2^+. As the He concentration is further increased to beyond 95 %, we observe both oxygen atoms and He* via the plasma emission. In this higher He dilution regime, the bonding in the film undergoes additional qualitative changes: (i) the dominant mode for H incorporation becomes SiH rather than SiOH; (ii) the films develop a suboxide character with evidence for SiSi bonds in addition to the SiO and SiH bonds; and (iii) the deposition rate drops. In this range of dilution, the atomic oxygen population increases dramatically, but the oxide deposition rate decreases. This means that the gas phase reaction for precursor formation and the surface CVD reaction cannot both involve atomic oxygen species. For example, if the pathway for precursor formation derives from a reaction between atomic oxygen and SiH_4, then the surface CVD reaction must then be based on a diatomic, or perhaps even a triatomic form of oxygen (ozone). We propose that the observation of SiH and SiSi bonding groups is driven by a competitive reaction involving silane fragmentation by He* and the generation of SiH_3 radicals. The SiH_3 radiacals then promote a deposition process that introduces an a-Si:H component to the deposited film. Note further that the SiH_3 radicals must also have a small cross section for reaction with atomic oxygen species, because a strong reaction between the SiH_3 radicals and atomic (or molecular) oxygen metastables would then open up an additional pathway for the incorporation of SiO bonding groups. This last observation parallels the same lack of reactivity that we have already noted between SiH_3 and the plasma generated nitrogen atoms.

We now consider the reaction chemistry that prevails for the deposition of nitride films using plasma excited mixtures of ammonia and He. Depending on the range of He dilution we observe at least four different species in plasma excited NH_3 and NH_3/He mixtures: (i) NH molecular fragments; (ii) helium metastables, He*; (ii) nitrogen atoms; and (iv) hydrogen atoms. We make the following addtional observations based on OES: (i) in pure ammonia and at He dilutions to about 80 % the dominant emissions in the plasma region are associated with NH molecular fragments and hydrogen atoms; (ii) at higher levels of He dilution the nitrogen atom concentration grows at the expense of the NH concentration but with hydrogen atoms present as well; and (iii) at the highest levels of He dilution, > 95 %, the nitrogen atom population decreases and the dominant species in the plasma are He*.

We propose the following pathway reactions. In the range of He dilution up to 80 %, the dominant active species being extracted from the plasma is the NH fragment which is similar in its chemistry to atomic oxygen. By analogy with oxygen atom/silane chemistry and with observations made on gas phase plasma reactions between AsH_3 and SiH_4 and PH_3 and SiH_4 [22], we propose that mixing NH and SiH_4 generates silazane precursors in the gas phase by a reaction of the type given below:

$$NH + SiH_4 \longrightarrow H_3Si\text{-}(NH)\text{-}SiH_3 + H$$

We further propose that these precursors then react with additional NH species at the film surface and generate 'silicon nitride' films with relatively high concentrations of NH groups (up to 30 at. % H). In the limit of saturation of the nitrogen sites with NH bonding groups, the film becomes an isoelectronic analog of SiO_2, silicon diimide with a stoichiometry corresponding to $Si(NH)_2$ [4,20]. This material is formed by complete replacement of the SiH_3 groups of the precursor with NH bonding groups, much the same way SiO_2 can be generated from disiloxane and oxygen atoms. However, for most of the deposition parameter space we have explored, we find diimide/nitirde alloys. This means that in addition to loss of H from SiH_3 groups of the precursor, there is also some loss of hydrogen atoms from the NH bonding group. Over the range of He dilution between pure NH_3 and 80 % He, the deposition rate increases from about 0.5 to 0.8 Angstroms/second and this increase simply reflects increases in the NH population, as monitored by OES.

For He dilution in excess in excess of about 80 %, we find both N and NH emission. In this regime two things occur: (i) the fraction of SiNH bonds in the films decrease at the expense of an increase in the fraction of SiN; and (ii) the deposition rate decreases; in addition, there are are no ir observable SiH absorptions. As the He dilution increases further deposition continues to drop and the fraction of NH bonding groups decreases. These changes correlate with decreases in the plasma emission associated with NH and increases in the emission associated with N. At the highest dilutions studied, about 98 % He, the films deposited above about 500°C are essentially stoichiometric Si_3N_4, and even though we observe He* in the plasma by OES, we do not observe any spectroscopic evidence by ir absorption for SiH groups in the deposited films.

APPLICATIONS OF REMOTE PECVD DIELECTRICS IN DEVICES

We discuss several different device configurations in which we have utilized Remote PECVD dielectric films. These studies were done in collaboration with R.J. Markunas and his group at the Research Triangle Institute, Research Triangle Park, N.C. [2,3,9]. Our first application was in an FET device based on (In,Ga)As. This was done at a very early stage of our research effort, and at a time when there was Na contamination present in our silicon dioxide films. The source of Na was from a pyrex tube that was used in the plasma excitation region. Replacement of this tube with a silica tube eliminated the Na problem. However, in order to compensate for the Na contamination, we used a three layer dielectric in our first device structures [2]: 100 Anstroms oxide/ 600 Angstroms nitride/ 100 Angstroms oxide. The two oxide layers blocked electron tunnelling into the nitride and the nitride layer served as a diffusion barrier for the mobile Na^+ ions. Prior to insulator deposition the (In,Ga)As surface was etched in bromine/methanol, and then subjected to an ammonia plasma discharge in the depostion chamber. Device structures formed in this manner displayed the following characteristics: (i) a small dc voltage offset < 1.0 V; (ii) a high FET gain, 75 Ms/mm; and (iii) good stability defined by less than 5 % drift over a 24 hour stress bias test.

Upon replacement of the pyrex tube in the deposition chamber by a fused silica tube, we were able to deposit SiO_2 and Si_3N_4 films with no Na^+ ion contanimation and incorporate them into device structures. Devices fabricated on both n and p Si substrates and using SiO_2 layers displayed MOS characteristics similar to devices fabricated with thermally grown SiO_2 layers (growth temperature > 1000°C) [23]. Prior to the insulator deposition, it was necessary to process the crystalline silicon surface to remove any surface contaminants, primarily oxygen and hydrocarbon, and to verify the surface was chemically clean and structurally in tact. This was done using standard etching techniques and then processing the cleaned and etched surface in a UHV environment. The result device structures exhibited breakdown fields in excess of 10×10^6 V/cm, and had integrated interface and bulk defect densities less than $5 \times 10^{10}/cm^2$.

In addition, it was demonstrated that both n and p-type germanium surfaces could be inverted at room temperature using tri-layer dielectric structures. Finally, we have also grown hydrogen free silicon oxynitride alloys on crystalline silicon and obtained MOS characteristics similar to those obtained using silicon dioxide dielectrics. The oxynitride compositions in this case correspond to solid soluitions of silicon dioxide and silicon nitride;i.e., the compositions lie along the join line from SiO_2 to Si_3N_4 on a ternary (three element: Si, O, N) phase diagram.

DISCUSSION

We first summarize the results we have presented in this paper, and then mention some of the extensions of the Remote PECVD process to other materials systems of interest in semiconductor device structures. We have found that the important requirements of: (i) maintaining compound stoichiometry in silicon oxides and nitrides; and (ii) controlling the nature and amount of bonded hydrogen films produced via PECVD processes are related to the way the silane reactant is utilized. By using a remote process in which the silane reactant is not subjected to a direct plasma discharge, but instead is either chemically consummed or selectively fragmented, we have been successful: (i) in growing stoichiometric dielectrics with no ir detectable bonded hydrogen; (ii) controlling the manner in which hydrogen is incorporated in hydrogenated amorphous silicon; and (iii) in producing 'device' or 'gate' quality dielectric materials.

We find that device quality silicon oxides, nitrides and oxynitrides can be grown, respectively from plasma excited and helium diluted oxygen, ammonia and oxygen/ammonia mixtures that are subsequently mixed with neutral silane. We have proposed a model for the reaction chemistry that is based on both gas phase precursor formation and surface reactions. The experimental basis for this model is derived mostly from comparisons with previously reported studies of plasma excited mixtures of silane and NO, and silane and arsine and phosphine [21-23]. We are presently studying the gas phase reactions with a deposition/analysis system that provides for: (i) optical emission spectroscopy; (ii) mass spectrometry; and (iii) thin film deposition on a heated and moveable substrate [11].

We are also extending our deposition approach to mixed oxide materials of the general form $A_xB_yO_z$, where A and B are metal or metalloid atoms. We have two applications in mind: (i) the deposition of chemically 'customized' oxides, such as (Si,Al)O, that may prove to be of use in compound semiconductor materials technologies that employ compounds with one or more of the mixed oxide metal or metalloid atoms, e.g., (Al,Ga)As; and (ii) the deposition of multicomponent oxides with special electronic properties such as ferroelectric materials with high dielectric constants, or superconducting oxides.

ACKNOWLEDGEMENTS

This work has been supported under ONR Contracts N00014-79-C-0133, and N00014-86-K-0760, NSF Grant DMR-8705633, and under ONR and SRC subcontracts to North Carolina State University from the Research Triangle Institute.

REFERENCES

1. W. Kern and V.S. Ban, in Thin Film Processes, Ed. J.L. Vossen and W. Kern (Academic Press, New York, 1978), Chap III.

2. P.D. Richard, R.J. Markunas, G. Lucovsky, G.G. Fountain, A.N. Mansour and D.V. Tsu, J. Vac. Sci. Technol. A3, 867 (1985).
3. G. Lucovsky, P.D. Richard, D.V. Tsu, S.Y. Lin and R.J. Markunas, J. Vac. Sci. Technol. A4, 681 (1986).
4. D.V. Tsu and G. Lucovsky, J. Vac. Sci. Technol. A4, 480 (1986).
5. G.N. Parsons, D.V. Tsu and G. Lucovsky, J Non-Cryst. Solids, (in press).
6. L. Meiners, J. Vac. Sci. Technol. 21, 655 (1982).
7. M.J. Helix, K.V. Vaidyanathan, B.G. Streetman, H.B. Dietrich and P.K. Chatterjee, Thin Solid Films 55, 143 (1978).
8. A.C. Adams, Solid State Technol. 26, 135 (1983).
9. G. Lucovsky and D.V. Tsu, J. Vac. Sci. Technol. A5, 2231 (1987).
10. J.P.M. Schmitt, J. Non-Cryst. Solids 59/60, 649 (1983).
11. D.V. Tsu and G. Lucovsky, J. Vac. Sci. Technol. A6 (1988), (in press).
12. S.S. Kim, D.V. Tsu and G. Lucovsky, J. Vac. Sci. Technol. A6 (1988), (in press).
13. G. Lucovsky, R.J. Nemanich and J.C. Knights, Phys. Rev. B19, 2064 (1979).
14. K. Tanaka and A. Matsuda, Materials Science Reports 2, 139 (1987).
15. J. Balanuta, M.F. Goldie and Y-S. Ho, J Chem. Phys. 79, 2822 (1983).
16. J.C. Knights, R.A. Street and G. Lucovsky, J. Non-Cryst. Solids, 35/36, 179 (1980).
17. G. Inoue and M. Suzuki, Chem. Phys. Letters 105, 641 (1984).
18. D.V.Tsu and G. Lucovsky, Mater. Res. Soc. Proc. (to be published).
19. G. Lucovsky, J. Yang, S.S. Chao, J.E. Tyler and W. Cubatyj, Phys. Rev. B28, 3225 (1983).
20. D.V. Tsu, G. Lucovsky and M.J. Mantini, Phys. Rev. B33, 7069 (1986).
21. P.A. Longeway, R.D. Estes and H.A. Weakliem, J. Phys. Chem. 88, 73 (1984).
22. W.L. Jolly, Adv. Chem. 80, 156 (1969).
23. R.A. Rudder, G.G. Fountain and P. Lindorme, Appl. Phys. Lett. (submitted for publication).

Chemical Vapor Deposition of Metals for VLSI Applications

Martin L. Green

AT&T Bell Laboratories
Murray Hill, New Jersey 07974

ABSTRACT

Chemical vapor deposition (CVD) of metals on VLSI circuits is a processing scheme that allows one to deposit low temperature, conformal and radiation-defect free films. For these reasons, a number of metals have been chemically vapor deposited for a variety of VLSI metallization applications. These applications are discussed in detail in this paper. CVD W and CVD Al, for example, have already shown great potential as via fill and interconnect metallizations for multilevel devices with small ($<1\,\mu$m) and high aspect ratio (>1) source and drain windows, and shallow ($<0.25\,\mu$m), easily damaged junctions. In addition, the selectivity exhibited by CVD W obviates the need for patterning of this layer. The success of CVD metallization processing will depend upon an understanding of the interactions between the substrate, the process gases and the deposited film. This paper emphasizes these reactions for the case of CVD W. Imminent applications of CVD W and CVD Al are discussed, and the advantages and potential disadvantages of each process are assessed.

Introduction

Metal films play vital roles in the operation of a VLSI circuit. Gate, contact, diffusion barrier, interconnect and via fill metallization applications bring various requirements to bear on these films and the way in which they are deposited. These films can be deposited using a variety of deposition techniques such as sputtering, evaporation or chemical vapor deposition. Each technique has its own merits and drawbacks, which have been discussed in detail.[1] This paper focusses on the chemical vapor deposition (CVD) of metals for VLSI applications, a field which is currently of great interest and whose applications are in their infancy compared to evaporation or sputtering. CVD metals processing offers solutions to VLSI metallization problems arising from the small ($\lesssim 1\,\mu$m design rules) dimensions typical of such devices.

CVD is the deposition of a solid film on a substrate, brought about by the reaction of one or more vapor phase species. Typical deposition reactions include pyrolysis, reduction, oxidation, hydrolysis, disproportionation, or combinations of these, and are usually catalyzed by the substrate. The ability to deposit a particular film depends only on favorable thermodynamics and kinetics for the chosen reaction, and the availability of vapor sources for the reactants.

The VLSI Metallization Challenge

The advantages of CVD processing are best discussed in the context of metallization applications on a VLSI circuit device. Figure 1 is a schematic diagram of an N-channel MOS device which features a multilevel metallization scheme. Each of the five metallization applications on this device, i.e., gate, contact, diffusion barrier, interconnect and via fill, represents a challenge in thin film processing that can be addressed by CVD. The materials and processing requirements for the various

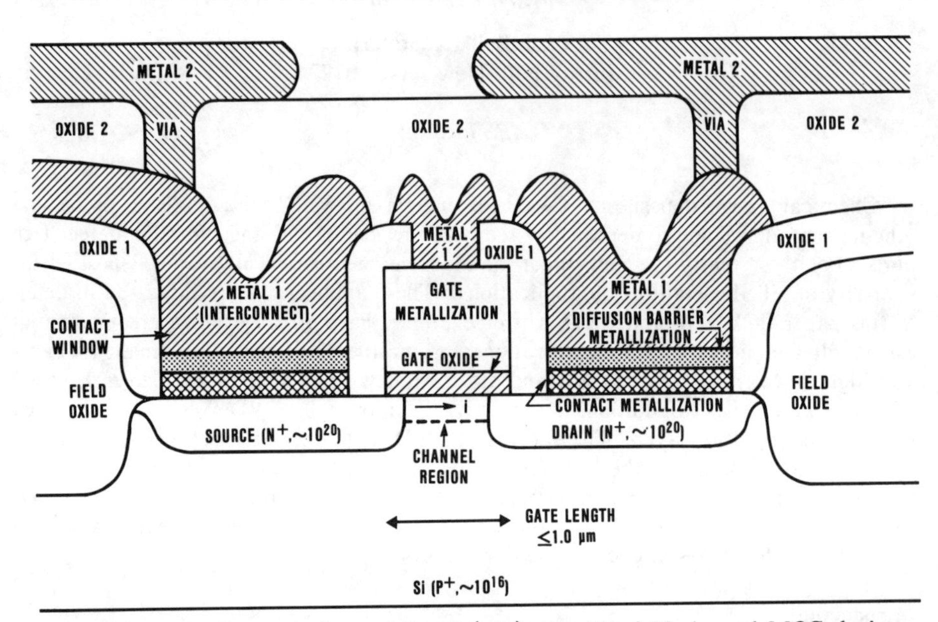

1) Schematic diagram of a multilevel (two) metallized N-channel MOS device. Dopant concentrations in the various Si regions are expressed in atoms/cm^3.

metallizations will now be summarized. Table I, which lists reported applications of CVD metals to VLSI processing, should be consulted in conjunction with the following discussion.

Gate Metallization

The gate is deposited early in the VLSI processing scheme, and will be exposed to subsequent high temperature processing steps ($\leq$950°C), sometimes in oxidizing atmospheres. Therefore, high melting temperature and oxidation resistance are important for this application. In addition, since the switching delay time of a transistor in the VLSI circuit decreases as the gate resistance decreases,[2] the gate should be a good conductor. Metal silicides are very often used as gates, due to their good oxidation resistance,[3] although their resistivities are approximately 5-10 times those of metals (~50 $\mu\Omega$-cm as opposed to 5-10 $\mu\Omega$-cm). CVD films of the refractory metals Mo,[4,5] Ru[6] and W[7] have been studied as gates, but their oxidation has to be prevented (not necessary for the case of Ru, whose oxide, RuO_2, is itself a good conductor[6]). Some refractory metal gates also have mid-bandgap (for Si) work functions, making them useful for a variety of CMOS applications.[8]

Contact Metallization

The contact between the metallization and the silicon substrate may be rectifying (i.e. a Schottky contact), or ohmic. Usually, the function of the contact metallization

TABLE I. Applications of CVD Metals to VLSI Processing

Application	Film	Source	Deposition Temperature, °C	Major Reasons for Selecting CVD Processing	Reference
Gate	Mo	$MoCl_5$	500-800	No radiation damage, high purity films.	4,5
	Ru	various Ru Organometallics	250-800	Low deposition temperatures, no radiation damage.	6
	W	WF_6	250-325	No radiation damage.	7
Contact (Schottky)	Cr	$(C_9H_{12})_2Cr$	350-520	Low deposition temperature, no radiation damage.	9
	Mo	MoF_6	300	Selective deposition on Si.	10,11
	V	VCl_4	1140-1300	High purity films.	12
	W	WF_6	325-400	High purity films, clean interfaces.	13
Contact (Ohmic)	Ag	AgF	80-600	Selective deposition on Si.	14
	Mo	$MoCl_5$	400-1350	High purity film, clean Mo/Si interface.	15,16
	Pt	$Pt(PF_3)_4$	200-300	Low temperature deposition, no radiation damage.	17
	W	WF_6	300-700	Low temperature deposition, conformal coverage, high purity films.	18,19
	W	WCl_6	700	Conformal coverage, high purity films.	20
Diffusion Barrier	Pt	see listing under Contact (Ohmic)			
	Ru	see listing under Gate			
	W	WF_6	290-550	Selective deposition, high purity films.	21-25
Interconnect Metallization	Al	i-$(C_4H_9)_3Al$	235-400	Conformal coverage, no radiation damage.	27-29
		AlCl	150-550	"	30
	Mo	$Mo(CO)_6$	$\geqslant$500	Conformal coverage, no radiation damage.	31,32
	W	WF_6	300-700	Conformal coverage, no radiation damage.	33-37
Via Fill or Planarization Metallization	W	WF_6	300-500	Selectivity, conformal coverage.	38-46

is to provide low resistance, ohmic contact between it and the Si and it and the interconnect metallization. However, CVD processing has been used to create both kinds of contacts. Schottky contacts have been formed with Cr[9], Mo[10,11], V[12] and W[13]. For the cases of Mo and W, the contacts were formed by selective deposition, i.e., the reactant gases only reacted with the exposed silicon, thereby obviating patterning. Low resistance ohmic contacts have been formed by CVD of Ag[14], Mo[15,16], Pt[17] and W[18-20]. Many CVD reactions automatically result in intimate contact between the deposited metal and the substrate due to the consumption of a thin layer of the substrate during deposition.

Diffusion Barrier Metallization

Aluminum, the interconnect metal of choice for sputtering or evaporation processes, generally establishes good ohmic contact with the silicon after proper heat treatment. However, depending upon processing after Al deposition, Al and Si can interdiffuse to cause "spiking". Spiking is intolerable in shallow junction ($\lesssim 0.25\ \mu m$) devices, since the presence of Al in the N^+ junction will ruin transistor action. Therefore, diffusion barriers are often used. If the diffusion barrier also provides good ohmic contact to the silicon, then no contact metallization film is needed. Diffusion barriers must be good electrical conductors and should be refractory materials, since in such materials diffusional activity will be limited at moderate ($\sim$300–600°C) temperatures. CVD Pt[17], Ru[6] and W[21-25] have been investigated as diffusion barriers. A wide variety of other refractory materials, deposited by physical deposition techniques, have been evaluated as diffusion barriers as well. However, the conformal coating typical of CVD processing, which insures the even coating of the source and drain window bottoms, results in the deposition of a more impervious barrier.

Interconnect Metallization

The primary requirements for interconnects are low resistivity and good electromigration resistance. Al has low resistivity ($2.7\ \mu\Omega$-cm), but poor electromigration resistance due to its low melting point. However, alloying improves this greatly.[26] Higher melting point metals such as W have excellent electromigration resistance, but higher resistivity than Al. Al,[27-30] Mo[31,32] and W[33-37] have all been evaluated as CVD interconnect metallizations, their major advantage being conformal coverage.

Via Fill or Planarization Metallization

Multilevel metallization gives rise to the need to fill vias that connect one level of interconnect metallization to another. CVD processing is well suited for this due to the conformal coverage that its films exhibit. Blanket deposition of CVD W followed by etchback[38-40] is one scheme for forming plugs in vias. Another technique involves the deposition of highly selective CVD W in the vias,[41-46] after which no etchback is required. This latter technique may not prove useful, however, for devices with vias that vary in depth. However, either approach can be used to plug source and drain windows to planarize first level interconnect metallization.

Advantages of CVD Processing

Table II compares the salient features of CVD processing with those of the physical vapor deposition techniques, sputtering and evaporation. Film characteristics such as

TABLE II. Comparison Between Chemical Vapor Deposition and Physical Vapor Deposition (Sputtering and Evaporation)

	CVD	Sputtering	Evaporation
Step Coverage	Conformal	Nonconformal (line-of-sight)	Nonconformal (line-of-sight)
Low Temperature Deposition	Yes	Yes	Yes
Radiation Damage	No	Yes	Yes (Electron-beam evaporation)
Selectivity	Possible	No	No
Film Purity	Good-Excellent (Process Dependent)	Good-Excellent (Gas Incorporation)	Excellent
Wafer Throughput	~100/hr	~60/hr	~30/hr
Relative Processing Cost per Wafer	Low	Medium	High

conformal coverage and high purity, and processing benefits such as low temperature and radiation-damage free deposition and deposition selectivity allow one to make high quality films by CVD processing. At the same time, the high throughput and correspondingly low cost per wafer associated with traditional CVD processing make it an economically attractive process.

Figure 2 is a schematic illustration of a typical hot-wall, horizontal LPCVD (low pressure) reactor. The reactants are pumped through the furnace, at which point they react to deposit a film on the substrates. The gaseous by-products are then pumped out of the system. Source materials, which are either gases, or liquids or solids with high vapor pressures (see Table I for examples), can be obtained in very pure form.

The reactor depicted in Fig. 2 can be used, for example, to deposit Al or W films. However, two current trends in CVD processing are the use of cold-wall reactors, Fig. 3, and the use of single-wafer reactors. Cold-wall reactors keep film contamination due to gas phase nucleation of particles to a minimum because only the wafers and the heater block on which they sit get hot. Cold-wall reactors have been found necessary to achieve the high degree of selectivity of CVD W that has recently been observed.[44] Single-wafer reactors are a variant of cold-wall reactors in which only one wafer at a time is processed. This is desirable mainly to exercise great deposition control over expensive, large size ($\geqslant$6") wafers. The theory of reactor design has been discussed in great detail.[47]

The advantages of CVD processing will now be considered further.

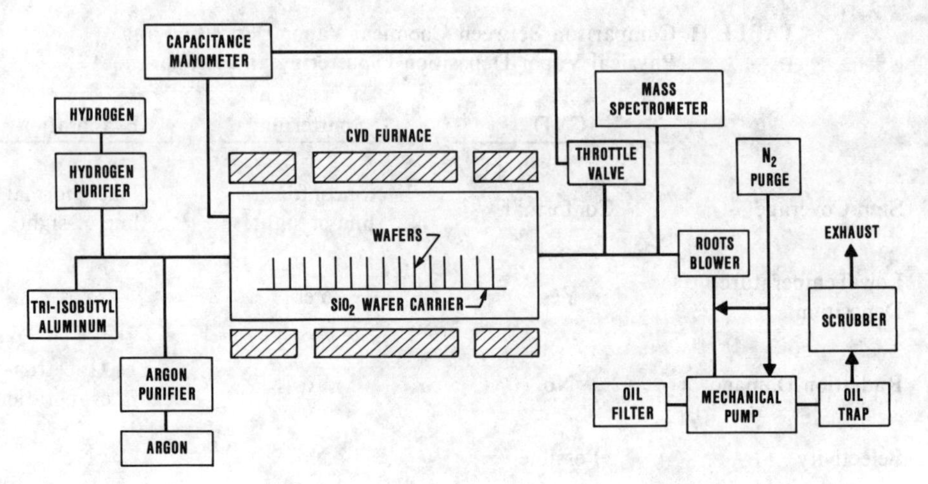

2) Schematic diagram of a hot-wall low pressure CVD system, such as might be used for the deposition of Al from tri-isobutylaluminum.

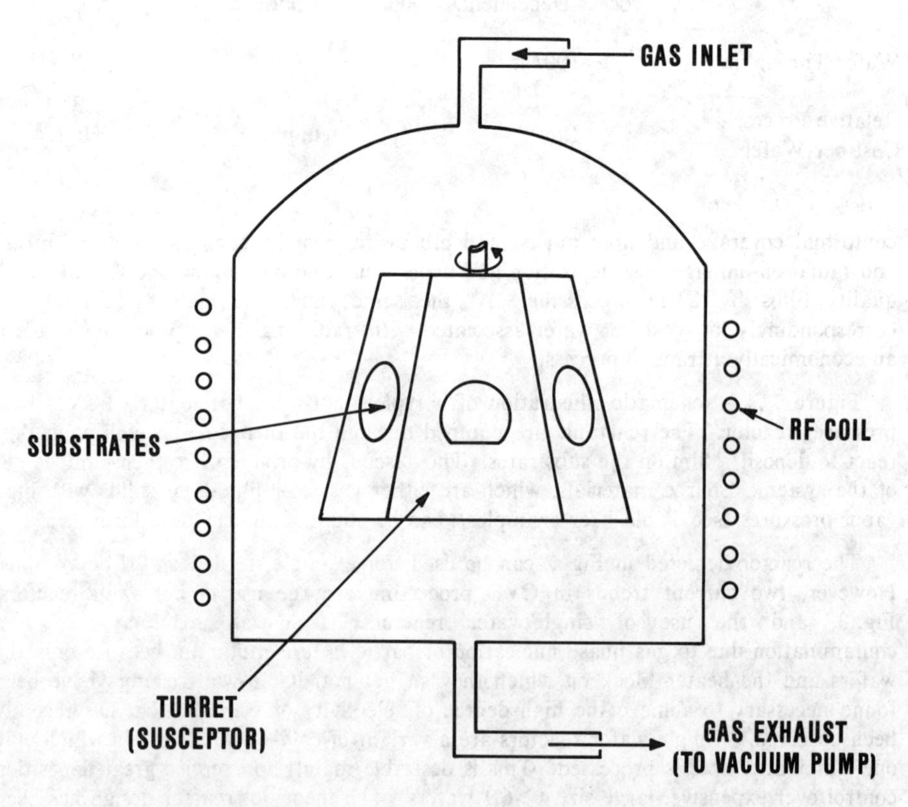

3) Schematic diagram of a cold-wall low pressure CVD system.

Conformal Coverage

Narrow, straight-walled windows and vias, re-entrant angles and overhangs are device features that can lead to poor conformal coverage with sputtered or evaporated films. Figure 4a illustrates the poor conformal coverage typical of physical deposition processing. During such line-of-sight deposition processes, shadowing effects due to device features leads to local film thickness nonuniformities. On the other hand, Fig. 4b illustrates the ideal conformal coverage that is typical of CVD processing. Because the growth of CVD films is usually surface catalyzed, nucleation and growth occurs on all surfaces, regardless of their orientation to the gas source. Therefore, by definition, uniform coverage is achieved. Uniform coverage can lead to "plugging" of windows and vias if the films are allowed to become thick enough, as is shown in Fig. 4c.

Low-Temperature Deposition

Many high melting point metals can be deposited at relatively low temperatures by CVD. For example, W can be deposited at 300°C, or 0.16 T_{mp}, by the H_2 reduction of WF_6. Such low homologous temperatures are useful in minimizing interdiffusion between the various layers of the device. Deposition of W films by evaporation, for example, would require heating of the W source to temperatures near its melting point by resistance heating or electron beam bombardment. The CVD apparatus is much simpler in that it operates at much lower temperatures.

Within the CVD field there has been a move towards the use of organometallic sources to achieve lower deposition temperatures. For example, Cr can be deposited from $CrCl_2$ at T = 1200-1325°C,[48] but from dicumene chromium, $(C_9H_{12})_2Cr$, at T = 320-545°C,[9] and V can be deposited from VCl_4 at T = 1140-1300°C,[12] but from $(C_5H_5)V(CO)_4$ at T = 325-500°C.[49]

Radiation-Damage Free Deposition

Sputtering can also deposit high temperature films at fairly low temperatures. However, sputtering is an energetic process that results in electron, x-ray and ion bombardment of the gate oxide and substrate. Defects called traps, produced at the gate oxide/silicon interface,[50] can affect device performance by altering threshold voltage characteristics, and must therefore be annealed out. Further trap formation may occur upon reactive-ion etching of Al films, in which case even CVD Al processing would not insure a defect free film. However, if lift-off, rather than reactive-ion etching processing techniques were employed for patterning, CVD Al processing would result in defect free films. It should also be mentioned that electron-beam evaporation processing of films can lead to trap formation.

Deposition Selectivity

CVD film formation is fundamentally different from sputtering or evaporation in that nucleation and growth are catalyzed by the deposition surface. Therefore, by deactivating certain surfaces, films will tend not to nucleate and grow on them; consequently, one can selectively deposit films. Figure 5 is an example of W selectively deposited on Si, to the exclusion of SiO_2. This is very important because patterning steps (photolithography and etching) are not necessary for these films. Selectivity can occur on the basis of choosing reactants such that reactions with only certain substrate

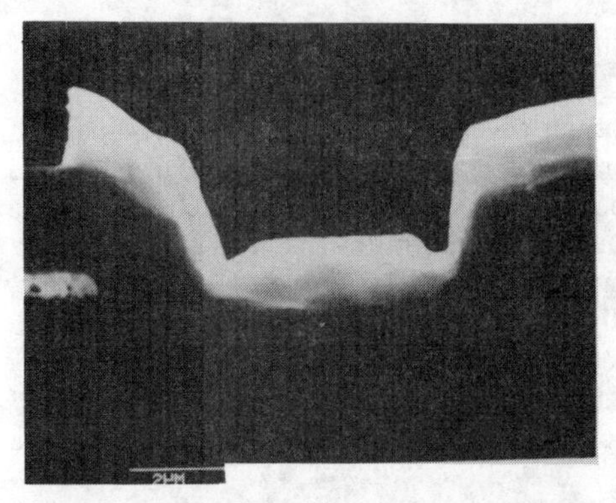

a

c

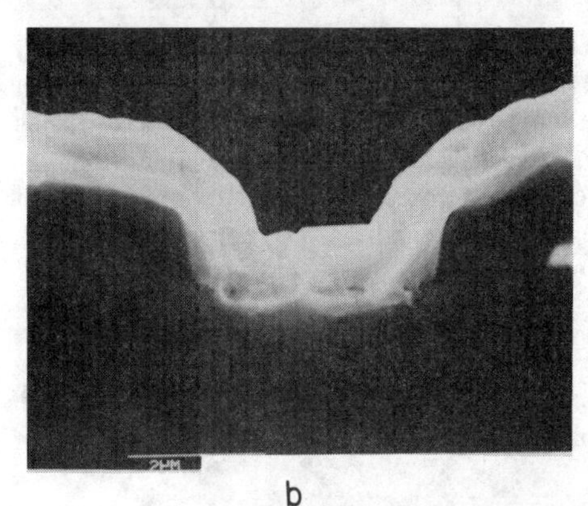

b

4) a) Evaporated Al film deposited on the contact window of a typical device, illustrating poor conformal coverage,
b) CVD Al film deposited on a similar contact window, illustrating nearly perfect conformal coverage, and
c) Thick CVD W film deposited on a small (~1μm) contact window, illustrating the filling, or "plugging", of the window.

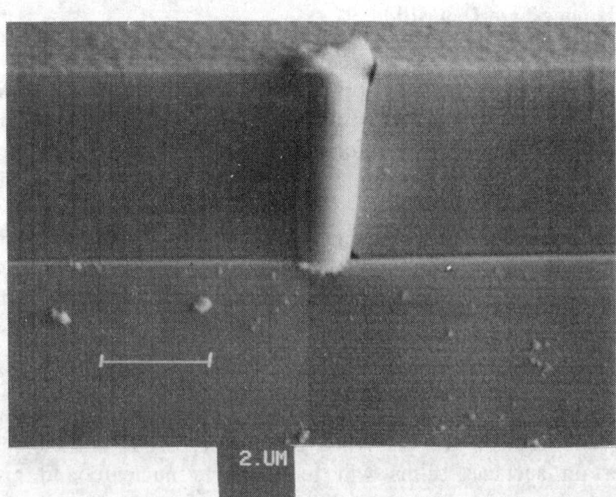

5) Thick selective plug of CVD W deposited in a window in SiO_2 to Si. Aspect ratio of window is 3:1. (from R. Wilson, GE Laboratories)

components will occur, as has been done with Ag,[14] Mo[10] or W[13,51] for deposition on Si to the exclusion of SiO_2, or by lowering the temperature and pressure so that film nucleation is easier on Si than on SiO_2, as has been done with W.[21-23]

The Current Status of CVD W and CVD Al as Integrated Circuit Metallizations

Of the various CVD metal processing schemes listed in Table I, CVD W, and to a lesser extent CVD Al, are closest to commercial acceptance. These two processes will now be discussed in detail.

CVD Al

As is illustrated in Fig. 4b, LPCVD Al, deposited by the pyrolysis of i-$(C_4H_9)_3Al$,[27-29] (tri-isobutylaluminum or TIBAL), exhibits excellent conformal coverage. This is the major reason for its intended application as an interconnect metallization for submicron VLSI circuits. Safer (nonpyrophoric) source materials might accelerate the introduction of CVD Al processing into commercial practice. These have yet to be identified.

Selective CVD Al has recently been demonstrated,[52] thereby opening CVD Al technology to many of the same applications for which selective CVD W is now being considered. In addition, whereas Si can react extensively with WF_6 during the selective CVD W process,[53,54] causing damage to devices, TIBAL does not interact with Si, leaving it undamaged.

Analysis of the CVD Al films[28] shows that they are pure Al, with only trace levels of oxygen, carbon and silicon. Consistent with this finding is the fact that the films have near bulk resistivities, varying between 2.8 and 3.5 $\mu\Omega$-cm.

An important characteristic of the CVD Al film is its rough surface. This roughness manifests itself by rendering the films milky-white in appearance, with low reflectivity,[28] which can give rise to photolithographic processing problems. Thick CVD W films, e.g. in Fig. 4c, also exhibit rough surfaces, and, thus far, patterning has been successful.

The electromigration resistance of the CVD Al films was studied in detail,[29] and was found to be typical of pure Al films. The addition of Cu, for example, would most probably improve the electromigration resistance, but this has yet to be done, for lack of volatile Cu sources.

CVD W

CVD W technology has undergone an interesting evolution, outlined in Table III, in which metallization applications have shifted as the capability of CVD W processing has expanded, and as more has been understood about the properties of the CVD W films. Due to recent intense interest in the commercialization of CVD W technology, its use in VLSI applications is imminent, although it is not clear now which application will be first. Table IV illustrates the relationships between the properties of CVD W films and their intended application. Not every application depends upon every property of the W film. Properties such as resistivity and stress, which are important to every application of CVD W, are well documented and are found to be well within the acceptable range.[23,58] However, the stress of CVD W films on substrates other than Si (i.e. SiO_2 (with various "glue" layers), metals, etc.) needs to be investigated.

TABLE III

Evolution of CVD W Applications

Date	Application	Major Benefit	Major Problem	Reference
1965	Schottky Barrier	selectivity		13
early 1970's – present	Interconnects	excellent conformal coverage, high electromigration resistance	poor adhesion to oxide	18,20
	Gate Electrodes	low resistivity, refractory	lack of oxidation resistance	7,19,20,55
late 1970's – present	Diffusion Barrier and Contact Metallization	selectivity	high contact resistance to P^+Si without silicide, high leakage current for junction depths $\lesssim 2500Å$.	22-25,41,56,57
early 1980's – present	Interconnects	excellent conformal coverage, high electromigration resistance	thick W films ($t \gtrsim 0.75$ μm), are very rough, with coarse, columnar grains. Adhesive layers needed on SiO_2.	33-37
1984 – present	Multilevel Metallization Applications (via fills, second level interconnect, planarization layer, etc.)	planarization (due to either highly selective film growth, or to etchback of nonselective layers)	coarse grain size	38-46

Further, the electromigration resistance of W is more than adequate for VLSI device requirements, due to its high melting point.

That CVD W is an effective barrier to Al/Si interaction is illustrated in Fig. 6. The RBS spectra show that whereas after a 450°C/30 min anneal the Si/W/Al layered structure is essentially the same as it was in the as-deposited state, a 550°C/30 min anneal leads to the destruction of the layers due to interdiffusion.[25] Although 450°C is a reasonable temperature limit for post-metallization integrated circuit processing, some processing schemes, notably those which involve multilevel metallization and intermediate dielectric layers, may necessitate process temperatures as high as 550°C. Diffusion barriers, other than W, may succeed in this temperature regime. CVD W films would probably also work in this higher temperature regime, if their grain size were not as small as it is, ~1000-3000Å.[23,58] Larger grain size films or "stuffed" grain-boundary films would undoubtedly be better barriers.

Electrical measurements, such as contact resistance and leakage current, are very sensitive indicators of the interactions between Si and the CVD W layer. Data from carefully characterized and controlled experiments have only recently become available. It is important to realize that contact resistance and leakage current characteristics will be important for applications other than the originally intended diffusion barrier application. Regardless of the processing, whether one deposits a thick selective window plug, a nonselective blanket layer or a thin selective diffusion barrier film, the electrical properties will be largely determined by the nature of the W/Si interface. Figure 7 illustrates the contact resistance (R_c) of CVD W to N^+ and P^+ doped Si source and drain areas, as a function of doping concentration.[25] These measurements were made on wet-etched, 2 μm diam. windows, into which W and then Al were deposited. The W was deposited with excess SiF_4 deliberately added to the gas mixture, to prevent junction erosion. In addition to the expected $1/\sqrt{N_d}$ dependence of R_c, it can be seen the the P^+ R_c values tend to be less than the N^+ values. P^+ R_c values should be lower than N^+ values, due to the lower barrier height of P^+ Si to W, although this has not been observed in the CVD W literature before, possibly due to the presence of extraneous layers at the W/Si interface. The use of SiF_4 has eliminated that problem, and in addition has result in R_c values which are stable to anneals up to 450°C. It can be further seen from Fig. 7 that reductions in R_c can be obtained by using silicide interlayers between W and Si. The trend in recent years has been to use silicide interlayers,[22] which may also serve as glue layers for the blanket deposition process.[33] The R_c values shown in Fig. 7, which are in the range of 3-30Ω

TABLE IV

Relationships between Properties and Intended Applications of CVD W Films

Application	Resistivity	Electromigration Resistance	Contact Resistance to Si	Leakage	Barrier Properties	Stress	Surface Roughness
Diffusion Barrier and Contact Metallization	x	–	x	x	x	x	–
Interconnect	x	x	x	x	x	x	x
Via fill, second level metal	x	x	–*	–	–	x	x
Gate Electrode	x	–	–	x	–	x	–

x This property is of primary importance for this application.

– This property is *not* of primary importance for this application.

* However, contact resistance to the first level metal is important.

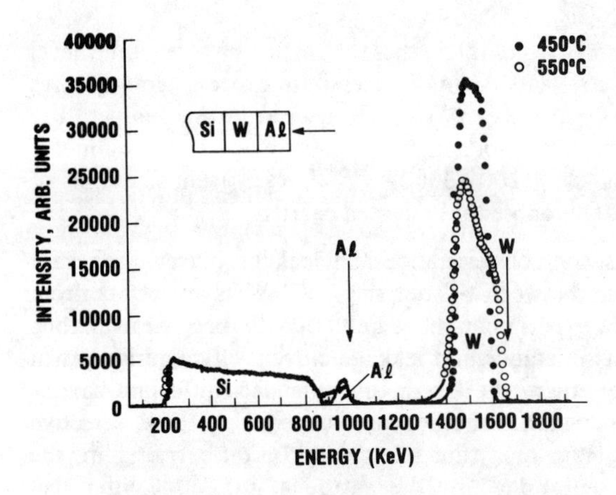

6) RBS spectra of an Al-W-Si structure (1500Å Al/1100Å W/Si substrate) after 30 min. anneals at 450°C and 550°C. Arrows mark the surface positions of the corresponding elements. The as-deposited spectrum, not shown, was identical to the 450°C spectrum.

8) Mean leakage current density vs. N^+ (As)/P junction depth in Si for unmetallized diodes (no W) and after 1-step or 2-step CVD W.

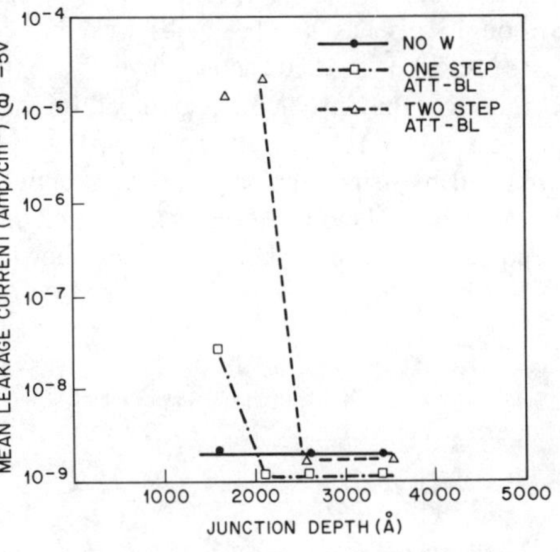

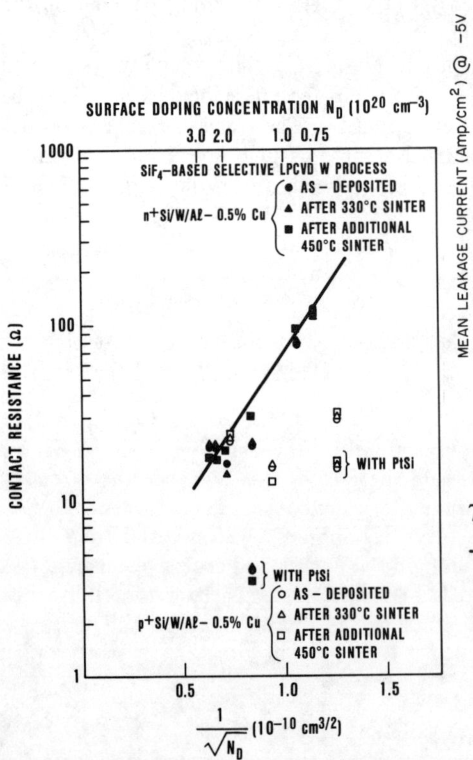

7) Contact resistance to N^+ and P^+ diffusions for 2.0 μm vias as a function of surface doping concentration. Data are shown for the SiF_4-based selective LPCVD W process after Al metallization, after a 330°C/30 min/H_2 anneal, and after an additional 450°C/30 min/H_2 anneal.

for N^+ contacts ($N_d(As) = 1.4 \times 10^{20} cm^{-3}$) and 15-30Ω for P^+ contacts ($N_d(B) = 0.6 \times 10^{20} cm^{-3}$), are compatible with high performance CMOS device requirements.

Figure 8 illustrates the effect of CVD W processing on leakage current, as a function of junction depth, for N^+(As) junctions. The same reference[57] also illustrates the case of CVD W on $CoSi_2/N^+$(As). It can be seen that a critical junction depth exists, shallower than which CVD W will give rise to unacceptably high leakage currents. The critical junction depth varies somewhat, depending upon the exact CVD W deposition conditions (i.e., one or two step deposition), but is less than or equal to about 2500Å. This may severely limit the use of CVD W as a metallization (interconnect, window plug or diffusion barrier) for next generation submicron junction (<1500Å) devices, unless another interlayer is introduced between the W and the Si. It has been determined that $CoSi_2$, and probably other silicides as well, are not barriers.[57] Therefore, other interlayers, which act as leakage barriers to the diffusion barrier (W), will have to be used with these devices. It should be emphasized that most CVD W leakage measurements up to now have been generated on devices with junction depths greater than 2500Å, so leakage current problems have been underestimated.

Thick blanket CVD W films (~1.0 μm) tend to have rough surfaces, a problem that they share with other CVD metal films.[28,34,58] The roughness is due to competitive columnar grain growth resulting from slight variations in local growth rates, which give rise to surface steps whose heights are proportional to the film thickness. As has been discussed in an early paper,[59] alloying, which can result in grain renucleation during growth, may potentially break up the columnar structure and result in smooth, thick films. Re, available from ReF_6, might be a good candidate for alloying with W.

$Si\text{-}H_2\text{-}WF_6$ thermochemistry has both positive and negative implications for CVD W processing. Although the interaction of Si and WF_6 is partially responsible for the deposition selectivity that is observed, it can also lead to problems in film deposition. Table V is a compilation of reaction free energies for the $Si\text{-}H_2\text{-}WF_6$ system.[60] One can readily see that the Si reduction of WF_6 has a much greater driving force than the H_2 reduction of WF_6, and so will always occur first. Therefore, the characteristics of the W/Si interface, and of many of the films most sensitive electrical properties, are determined by reaction (1). Furthermore, although much has been said about the thin W films (~100Å) that form as a result of the self-limiting nature of reaction (1), there is now ample evidence to show that this reaction is self-limiting, but not always at small W coverages. In fact, Si reduced W films as thick as 1 μm have been reported.[61,62] Of course, under such circumstances, large amounts of Si are consumed, and devices are destroyed.[25] Among the factors that influence the extent of the Si reduction are Si native oxide thickness,[63] damage and dopant effects due to wafer implantation,[53] damage due to dry etching of wafers, and deposition temperature.[54] Although attempts have been made to understand excessive consumption of Si,[54] reaction (1) is not well understood. However, such deleterious efforts can usually be avoided by conservative processing, especially the use of excess SiF_4.[25]

Another manifestation of excessive erosion of Si by WF_6 is a phenomenon called encroachment.[23,43] In fact, encroachment may be the consumption of Si by WF_6 that was alluded to before, with the interfacial stress between Si and SiO_2 (or any other layer), providing an increased driving force for reaction (1). Figure 9 illustrates the

TABLE V

Reaction Free Energies (ΔG) in the $Si-H_2-WF_6$ System

Reaction	T = 600°K (327°C)	T = 800°K (527°C)
(1) $2WF_6 + 3Si \rightarrow 2W + 3SiF_4\uparrow$	−147	−153
(2) $WF_6 + 3H_2 \rightarrow W + 6HF\uparrow$	−15	−33
(3) $4HF + Si \rightarrow SiF_4 + 2H_2\uparrow$	−88	−80

reactions (1), (2) : Kcal/mole of W

reaction (3) : Kcal/mole of Si

"unzipping" of the Si/SiO_2 interface due to encroachment.

Notice from Table V that the H_2 reduction of WF_6, reaction (2), which is the most desirable W deposition reaction in that no Si is consumed, has the smallest driving force. The by-product of this reaction, HF, might be implicated in the formation of another defect, called a "wormhole", or tunnel defect, according to reaction (3). Figure 10 illustrates these wormholes, which always have a W particle at their ends. This W particle probably catalyzed reaction (3), which then caused it to be drilled into the Si.[25,64] Similar phenomena have been reported previously for the catalytic gasification of other materials.[65,66] The wormholes tend to form around the periphery of the junction areas (Fig. 10), because the annular space between the oxide window wall and the W layer provides access for the HF to reach the underlying Si. Wormholes are reliability threats and should be eliminated.

Conclusions

The advantages of CVD processing, which include conformal coverage, low temperature and radiation-free deposition, the possibility of deposition selectivity, and high purity film formation are certain to insure the use of CVD metallization in future VLSI processing schemes. At this point, CVD W, and to a lesser extent CVD Al

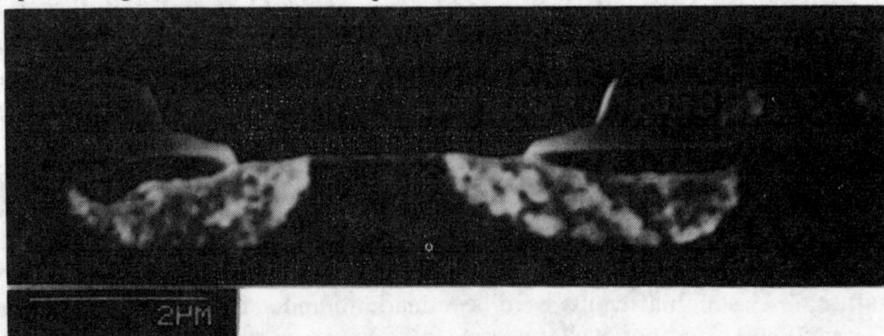

9) Encroachment phenomenon observed in a source/drain window after CVD W deposition. The Si at the Si/SiO_2 interface has been consumed by WF_6. Notice that a thin, self-limiting film of W exists at the center of the window. Deposition temperature = 290°C. To exaggerate the encroachment effect, no H_2 was used in this experiment.

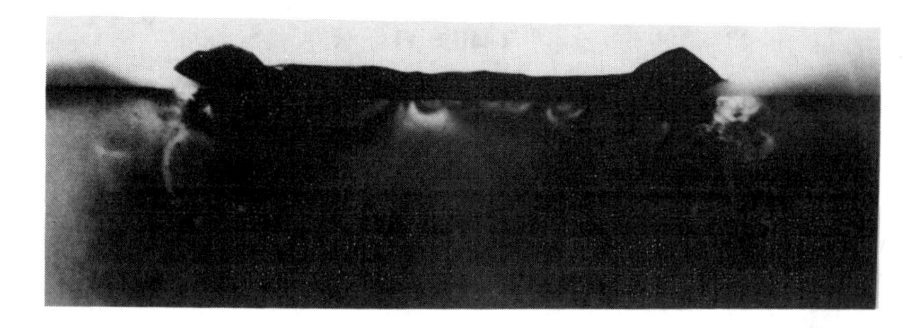

10) Transmission electron photomicrograph illustrating "wormhole" defects originating at the W/Si interface and penetrating into the Si bulk.

technologies are closest to commercial realization. The major applications of CVD W might be as follows:

a) As a multilevel metal via fill: deposited either selectively or nonselectively, the CVD W would not be in contact with Si, so potential problems such as Si consumption and high leakage current would not arise. However, the CVD W would have to be deposited on first level metal, probably Al, requiring the use of a silicide, e.g., interlayer, due to the native Al_2O_3. Furthermore, the problem of surface roughness would have to be solved, perhaps by depositing W alloys rather than W.

b) As an interconnect or diffusion barrier or plug, but only if the WF_6/Si interactions can be avoided, perhaps through the use of a protective, conductive layer such as CVD TiN on top of the Si.

The potential applications of CVD W may be limited by the fact that WF_6 attacks Si too aggressively (i.e., ΔG is too large), thereby creating defects in the Si which deleteriously effect device characteristics. Therefore, a possible future direction for CVD W research might be the development of new W sources which do not interact as strongly with Si. Of course, this may mean that selectivity is compromised. Table VI summarizes some reaction free energy data for alternative sources of W, and for other refractory metals as well. It can be seen that switching to other W halide sources, or even going to other metals, may not offer any obvious advantage. However, classes of W compounds such as the carbonyls, and perhaps W organometallics, may be developed which offer more promising chemistries.

Finally, the use of cold wall reactors to deposit highly selective CVD films greatly increases the value of CVD W processing by opening up the multilevel metallization field. It is also possible that in cold wall reactors, the detrimental effects of WF_6/Si interactions may be alleviated, due to their lower thermal budget deposition cycle.

Acknowledgements

Many thanks to R. A. Levy, V. V. S. Rana and R. Wilson (G.E.) for their contributions to this paper.

TABLE VI

Reaction Free Energies (ΔG) for Alternate Chemistries

Alternate CVD W Chemistries

			T = 600°K (327°C)	T = 800°K (527°C)
WCl_6:				
$WCl_6 + 3H_2$	→	$W + 6HCl$	−99	−108
$2WCl_6 + 3Si$	→	$2W + 3SiCl_4$	−144	−144
$4HCl + Si$	→	$SiCl_4 + 2H_2$	−30*	−24*

− no obvious advantage over WF_6

− WCl_6 is an inconvenient (solid) source

$W(CO)_6$:

$W(CO)_6$	→	$W + 6CO$	pyrolyzes at T > 250°C

\+ lower dep. temp. possible

\+ chemically benign environment (no Si interaction)

− no selectivity

− film must be annealed at T ~ 700°C to yield low resistivity

Alternate Metals

MoF_6, $MoCl_6$:

$MoF_6 + 3H_2$	→	$Mo + 6HF$	−27	−42
$2MoF_6 + 3Si$	→	$2Mo + 3SiF_4$	−159	−162
$MoCl_6 + 3H_2$	→	$Mo + 6HCl$	−99	−114
$2MoCl_6 + 3Si$	→	$2Mo + 3SiCl_4$	−144	−150

− no obvious advantage

− Mo deposition by MoF_6 does not self-limit (continuous Si consumption[67])

ReF_6:

$ReF_6 + 3H_2$	→	$Re + 6HF$	−17	−36
$2ReF_6 + 3Si$	→	$2Re + 3SiF_4$	−156	−156

− no obvious advantage

All ΔG values in units of Kcal/mole of metal
* Kcal/mole of Si

REFERENCES

1. J. L. Vossen and W. Kern, *Thin Film Processes*, (Academic Press, New York, 1978).

2. A. N. Saxena and D. Pramanik, Solid State Technology , *27*, 93 (1984).

3. S. P. Murarka, *Silicides for VLSI Applications*, (Academic Press, New York, 1983).

4. K. Yasuda and J. Murota, Jap. J. Appl. Phys., *22*, L615 (1983).

5. D. M. Brown, W. R. Cady, J. W. Sprague and P. J. Salvagni, IEEE Trans. Elec. Dev., *ED-18*, 931 (1971).

6. M. L. Green, M. E. Gross, L. E. Papa, K. J. Schnoes and D. Brasen, J. Electrochem. Soc., *132*, 2677 (1985).

7. V. Lubowiecki, J. L. Ledys, C. Plossu and B. Balland, in *Tungsten and Other Refractory Metals for VLSI Applications II*, (MRS, Pittsburgh, 1987) p. 169.

8. H. B. Michaelson, IBM J. Res. Dev., *22*, 72 (1978).

9. N. G. Anantha, V. Y. Doo and D. K. Seto, J. Electrochem. Soc., *118*, 163 (1971).

10. G. G. Pinneo, in *Proceedings of the 3rd International Conference on CVD*, (ANS, Hinsdale, Ill., 1972) p. 462.

11. S. S. Simeneov, E. J. Kafedjiiska and A. L. Guerassimov, Thin Solid Films, *115*, 291 (1984).

12. K. J. Miller, M. J. Grieco and S. M. Sze, J. Electrochem. Soc., *113*, 902 (1966).

13. C. R. Crowell, J. C. Sarace and S. M. Sze, Trans. Met. Soc. AIME, *233*, 478 (1965).

14. R. J. H. Voorhoeve and J. W. Merewether, J. Electrochem. Soc., *119*, 364 (1972).

15. J. J. Casey, R. R. Verderber and R. R. Garnache, J. Electrochem. Soc., *114*, 201 (1967).

16. T. Sugano, H. Chow, M. Yoshida and T. Nishi, Jap. J. Appl. Phys., *7*, 1028 (1968).

17. M. J. Rand, J. Electrochem. Soc., *120*, 686 (1973).

18. J. M. Shaw and J. A. Amick, RCA Review, 306 (1970).

19. N. E. Miller and I. Beinglass, Solid State Tech., *25*, 85 (1982).

20. C. M. Melliar-Smith, A. C. Adams, R. H. Kaiser and R. A. Kushner, J. Electrochem. Soc., *121*, 298 (1974).

21. N. E. Miller and J. Beinglass, Solid State Tech., *23*, 79 (1980).

22. S. Swirhun, K. C. Saraswat and R. W. Swanson, IEEE Electron Dev. Lett. *EDL-5*, 209 (1984).

23. M. L. Green and R. A. Levy, J. Electrochem. Soc., *132*, 1243 (1985).

24. T. Hara, S. Enomoto, N. Ohtsuka and S. Shima, Jap. J. Appl. Phys., *24*, 828 (1985).

25. R. A. Levy, M. L. Green, P. K. Gallagher and Y. S. Ali, J. Electrochem. Soc., *133*, 1905 (1986).

26. F. M. d'Heurle and P. S. Ho, in *Thin Films-Interdiffusion and Reactions*, (J. Wiley and Sons, New York 1978) p. 243.

27. M. J. Cooke, R. A. Heinecke, R. C. Stern and J. W. C. Maes, Solid State Tech., *25*, 62 (1982).

28. M. L. Green, R. A. Levy, R. G. Nuzzo and E. Coleman, Thin Solid Films, *114*, 367 (1984).

29. R. A. Levy, M. L. Green and P. K. Gallagher, J. Electrochem. Soc., *131*, 2175 (1984).

30. R. A. Levy, P. K. Gallagher, R. Contolini and F. Schrey, J. Electrochem. Soc., *132*, 457 (1985).

31. L. H. Kaplan and F. M. d'Heurle, J. Electrochem. Soc., *117*, 693 (1970).

32. W. E. Engeler and D. M. Brown, IEEE Trans. Elec. Dev., *ED-19*, 54 (1972).

33. W. A. Metz and E. A. Beam, in *Tungsten and Other Refractory Metals for VLSI Applications*, (MRS, Pittsburgh, 1986) p. 249.

34. C. Fuhs, E. J. McInerny, L. Watson and N. Zetterquist, ibid, p. 257.

35. D. W. Woodruff, R. H. Wilson and R. A. Sanchez-Martinez, ibid, p. 173.

36. K. C. Ray Chiu and N. E. Zetterquist, in *Tungsten and Other Refractory Metals for VLSI Applications II*, (MRS, Pittsburgh, 1987) p. 177.

37. V. V. S. Rana, J. A. Taylor, L. H. Holschwandner and N. S. Tsai, ibid, p. 187.

38. G. C. Smith, in *Tungsten and Other Refractory Metals for VLSI Applications*, (MRS, Pittsburgh, 1986) p. 323.

39. R. J. Saia and B. Gorowitz, in *Tungsten and Other Refractory Metals for VLSI Applications II*, (MRS, Pittsburgh, 1987) p. 349.

40. C. H. Chen, L. C. Watson and D. W. Schlosser, ibid, p. 357.

41. T. Moriya, S. Shima, Y. Hazuki, M. Chiba and M. Kashiwagi, *Proc. 1983 IEDM Meeting*, Paper 25.3, 550 (1983).

42. R. S. Blewer and V. A. Wells, *Proc. 1st IEEE VLSI Multilevel Interconnection Conf.*, 153 (1984).

43. H. Itoh, R. Nakata and T. Moriya, *Proc. 1985 IEDM Meeting*, Paper 25.6, 606 (1985).

44. R. H. Wilson, R. W. Stoll and M. A. Calacone, in *Tungsten and Other Refractory Metals for VLSI Applications*, (MRS, Pittsburgh, 1986) p. 35.

45. D. C. Thomas and S. S. Wong, *Proc. 1986 IEDM Meeting*, Paper 12.8, 811 (1986).

46. N. Tsuzuki, M. Ichikawa, K. Kurita, K. Watanabe and K. Inayoshi, in *Tungsten*

and Other Refractory Metals for VLSI Applications, (MRS, Pittsburgh, 1987) p. 257.

47. K. F. Jensen, Chem. Eng. Sci., *42*, 923 (1987).

48. H. M. J. Mazille, Thin Solid Films, *65*, 67 (1980).

49. B. A. Macklin, Technical Report AFML-TR-68-9 (Wright-Patterson AFB) February 1968.

50. J. M. Aitken, J. Non-Crystall. Solids, *40*, 31 (1980).

51. K. Y. Tsao and H. H. Busta, J. Electrochem. Soc., *131*, 2702 (1984).

52. G. S. Higashi and C. G. Fleming, Appl. Phys. Lett., *48*, 1051 (1986).

53. M. L. Green, Y. S. Ali, B. A. Davidson, L. C. Feldman and S. Nakahara, in *Mat. Res. Soc. Symp. Proc.*, Vol. *54* (MRS, Pittsburgh, 1986) p. 723.

54. M. L. Green, Y. S. Ali, T. Boone, B. A. Davidson, L. C. Feldman and S. Nakahara, J. Electrochem. Soc., *134*, 2285 (1987).

55. W. A. Metz, J. E. Mahan, V. Malhotra and T. L. Martin, Appl. Phys. Lett., *44*, 1139 (1984).

56. P. A. Gargini, Ind. Res. Dev., *25*, 141 (1983).

57. G. Georgiou, J. M. Brown, M. L. Green, R. Liu, D. S. Williams and R. S. Blewer, in *Tungsten and Other Refractory Metals for VLSI Applications II*, (MRS, Pittsburgh, 1987) p. 227.

58. T. I. Kamins, D. R. Bradbury, T. R. Cass, S. S. Laderman and G. A. Reid, J. Electrochem. Soc., *133*, 2555 (1986).

59. W. R. Holman and F. J. Huegel, in *Proc. Chem. Vap. Dep. Ref. Metals, Alloys and Comp.*, (AIME, New York, 1967) p. 127, 427.

60. T. P. Reed, *Free Energy of Formation of Binary Compounds*, (MIT Press, Cambridge, 1971).

61. C. Morosanu and V. Soltuz, Thin Solid Films, *52*, 181 (1978).

62. R. J. Mianowski, K. Y. Tsao and H. A. Waggener, *Tungsten and Other Refractory Metals for VLSI Applications*, (MRS, Pittsburgh, 1986) p. 145.

63. H. H. Busta and C. H. Tang, J. Electrochem. Soc., *133*, 1195 (1986).

64. W. T. Stacy, E. K. Broadbent and M. H. Norcott, J. Electrochem. Soc., *132*, 444 (1985).

65. R. T. K. Baker, J. A. France, L. Rouse and R. J. Waite, J. Catal., *41*, 22 (1976).

66. D. J. Coates, J. W. Evans, A. L. Cabrera, G. A. Somorjai and H. Heinemann, J. Catal., *80*, 215 (1983).

67. N. Lifshitz, J. M. Brown and D. S. Williams, in *Tungsten and Other Refractory Metals for VLSI Applications II*, (MRS, Pittsburgh, 1987), p. 215.

NON-SELECTIVE TUNGSTEN CHEMICAL-VAPOR DEPOSITION USING TUNGSTEN HEXACARBONYL

J. R. Creighton
Sandia National Laboratories, Albuquerque, NM 87185.

ABSTRACT

We have used tungsten hexacarbonyl to deposit thin (<1000 Å) non-selective tungsten films on silicon and silicon dioxide at 550°C. Thicker (≥ 1 micron) tungsten films were then deposited using conventional H_2 reduction of WF_6 at 470°C using the non-selective film as an adhesion layer. Films grown in this manner have excellent adhesion to SiO_2, essentially 100% step coverage, and good resistivity (7.5-14 μΩ-cm). Samples could be transferred under vacuum from the deposition chamber to a UHV chamber equipped with Auger spectroscopy, thus allowing surface and interface properties of the tungsten films to be studied at the initial stages of growth. No evidence was found for a stoichiometric tungsten oxide or tungsten silicide at the W/SiO_2 interface.

INTRODUCTION

Non-selective (blanket) tungsten deposition has a number of potential uses for via filling and interconnect formation in microelectronic circuits [1,2]. Unfortunately, tungsten deposition on insulators by the typical hydrogen reduction of WF_6 usually suffers from nucleation difficulties and poor adhesion. One solution to these problems is to first deposit an adhesion ("glue") layer on the oxide surface, and then deposit tungsten by conventional CVD. The adhesion layers that have been tried with varying degrees of success have included CVD tungsten silicide [1,3-6], CVD polysilicon and sputtered Cr, Ti, W, and Mo [1,2,7], as well as sputtered Al and TiN [8]. One method of simplifying the etch back process needed for via fill applications would be to use an adhesion layer with an etch rate similar to the CVD tungsten etch rate. Obviously this would best be accomplished if the glue layer itself were tungsten. Interestingly, some authors have rationalized, using thermodynamic arguments, that the tungsten-silicon dioxide interface is inherently weak [8], while others have reported good adhesion using sputtered tungsten as an adhesion layer [2,7]. We have evidence [9] that suggests the poor adhesion of tungsten deposited by the hydrogen reduction of WF_6 is due to a replacement of oxygen with fluorine at the SiO_2 surface which prevents a good linkage of the metal with the substrate. In other words, poor adhesion may not be an intrinsic property of the W/SiO_2 interface and it is, therefore, worth considering other methods of tungsten deposition. We chose to investigate tungsten hexacarbonyl, $W(CO)_6$, as a potential source for depositing tungsten adhesion layers because of the inherent advantages of a CVD process and the lack of halide (F, Cl) chemistry which may be detrimental for adhesion.

The electrical and structural properties of tungsten deposited using tungsten hexacarbonyl have been reported by Kaplan and d'Heurle [10] and their results served as a useful guideline for our study. To briefly summarize their results, tungsten with a resistivity as low as twice the bulk value (5.5 $\mu\Omega$-cm) could be deposited at temperatures above 450°C. The film resistivity increased dramatically for deposition temperatures below 450°C, apparently due to an increase in the carbon incorporation.

EXPERIMENTAL

These experiments were performed in a dual chamber vacuum system composed of a reaction chamber and a UHV analysis chamber (see Fig. 1). The analysis chamber was equipped with a cylindrical mirror analyzer for Auger electron spectroscopy (AES) and a 4 keV ion gun for sputter cleaning. This chamber was pumped with a turbomolecular pump to a base pressure of less than 2×10^{-10} torr. The reaction chamber was also pumped by a turbomolecular pump to a base pressure of $\simeq 5 \times 10^{-9}$ torr. However, due to the number of samples studied and the slow outgassing of adsorbed reactants and products, the working base pressure was higher; normally ≈2 × 10^{-7} torr. Because the room temperature vapor pressure of $W(CO)_6$ is only ≈100 mTorr, it was difficult to obtain high flow rates and pressures in the reaction chamber without the aid of carrier gases and/or a heated gas-handling system. Since neither option was immediately available we chose to use low flow rates by throttling the turbopump in the reaction chamber to give a nitrogen pumping speed of ≈0.7 l/sec. With this arrangement we could achieve a 1 mTorr hexacarbonyl pressure at an approximate flow rate of 0.014 sccm. Pressures in this range were measured with a capacitance manometer.

Two types of samples were used in this study. For adhesion tests, resistivity determinations, and studies of the interface chemistry, unpatterned oxidized silicon substrates were used. These samples were 2-cm x 1-cm x 0.5-mm rectangular slices of 0.4 Ω-cm Si(100) with 400 Å of thermally grown silicon dioxide. The second samples tested were patterned silicon wafers covered with 0.9 micron high oxide lines of various widths and spacings. Both types of samples were etched for 15 seconds in a 20:1 NH_4F/HF solution and rinsed with distilled water shortly before installation into the vacuum system. The samples were mounted on a long-throw manipulator by means of tantalum clips which allowed direct resistive heating to above 600°C. For temperature measurements, a chromel-alumel thermocouple was attached to the top edge of the samples with a small drop of ceramic cement (Aremco #516). By opening a straight-through valve, a sample could be moved from the reaction chamber into the UHV chamber for surface analysis within a few minutes after deposition was terminated.

RESULTS AND DISCUSSION

The first set of experiments was designed to examine the properties of thin tungsten films deposited on blanket oxide surfaces using $W(CO)_6$. Tungsten films with thicknesses ranging from 600-900 Å were deposited at a substrate temperature of 550±5°C and a tungsten hexacarbonyl pressure of ≈1 mTorr. Film thicknesses were measured by profilometry over a region of the sample masked by the supporting tantalum clips and also by profilometry over tungsten mesas formed by masking and etching back with hydrogen peroxide. Thicknesses measured by both methods agreed within experimental error of ±100Å. The growth rates achieved ranged from 12 to 18 Å/min with the probable source of fluctuation being the difficulty in maintaining a constant $W(CO)_6$ pressure. Film resistivities, measured with a four point probe and using the measured thicknesses, ranged from 23 to 33 μohm-cm. All films adhered to the oxide, passing the tape pull test.

In order to determine if the adhesion of the tungsten films is due to an interfacial compound formation, such as WSi_2 or WO_3, we terminated several deposition experiments at the very early stages (i.e., thickness <10 Å) of growth and examined the surface with Auger spectroscopy. A representative data set is outlined in Fig. 2. First, an Auger spectrum of the initial SiO_2 surface (curve (a), Fig. 2) was measured as a benchmark. The sample was then transferred into the reaction chamber and a very short deposition (≃30 sec) was performed using the conditions described in the previous paragraph. The sample was cooled, the reaction chamber pumped out, and the sample returned to the analysis chamber. After the deposition, the Auger spectrum (curve (b), Fig. 2) shows the presence of tungsten but the silicon peak has not been completely attenuated. Assuming a uniform tungsten film and using the inelastic mean free path for electrons at the silicon Auger electron energy [11], we can use the silicon peak attenuation to calculate an approximate tungsten film thickness of ≃6 Å. This value, though approximate, is in good agreement with the predicted thickness of 7.5±2.0 Å based on the growth rates of thicker films measured by profilometry. Also, the silicon Auger peak (at 76 eV) is indicative of SiO_2 and no features due to elemental silicon or silicon in tungsten silicide, which would show up clearly at higher energy (≃90 eV), were detected. In order to determine the chemical state of the tungsten, it is necessary to quantify the amount of adsorbates, such as oxygen, which are chemically bound to the tungsten. This is difficult because the oxygen AES spectrum from the underlying SiO_2 overlaps with the AES spectrum of oxygen associated with tungsten. The oxygen AES

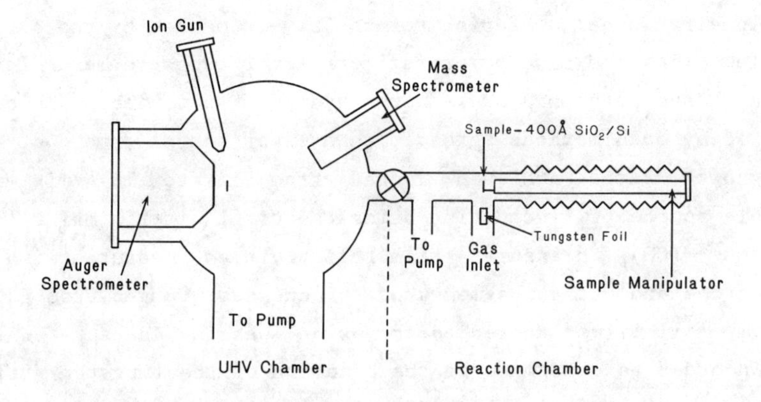

Fig. 1. Schematic of dual chamber vacuum system.

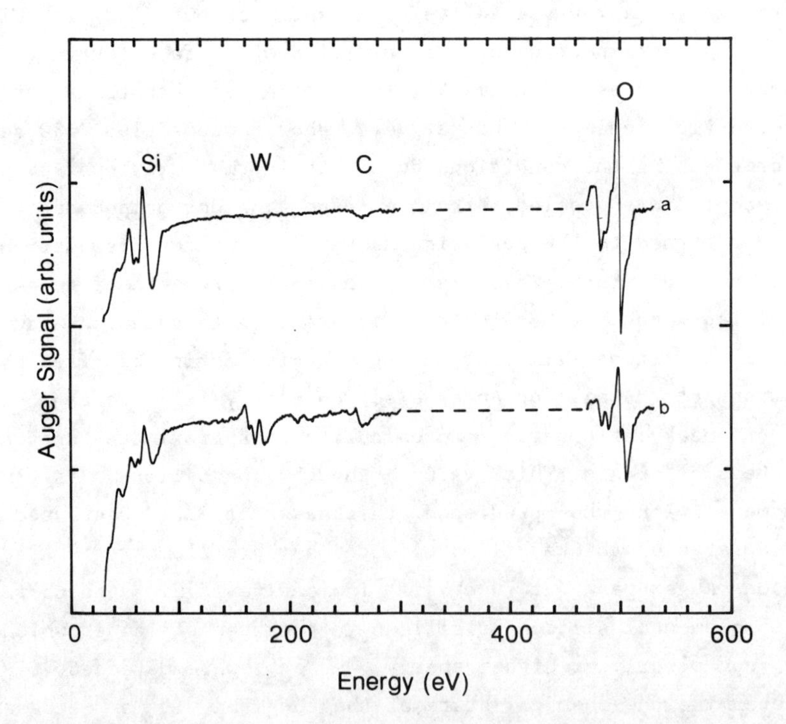

Fig. 2. Auger spectra of: (a) initial SiO_2 surface, (b) surface after deposition of a thin (≃6 Å) tungsten layer using tungsten hexacarbonyl.

lineshape in curve b, Fig. 2, is due to a combination of these two types of oxygen. Despite this difficulty, we can get an upper limit on the O/W AES peak height ratio of 2.9. This compares to a O/W AES peak height ratio of 7.1 measured in our laboratory for bulk WO_3. Using this latter value for calibration, we calculate an approximate upper limit to the stoichiometry of the WO_x observed in curve b, Fig. 2, and find $X \leq 1.2$. These Auger results clearly show that no elemental silicon, tungsten silicide, or stoichiometric tungsten oxide are present at the W/SiO_2 interface and therefore these compounds cannot be responsible for the good tungsten film adhesion.

Step coverage was examined by depositing thicker tungsten films on the patterned substrates. The TEM cross section of such a film, as displayed in Fig. 3, shows the excellent conformality of the tungsten over the oxide line. In addition, the tungsten-silicon interface is smooth and there is no evidence of silicon consumption or wormholes. This is to be expected since there is no fluorine or HF present which are believed responsible [12-14] for the deleterious effects observed when using WF_6. The dark line in the silicon substrate, most obvious under the oxide lines, is due to ion implant damage which occurred during fabrication of the patterned oxide substrates. There are some noticeable differences between the tungsten film on the oxide as compared to the film on the silicon. The tungsten grain size on the oxide is larger than on the silicon (≃1000 Å vs. ≃500 Å) and the apparent film thickness is also greater (≃1500 Å on oxide vs. ≃1200 Å on silicon). We do not, as yet, have an explanation for these differences. Selected area electron diffraction indicates that the films deposited on the oxide and on the silicon are composed of α-tungsten, and no other diffraction features due to tungsten oxide or silicide are detected.

Since the primary focus of this study was to determine the feasibility of using a thin tungsten layer deposited from the hexacarbonyl as an adhesion layer, a two-step deposition process was then used. First, a thin tungsten layer was deposited at 550°C using $W(CO)_6$. Then, a thicker (≃1μm) tungsten film was deposited by the hydrogen reduction of WF_6. Tungsten from the hexacarbonyl was deposited to thicknesses ranging from ≃50-500 Å using the procedure

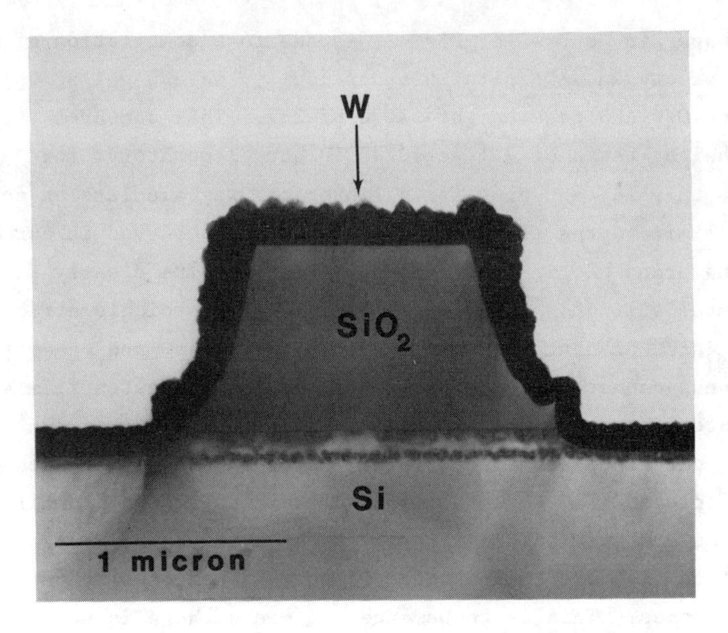

Fig. 3. Cross-section TEM of patterned oxide sample with tungsten thin film deposited using tungsten hexacarbonyl.

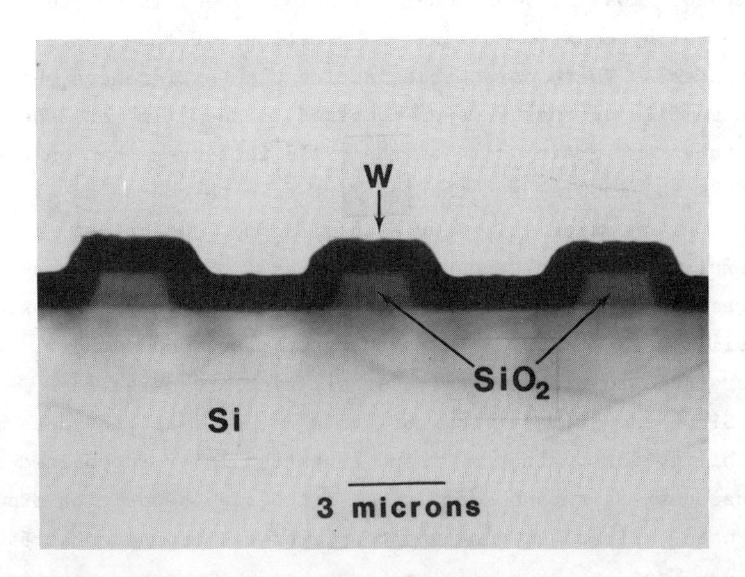

Fig. 4. Cross-section TEM of patterned oxide sample with tungsten film deposited from two-step method (described in text) using a thin (≃75 Å) tungsten adhesion layer deposited using tungsten hexacarbonyl.

described previously. Samples remained in the vacuum system for both steps of the deposition process so film thickness after the first deposition step was estimated from the average growth rate determined previously. The samples were then exposed to the following conditions for twenty minutes; WF_6 partial pressure = 5 mTorr, H_2 partial pressure = 100 mTorr, total flow rate ≈ 40 sccm, substrate temperature = 470±5°C. Total tungsten film thickness was then measured by profilometry and ranged from 1.1 to 1.8 microns. The temperature uncertainties may account for ±0.1 micron of the thickness variation as calculated using a 0.71 eV activation energy for the growth kinetics [15]. The additional source(s) of the thickness variation have not yet been determined. Resistivity ranged from 7.5 to 14 μΩ-cm. Differences in the resistivity did not correlate with the tungsten film thickness deposited in the first step, i.e. from the hexacarbonyl. All films passed the tape pull test and an attempt to be more quantitative was made by using a calibrated pull tester. In all cases, the silicon substrates broke before the tungsten films delaminated from the surface. The lower limit on the adhesion strength of the tungsten films to the SiO_2 surface was $\simeq 10^8$ dynes/cm^2.

Step coverage of tungsten films grown by the two-step process was examined using the patterned oxide substrates. A cross-section TEM micrograph of such a film is displayed in Fig. 4. This film was grown by depositing ≃75 Å of tungsten from the hexacarbonyl and then 9000 Å of tungsten using a 15 minute H_2+ WF_6 deposition. Although the step coverage is excellent, essentially 100%, close examination of the tungsten-silicon interface did reveal the presence of some wormholes. We believe that the initial 75 Å tungsten film may have been porous or discontinuous which allowed interactions to occur at the tungsten-silicon interface during the H_2+ WF_6 deposition step.

SUMMARY

Tungsten hexacarbonyl was used to deposit thin α-tungsten films that adhere well to SiO_2. Coverage was conformal on patterned oxide substrates. The resistivity of these films was low enough (≤ 35

μohm-cm) to be considered acceptable for some applications such as diffusion barriers. These films were used as underlying adhesion layers for thick (≃1.5 micron) tungsten films grown using the conventional hydrogen reduction of WF_6 deposition process. With the adhesion layers, the thick films adhered to SiO_2, passing the tape pull test. The resistivities of the thick films ranged from 7.5 to 14 $\mu\Omega$-cm. Two-step deposition of tungsten on patterned oxide substrates yielded films with essentially 100% step coverage. Some wormholes were found near the tungsten-silicon interface, perhaps due to porosity of the very thin (≃75 Å) adhesion layer used which allowed some exposure of the silicon surface during the H_2+ WF_6 deposition step. Determination of the minimum thickness of the tungsten adhesion layer (from tungsten hexacarbonyl) which prevents wormhole formation during the conventional tungsten CVD step will be the subject of future work.

ACKNOWLEDGMENTS

It is a pleasure to acknowledge T. J. Headley for his expedient and flawless TEM analysis.

This work was performed at Sandia National Laboratories supported by the U.S. Department of Energy under contract # DE-AC04-76DP00789 for the Office of Basic Energy Sciences.

REFERENCES:

1. N. Kobayashi, S. Iwata, N. Yamamoto, and N. Hara, in Tungsten andOther Refractory Metals for VLSI Applications II, edited by E. K. Broadbent (MRS, Pittsburgh, PA, 1987), p. 159.
2. V. Lubowiecki and J. Ledys, in Tungsten and Other Refractory Metals for VLSI Applications II, edited by E. K. Broadbent (MRS, Pittsburgh, PA, 1987), p. 169.
3. K. C. R. Chiu and N. E. Zetterquist, in Workshop on Tungsten and Other Refractory Metals for VLSI Applications II, edited by E. K. Broadbent (MRS, Pittsburgh, PA, 1987), p. 177.

4. W. A. Metz and E. A. Beam, in Tungsten and Other Refractory Metals for VLSI Applications, edited by R. S. Blewer (MRS, Pittsburgh, PA, 1986), p. 249.
5. C. Fuhs, E. McInerney, L. Watson, and N. Zetterquist, in Tungsten and Other Refractory Metals for VLSI Applications, edited by R. S. Blewer (MRS, Pittsburgh, PA, 1986), p. 257.
6. S. Suresh and S. D. Mehta, in Tungsten and Other Refractory Metals for VLSI Applications, edited by R. S. Blewer (MRS, Pittsburgh, PA, 1986), p. 161.
7. D. W. Woodruff, R. N. Wilson, and R. A. Sanchez-Martinez, in Tungsten and Other Refractory Metals for VLSI Applications, edited by R. S. Blewer (MRS, Pittsburgh, PA, 1986), p. 173.
8. V. V. S. Rana, J. A. Taylor, L. H. Holschwander, and N. S. Tsai, in Tungsten and Other Refractory Metals for VLSI Applications II, edited by E. K. Broadbent (MRS, Pittsburgh, PA, 1987), p. 187.
9. J. R. Creighton, unpublished results.
10. L. H. Kaplan and F. M. d'Heurle, J. Electrochem. Soc. 117, 693 (1970).
11. B. Lang, P. Scholler, and B. Carriere, Surface Sci. 99, 103 (1980).
12. W. T. Stacy, E. K. Broadbent, and M. H. Norcott, J. Electrochem. Soc. 132, 444 (1985).
13. D. C. Paine, J. C. Bravman, and K. C. Saraswat, in Tungsten and Other Refractory Metals for VLSI Applications, edited by R. S. Blewer (MRS, Pittsburgh, PA, 1986), p. 117.
14. M. L. Green, Y. S. Ali, T. Boone, B. A. Davidson, L. C. Feldman, and S. Nakahara, in Tungsten and Other Refractory Metals for VLSI Applications II, edited by E. K. Broadbent (MRS, Pittsburgh, PA, 1987), p. 85.
15. E. K. Broadbent and C. L. Ramiller, J. Electrochem. Soc. 131, 1427 (1984).

A MOLECULAR BEAM STUDY OF THE REACTION OF WF_6 ON Si(100)

Ming L. Yu, Benjamin N. Eldridge, and Rajiv V. Joshi
IBM T. J. Watson Research Center, Yorktown Heights, NY 10598, USA

ABSTRACT

We have studied the reduction of WF_6 on Si(100) for the selective deposition of tungsten on silicon from 300°C to 700°C. We simulated the process by impinging a pulsed beam of WF_6 on Si(100) surfaces and used time-resolved mass spectrometry to monitor the evolution of the reaction products. While SiF_4 is the major reaction product below 400°C, SiF_2 becomes the dominant species above 500°C with a gradual transition as the temperature is raised to 500°C. Both SiF_2 and SiF_4 have a fast and a slow component. It is very likely that the fast component, which is dominant at high temperatures, may be produced by the immediate reaction of WF_6 with Si as the molecule strikes the Si(100) surface. The slow component, which can have time scales in the order of tens of seconds, is likely related to the regrouping of fluorine atoms on the silicon surface to form the volatile products. We also found that Si atoms, which diffuse through the W overlayer, can sustain the WF_6 reduction reaction. The production of SiF_2 and SiF_4 from such a Si layer on W is qualitatively similar to that from the bare Si surface.

INTRODUCTION

The reduction of tungsten hexafluoride (WF_6) by silicon is often used for the selective chemical vapor deposition (CVD) of tungsten on silicon[1]. The overall reaction is believed to be

$$2WF_6 + 3Si \rightarrow 2W + 3SiF_4\uparrow . \qquad (1)$$

The chemical inertness of SiO_2 to WF_6 results in high selectivity in deposition. Interestingly the reaction does not stop at a monolayer coverage of tungsten. The W growth rate is initially large. Then it decreases rapidly with time as if the growth is self-limiting[1–3]. The film thickness can reach a few hundred Å. We want to examine whether Eq. (1) is actually the reaction responsible for the tungsten deposition. We also want to explore the possible differences between the initial reaction of WF_6 on bare silicon, and the reaction during the growth of the tungsten film.

We have studied this reduction reaction by impinging a pulsed beam of WF_6 molecules on Si(100) surfaces in an ultrahigh vacuum, and using time-resolved mass spectrometry to monitor the evolution of the reaction products. Our experiments show that Eq. (1) is not the only reaction pathway. Both SiF_2 and SiF_4 can be the dominant reaction products, depending on the reaction temperature. In this paper, we shall focus on the chemical steps in the tungsten deposition process.

EXPERIMENT

We have used Si(100) surfaces throughout the experiment. The samples were boron doped, p-type, 1 ohm-cm Si wafers cut into 1.9 cm diameter, 0.38 mm thick, discs. A tantalum heater of similar dimension provided uniform heating for the samples. A 0.003 inch W-5%Re/W-26%Re thermocouple was pressed against the back of the Si wafer to measure the sample temperature. Clean and ordered p(2x1) Si(100) surfaces were prepared by Ar^+ sputtering and 1000°C thermal anneal. The condition of the surfaces was monitored in situ by Auger spectroscopy (AES) and low energy electron diffraction (LEED).

A collimated, 3 mm diameter, molecular beam of WF_6 was produced by the supersonic expansion of WF_6 at about atmospheric pressure through a 0.5mm diameter nozzle. Short pulses of WF_6 molecules were generated by the combination of a pulsed valve and a synchronized high speed chopper. The gas pulses impinged on the Si(100) surface at 45°. With three stages of differential pumping, the vacuum in the reaction chamber was maintained at about 1×10^{-10} Torr. The reaction products were detected through a 4.8 mm diameter aperture by a differentially pumped quadrupole mass spectrometer located at right angles to the beam direction. The distance between the sample and the ionizing volume was about 4 cm. The signal was detected by pulse counting and recorded by a multichannel scalar. A detailed description of the apparatus has been published[4].

Certain specifics are relevant to this experiment. As will be shown below, fluorine atoms from WF_6 molecules have very long residence time on the Si surface, sometimes extending into many seconds. We have used 250 μs wide WF_6 pulses and 50 μs or longer per channel in the detector to obtain reasonable signal to noise ratio. Transit times of reaction products from the sample to the mass spectrometer are negligible in this time scale. Signal to noise consideration also forced us to use He seeding (19% WF_6 in He) to improve the intensity of the WF_6 beam. The resulting beam has a translational energy of about 0.4 eV. We have used beams formed from pure WF_6 on several occasions and found that the data is not qualitatively different from those acquired with the He seeded beam. All the molecular beam data reported here was obtained with the He seeded beam for the better signal to noise ratio. We do not have a direct measure of the quantity of WF_6 molecules in each pulse since our mass spectrometer was not calibrated for WF_6. Our previous experience with other gases suggests that there should be about 10^{10} molecules per pulse. Hence the exposure from each pulse is far less than a monolayer. We also used x-ray photoemission (XPS) and LEED to study the static chemisorption state of WF_6 on silicon in a separate UHV chamber. The XPS spectra was obtained with a Leybold EA-11 hemispherical analyzer with an Al Kα source. An exit angle of 85° was often used to improve the surface sensitivity.

RESULTS

(i) Si(100)

WF_6 has a high sticking coefficient on Si (100). Exposure of 30 Langmuir ($1L = 1 \times 10^{-6}$ Torr-sec) at room temperature saturates the surface. To determine the initial sticking coefficient (reaction probability) s_0, we exposed the Si (100) surface to short pulses of WF_6 The unreacted WF_6 molecules that got reflected into the mass spectrometer were monitored by the mass 184 (W^+) peak. The sticking coefficient s_0 is given by:

$$s_o = (r_{sat} - r_o)/r_{sat}$$

where r_0, r_{sat} are the reflected W^+ signals from the clean and the WF_6 saturated Si (100) surfaces respectively. The assumption here is that the angular distributions of the reflected WF_6 molecules are the same in both cases. s_0 is about 0.3 $\pm$ 0.04 and is not sensitive to temperature variations from 200°C to 700°C. The saturation WF_6 coverage at room temperature estimated by the W4f XPS peak (corrected for F 2s) is about $1.5 \times 10^{14}/cm^2$. This is much smaller than the number of first layer Si atoms on Si (100): $6.8 \times 10^{14}/cm^2$. It suggests that each WF_6 molecule reacts with several Si atoms. The shape of the Si 2p XPS peak (Fig. 1)

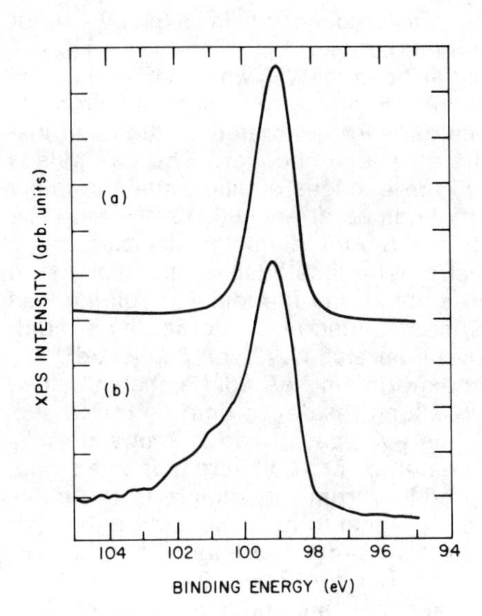

Fig 1. Si2p XPS spectra for (a) clean, (b) WF_6 saturated Si(100).

reveals the formation of a range of silicon fluorides $SiF_x(x = 1, 2, 3)$. The Si2p peak shifts towards higher binding energy by 1, 2, and 3 eV for x = 1, 2, and 3 respectively[5]. The WF_6 layer reduced the intensity of the p(2x1) LEED pattern of Si (100), but there was no evidence of the formation of an ordered WF_6 layer.

While the fluorine atoms are quite stable on the Si surface at room temperature, the desorption of silicon fluoride products were observed at elevated temperatures. The major peaks in the mass spectrum related to the reaction products are SiF^+, SiF_2^+, and SiF_3^+. While the SiF_3^+ signal is well documented[6] to be the major cracking product of SiF_4 in the mass spectrometer ionizer, we found that the SiF^+ and SiF_2^+ signals came from a different origin. Figure 2 shows the time evolution of (a) W^+, (b) SiF^+, (c) SiF_2^+, and (d) SiF_3^+ signals during the reaction of 250 μs wide pulses of WF_6 on Si (100) at 500°C. The W^+ signal represents the unreacted portion of the WF_6 pulse after impinging on the Si (100) surface. The width of the W^+ signal reflects the width of the incident WF_6^+ pulse. The SiF^+ and SiF_2^+ signals have practically the same time variation while the SiF_3^+ signal is appreciably sharper. The similarity of the SiF^+ and SiF_2^+ waveforms and their differences from the SiF_3^+ waveforms were invariably observed. That leads us to postulate that SiF_2 is also a major reaction product, and the SiF^+ signal is from the cracking of SiF_2 in the ionizer. We should note that SiF_2 was also found[6] to be a reaction product during the reaction of F on Si. In the rest of the paper we shall assume that the SiF^+ signal is proportional to the production of SiF_2, and the SiF_3^+ signal represents the production of SiF_4.

To further verify the presence of a second reaction pathway:

$$WF_6 + 3Si \rightarrow W + 3SiF_2\uparrow \qquad (2)$$

in addition to Eq. (1), we monitored the time averaged SiF^+ and SiF_3^+ signals during the reaction of Si (100) with the WF_6 pulses at various temperatures. Figure 3(a) and (b) show the variations of these signals with the reaction temperature. At 300°C, the reaction pathway (1) dominates and SiF_4 is the major reaction product. At 400°C and above, the reaction pathway (2) becomes increasingly more important and finally SiF_2 is the major reaction product at 700°C.

The rates at which the reaction products leave the Si (100) surface decreases rapidly with decreasing reaction temperature. Figure 4(a) shows the time evolution of the SiF^+ signal at 500°C. Aside from the much longer time scale when compared to Fig. 2b for 550°C, the signal can roughly be divided into two components. The fast component was practically instantaneous with the incident WF_6 pulse. While its magnitude decreased rapidly with decreasing temperature, it was always observed. The slower component usually has a time scale in order of seconds. It does not have a simple exponential dependence on time. Hence the production of SiF_2 is not first order.

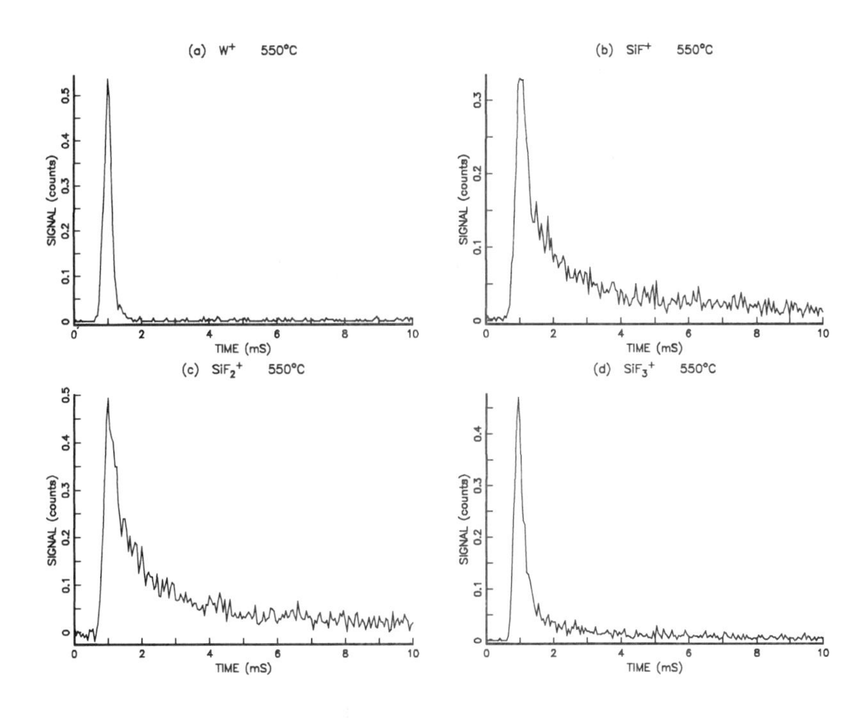

Fig. 2 Product waveform of (a) W^+, (b) SiF^+, (c) SiF_2^+, and (d) SiF_3^+ at 550°C.

(ii) Si on W

The evolution of SiF_4(SiF_3^+) is always faster than that of SiF_2(SiF^+) as shown in Fig. 2(d) and Fig. 4(b). But careful examination of the waveform always reveals the existence of a slow component. This is clearly demonstrated by Fig. 4(c). The slow component of the SiF_3^+ signal at 350°C extends beyond 50 seconds.

The silicon reduction of WF_6 does not stop after a monolayer of tungsten is deposited onto the silicon surface. Tungsten films up to hundreds of Å in thickness has been grown by this reaction[1–3]. We have studied one possible mechanism for this continual growth of tungsten. A tungsten film was grown in our vacuum system by heating a Si (100) sample to 450°C in one Torr of WF_6 for five minutes. It has been reported that the W film should reach the "self-limiting" thickness under this growth condition[1–3]. In situ XPS did not detect above the noise level the presence of Si on the tungsten surface. However, the film growth can resume by simply heating the sample to 700°C for 5 minutes. XPS revealed the segregation of $1.9 \times 10^{15}/cm^2$ of Si atoms on and near the surface of tungsten as shown by the Si2p peak (Fig. 5a). These Si atoms react with WF_6 readily. Figure 5(b) shows the formation of silicon fluorides SiF_x ($x = 1,2,3$) after this Si on W layer was exposed to 60L of WF_6 at room temperature. Hence, W film growth can proceed by the thermal diffusion of Si atoms from the Si substrate to the surface of the W film. We also observed that a tungsten surface can adsorb WF_6.

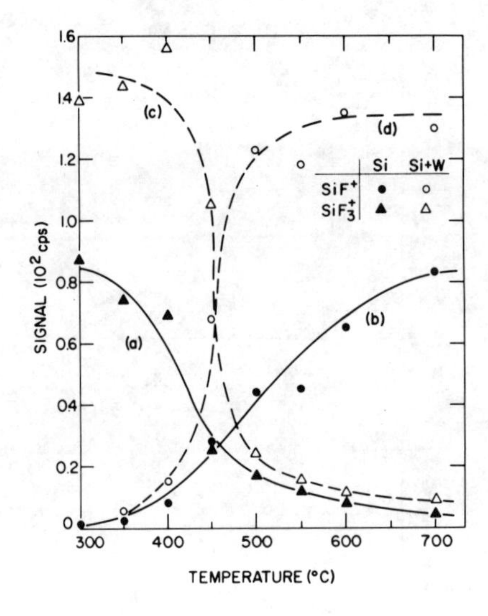

Fig. 3 Temperature dependences of the SiF^+ and SiF_3^+ signals from the Si(100) (a and b), and the Si covered W surfaces (c and d).

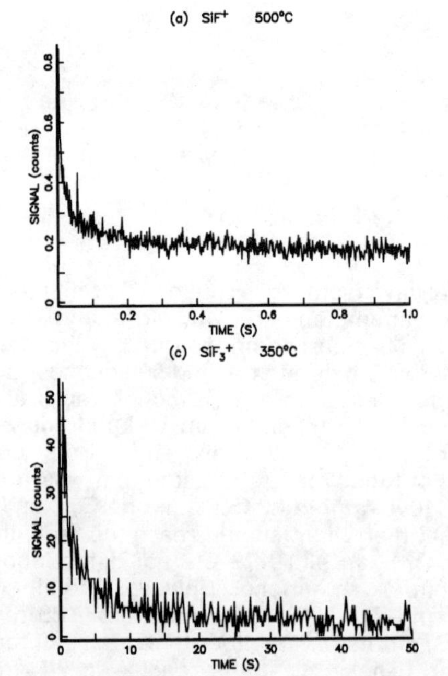

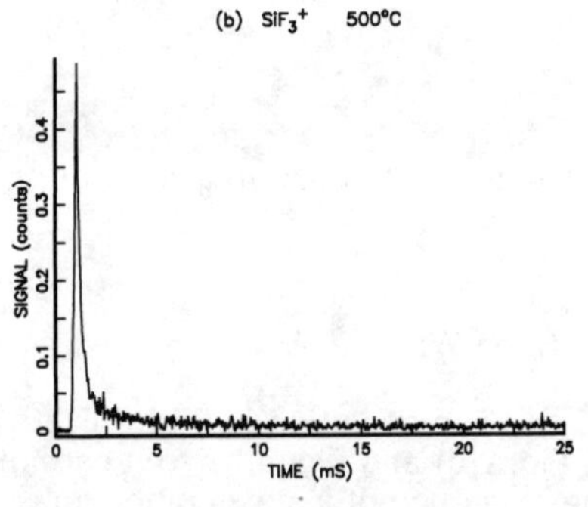

Fig. 4 Product waveform of (a) SiF^+ at 500°C, (b) SiF_3^+ at 500°C and (c) SiF_3^+ at 350°C.

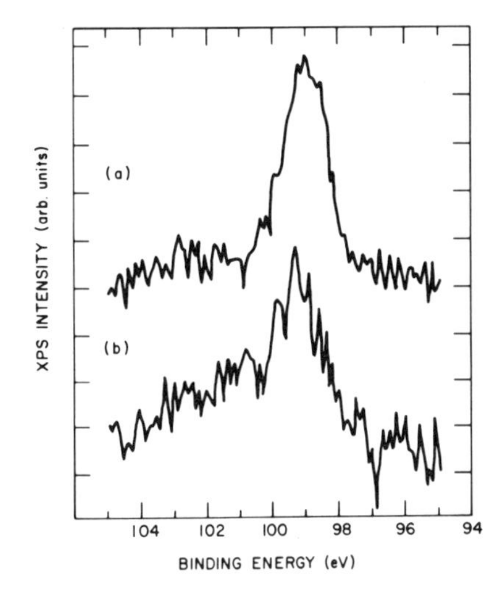

Fig. 5 Si2p XPS spectra for (a) Si segregated on W, (b) after WF_6 adsorption on the same surface.

60L of WF_6 at room temperature produces a saturated surface with about $1.1 \times 10^{14}/cm^2$ of WF_6 molecules, as determined by the magnitude of the F1s XPS peak. Hence, fluorine atoms are readily available to react with the silicon atoms that are able to diffuse to the tungsten surface.

The chemical properties of these segregated Si atoms on W are similar to those of the Si (100) surface. We have repeated our molecular beam study on these Si overlayers on W. Although we did not have a control over the Si coverage since the surface is at a dynamic state at the elevated temperatures, the qualitative comparison with Si (100) is still instructive. It is also more relevant to the film growth process. We shall use "Si + W" to label such a surface. We always started with a five minute anneal at 700°C. Hence, we believe that the Si coverage should be close to a monolayer.

The "Si + W" surface was actually more reactive than the Si (100) surface. The sticking coefficient (reaction probability) s_0 is about $0.48 \pm .05$ with no obvious temperature dependence. The WF_6 reaction also follows both reaction pathways (Eqs. 1 and 2). Curves (c) and (d) in Fig. 2 show the temperature dependences of the SiF_2 (SiF^+) and SiF_4 (SiF_3^+) production. The signals are larger than those from Si(100) because of the higher value of s_o. Again SiF_4 is the dominant reaction product at lower temperatures. However, the transition to the second reaction pathway is more rapid. The production of SiF_2 dominates even at 500°C.

With a 250μs wide WF_6 pulses, we monitored in real time the evolution of SiF_2 and SiF_4. Curve (a) in Fig. 6 shows the SiF^+ signal as a function of time at 600°C. Again the SiF_2 production has a fast component comparable to the incident pulse width, and a slow component. The evolution of SiF_4 (SiF_3^+) at the same temperature is more rapid as shown by curve (b) in Fig. 6. However, SiF_4 has a large slow component which lasts into many tens of seconds at lower temperatures as shown in curve (c) for 350°C.

DISCUSSIONS

We have identified two reaction pathways for the reaction of WF_6 with both Si(100) and "Si + W" surfaces. At lower temperatures, SiF_4 production is the dominant reaction while SiF_2 production is favored at higher temperatures. The productions of SiF_2 and SiF_4 have different kinetics and are not first order reactions. We have no quantitative model for these reactions at present. We speculate that the time evolution of the reaction products may be governed not only by the desorption activation energies, but also by the diffusion of fluorine atoms on the surface. Since the WF_6 molecule has six fluorine atoms, it is conceivable that SiF_2 and SiF_4 can be produced immediately after the dissociation of WF_6 on the surface. This picture is consistent with the appearance of the "fast" components of the reaction products. But once the fluorine atoms diffuse around on the surface, the formation of SiF_2 and SiF_4 would require the regrouping of two

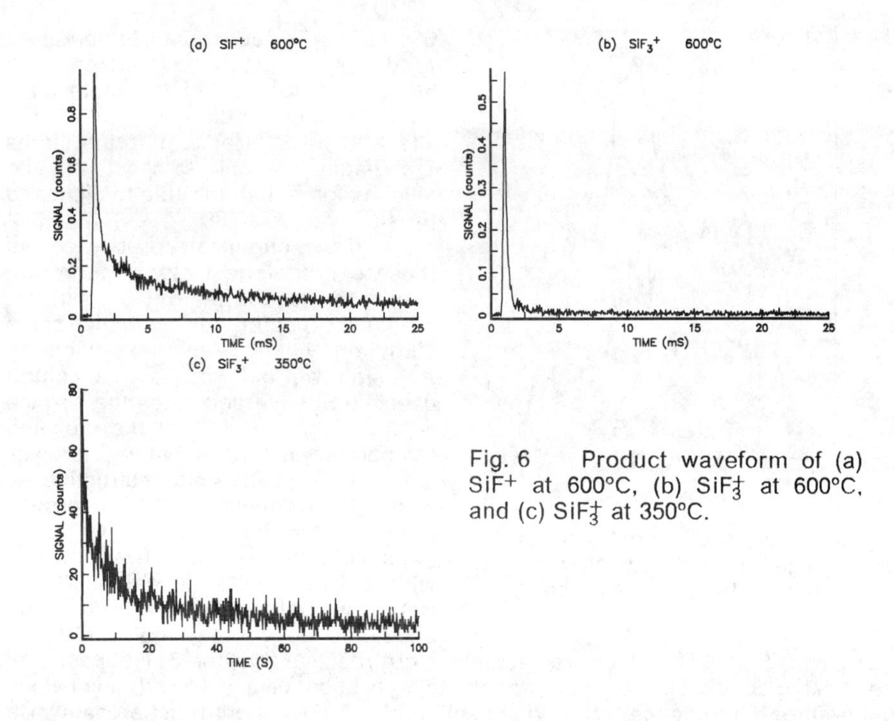

Fig. 6 Product waveform of (a) SiF^+ at 600°C, (b) SiF_3^+ at 600°C, and (c) SiF_3^+ at 350°C.

and four fluorine atoms respectively, which would be strongly controlled by the diffusion process. The desorption kinetics would not be first order, and this is consistent with our experimental observations. We should caution that the separation into the "fast" and "slow" components are only qualitative. The correct model should predict both behaviors self-consistently.

We have also identified that the diffusion of Si to the tungsten surface can be a mechanism to sustain the growth of the tungsten film. In this mechanism, the growth rate is limited by the flux of Si atoms through the W film, which decreases rapidly with film thickness. Qualitatively, the growth rate is large at the initial stage, but decreases rapidly with time. This is consistent with experimental observations[1–3]. Since the reaction probability s_o of WF_6 is relatively insensitive to the reaction temperature, the temperature dependence of the W film growth process would be controlled by that of the Si diffusion. We are in the process of determining whether this mechanism dominates the film growth process.

ACKNOWLEDGEMENT

The authors would like to acknowledge the useful discussions with Dr. F. R. McFeely, and Dr. K. Ahn, and a critical review of the manuscript by Dr. W. Reuter. Research is partially supported by the Office of Naval Research.

REFERENCES

1. T. Moriya and H. Itoh, Proc. of Workshop on Tungsten and other Refractory Metals for VLSI Applications 1984 and 1985, Ed. R. S. Blewer, (Materials Research Society, 1986) p. 21.
2. M. L. Green and R. A. Levy, J. Electrochem. Soc. 132, 1243 (1985).
3. E. K. Broadbent, Proc. of Workshop on Tungsten and other Refractory Metals for VLSI Applications 1984 and 1985, Ed. R. S. Blewer, (Materials Research Society, 1986) p. 365.
4. B. N. Eldridge and M. L. Yu, Rev. Sci. Instrum. 58, 1014 (1987).
5. F. R. McFeely, J. F. Morar, and F. J. Himpsel, Surf. Sci. 165, 277 (1986).
6. M. J. Vasile and F. A. Stevie, J. Appl. Phys. 53, 3799 (1982).

CHEMICAL VAPOR DEPOSITION OF TUNGSTEN ON SILICON AND SILICON OXIDE STUDIED WITH SOFT X-RAY PHOTOEMISSION

J.A. Yarmoff* and F.R. McFeely
IBM T.J. Watson Research Center, Box 218, Yorktown Heights, NY 10598

ABSTRACT

The growth of tungsten films on silicon and oxidized silicon surfaces via the silicon reduction of WF_6 was studied with soft x-ray photoemission. The films were grown in ultra-high vacuum and analyzed in situ. It was found that the growth on clean Si proceeds via diffusion of Si atoms through the W film to the surface, so that the silicon atoms become available for the reduction reaction. Post-fluorination of these films via XeF_2 was performed in order to ascertain the structural details and to investigate further the role of fluorine in the growth process. Silicon oxide surfaces were prepared and subsequently exposed to WF_6. It was found that a fully formed SiO_2 surface is inert with respect to WF_6, but that a partially formed oxide will permit partial WF_6 dissociation.

I. INTRODUCTION

The growth of metallic films on semiconductors via Low Pressure Chemical Vapor Deposition (LPCVD) has many advantages over traditional methods, such as evaporation or sputter deposition. One of the major advantages of LPCVD is that the chemical nature of the process allows for the ability to produce spatially selective deposits. In the present case, tungsten films are grown on silicon by exposure of silicon wafers to tungsten hexafluoride. This process is selective in that a tungsten film will grow on a bare Si substrate, but no growth will take place on silicon oxide.[1,2]

When making bulk tungsten deposits, i.e. for electronic applications, a two-step procedure is employed. The first step involves the exposure of the substrate to pure WF_6 vapor, in which case the molecule reacts via the silicon reduction of WF_6 to form a 'seed layer'. In the technological environment, a seed layer thickness will typically reach a limit of 200-900 Å, which is dependent on the initial surface condition.[3] To grow a thicker film, hydrogen is mixed with the WF_6, which promotes W growth via the hydrogen reduction of WF_6. A typical growth temperature is 450°C.

In the present work, LPCVD is performed entirely in ultra-high vacuum (UHV) and the films are analyzed by soft x-ray core-level photoemission. The mechanism for the growth of the seed layer is found to involve the diffusion of a small number of silicon atoms to the surface of the seed layer, where they are available for the silicon reduction reaction. In present work, no evidence of a limiting film thickness was found, thus suggesting that the thickness limit seen in previous non-UHV studies may be related to small amounts of contaminants.

The use of surface science techniques for the study of LPCVD reactions offers insights into the reaction mechanisms that cannot be obtained in other ways. This fact is readily apparent in the present case, as it is the reaction of an incoming WF_6 molecule with a surface Si atom that is responsible for the growth of the seed layer. To directly observe this reaction an extremely surface sensitive technique was necessary. Si 2p core levels collected with 130 eV photons produce electrons with an escape length of ~5Å.[13] A previous comparison of the relative intensities of the Si 2p to the W 4f core levels of the seed layer to a WSi_2 reference sample showed that there is approximately 25% silicon within the top 5Å of the surface,[14] and this silicon is concentrated at the surface.

To more fully understand the nature of the chemical selectivity and the role of a thin oxide layer in inhibiting the W growth, exposures were made on silicon surfaces that were previously oxidized in a controlled manner. These results show that a thin thermally annealed oxide layer completely inhibits the WF_6 from reacting with the surface, while an oxide layer deposited at room temperature, which is similar to a native oxide, serves to hinder the dissociation of the WF_6 molecule, but still allows W to grow.

II. EXPERIMENTAL

The measurements were performed on beamline UV-8 at the National Synchrotron Light Source using either a 3m or a 6m toroidal grating monochromator as the photon source. Si (111) wafers were cleaned with standard techniques, and photoemission was used to judge cleanliness.[4] WF_6 exposures and oxidations were performed entirely in UHV, with a sample transfer system employed to move the samples from a dosing chamber into the spectrometer chamber. The Si wafers were heated resistively, and IR and optical pyrometry were used to measure the temperature. W foil and WSi_2 were employed as reference materials.

III. RESULTS AND DISCUSSION

W 4f, W $5p_{3/2}$, and F 2s core levels collected with a 90 eV photon energy are shown in Fig. 1. Fig 1a shows the results obtained by exposure of a clean Si (111) wafer to WF_6 at room temperature and Fig. 1b shows a sample which was exposed to WF_6 at a temperature of 450°C. The peak shapes shown in Figs. 1a and 1b were found to be virtually independent of exposure, from 10L to 10000L, and the W core level positions in these spectra are indistinguishable from metallic W. The major difference between the 90 eV photoemission spectra collected after exposure at the two temperatures was in the F 2s that was observed following the room temperature exposure. There is overlap between the W 4f levels

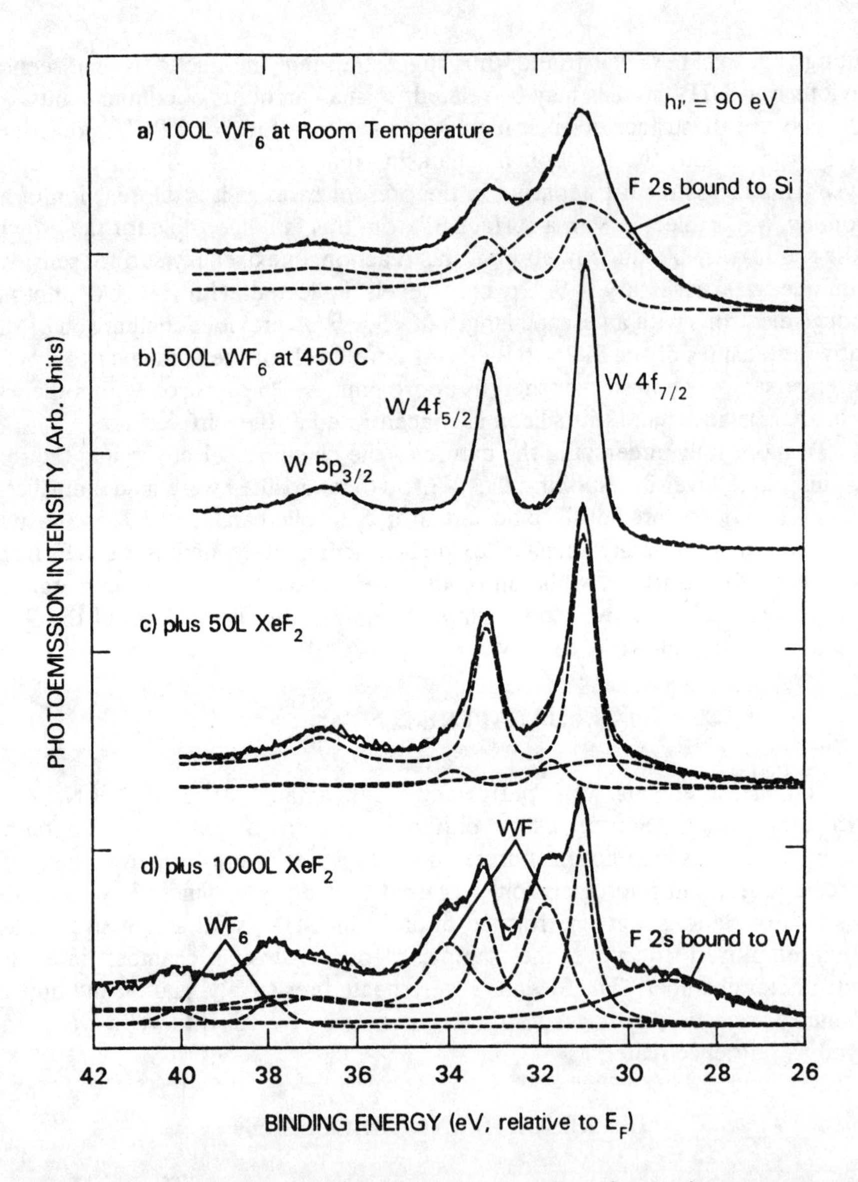

Figure 1. Photoemission spectra of a Si (111) wafer after various treatments using a 90 eV photon energy. This region shows the W 4f, $5p_{3/2}$ and F 2s levels for a) a sample exposed to 100L of WF_6 at room temperature, b) a sample exposed to 500L of WF_6 at 450°C, c) a sample, such as the one shown in Fig. 1b, after exposure to 50L of XeF_2, and d) the sample from c) after a 1000L exposure of XeF_2. The solid lines show the raw data and the dashed lines are a numerical fit (Gaussian broadened Lorentzian lineshapes) to the data using a composite W lineshape obtained from reference samples.

and the F 2s level, but the contributions from each of the core levels can be distinguished by using a binding energy for the F 2s level of F bound to Si obtained from measurements of XeF_2 chemisorption on Si.[5]

Figure 2 shows photoemission spectra of the Si $2p_{3/2}$ core level, collected with 130 eV photons, of the same surfaces as Fig. 1. The distribution of the silicon fluorides resulting from a room temperature exposure to WF_6, shown in Fig. 2a, is almost identical to that previously observed for a 50L exposure of Si (111) to XeF_2.[5] This result, in conjunction with the observance of only metallic W in Fig. 1a, demonstrates that the dissociation of WF_6 on a silicon surface is complete, even at room temperature, and all the fluorine remaining on the sample is bound to silicon.

As the seed layers are typically grown at 450°C, the spectra in Figs. 1b and 2b are representative of a seed layer. As can be seen in Fig. 2b, there is silicon present in the surface region of these samples even after large exposures to WF_6. The fact that Si is always seen on the surface suggests a mechanism for the growth of the seed layer in which Si atoms diffuse through the layer and collect at the surface, thereby propagating the growth. This implies that the Si atoms which are observed in the 130 eV Si 2p spectra are most likely concentrated at the surface, and depleted from the bulk, of the seed layer. A confirmation of the surface segregation of the Si in the seed layer was given by a comparison of bulk sensitive to surface sensitive photoemission.[14] Additionally, since Si is seen on the surface after very large WF_6 exposures, and since there was never any seed layer film grown in the present study that did not show this surface Si, it is presumed that the growth of the seed layer in the UHV conditions of the present apparatus did not reach the limiting thickness observed under conventional LPCVD conditions.

In an effort to further elucidate the role of fluorine in the reaction mechanism, and to further understand the nature of the seed layer, a surface such as the one shown in Figs. 1b and 2b, was prepared by a 500L exposure of clean Si (111) to WF_6 at 450°C and subsequently exposed to XeF_2. XeF_2 has been successfully employed as a source of fluorine atoms for reaction with both Si and W.[8,9] These post-fluorination experiments indicated that the additional fluorine atoms at first became bonded to silicon atoms. After further exposure to XeF_2, the silicon atoms were preferentially etched, as evidenced by a loss of intensity in the Si core levels, and fluorine began to bond to W.

Figure 1c shows the W core levels after a 50L post-fluorination. This spectrum shows basically metallic W, with a small amount of monofluoride, shifted 0.72 eV from the metallic W peak, becoming visible. (The core level shift for tungsten monofluoride was not obtained from this spectrum, as the intensity of the shifted peak is very low. Rather, it was obtained from the larger XeF_2 exposures discussed below.) Figure 1d shows the same surface after a 1000L XeF_2 exposure. After this dose, a strong monofluoride peak is evident, and a peak shifted 6.9 eV towards higher binding energy appeared. This new peak is tenta-

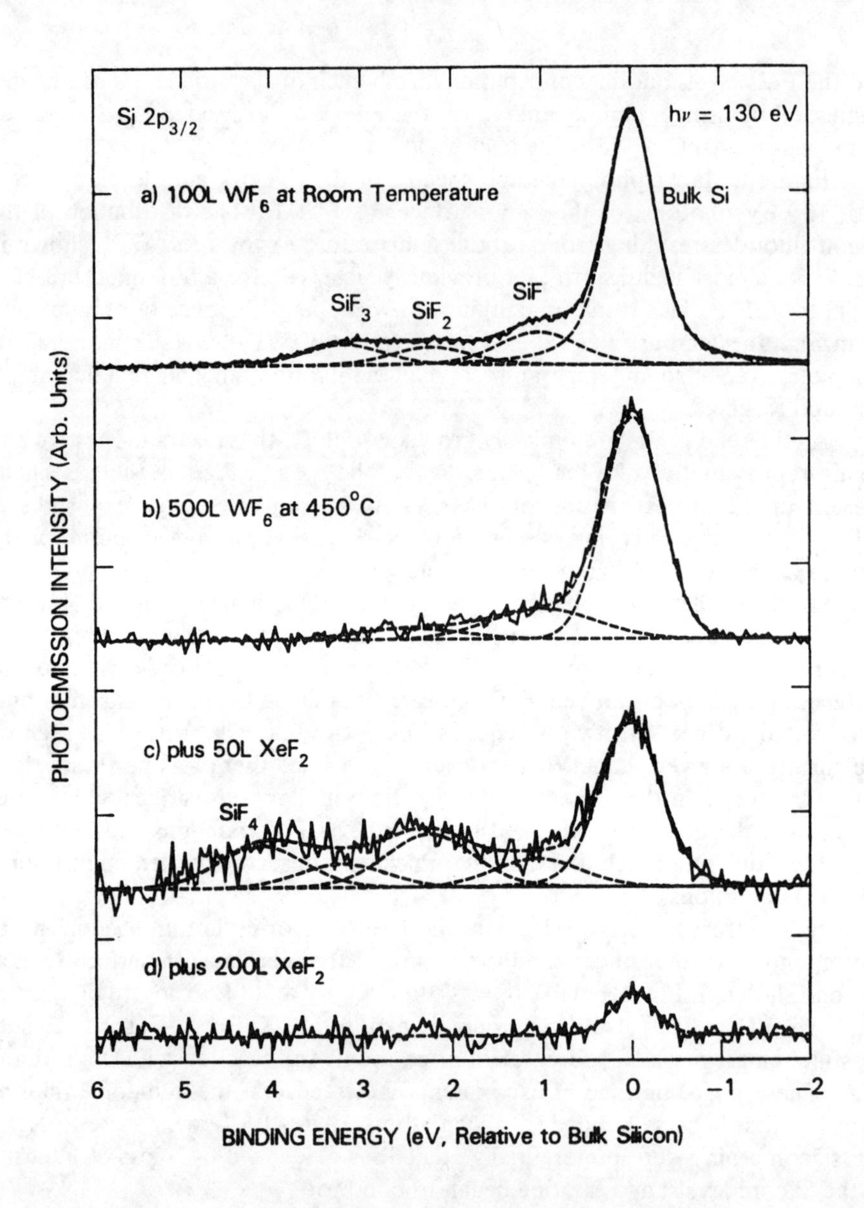

Figure 2. Photoemission spectra of a Si (111) wafer after various treatments collected with a 130 eV photon energy. This region shows the Si $2p_{3/2}$ core level for the same samples as shown in Fig. 1. The $2p_{3/2}$ component, shown as a solid line, was obtained by subtracting a background from the raw data and then numerically removing the $2p_{1/2}$ component. The dashed lines are a numerical fit to the data using F-induced shifts of the Si 2p obtained from XeF_2 chemisorption on Si.[5,6]

tively identified as WF_6 molecules, formed by the etching of the seed layer by XeF_2, which have become trapped in the layer. WF_6 has been observed to be the major reaction product resulting from the XeF_2 etching of tungsten,[9] and is thus expected as a reaction product in the etching of the seed layer. In addition, it should be noted that the F 2s core level for the F atoms bound to W has a lower binding energy than the F 2s core level shown in Fig. 1a for F bound to Si.

The corresponding Si core levels for the post-fluorination of the seed layer are shown in Figs. 2c and 2d. The numerical fit shown in Fig. 2c was performed by using the known chemical shifts of the Si 2p core level obtained from measurements of Si wafers etched with XeF_2.[5] This spectrum shows a relatively large amount of SiF_4 trapped in the layer. SiF_4 has been identified as the major reaction product during XeF_2 etching of silicon,[15] and has been previously observed as a trapped reaction product in etched Si wafers.[6] The relative amount of SiF_2 in this system is larger, however, and the amount of SiF_3 is much smaller than is seen in the XeF_2 etching of pure Si.[6] Figure 2d shows the Si 2p obtained after a 200L exposure of the seed layer to XeF_2. Although the data is not shown to scale in these figures, the amount of Si remaining at this point was very small, and hence only a single peak was observed. After the 1000L exposure, this peak was barely detectable.

The utility of these post-fluorination experiments lies in the information concerning the structure of the seed layer that is indicated by the results. This structural information is not available from an analysis of the photoemission spectra of the seed layer alone. For example, the fact that a Si signal is observed from the seed layer does not necessarily indicate that this Si is uniformly distributed on the surface. Two other possibilities exist, both of which can be ruled out by virtue of the results of the post-fluorination experiments.

The first of these possibilities is that the Si is grouped into islands on an otherwise clean W substrate. If this were the case, then areas of bare W would have been just as readily available for fluorination as was the Si. A comparable dose of 50L of XeF_2 on bare W produces a measurable amount of shifted W.[10] In addition, more shifted Si relative to unshifted (bulk) Si was seen for the 50L post-fluorination of the seed layer than is seen after a 50L exposure of a clean Si (111) wafer to XeF_2.[5] This is further evidence of the fact that the morphology of this Si is different from bulk Si. Most likely, the silicon is contained in a thin layer at the seed layer surface, as there is little Si substrate underneath this layer to contribute to the bulk Si peak.

The second possible explanation to the Si signal observed from the seed layers is islanding of the W growth, whereby areas of the Si substrate remain exposed. If this were the case, then these areas of bare bulk Si would be responsible for the observed Si 2p signal, and the post-fluorination would be expected to behave, with respect to the Si core level, in the same manner as on a clean Si (111) substrate. The possibility that the W exists in islands can be directly eliminated by the post-fluorination results, however, via the following analysis. When a bare Si surface is subjected to XeF_2 etching, SiF_3 intermediate species are formed in abundance.[6] In the present case, post-fluorination of the seed layer produced a surface containing considerably more SiF_2 than SiF_3. This result implies that the morphology of the Si in the seed layer is different than that of bulk Si, making it unlikely that there are regions of bare Si substrate on these seed layer surfaces. The build-up of a large number of SiF_2 species may be indicative of the formation of a polymeric Si-F structure, perhaps as a result of fluorination of a thin layer of silicon atoms residing atop a W substrate.

Another structural fact concerning the nature of the seed layer is evident from the 1000L post-fluorination, after which WF_6 molecules were observed to be trapped in the seed layer. As a comparison, no trapped WF_6 is observed on bare W metal, even after much larger exposures to WF_6.[10] This indicates that the structure of the seed layer is more open than the structure that exists in W metal, which allows for the WF_6 trapping. The openness of the seed layer structure is most likely in the form of large grain boundaries. These grain boundaries may have some connection with both the ability of Si to diffuse through the layer, and the fact that the resistivity of the CVD-grown W films is not quite as low as that of W metal.[2,3]

To gain an understanding of the detailed chemistry responsible for the selectivity of the WF_6 LPCVD process, Si wafers were oxidized in a controlled fashion, and then exposed to WF_6. Figure 3 shows Si $2p_{3/2}$ photoemission of two types of oxidized surfaces that were employed for these purposes. The surface shown in Fig. 3a was prepared by exposure of a clean Si (111) wafer to $5X10^{-5}$ torr of O_2 at 750°C for 10 minutes. This produces a fully formed SiO_2 layer approximately 11Å thick.[16] Exposure of this fully formed oxide to WF_6 produced no measurable W growth at any temperature below the oxygen desorption temperature itself (approximately 850°C). Since no growth was seen on this thermal oxide layer, it is suggested that the oxide layer forms a diffusion barrier to the Si atoms residing in the substrate.

A partial oxide layer was grown by exposure of a clean Si (111) wafer to 200L of O_2 at room temperature. The Si $2p_{3/2}$ spectrum of this sample is shown in Fig. 3b. This surface had only a small amount of Si^{4+} (i.e. fully formed SiO_2) and contained mostly partially oxidized Si atoms on the surface. Exposure of this sample to 1000L of WF_6 at room temperature did result in chemisorption of WF_6. The exposure was done at room temperature and not at a normal growth temperature because the elevated temperature would have had the effect of both altering the sub-oxide distribution and of thermally desorbing fluorine from the surface. In this case, the fluorine remaining on the surface was useful for monitoring the reaction chemistry.

Following the room temperature WF_6 exposure of the partially oxidized Si surface, a distribution of tungsten fluorides was observed that indicates an incomplete dissociation of the chemisorbed WF_6 molecules. This distribution can be seen in Fig. 4, which is a 150 eV photoemission spectrum. A photon energy

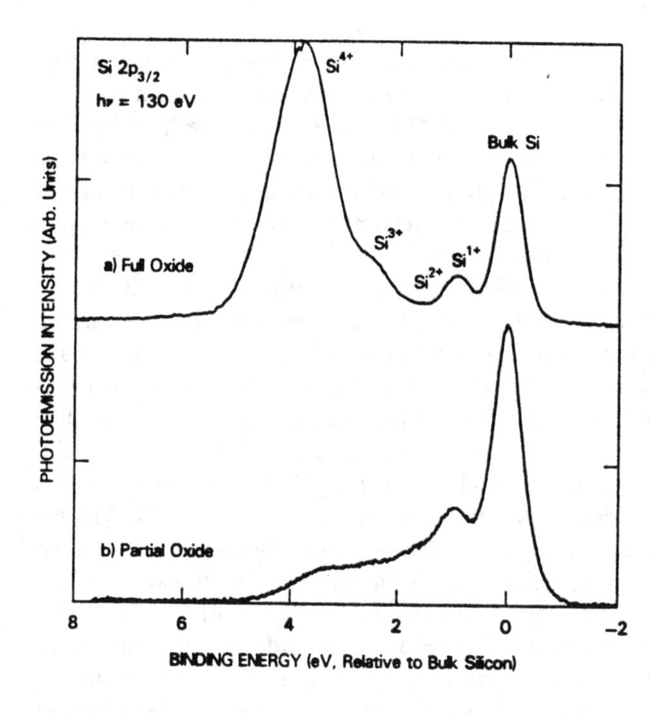

Figure 3. Si $2p_{3/2}$ photoemission spectra of a Si (111) wafer after oxidation prior to WF_6 exposure. a) Full oxide which did not allow any tungsten growth. The oxide was prepared by exposing a clean Si wafer to $5X10^{-5}$ torr of O_2 for 10 minutes at 750°C. b) The partial oxide layer which was prepared by exposure to 200L of O_2 at room temperature.

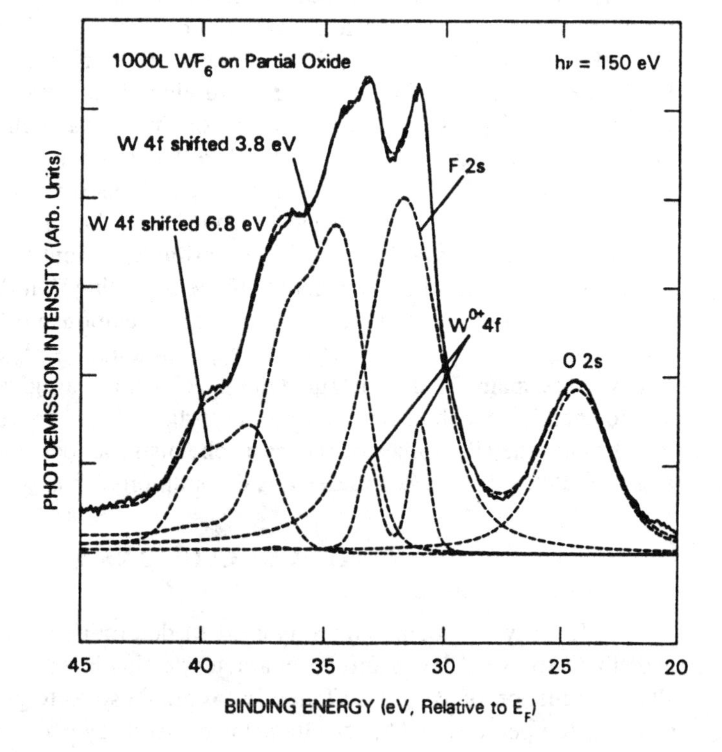

Figure 4. 150 eV photoemission spectrum of the partial oxide surface shown in Fig. 3b after exposure to 1000L of WF_6 at room temperature. The solid line is the raw data and the dashed lines are a numerical fit to the data.

of 150 eV, as compared to the 90 eV energy used for the spectra shown in Fig. 1, was used to reduce the relative photoemission cross sections of both the F 2s and the W $5p_{3/2}$ levels with respect to the W 4f level, in order to reduce the congestion of the spectrum. Even so, there is still a large contribution from the F 2s, some of which comes from F atoms still bound to W while some may result from F that was liberated from WF_6 and has now become bound to Si. Additionally, the O 2s peak is visible in this region of the spectrum.

The W 4f distribution shown in Fig. 4 has a small amount of metallic W and a large amount of W 4f intensity shifted towards higher binding energy from the metallic W 4f. The shifted W 4f intensity was divided into two separate peaks for the numerical fitting shown in Fig. 4 because of the appearance of two features in the spectrum indicative of the spin-orbit splitting of the W 4f level. The use of only two shifted W 4f peaks was also the only manner found in which the fitting procedure was able to mathematically converge. One of these regions was shifted 3.3 eV from the metallic W 4f and the other was shifted 6.8 eV. The regions that resulted from the fitting procedure were quite broad, which is most likely indicative of the inhomogeneity of the sample. The identification of the actual surface species represented by these core level shifts is difficult, however, as it is possible that some of the species are W atoms bound both to fluorine and oxygen, for example an WF_xO unit which is bound to Si through the O atom.

These data demonstrate that even a small amount of oxygen deposited at room temperature can have a large effect on WF_6 chemisorption, and thus can also affect the results of an LPCVD process. By performing LPCVD entirely in UHV, small amounts of contaminants are eliminated. Previous studies of the Si reduction of WF_6 have involved cleaning of silicon via chemical etching methods in air, then inserting the wafer into the LPCVD reactor. After a film was grown, the sample was then removed, again to air, and measured in some fashion. This method may be practical in device manufacture, but in order to obtain better control of the film growth, it would be desirable to limit the amount of contaminants present in an LPCVD reactor. Since a limiting film thickness was not observed in the present studies, there may be a relationship between the limiting film thickness and the contaminants present in a non-UHV environment. In addition, since many of the intermediate species that can give clues to the microscopic chemical mechanisms that are important to the process will not survive a trip through the air, studies of the basic chemistry associated with LPCVD processes should be carried out under UHV conditions.

IV. CONCLUSIONS

Soft x-ray photoemission was used to determine a mechanism for the the growth of the seed layer and to characterize this layer. It was found that the silicon reduction of WF_6 is sufficiently favorable so as to go to completion even at room temperature. The mechanism for seed layer growth involves the diffusion of Si atoms to the surface of the seed layer, where they become available for the reduction reaction. Post-fluorination of the seed layer indicated that the Si atoms at the surface are uniformly mixed with W, and that the seed layer structure is more open than W metal. A thin thermal oxide layer (11 Å) was sufficiently unreactive so that no W deposition occurred. Reaction of WF_6 with a partially formed oxide layer resulted in partial dissociation of the WF_6 molecules.

V. ACKNOWLEDGEMENTS

This research was carried out at the National Synchrotron Light Source, Brookhaven National Laboratory, which is supported by the U.S. Department of Energy, Department of Material Sciences. The authors wish to acknowledge the technical assistance of A. Marx, J. Yurkas and C. Costas.

*Present address: National Bureau of Standards, Surface Science Division, Gaithersburg, MD 20899

REFERENCES

1. E.K. Broadbent and C.L. Ramiller, J. Electrochem. Soc. **131**, 1427 (1984).
2. E.K. Broadbent and W.T. Stacy, Solid State Tech. **28**, 51 (December 1985).
3. K.Y. Tsao and H.H. Busta, J. Electrochem. Soc. **131**, 2702 (1984).
4. F.J. Himpsel, D.E. Eastman, P. Heimann, B. Reihl, C.W. White and D.M. Zehner, Phys. Rev. **B24**, 1120 (1981).
5. F.R. McFeely, J.F. Morar, N.D. Shinn, G. Landgren and F.J. Himpsel, Phys. Rev. **B30**, 764 (1984).
6. F.R. McFeely, J.F. Morar and F.J. Himpsel, Surf. Sci. **165**, 277 (1986).
7. F.J. Himpsel, J.F. Morar, F.R. McFeely, R.A. Pollak and G. Hollinger, Phys. Rev. **B30**, 7236 (1984).
8. H.F. Winters and J.W. Coburn, Appl. Phys. Lett. **34**, 70 (1979).
9. H.F. Winters, J. Vac. Sci. Technol. **A3**, 700 (1985).
10. F.R. McFeely and J.A. Yarmoff, unpublished.
11. E. Grossman, A. Bensaoula and A. Ignatiev, Surf. Sci., in press.
12. G. Hollinger and F.J. Himpsel, Appl. Phys. Lett. **44**, 93 (1984).
13. F.J. Himpsel, P. Heimann, T.-C. Chiang and D.E. Eastman, Phys. Rev. Lett. **45**, 1112 (1980).
14. J.A. Yarmoff and F.R. McFeely, J. Appl. Phys., submitted.
15. H.F. Winters and F.A. Houle, J. Appl. Phys. **54**, 1218 (1983).
16. G. Hollinger and F.J. Himpsel, Appl. Phys. Lett. **44**, 93 (1984).

(about 0.5) atoms in the surface of the thin layer, whereas the [illegible] for the adsorbate reaction. Post-[illegible] [illegible] that the Si(100) at the surface [illegible] with [illegible] structure [illegible] a [illegible] (1×1) [illegible] apparently [illegible] W deposition [illegible] a partially formed [illegible] resulting in [illegible] formation of the W [illegible].

ACKNOWLEDGMENTS

This research was conducted at the National Synchrotron Light Source, Brookhaven National Laboratory, which is supported by the U.S. Department of Energy, Department of Materials Sciences. The authors [illegible] the technical assistance of [illegible].

Present address: [illegible] Gaithersburg, MD 20899.

REFERENCES

1. E.K. Broadbent and C.L. Ramiller, J. [illegible]
2. E.K. Broadbent and [illegible]
3. [illegible] and H. [illegible]
4. [illegible] Himpsel, [illegible] and D.M. [illegible] (1984).
5. [illegible] McFeely, [illegible]
6. [illegible]
7. [illegible] Rev. B30 [illegible] (1984).
8. [illegible] (1979).
9. [illegible] (1985).
10. [illegible] unpublished.
11. [illegible]
12. [illegible]
13. [illegible] 45 1112 (1980).
14. [illegible]
15. [illegible]
16. [illegible]

CHAPTER V
ALTERNATIVE GROWTH TECHNIQUES AT THE FOREFRONT I

LOW TEMPERATURE SILICON EPITAXY BY PHOTO- AND PLASMA-CVD

A. Yamada, A. Satoh, M. Konagai and K. Takahashi
Department of Electrical and Electronic Engineering
Tokyo Institute of Technology
2-12-1, Ohokayama, Meguro-ku, Tokyo 152, JAPAN

ABSTRACT

Novel techniques for the epitaxial growth of silicon have been developed using mercury-sensitized photochemical vapor deposition and plasma chemical vapor deposition. Specular epitaxial silicon films were grown on (100)-oriented Si substrates at growth temperatures of between 100°C to 300°C by the photochemical or glow discharge decomposition of a gas mixture consisting of either Si_2H_6+SiH_2F_2+H_2 or SiH_4+SiH_2F_2+H_2. It was found that an addition of SiH_2F_2 gas to the reactant gases and a high dilution ratio of Si_2H_6 (or SiH_4) to H_2 were essential for this very low-temperature silicon epitaxy. Furthermore, silicon epitaxy at between 600-700°C has been demonstrated by the ArF or XeF excimer laser induced photochemical vapor deposition technique. Improvements in both the film crystallinity and electrical properties were produced by laser irradiation.

I. INTRODUCTION

A further reduction in the dimensions of high performance silicon devices is limited because conventional device fabrication methods are carried out at high processing temperatures of greater than 1000°C. For the development of future bipolar and complementary metal-oxide-semiconductor (CMOS) devices, silicon epitaxial growth will be one of the key technologies since epilayers of arbitrary conduction type can be readily grown.

The silicon epitaxial growth has been used as an essential technique for the fabrication of bipolar devices. Recently, in CMOS fabrication, epilayers grown on highly doped substrates have been used to avoid latch-up phenomena due to fine device isolation and to prevent soft errors caused by alpha rays. However, epitaxial growth is one of the highest temperature processes in the fabrication of very large scale integrated (VLSI) circuits. Thus the dimensions of device layers deposited by silicon epitaxy are fixed at values greater than the diffusion length of dopants out of the substrate, where the device dimensions are restricted to several microns under typical high-temperature conditions. For this reason, the development of low-temperature processes for fabricating VLSI chips becomes important for the continuing development of smaller and faster silicon devices and circuits.

In this paper we report on three methods for low-temperature silicon epitaxy. They are mercury-sensitized photochemical vapor deposition (photo-CVD), plasma-CVD and excimer laser induced chemical vapor deposition (laser-CVD). We begin in Sec.II by describing the recent advances in low-temperature silicon epitaxial growth. Next, in Secs.III and IV, we present our results on low-temperature silicon epitaxial growth by photo-CVD and plasma-CVD, respectively. Then, in Sec.V, a new technique for cleaning the substrate prior to the growth by plasma etching will be described. Following this, in Sec.VI, low-temperature silicon epitaxy by laser-CVD is demonstrated. And finally, we conclude our results.

II. RECENT ADVANCES IN LOW-TEMPERATURE SILICON EPITAXIAL GROWTH

The most popular technique for the growth of silicon epitaxial films is thermal-CVD at between 1050-1150°C by SiH_4, SiH_2Cl_2 or $SiCl_4$ gas decomposition. However, it is difficult to deposit a slightly doped epilayer on a highly doped substrate using the thermal-CVD method, because of the unavoidable diffusion of dopants into the epilayer that occurs due to various thermodynamic phenomena.

The diffusion mechanisms are as follows:-

1. solid-state diffusion effects due to the thermodynamic behavior of atoms.
2. auto-doping effects due to vaporization of dopants from the substrate or the growth chamber.

The most important parameter, which affects the diffusion and the vaporization processes, is the substrate temperature during growth. Therefore, the abrupt dopant transition from the substrate to the epilayer could be obtained by decreasing the growth temperature which would result in both a shortening of the dopant diffusion length and a lowering of the dopant vapor pressure. Calculations which considered the diffusion coefficient of dopants such as boron, arsenic and phosphorus show that the preferable growth temperature is under 800°C for VLSI fabrication.

Of the well known thin film growth methods, silicon molecular beam epitaxy (MBE) has been widely studied[1),2)] and has shown potential for Si epitaxial growth at low temperatures. From the research of Si-MBE, the controlled use of ion-sources has been shown to be effective in lowering the growth temperature since low energy ion bombardment on the substrate and the growing film could possibly enhance the element sticking probabilities and the surface reaction rates. Epitaxial growth of silicon was observed at a substrate temperature of 400K under UHV conditions (10^{-7}Pa) using single charged Si^+ beams at an acceleration energy of 50eV[3)]. However, Si-MBE has some disadvantages such as low throughput and complexity of apparatus.

Therefore one suitable method for VLSI fabrication might be the chemical vapor deposition technique. For example, the low pressure CVD (LPCVD) technique[4)] and the plasma enhanced CVD (PECVD) technique[5)] have been proposed in the literature. Using the PECVD process, epitaxial silicon film was deposited at between 750-800°C by Reif. They have also succeeded in the fabrication of MOSFET's with their method[6)]. Despite this successful result, the plasma process has an implicit disadvantage of radiation-induced damage from high-energy charged particles, which might exist to some extent in the glow-discharge plasma.

In contrast to these methods, photo-assisted epitaxy has been in the spotlight as a novel low-temperature epitaxial growth technique. Using the photo-CVD technique, the growth temperature could be drastically reduced, because the reactant gases are not decomposed thermally but photochemically. In addition, the process seems to make it possible to control the epitaxial growth by UV radiation, since reactive species are only produced from source gases when the UV photon energy matches their absorption bands. Moreover, photo-assisted processes could be applicable not only for epitaxial growth but also for other VLSI fabrication techniques such as etching, oxidation and doping.

The concepts of photo-epitaxy have been proposed by Nishizawa[7] and Frieser[8] in the 1960's. Recent advances in the photo-epitaxy of silicon is summarized in Table I. Examples 1 and 2 were the first reports on photo-epitaxy. From the results of Example 1, a decrease in the lower limit of the growth temperature by 40°C was observed with the UV irradiation using a high pressure mercury lamp. In Example 2, it was shown that the improvement of the film orientation was produced by illumination of the substrates with a mercury vapor lamp during deposition. Examples 3 and 4 reported on the effects of UV irradiation and found that irradiation with a mercury-xenon (Hg-Xe) lamp during deposition was essential for obtaining an epitaxial film at a low substrate temperature. It was demonstrated in Example 5 that pulsed-laser irradiation of the film could be used to sequentially melt and resolidify thin layers of as-deposited material and that amorphous layers could be converted to polycrystalline layers or single crystals under the proper conditions. Thus, there has been an increase in the active on silicon photo-epitaxy primarily because of its great possibilities for VLSI fabrication processes. In the following text, we will discuss several techniques for silicon epitaxial growth using UV or laser irradiation. In addition, we will demonstrate the plasma-CVD method for low-temperature silicon epitaxy as a reference technique for photo-CVD and for clarification of the growth mechanism.

Table I. Recent advances in the photo-epitaxy of silicon

Ex.	Gases	Lamp(Laser)	Tsub(°C)	Ref.
1	$SiCl_4/H_2$	H.P.Hg	860	7
2	Si_2Cl_6/H_2	L.P.Hg	700-1000	8
3	Si_2H_6/H_2	Hg-Xe	630-1000	9
4	SiH_2Cl_2/H_2	Hg-Xe	730	10
5	SiH_4/Ar	KrF	535-650	11
6	$Si_2H_6,SiH_2F_2/H_2$	L.P.Hg	200	12

III. MERCURY-SENSITIZED PHOTOCHEMICAL VAPOR DEPOSITION

The details of the deposition system and experimental procedure have been reported elsewhere[12]. Briefly, the deposition system consists of two chambers, one is used for photo-CVD and the other for plasma-CVD, both being accessible by load-lock entry. (100)-oriented Si wafers were used as substrates and cleaned in organic solvents, followed by a HF(2.5%) dip. After these substrate treatments, the photo-epitaxy was carried out using a low-pressure mercury lamp as a UV irradiation source. The typical preparation conditions are summarized in Table II.

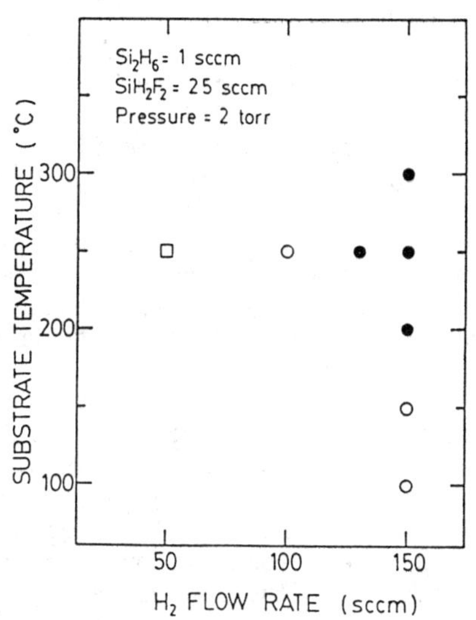

Fig.1 The film crystallinity as a function of the T_{sub} and the H_2 flow rate

In our experiments, the high dilution ratio of Si_2H_6 to H_2 and the addition of SiH_2F_2 to the gas system are key points for low-temperature silicon photo-epitaxy. Figure 1 shows the film crystallinity determined by reflective high-energy electron diffraction (RHEED) as a function of the substrate temperature and the H_2 flow rate. In this figure, the open square indicates a halo pattern which implies that the grown epilayer was amorphous. The open and closed circles indicate the distinct spotty patterns with slightly streaked lines, and elongated streak patterns respectively, which suggests that single-crystal Si films were obtained and that the latter samples had better crystalline quality than the former ones. It can be seen from the figure that the film crystallinity is very sensitive to the H_2 flow rate. At the substrate temperature of 250°C, the crystalline structure of the layer varied from amorphous to single crystal with increasing H_2 flow rate from 50sccm to 150sccm. This tendency, for an

Table II. The typical preparation conditions

T_{sub}(°C)	Si_2H_6(sccm)	SiH_4(sccm)	SiH_2F_2(sccm)	H_2(sccm)	P_{tot}(Torr)
100-300	1		20-30	150	2
250-350		5	25	50	2

improvement of the film crystallinity with the H_2 flow rate was also observed in the deposition of microcrystallized amorphous silicon[13].

The crystallinity is more sensitive to the SiH_2F_2 flow rate. Figure 2 represents the variation of RHEED patterns of epitaxial layers grown at 200°C with the SiH_2F_2 flow rate, using Si_2H_6 as a reactant gas. Without the SiH_2F_2 gas (a), narrow ring pattern which is a typical pattern of microcrystalline material, was obtained. With the addition of 15sccm of SiH_2F_2, slightly streaked lines appeared in the pattern (b). When the SiH_2F_2 flow rate exceeded 20sccm, the layer became epitaxial (c), and the most smooth surface of epitaxial layers was obtained at 35sccm (d). Using SiH_4 as a reactant gas, the RHEED pattern showed elongated streaks and clear Kikuchi lines.

The improvement of the crystallinity brought about by the addition of SiH_2F_2 gas suggests that fluorine or fluoride radicals may play an important role in the epitaxial growth. Similar to the deposition of μc-Si[14], it might be that the hydrogen radicals reduce the barrier height associated with the surface migration of the precursors by covering the g rowing surface, and that the fluorine or fluoride radicals, introduced by the addition of SiH_2F_2, serve to extract the excess bonded hydrogen on the growing surface. Another possible effect of these radicals might be to remove the native oxide layer on the growing surface. To verify these models, we checked the existence of the fluorine or fluoride radicals by mass spectrometric (MS) measurements using a quadrupole mass spectrometer (QMS). The MS measurements were performed under identical growth conditions so that sampling of

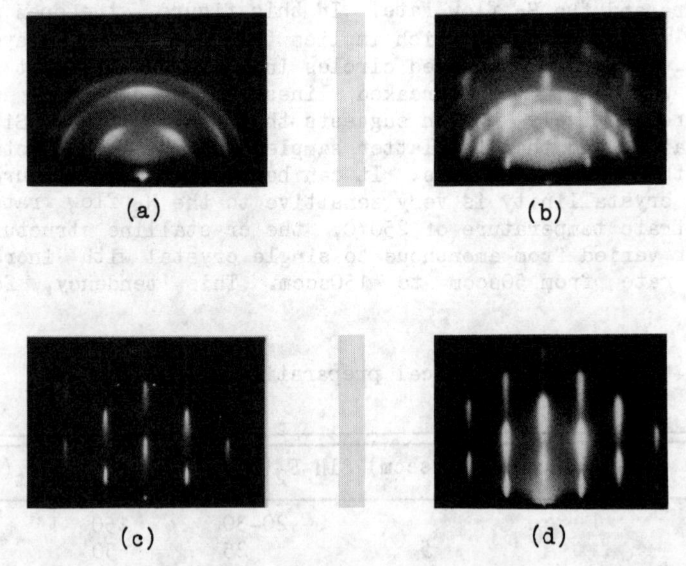

Fig. 2 The variation of RHEED patterns of epitaxial layers grown using the SiH_2F_2 flow rate

reactant gases was carried out by a differential evacuation system and the sample gas was introduced into the QMS analyzing tube. Figure 3 shows the time dependence of the mass line intensities for each of the species H_2(m/e=2), SiH_2(30), SiH_2F(49), $SiHF_2$(67) and SiF_3(85) when the UV lamp was turned on. In the figure, each intensity line was normalized with respect to its intensity when the lamp was off to cancel the decomposition in the QMS analyzing tube. As shown in Fig.3, the intensities of the fragments from SiH_4 and SiH_2F_2 decreased with increasing time because photolysis of SiH_4 and SiH_2F_2 occurred due to the UV irradiation. However, the intensity of SiF_3(85), which might be produced in a secondary process, increased with increasing time. At present, it is not clear whether this fragment was produced in gas phase reactions or in reactions of solid-phase silicon with the fluorine or fluoride radicals as suggested in our model. However, it is clear that the fluorine or fluoride radicals were certainly present in the gas phase during the photoepitaxial growth of silicon.

Further investigation of the crystallinity of the epitaxial layers was carried out by TEM and Raman scattering measurements. A Raman spectrum of the layer grown at 250°C using Si_2H_6 gas exhibited a sharp peak at 520cm^{-1}, identical to that of single crystal silicon, and the full width at half maximum (FWHM) of the spectrum was estimated to be about 6cm^{-1}. The FWHM of the spectrum was further improved by using SiH_4 gas during growth. The Raman spectrum of the epitaxial layer grown using SiH_4 gas also

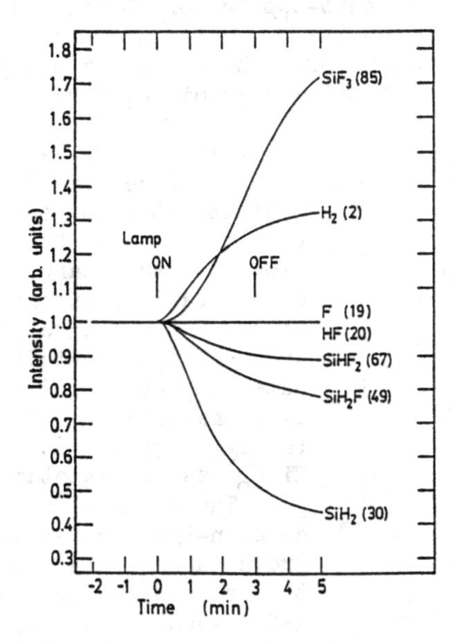

Fig.3 The time dependence of the mass line intensities

Fig.4 A dark TEM image of the sub-epi interface (x152,000)

exhibited a sharp peak at $520cm^{-1}$ with a FWHM of $4cm^{-1}$. This value is slightly different from the value of $3cm^{-1}$ obtained for the silicon substrate. Figure 4 shows a dark TEM image taken at 1000kV, of the interface of the photo-CVD epitaxial layer and the silicon substrate. The electron beam was parallel to the <011> direction. As shown in this photography, the epitaxial nature of the layer is confirmed by the fact that no grain boundaries are observed. However, many dislocations are observed in the epitaxial layer, which cause broadening of the Raman scattering peak of the layer due to structure fluctuations.

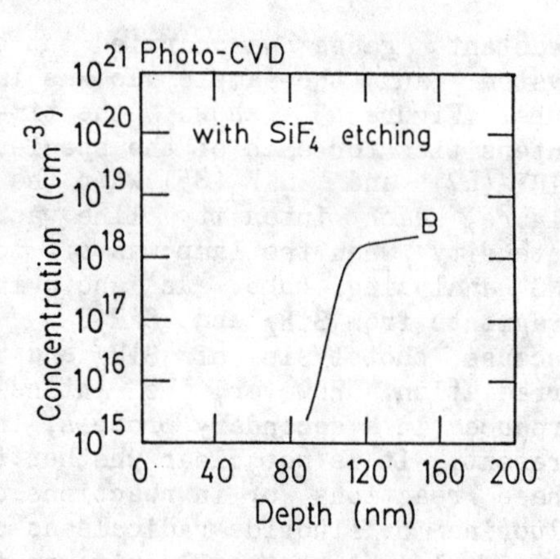

Fig.5 A boron profile as measured by SIMS

Figure 5 shows a boron profile as measured by secondary ion mass spectroscopy (SIMS). The photo-epitaxial layer was grown at 200°C on a p^+ Si substrate with a boron concentration of $5x10^{18}cm^{-3}$. The boron concentration in the epitaxial layer was about $1.7x10^{15}cm^{-3}$, which is the detection limit of the SIMS system, and it was found that the auto-doping of boron was negligible. The transition distance from the sub-epi interface to the epitaxial layer surface was less than 150Å. The abrupt impurity profile of the sub-epi interface is potentially promising for the realization of VLSI devices.

The electrical properties of the epitaxial layers were evaluated by Hall measurements. Figure 6 shows the free carrier mobility at room temperature as a function of the carrier concentration. In this figure, the circles and the triangle indicate the n-type and p-type material respectively, grown using Si_2H_6 as a reactant gas. The squares indicate n-type material grown using SiH_4 as the reactant gas. No difference in the electrical properties was observed between the samples prepared from Si_2H_6 and the

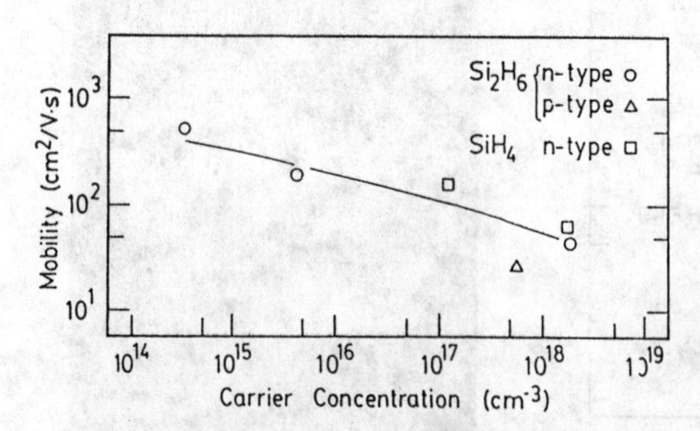

Fig.6 The free carrier mobility as a function of the carrier concentration

samples prepared from SiH_4. The electron mobility was seen to be the large value of $520cm^2/Vs$ with a low electron carrier concentration of $3.2x10^{14}cm^{-3}$. However, all values are less than that of bulk material. This is because the carriers were mainly scattered by the dislocations, which were confirmed to be in the epitaxial layers by TEM measurements.

IV. PLASMA CHEMICAL VAPOR DEPOSITION

The plasma-CVD system was a conventional diode-type glow-discharge reactor. Additional details of the system and the deposition procedure are available elsewhere[15]. The deposition conditions were as follows:- The gas flow rates were 1sccm for SiH_4, 0-15sccm for SiH_2F_2 and 100sccm for H_2. The total pressure during deposition was between 1-5Torr. The substrate temperature was 250°C and the RF power ranged from 10 to 100W. The epitaxial growth rate during plasma-CVD was 70-80A/min which is consider-ably greater than the low-temperature photo-CVD growth rate of 16A/min. The reason is probably that growth radicals are produced more easily in plasma-CVD than in mercury-sensitized photo-CVD.

Figure 7 shows the growth rate and crystallinity (from RHEED) of the epitaxial Si films as a function of the SiH_2F_2 flow rate. The film crystallinity was seen to be critically sensitive to the SiH_2F_2 flow rate as shown in Fig.7. This tendency is very similar to that of silicon epitaxy by photo-CVD. Thus, in plasma-CVD

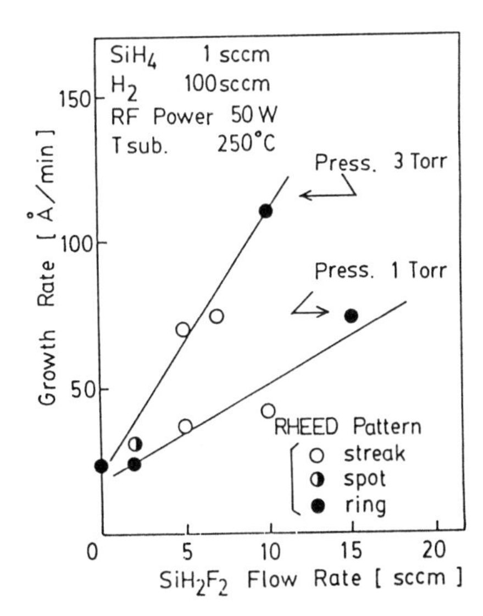

Fig.7 The growth rate of the epitaxial silicon as a function of the SiH_2F_2 flow rate

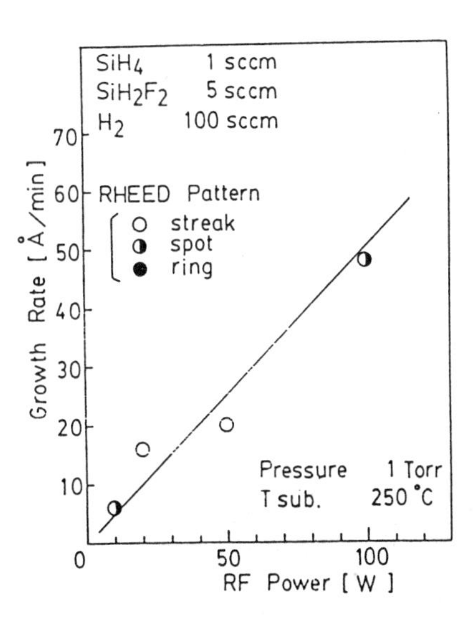

Fig.8 The growth rate of the epitaxial silicon as a function of applied plasma RF power

epitaxial growth, it might be that the hydrogen radicals and the fluorine or fluoride radicals could act in a similar manner as in photo-CVD growth. Hence, the epitaxial process proceeds at this low temperature.

Figure 8 shows the growth rate of the epitaxial silicon and its film crystallinity as a function of applied plasma RF power. The RHEED pattern showed fine streaked lines at powers ranging from 20 to 50W. The morphology of the epitaxial surface was also measured by optical microscopy and no roughness was observed. These features indicate that the epitaxial films had a smooth surface. At powers above 100W, the RHEED pattern changed to the ring pattern which indicated that crystallite growth was hindered by ion bombardment in the plasma at such high powers.

The transition distance from the sub-epi interface to the epitaxial layer surface defined by the boron profile was found to be 150A, which shows that the abruptness of sub-epi interface produced by plasma-CVD is comparable to that obtained by photo-CVD. However, in hydrogen evolution measurements, samples produced by both techniques showed different trends. From IR absorption measurements, it was found that our samples had a high hydrogen content with an average concentration of $\sim 10^{21} cm^{-3}$. This hydrogen might be concentrated in the dislocation region as observed by TEM measurements and is easily desorbed by raising the sample temperature. Figure 9 shows a set of hydrogen evolution spectra as a function of temperature. The hydrogen evolution occurred above 400°C with the sample prepared by plasma-CVD and exhibited an evolution peak at 430°C. In contrast to this feature the sample prepared by photo-CVD exhibited a flat profiled spectrum. This result implies that epilayers grown by plasma-CVD include a weak bonded Si-H structure in the dislocation region and these weak-bonded structures could hinder the crystal growth. Indeed, we have observed twin structure instead of single-crystalline structure in a thick layer whose thickness exceeded 5000A grown by plasma-CVD.

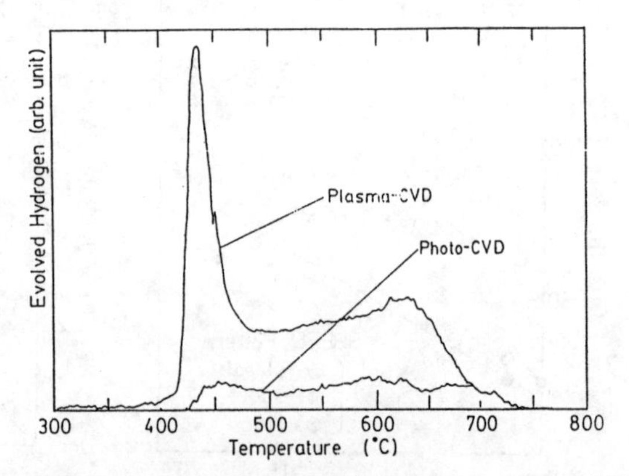

Fig.9 Hydrogen evolution spectra as a function of temperature

In conclusion to the previous two sections, it was shown that the presence of fluorine or fluoride radicals is absolutely necessary for the epitaxial growth of silicon at this low temperature of 200°C.

V. A METHOD FOR CLEANING Si SUBSTRATES AT LOW TEMPERATURE

The low-temperature epitaxy of silicon has been successfully demonstrated in the preceding sections by the photo- and plasma-CVD techniques. However, several undesirable properties of our epilayers were observed:-

1) high dislocation density in the layer.
2) poor electrical properties of the layer.
3) pileup of impurities at the sub-epi interface (as shown in Fig.10).

Figure 10 shows impurity (carbon, oxygen and nitrogen) profiles as measured by SIMS ((a) photo-CVD, (b) plasma-CVD). In Fig.10, the broken line indicates the impurity concentration when the substrate was cleaned by the routine method described in the previous sections, and the solid line indicates the impurity concentration when the substrate was etched by the SiF_4 plasma prior to the growth, which is a main subject of this section, and possibly be a solution to problem 3. To solve the problems 1 and 2, it was planned to raise the epitaxial growth temperature.

We investigated the plasma etching with SiF_4 gas using the same reactor as plasma-CVD. The substrate was a (100)-oriented Si wafer cleaned by organic solvents, followed by a HF(2.5%) dip. SiF_4 (100%) was used as an etching gas, because carbon contamination might occur by plasma etching using conventional halogenated-carbon etchants such as CF_4 and CF_3Br. Etching conditions were as follows:- SiF_4 flow rate=10sccm; $T_{sub.}$=250°C; RF power=20-100W; and total pressure=0.3Torr.

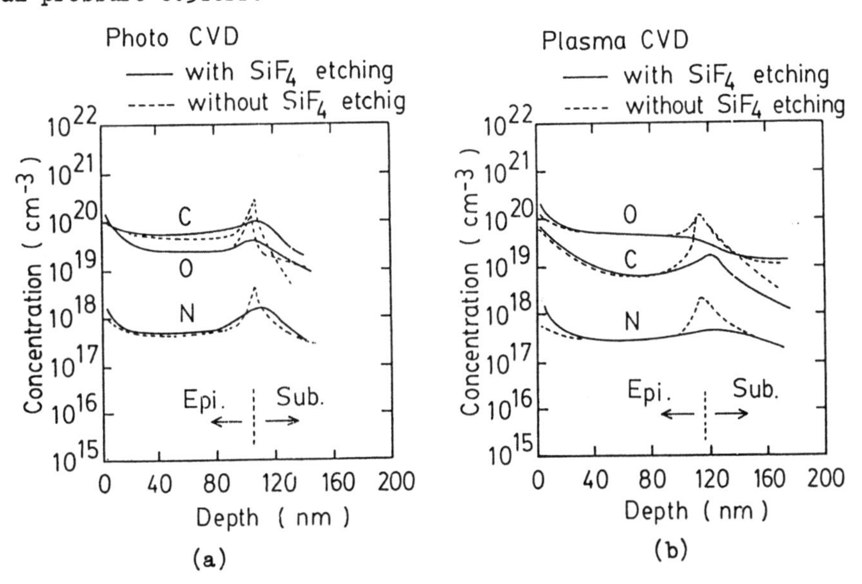

Fig. 10 Impurity profiles as measured by SIMS (a) photo-CVD, (b) plasma-CVD

Figure 11 shows the dependence of the etching rate as a function of the RF power. A silicon dioxide layer prepared by thermal oxidation at 1100°C was also etched by the SiF_4 plasma at a RF power of 50W with an etch rate of 18A/min. The etch rate became constant in the high-power region. The highest etch rate of 67A/min was obtained at a RF power of 50W.

The MS measurement was carried out to clarify the etching mechanism. Figure 12 shows the dependence of the mass line intensity for each of the species such as F (open circle), SiF (closed rectangle), SiF_2H (open triangle), SiF_3 (open rectangle) and SiF_4 (closed circle) as a function of the RF power. In this figure, the dependence of the mass line intensity of SiF and SiF_3 on the RF power is similar to that of SiF_4, which implied that both SiF and SiF_3 species were produced in the primary reaction process of the glow-discharge decomposition of SiF_4 gas. On the other hand, the dependence of the mass line intensity of SiF_2H on the RF power closely corresponds to the dependence of the etching rate on the RF power. This result suggested that the SiF_2H species was produced in the reaction of solid-phase silicon with fluoride radicals, which is in agreement with the results of Flamm[16], that is, in silicon etching phenomena with F atoms the formation of SiF_2 is the rate-limiting step which controls the gasification of Si. The morphology of the etched layer was specular and no roughness was observed by scanning electron microscopic (SEM) measurement.

Thus, the plasma etching of silicon was confirmed using SiF_4 gas as described above. We employed this technique for cleaning a silicon substrate. The cleaning of the silicon substrate and deposition procedure were conducted in the following way. Firstly, the substrate was cleaned using conventional organic solvents,

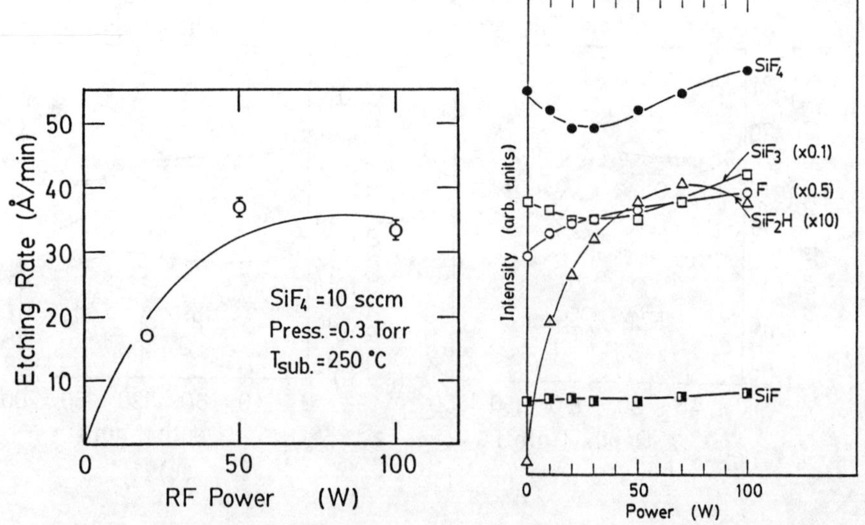

Fig.11 The power dependence of the etching rate

Fig.12 The power dependence of the mass line intensity

followed by a chemical cleaning using the RCA process and then it was loaded into a reactor with a thin native oxide layer preventing impurity contamination. Following this the plasma etching was carried out to etch the native oxide and silicon substrate (approximately 50A), which was immediately followed by a photo or plasma deposition. The resulting samples were evaluated by SIMS analysis. A typical SIMS scan for an epitaxial film cleaned by this method is shown in Fig.10. The concentration of the impurities at the sub-epi interface was reduced by about one order of magnitude due to the plasma etching. Thus it was found that the plasma etching is very effective in the cleaning of silicon substrates prior to growth.

VI. LASER-ASSISTED CHEMICAL VAPOR DEPOSITION

Figure 13 shows a schematic diagram of the laser-CVD system that was use for the silicon epitaxial growth. The deposition system consists of a single-wafer reactor that has a load-lock entry and a turbomolecular pump to produce a hydrocarbon-free, low base pressure environment ($\sim 10^{-7}$Torr). The substrates used for the the epitaxial depositions were (100)-oriented silicon wafers, which were chemically cleaned using Shiraki's procedure[17] and pre-annealed at 800°C under high vacuum conditions ($\sim 10^{-7}$Torr). An ArF(193nm) or XeF(351nm) excimer laser(100pps, 50-70mJ/cm^2) was used as a UV irradiation source. The laser beam was kept normal to the substrate throughout this work. The reactant gas was 10% Si_2H_6 diluted with H_2, whose fundamental absorption edge is at 200nm[18]. Hence, direct photodecomposition of the reactant gas is caused by ArF laser irradiation, whereas it does not occur by XeF laser irradiation. The inner surface of the quartz window was coated with FOMBLIN oil and purged with H_2 gas to maintain its transparency. The detailed deposition conditions were as follows:- $T_{sub.}$=550-750°C; Si_2H_6(10%) flow rate=1sccm; H_2 flow rate=9sccm; and total pressure=0.15Torr. Epitaxial nature of the obtained films were confirmed by RHEED measurements and it was found that the films prepared by thermal-CVD above 650°C and the films prepared by laser-CVD above 600°C were single crystals.

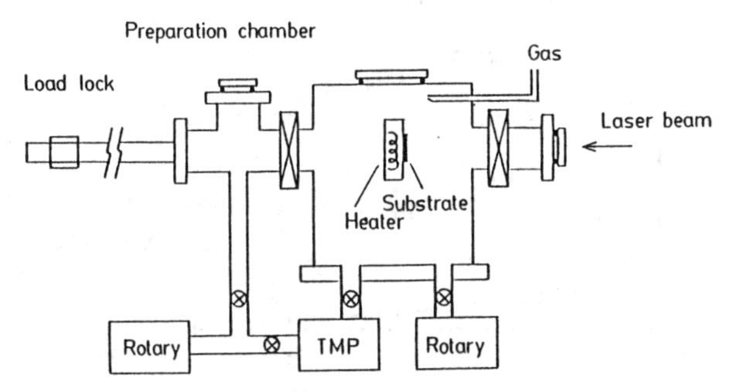

Fig.13 A schematic diagram of the laser-CVD system

The dependence of growth rate on the substrate temperature is shown in Fig.14. From this figure, the growth rates of the thermal-CVD and laser-CVD using XeF irradiation became constant in the high-temperature region above 700°C. In contrast to this result, the growth rate of laser-CVD with ArF irradiation did not become constant. The reason might be that only thermal decomposition contributed to the growth rate and the epitaxial growth was limited by the supply of the reactant gases in the high-temperature region in the former case, but on the other hand, the photodecomposition of the reactant gases due to ArF irradiation contributed to the growth rate and the enhancement of growth rate occurred in the latter case. The effect of the laser irradiation is more apparent in the rate-limiting region less than 700°C. In this region, laser irradiation of the growing films was found to have a strong affect on the growth rate, which was 3 times greater with the laser irradiation. Moreover, the activation energy of the epitaxial growth decreased to 39kcal/mol (ArF irradiation) or 37kcal/mol (XeF irradiation) from 45kcal/mol (thermal decomposition). The

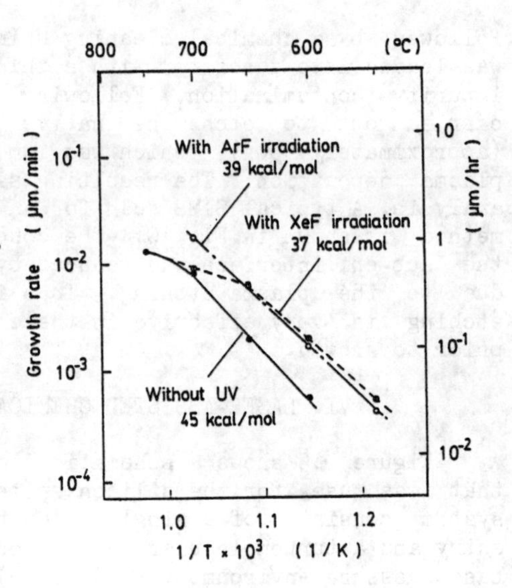

Fig.14 The dependence of growth rate on the substrate temperature

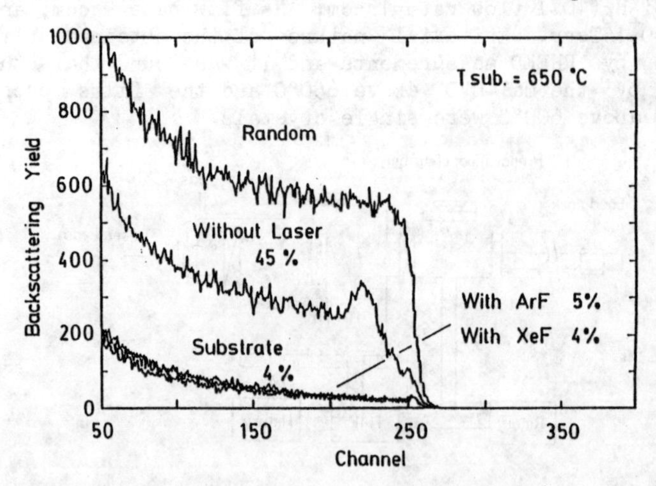

Fig.15 A typical set of backscattering spectra

reason of these effects are not known, however, it might be probable that laser irradiation makes it possible to activate both gas-phase and surface reactions and to enhance the decomposition of the reactant gases. Consequently, the enhancement of growth rate and the reduction of activation energy was realized by laser irradiation.

The improvement of the layer crystallinity with laser irradiation is more pronouced than the effect on the growth rate. To evaluate the crystallinity, epitaxial layers were examined by Rutherford backscattering and channeling measurements. A typical set of backscattering spectra are shown in Fig.15 for the epitaxial layer grown by laser-CVD and thermal-CVD at 650°C on a silicon substrate. Crystallinity was measured by comparing channeled to random yields (Y_{chan}/Y_{rand}). An ideal crystal should give a Y_{chan}/Y_{rand} value of ∿3.0%. This value is close to that obtained from a silicon substrate used as a control sample for our measurement system. It is evident from the figure that smooth growth is achieved for the samples irradiated with laser beam without a major sacrifice of crystalline quality. However, it is significant that for the sample grown only by thermal energy Y_{chan}/Y_{rand} value is very large indicating poor crystalline quality. Furthermore, the thermally prepared sample shows a peak of backscattered yield at a channel number of 220, which suggests an existence of structure distortion at the sub-epi interface. Such superiority was also observed in the electrical properties of the epitaxial films.

The electrical properties of the epitaxial layers were evaluated by Hall measurements. The undoped samples were deposited on 1kΩcm Si substrates. The results are summarized in Table III. The high carrier concentration of the epitaxial layers (∿10^{17}cm^{-3}) means that In atoms were incorporated into the epitaxial layer during the growth because the silicon substrate was mounted on the substrate holder using In solder. This consideration was also supported by the activation energy of the carrier concentration (∿170meV, the ionization energy of In is 160meV[19]). Significant differences could not be observed between the two samples deposited with laser irradiation. However, mobility of the samples deposited by thermal-CVD was inferior to that of the samples deposited by laser-CVD, correlating well with the backscattering results. The reason for the enhancement of the crystallinities and the elec-

Table III. The electrical properties of the epitaxial layers

	ArF	XeF	Thermal
Resistivity (Ω•cm)	0.75	0.72	2.6
Mobility (cm^2/Vs)	90	86	35
Carrier Con. (cm^{-3})	9.4x10^{16}	1.0x10^{17}	6.7x10^{16}
Thickness (μm)	0.68	0.58	1.35

trical properties of the epitaxial layers are now being studied.

CONCLUSIONS

We have grown silicon epitaxial layers at a temperature as low as 200°C and also succeeded in doping the layers n and p type using the mercury-sensitized photochemical vapor deposition and plasma-CVD techniques. It was found that the presence of SiH_2F_2 in the gas phase is essential for the growth of Si epitaxy by both methods at this low temperature. Furthermore, silicon epitaxy at 600°C has been demonstrated by the ArF and XeF excimer laser-induced photo-CVD technique and found that laser irradiation on the substrate was very effective for the enhancement of the layer crystallinities and electrical properties. However, systematic characterization of the Si films in terms of defect density is necessary in evaluating the usefulness of low-temperature epitaxy proposed in this study.

ACKNOWLEDGMENT

The authors wish to thank Mitsui Toatsu Chemicals, Inc. for the supply of Si_2H_6 and SiH_2F_2 gases. This work was supported in part by scientific research grant-in aid No.62850053 for Developmental Scientific Research from the Ministry of Education, Science and Culture of Japan.

REFERENCES

1) Y.Ota, J.Electrochem.Soc., 126, 1761 (1979).
2) G.E.Becker and J.C.Bean, J.Appl.Phys., 48, 3395 (1977).
3) P.C.Zalm and L.J.Beckers, Appl.Phy.Lett., 41, 167 (1982).
4) M.J.P.Duchemin, M.M.Bonnet and M.F.Koelsch, J. Electrochem. Soc., 125, 637 (1978).
5) T.J.Donahue and R.Reif, J.Appl.Phys., 57, 2757 (1985).
6) W.R.Burger and R.Reif,IEEE Electron Device Lett.,EDL-7,206(1986)
7) M.Kumagawa, H.Sunami, T.Terasaki and J.Nishizawa, Jap. J. Appl. Phys., 7, 1332 (1968).
8) R.G.Frieser, J.Electrochem.Soc., 115, 401 (1968).
9) T.Yamasaki, T.Ito and H.Ishikawa, Symp.VLSI Technol., 56 (1984).
10) A.Ishitani, M.Kanamori and H.Tsuya, J.Appl.Phys.,57,2956(1985).
11) K.Suzuki, D.Lubben and J.E.Greene, J.Appl.Phys., 58, 979(1985).
12) S.Nishida, T.Shiimoto, A.Yamada, S.Karasawa, M.Konagai and K.Takahashi, Appl.Phys.Lett., 49, 79 (1986).
13) S.Nishida, H.Tasaki, M.Konagai and K.Takahashi, J.Appl.Phys., 58, 1427 (1985).
14) A.Matsuda, J.Non-Cryst.Solids, 59&60, 767 (1983).
15) K.Nagamine, A.Yamada, M.Konagai and K.Takahashi, Jap. J. Appl. Phys, 26, L951 (1987).
16) D.L.Flamm, V.M.Donnelly and J.A.Mucha,J.Appl.Phys.,55,3633(1981)
17) A.Ishizaka and Y.Shiraki, J.Electrochem.Soc., 133,666 (1986).
18) H.J.Emeleus and K.Stewart, Trans.Farady Soc., 22,1577 (1936).
19) S.M.Sze, Physics of Semiconductors Devices, 2nd ed., Wiley, New York, 1981.

DEPOSITION OF THIN FILMS BY ION BEAM SPUTTERING : MECHANISMS AND EPITAXIAL GROWTH

C. Schwebel, G. Gautherin
Institut d'Electronique Fondamentale, Unité Associée au CNRS
(UA 22) Université Paris XI, Bât. 220 - 91405 ORSAY-CEDEX (F)

ABSTRACT

In this paper, the capabilities of ion beam sputtering (IBS) to prepare epitaxial layers of various materials are investigated. Thin films (W < 0,5 μm) of silicon, yttria stabilized zirconia (YSZ), tungsten and nickel were deposited using an UHV apparatus equipped with surface analytical tools (RHEED, Auger electron spectrometer). Rare gas ions (Ar, Kr, Xe) of 20 keV energy were used. Silicon homoepitaxial films deposited at temperatures above 700°C showed structure and electrical properties close to the bulk. Layers of YSZ on Si substrates were deposited by sputtering of a $(Y_2O_3)_{0.23}(ZrO_2)_{0.77}$ target. Films of good epitaxial quality were grown on Si(100) substrates under partial pressure of oxygen at temperatures in the range 700°C-800°C. Tungsten silicide and Ni_xGaAs films were prepared by pure metal deposition on heated silicon and GaAs substrates respectively. Structure and electrical properties of these films have been determined as a function of the deposition temperature. In order to improve the thin film properties, sputtering-related phenomena were studied.

INTRODUCTION

In very large scale integration technology, device performances are strongly dependent on the interface quality between the different active layers. Consequently, the improvement of such device characteristics requires, in particular, a lowering of the process temperature to minimize autodoping and outdiffusion phenomena. What are the possible means to reach this goal in the epitaxy field ? To answer this question it is necessary to examine the process of accommodation between the condensing particles and the substrate surface at temperature T. If this temperature is so low that the adsorption of incident atoms onto the surface occurs at the impingement point, a disordered structure i.e. amorphous grows. At higher substrate temperature and if the growth rate is not too high, the condensed atoms diffuse over the surface leading to their arrangement. In this case a more or less ordered structure i.e. monocrystalline, polycrystalline, may be obtained. If the condensing atoms own higher energy due to either emission process or momentum transfer from fast particles impinging the growing film one can assume that the same surface mobility may be obtained at lower substrate temperature. However it is important to note that an energy too large could lead to the creation of defects in the growing films.

In conclusion one can expect, that deposition of superthermal particles will lead to a lowering of the epitaxial growth temperature. So the problem consists of creating free particles with adequate energy. This has been realized either by sputtering a target or by accelerating ionized species in an electrostatic field towards a substrate. Two techniques based on the second principle are already developed : ion beam deposition (IBD) [1, 2] and ionized cluster beam deposition (ICBD)[3]. The most important limitation of these techniques are the difficulty to deposit less volatile material. However these techniques are very interesting because the particle energy can be theoretically selected by adjustment of the voltage applied between the region of ionization and the substrate. Due to space charge effect, difficulties are encountered experimentally with IBD to transport intense beams at low energy.

The sputtering process creates atoms with most probable energy of the order of $U_o/2$ [4] (a few eV), where U_o is the binding energy for target surface atoms. It is important to note that, contrary to evaporation, vapor pressure problems do not exist with a sputtering process. Among the sputtering techniques, the ion beam sputtering deposition (IBS) is quite attractive since the layer growth is isolated from the ion generation process.

In this paper, we report only our results on IBS in the epitaxy field. The reader will find elsewere, in a recent article, a review of the results obtained by the other techniques [5]. We have investigated the epitaxial growth of silicon, zirconia, tungsten silicide and Ni_xGaAs by IBS. Silicon has been selected because it is a good test vehicule in the microelectronics field ; other materials were chosen because the IBS technique is well adapted to deposition of material of low vapor pressure.

EXPERIMENTAL

A) Experimental apparatus

The experimental apparatus (Fig. 1), previously described in more detail [6], consists of three main parts : the ion source, the intermediate chamber and the deposition chamber. The duoplasmatron ion source used has been constructed by our group, using UHV technology. In order to minimize ion beam contamination, the internal electrode of the source which undergoes erosion has been partly made of the same material as the target. The discharge composition is continually checked by mass analysis of the gas present in the anodic part of the ion source.

The intermediate chamber, situated between the ion source and the deposition chamber, contains the extractor and two Einzel lenses, and is pumped by a cryogenerator (1500 l/s for Ar).

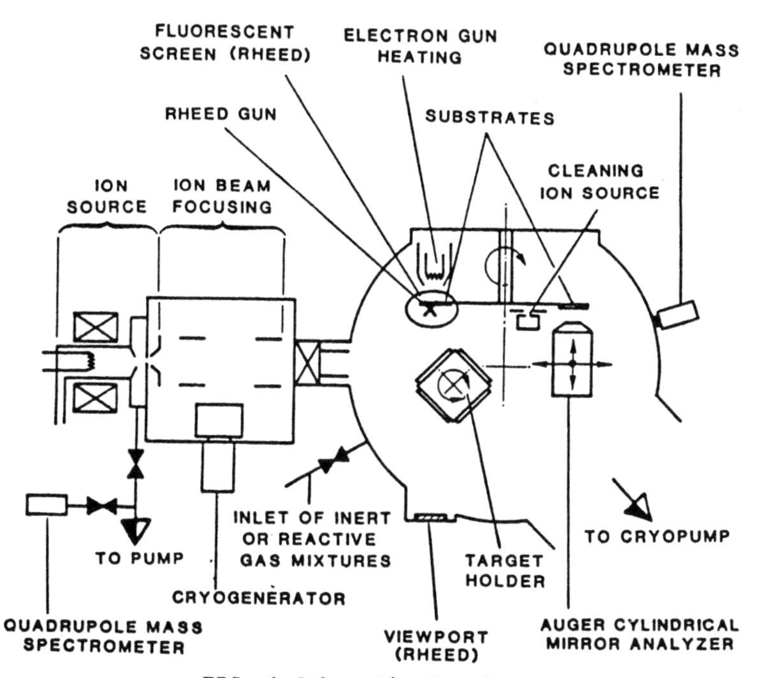

FIG. 1 Schematic drawing of the ion-beam sputter deposition system.

When the ion source is operating, the rare gas pressure in this chamber, caused by the neutrals coming from the source, is 3×10^{-4} Pa.

The deposition chamber is connected to the preceding one by a tube of low conductance (C = 70 l/s) allowing the passage of the ion beam, and is pumped by a He liquid cryopump giving an ultimate pressure of 10^{-7} Pa. During ion beam sputter deposition, the pressure is 2×10^{-5} Pa and consists mainly of rare gas (99%).

The deposition chamber contains a water cooled copper target holder and a carousel-type susbtrate holder in which ten samples can be inserted. The substrate at the deposition position can be heated directly on its back side by a hot tungsten filament. To obtain temperature higher than 550°C, this filament is negatively polarized relatively to the substrate and acts like an e-gun. With this heating technique a temperature of 1200°C is reached for a total power received by the substrate of 45 W. The substrate temperature is obtained using a sheet resistor deposited on a silicon substrate and previously calibrated in an oven. In order to allow sputtering of insulating materials, a neutralizer filament implanted near the target was used.

Additional equipment attached to the deposition chamber

includes a residual gas analyzer, a thickness sensor, a reflection high energy electron diffraction system (RHEED), an Auger electron spectrometer (AES), and a secondary ion mass spectrometer (SIMS) for sputtered ion analysis.

B) **Déposition parameters**

1. Nature of the primary ions

When an ion strikes a solid surface it can be either trapped or backscattered. The second mechanism leads to the emission by the target of particles having an energy similar ot the primary ion energy. Consequently, in thin film deposition involving sputtering, backscattering will induce defects in the growing film. Nevertheless, a higher M_1/M_2 ratio, where M_1 and M_2 are the mass of ion and target atoms, respectively, lowers the probability of one collision processes [7]. Consequently, all of our experiments are carried out with primary ions heavier than target atoms.

2. Deposition rate

The deposition rate is given as a function of the beam characteristics as follows :

$$V = \int_{So} Y(E_1)\ j(x, y)\ f(\theta)\ s_A\ (E_2, T)\ \Omega\ (x, y)\ dx\ dy, \quad (1)$$

where So is the surface of the target covered by the beam, $Y(E_1)$ is the sputtering yield of the target material for primary ion of energy E_1, $j(x, y)$ is the ion beam density at point (x, y) of the target, $f(\theta)$ is the angular distribution of sputtered atoms emitted in direction θ, $s_A\ (E_2, T)$ is the sticking coefficient of a sputtered A particle having an energy E_2 on a substrate at temperature T, $\Omega\ (x, y)$ is the solid angle under which the substrate is seen from point (x, y). The function $f(\theta)$ has been determined by thickness measurement of films deposited on ten substrates set on a cylindrical collector centered at the point where the beam axis intercepts the target (Fig. 2). The intensity distribution of the beam is measured using a movable multicollector probe. A Gaussian distribution can be obtained by adjustment of the lens voltage. For example, taking into account such a beam profile, the angular distribution of sputtered silicon was determined for different deposition conditions. Whatever the nature (Ar, Kr, Xe), the energy

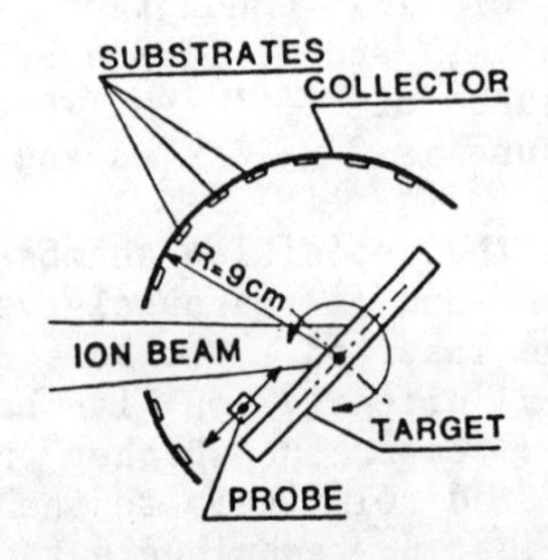

FIG. 2 Schematic drawing showing arrangement of the target, collector and probe for angular distribution studies.

(10-20 keV) of the primary ion and the mean current density of the beam (70-450 μA/cm^2) we have used, the angular distribution observed could always be fitted well by a $\cos^2 \theta$ function.

C) Substrate preparation

Before their introduction in the deposition chamber the substrates were first degreased and then chemically treated using the process of Kern and Puotinen for the silicon, and HCl : H_2O (1:1) solution for the gallium arsenide. Prior to the deposition the surface oxide was removed in situ under ultra high vacuum (p = 2 x 10^{-7} Pa) by flashing the substrates to 1100°C for 180 s and 650°C for 30 s for Si and GaAs respectively. These substrate preparations were checked in situ by means of Auger electron spectrometry (AES).

FILM CHARACTERISTICS

A) Silicon homoepitaxy

Silicon films were deposited with 1.5x10^{-2} to 15x10^{-2} nm/s growth rates. This deposition parameter was found to be independent of the substrate temperature in the range 20-900°C. This result confirms that the sticking coefficient s_{Si} (E_2, T) is also independent of temperature in this range.
As expected the crystallographic nature of films deposited on Si (100) subtrates is a function of the growth temperature. The beginning of single crystal growth occurs at 250°C (see Fig. 3). Above 700°C, good crystalline structure was obtained and the RHEED patterns show Kikuchi lines. Rutherford backscattering (RBS) and channeling of MeV He ions were used to study lattice disorder. The value of channeling yield χ_{min} was 2% for a thin film (W = 500 nm) deposited at 850°C. This shows that the structure of this film is perfectly crystalline.

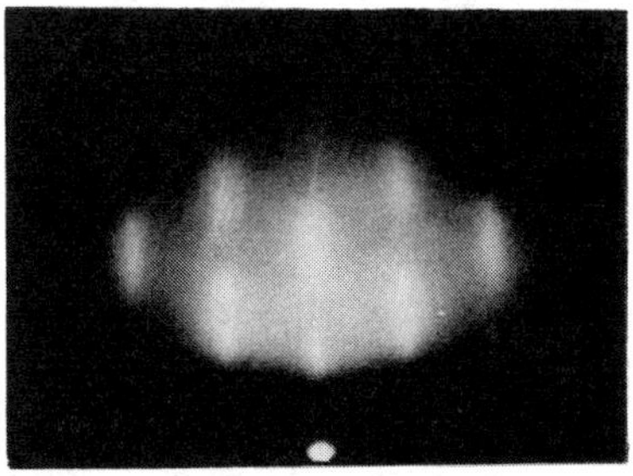

FIG. 3 RHEED pattern of a Si film deposited at 250°C.

The surface roughness of the deposited layer was investigated by a light scattering technique [8]. From the first stage of growth, the roughness is quite small ; at thicknesses above 30 nm, the value obtained (1 nm) is of the order of the resolution of the measurement method. These results are corroborated by the examination of the film with scanning electron microscopy (G = 10 000 X, resolution 5 nm) which showed that the surface was featureless for a 9 nm thick layer.

The electrical properties of a semiconductor film depend on

the concentration of both desired doping impurities and contamination impurities, as well as the electrical levels of these impurities in the band gap. The second kind of impurities may have two origins : accidental contamination due to a wrong technology and rare gas incorporation during the film growth. Rutherford backscattering (RBS) of MeV He ions and secondary ion mass spectrometry (SIMS) analysis were used to detect the concentration and nature of impurities. Rare gas concentration in the films was always found to be less than 1 at .%. The lowest value was obtained for heavier ions [9]. Possible mechanisms of rare gas incorporation will be analyzed in a later paragraph. In order to reduce the probability of accidental contamination, the ion source and beam transport system were specially designed. As a result, no metallic impurities were detected.

The doping of the layer was achieved by sputtering doped targets. Since at steady state the composition of the flux of sputtered particles is the same as that of the target, the transfer efficiency R of the dopant from the target to the film is

$$R(\theta, T) = \frac{f_B(\theta)\ s_B(T, E)}{f_{Si}(\theta)\ s_{Si}(T, E)}, \quad (2)$$

where $f(\theta)$ and $s(T,E)$ are the functions defined above (eq. (1)), and the subscripts B and Si denote respectively boron and silicon. Let us note that the energy E of the sputtered particle may be a function of the emission direction θ. Figure 4 shows that R is constant and of the order of 50%. We found that R is not affected significantly by the doping level (10^{16} - 5×10^{18} cm^{-3}), ion nature (Ar^+, Xe^+) or substrate temperature (T = 20-900°C). The distribution of the dopant determined by SIMS analysis was found homogeneous within a layer deposited at 700°C. As expected at this deposition temperature, a steep doping profile was observed at the layer-substrate interface[10]. The hole concentration and profile have been examined by capacitance-voltage measurements on mercury Schottky diodes. The C(V) measurements give results in agreement with the doping level and profile determined by SIMS analysis. This shows that the doping impurities are electrically active and therefore in a substitional

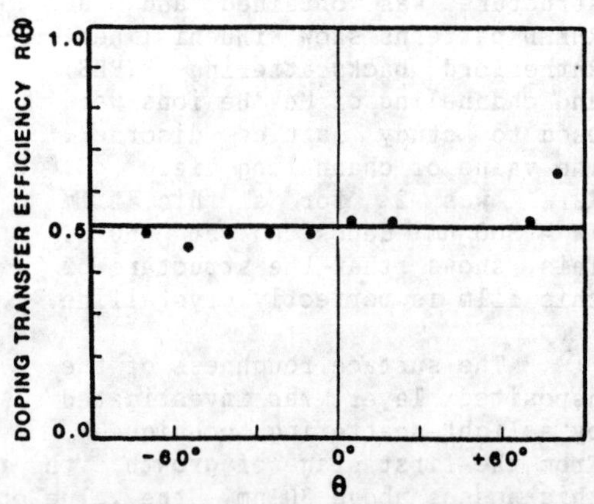

FIG. 4 Doping transfer efficiency as a function of the emission direction of sputtered particles.

FIG. 5 Hall mobility vs temperature of a 500 nm Si epitaxial film deposited at 710°C ($p = 2 \times 10^{16} cm^{-3}$).

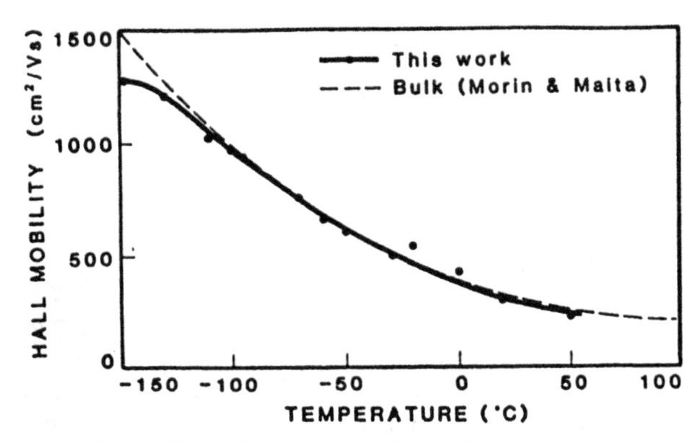

site in the silicon crystal. The free carrier density versus temperature suggests the presence of an acceptor level at E_v + 0.05 eV which can be attributed to the boron dopant. Figure 5 shows a typical curve of Hall mobility as a function of temperature for a 500 nm thick film grown at 700°C. For films grown at 600-900°C, room temperature Hall mobilities fell within 20% of the bulk value at the same doping level. However, at low temperature the mobility diverges from the bulk mobility reported by Morin and Maita [11].

B) Heterostructures

In this part we report our initial results on epitaxial growth of yttria stabilized zirconia (YSZ) on silicon, tungsten silicide on silicon and Ni_x GaAs on gallium arsenide by IBS.

1. YSZ/Si

YSZ thin films (W ≈ 150 nm) were deposited on Si (100) substrates by sputtering of a $(Y_2O_3)_{0.23}(ZrO_2)_{0.77}$ target. The deposition rate was 8×10^{-2} nm/s corresponding to 20 keV Xe^+ and beam current of 1.5 mA.

The crystallographic properties of the films were investigated in situ by RHEED and ex situ by X ray diffraction (XRD) using CuKα radiation. The film composition was checked indirectly by this latter technique and by AES. As expected, thin films prepared in UHV conditions were oxygen deficient. In order to compensate the oxygen loss, the majority of the films was grown under oxygen partial pressure. YSZ layers deposited at temperatures within the range 700°C-800°C under an oxygen pressure of the order of 10^{-4} Pa, exhibited monocrystalline cubic phase. Fig. 6 presents the diffraction pattern of such a layer -deposition temperature of 700°C, oxygen deposition pressure 10^{-4} Pa-. One can observe (200) and (400) peaks of YSZ as well as the strong (400) peak of Si. The lattice parameter deduced from the X ray spectrum was found to be 0.518 nm. This values corresponds to an yttria content in the

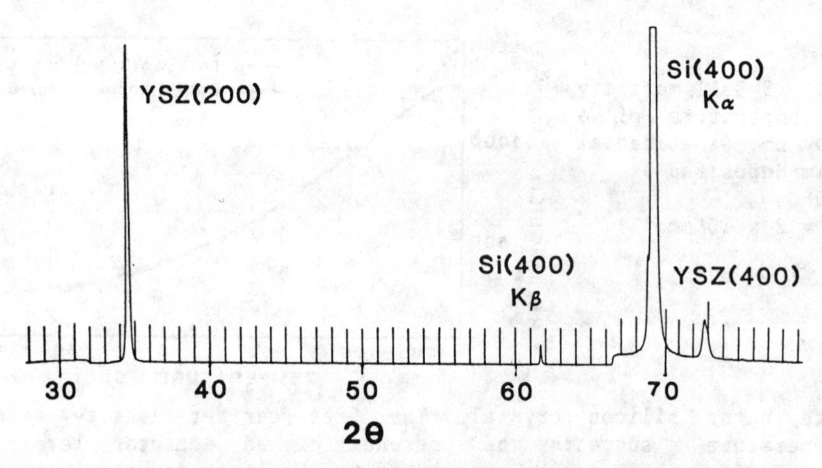

FIG.6 X Ray Diffraction pattern of 150 nm YSZ film grown at 700°C under an 1.2×10^{-4} Pa oxygen pressure on Si(100) substrate

film close to the target one. This was can corroborated by AES examination of the film (Fig. 7). Indeed, according to Salomon investigations on the quantification of yttria in YSZ [12], the Y_{MNN}/Zr_{MNN} peak to peak height ratio of 0.67, confirms this yttria content. RHEED pattern of Figure 8 gives a qualitative indication of the degree of crystallinity of films deposited at 750°C under 1.2×10^{-4} Pa oxygen pressure.

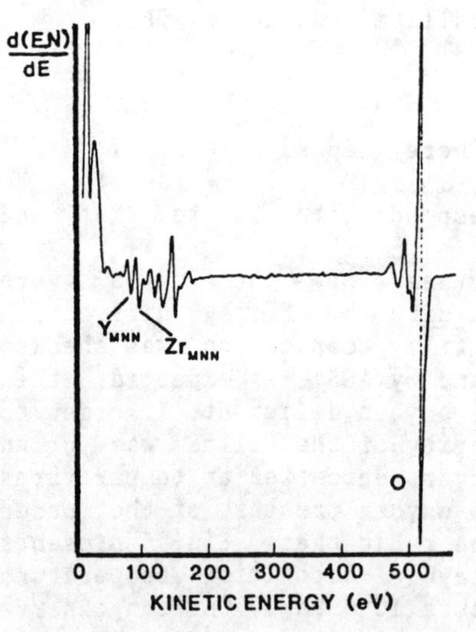

FIG. 7 Auger spectrum of YSZ film grown at 700°C under an 1.2×10^{-4} Pa oxygen pressure

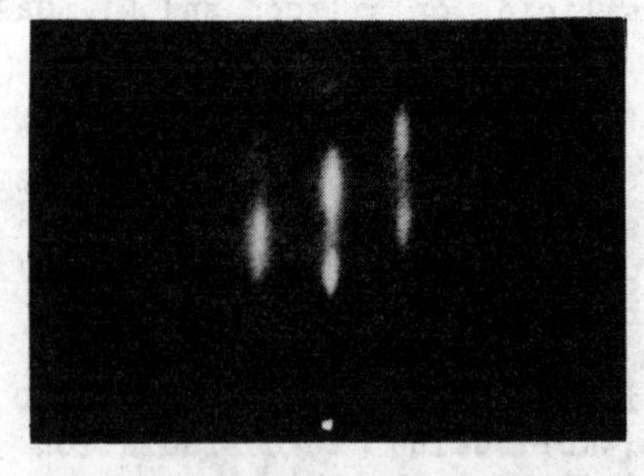

FIG. 8 RHEED pattern of 150 nm YSZ film grown at 750°C under an 1.2×10^{-4} Pa oxygen pressure on Si(100) substrate.

2. *Tungsten silicide*

Tungsten silicides were obtained by pure metal deposition on heated (100) and (111) silicon substrates [13]. W layers of 8nm were deposited at 4.5 x 10^{-2} nm/s growth rate. At deposition temperatures above 450°C the condensing material reacts with the substrate to lead to a silicide formation. Under this conditions the silicide thickness measured by ellipsometry was of the order of 19 nm. SEM examination of these silicide films revealed rough surfaces. The crystalline structure of the layers was investigated by RHEED. At a deposition temperature as low as 450°C, hexagonal WSi_2 structure was observed (Fig. 9a). The tetragonal structure began to appear at 500°C , while only this phase was detected above 650°C (Fig. 9b).

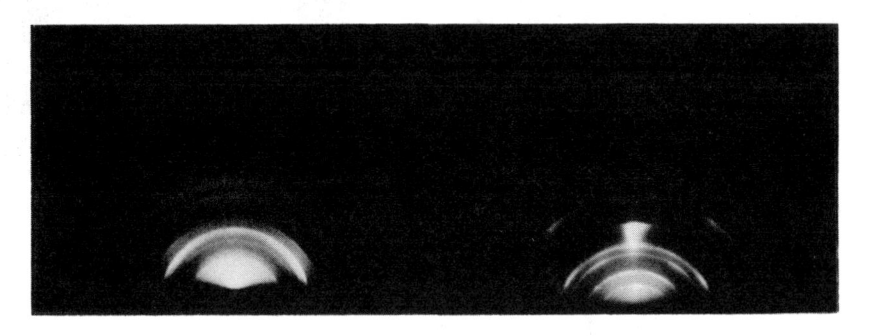

FIG. 9 RHEED patterns of tungsten silicide films prepared at a) 450°C and b) 710°C

The electrical films characteristics involving resistivity on Schottky barrier height ϕ_B were determined by four probe and I-V measurements respectively. The WSi_2 layer prepared at 450°C deposition temperature had resistivity of 400 µΩcm. With the increase of deposition temperature up to 500°C, the resistivity increased up to a maximal value of 800µΩcm. At deposition temperatures above 500°C i.e. when the tetragonal phase began to appear, the resistivity decreased as the temperature was increased. At 700°C deposition temperature, WSi_2 films had a resistivity of 250 µΩcm. In order to assist in the top contact formation for I-V measurement, approximatively 100 nm of W were deposited at room temperature on the top of the silicide layers. WSi_2 layers prepared at 450°C on Si n type (4 x 10^{15}cm^{-3} P) substrates had a Schottky barrier height of 0.65 eV. These electrical characteristics are in agreement with those reported in literature (see for example [14]).

3. Ni_x GaAs/GaAs

Ni_x GaAs films were formed by nickel deposition on heated GaAs n type ($8 \times 10^{15} cm^{-3}$ Si) substrates. Ni layers of 15 nm were deposited at 8×10^{-2} nm/s growth rate. Films grown at room temperature had resistivity of 16 µΩcm. At 200°C Ni_2GaAs hexagonal structure was observed. In this case, the equivalent thickness of the composite films was estimated at 22nm. These films had a resistivity of 50 µΩcm, Schottky barrier height ϕ_B of 0.8 eV and an ideality factor n = 1.04. Films formed at 500°C deposition temperature exhibited a complex structure with a mixing of Ni_2GaAs hexagonal phase and other undefined phases (Fig. 10). Such films were characterized by a resistivity of 10 µΩcm, Schottky barrier height ϕ_B of 0.76 eV and an ideality factor n = 1.14. These electrical parameters are close to those reported by Lahav et al.[15] for films prepared by post-deposition annealing.

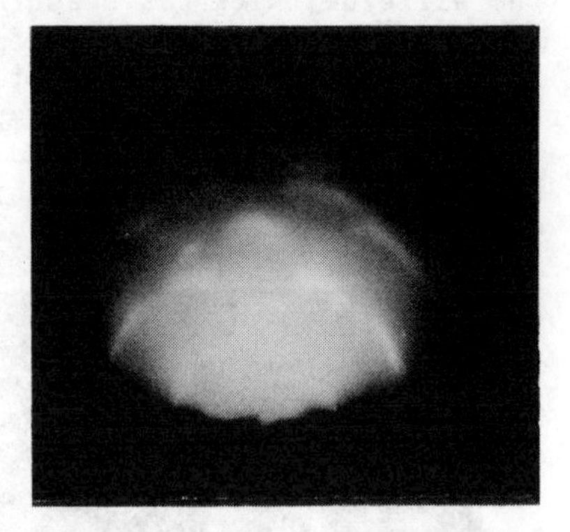

FIG. 10 RHEED pattern of Ni-GaAs film grown at 500°C.

DISCUSSION

The present results show that IBS films have properties very similar to those prepared by other techniques at equal or higher deposition temperature. However, thorough examination of the electrical properties sometimes reveals anomalies. For example, as we have previously noted, homo-epitaxial silicon films had low temperature mobility less than that of the bulk. This can be explained by carrier scattering on structural defects -point defects, dislocations-. These defects could be ascribed to a) the presence of rare gas in the films $-n_{Xe} = 10^{-2}$ at. % at 750°C deposition temperature-, b) the interaction of fast rare gas or/and energetic sputtered Si particles with the growing film. With the goal of determining the relative responsibility of each process, study of rare gas incorporation mechanisms in growing film was carried out.

When an energetic ion strikes a solid surface, it can be either trapped in the target or backscattered. In the first case, the implanted rare gas atoms can leave the target by two processes : sputtering and more likely out diffusion. Since thermal rare gas atoms could only be weakly physisorbed on the silicon surface at room temperature none of the mechanisms involving such species could participate in rare gas incorporation. Therefore the trapped rare gas atoms in our films could only result from sputtering of atoms implanted in the target or from backscattering of the primary ions on the target surface. In order to define the right hypothesis, angular distribution of the energetic rare gas emitted from the target was determined. It was calculated from the silicon one (§B2) by using the following relation :

$$I_{R.G.}(\theta) = I_{Si}(\theta) \times C_{R.G./Si}(\theta)$$

where $I_{R.G.}(\theta)$ and $I_{Si}(\theta)$ are emission intensities in direction θ for rare gas and silicon respectively, and $C_{R.G./Si}(\theta)$ the rare gas concentration in a film grown at direction θ. Typical angular distribution is shown on figure 11a for 20 keV argon ions and 45° beam incidence.

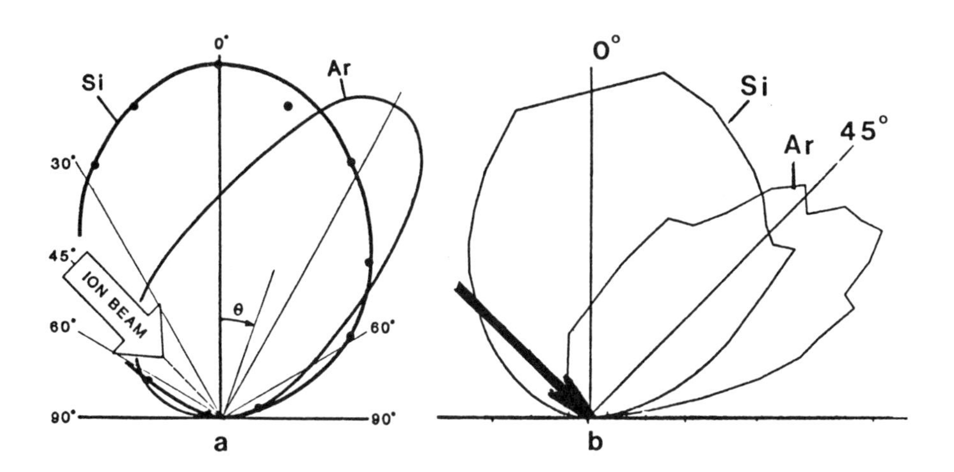

FIG. 11 Angular distributions of rare gas and sputtered silicon emitted from silicon target bombarded by 20 keV Ar^+ at oblique incidence (θ = 45°) a) measured b) calculated.

Contrary to the silicon angular distribution, the rare gas one is not symmetric with respect to the target normal. The sharp shape distribution points in a forward direction near the specular reflection direction of the beam. Similar results were obtained for Xe^+/Si and Kr^+/Si [9]. This is proof that the major part of the incorporated rare gas results from backscattering by a multiple collisions process.

The results of this study are confirmed by those obtained from a computer simulation of sputtering and backscattering processes using our experimental parameters. This work was carried out by W. Eckstein using the Monte Carlo Program TRIMSP [16]. Good agreement between experimental and calculated results was obtained

(Fig. 11b). The backscattering yield of the primary ions deduced from this calculation was 2.5 %.

This work is an important stage towards understanding of mechanisms which govern rare gas incorporation in film. Nevertheless, it does not still enable us to conclude categorically about the role played by the rare gas in film properties. Knowledge of the energetic distribution function of backscattered rare gas may be a decisive element in this research. We are performing experiments to define it.

CONCLUSION

This study on the potentialities of IBS has shown that epitaxies of good quality can be grown at low temperature. In addition it is hoped that the study of rare gas incorporation process will result in an improvement of the properties of these films and eventually a lowering of epitaxial temperature.

ACKNOWLEDGMENTS

The authors wish to thank their collaborators F. Meyer, C. Pellet, P. Legagneux, and W. Eckstein from Max Planck Institut für Plasma Physik, Garching - Germany - for the computer simulations. They are also grateful to A. Chabrier and F. Fort for their technical assistance.

REFERENCES

1. G.E. Thomas, L.J. Beckers, J.J. Vrakking and B.R. de Koninge, J. Cryst. Growth 56, 557 (1982).
2. H. Yamada and Y. Torii, Appl. Phys. Lett. 50, 386 (1987).
3. T. Takagi, K. Matsubara and M. Takaoka, J. Ap. Phys. 51, 5419 (1981).
4. P. Sigmund, Phys. Rev. 184, 283 (1969).
5. G. Gautherin, D. Bouchier and C. Schwebel from "Thin Films from Free Atoms and Particles" edited by K.J. Klabunde - Academic Press (1985) p. 203.
6. G. Gautherin, C. Schwebel and C. Weissmantel in Proceedings of the 7th Int. Vacuum Congress and the 3rd Int. Conf. on Solid Surface edited by R. Dobrozemsky (Vienna 1977) p. 1579.
7. J. Bottiger, J.A. Davies, P. Sigmund and K.B. Winterbon, Radiation Effects 11, 69 (1971).
8. P. Groce and L. Prod'homme, Nouvelle Revue d'Optique 7'121 (1976).
9. C. Schwebel, C. Pellet and G. Gautherin, Nucl. Instr. and Meth. B18, 525 (1987).
10. C. Schwebel, F. Meyer, G. Gautherin and C. Pellet, J. Vac. Sci. Technol. B4(5) 1153 (1986).

11. F.J. Morin and J.P. Maita, Phys. Rev. 96, 28 (1954).
12. J.S. Solomon and J.T. Grant, J. Vac. Sci. Technol. A3(2), 373 (1985).
13. F. Meyer, C. Pellet and C. Schwebel. Le Vide - Les Couches Minces 236, 207 (1987).
14. S.P. Muraka, J. Vac. Sci. Technol. 17(4), 775 (1980).
15. A. Lahav, M. Eizenberg and Y. Komem, J. Appl. Phys. 60(3), 991 (1986).
16. D.E. Harrison Jr, Radiat. Eff. 70, 1 (1983).

ELECTRICAL AND STRUCTURAL CHARACTERISTICS OF LASER-DEPOSITED Zn ON GaAs

T. J. Licata, D. V. Podlesnik, R. E. Colbeth, R. M. Osgood, Jr.
Microelectronics Sciences Laboratories,
Columbia University, New York, NY 10027-6699

C. C. Chang
Bell Communications Research
Red Bank, NJ 07701

INTRODUCTION

Laser processing of materials has been developed for microelectronics applications including the deposition of metals and semiconductors, etching and oxidation of semiconductors.[1-4] Lasers are used to accomplish low temperature photolytic processing or to produce elevated temperatures in a localized area. The low temperature treatments are important, for example, in the processing of compound semiconductors which often contain a volatile component. In this paper we report on the room temperature laser deposition of Zn on GaAs substrates. Both Zn and GaAs are materials that are sensitive to elevated temperatures. The high vapor pressure of Zn makes it unsuitable for use in high vacuum systems[5] while loss of arsenic during thermal processing of GaAs is a problem that has been intensively studied.[6] Here, good quality conductive Zn lines were written on GaAs and used as wiring for laser written devices. In addition, large area excimer laser deposited Zn/GaAs Schottky contacts were investigated. Finally, the ability of laser irradiation to modify the Zn/GaAs interface during the deposition process was examined.

EXPERIMENTAL TECHNIQUES

Photochemical depositions of micrometer structures are accomplished by focusing the 257-nm beam from a frequency-doubled argon-ion laser tuned to 514 nm onto a GaAs sample mounted in a stainless steel cell containing dimethylzinc gas (DMZn). The cell can be moved by computer-controlled stages with an accuracy of 0.25 μm. The cell volume is 1.8 cm^3 and is statically filled with typically 5-10 Torr of organometallic gas as measured by a capacitance manometer. The base pressure before introduction of the reactive gas is $\sim 10^{-6}$ Torr as measured by an ionization gauge. The laser beam is focused through a quartz window to a $\sim$3 μm spot on the sample surface using a 10X microscope objective (N.A. = 0.2). An automated data acquisition system is employed to measure focused laser spot sizes by scanning a razor blade beneath the focused beam while measuring laser power beneath the blade. The spot sizes reported represent the widths of the gaussian beam profiles at the 1/e points. The incident laser power was measured using an ultraviolet photodiode positioned beneath the microscope objective and ranged from 0.25 to 1.5 mW, corresponding to laser power densities from 3.5 to 21 kW/cm^2. For this range of laser power densities, the calculated temperature rise at the semiconductor surface is ≤ 5° C as obtained from the steady state solution of the nonlinear heat conduction equation.[7] A reflectivity of 0.58 was assumed for the 257-nm beam[8] while the temperature dependent thermal conductivities[9] were fitted to a 1/T dependency. A thermal component was added in several experiments by introducing the higher intensity 514-nm beam either simultaneously or sequentially to investigate curing or thermal diffusion of Zn into GaAs. In the large area deposition experiments, the 248-nm beam from an excimer laser charged with KrF was used. The laser beam was apertured to irradiate the GaAs sample in an area of about 0.5 cm^2. The laser intensity was regulated to 5-150 mJ/cm^2 using an external attenuator.

The incident ultraviolet beam will photodissociate molecules in the gas phase and in the adlayers. Since some of our interest is in maskless direct writing of micrometer structures, it is important to enhance the adlayer component of the deposition and suppress the gas phase component, which is less spatially restricted. Some control over the relative contributions of the two components is achieved through manipulation of ambient temperature and laser spot size. While the adlayer thickness increases with lower temperatures, room temperature is a convenient ambient temperature yielding sufficient adlayer coverage. The effect of laser spot size has been documented previously.[10] Specifically, it has been shown that within the diffusive regime, the gas phase deposition rate scales inversely with focused laser spot size, while the adlayer deposition rate scales inversely with spot size squared. The ~3 μm spot size used in the writing of micrometer structures favors the adlayer component of the deposition. Despite the fact that at pressures approaching the vapor pressure, the number of atoms in the adlayers increases faster than the number of atoms in the gas phase, the range of reactive gas pressures was restricted to 2-10 Torr (0.7 - 3.3% of the DMZn vapor pressure) in order to limit deposition on the cell's quartz window and resulting attenuation of the ultraviolet beam. Note that the beam waist on the interior window surface was ~1,000 μm.

Photochemical Zn deposition has been achieved previously using both DMZn and diethylzinc (DEZn). In our experiments, DMZn was chosen over DEZn as the parent gas since the ratio of gas phase to physisorbed layer ultraviolet absorption cross-sections is lower for DMZn than for DEZn. This is because the absorption cross-sections for the DMZn gas phase, chemi- and physisorbed layers occur at shorter wavelengths than do those for DEZn.[10,11] For many of the metal dialkyls studied thus far, especially those involving Zn or Cd, the absorption spectra exhibit broader tails for the physisorbed layers than for the gas phases.[12] The 257 nm photons access the broad tail in the DMZn physisorbed spectrum while avoiding much of the narrower gas phase absorption peak that would be accessed at 257 nm using DEZn. The result is a more localized deposition with greater spatial resolution for DMZn. The 257-nm beam corresponds to a photon energy of about 5 eV, which is sufficient to photodissociate the DMZn.[13]

EXPERIMENTAL RESULTS

The deposition conditions of 1.5 mW ultraviolet, 10 Torr DMZn and about 2 μm/sec sample translation speed were chosen to yield optimal spatial resolution. SEM images reveal well defined lines with a particle size of about 0.15 μm, as shown in Fig. 1. It has not yet been determined conclusively whether or not each particle is an individual grain. These lines have been characterized in terms of their usefulness as conductors. The resistivity of deposited metal is a reliable measure of both material purity and microstructural integrity. To evaluate the Zn conductivity, lines typically 300 μm long, 3 μm wide and 1500 Å thick were used as interconnects between chip circuit elements. Resistivity ratios between our deposits and the literature value[14] for bulk Zn ($6 \times 10^{-6} \Omega$-cm) are typically 7 for metal lines deposited with no thermal treatment. This compares well with results for other laser photodeposited metals cited in the literature, which generally exhibit ratios of from 4 to 100.[15-17]

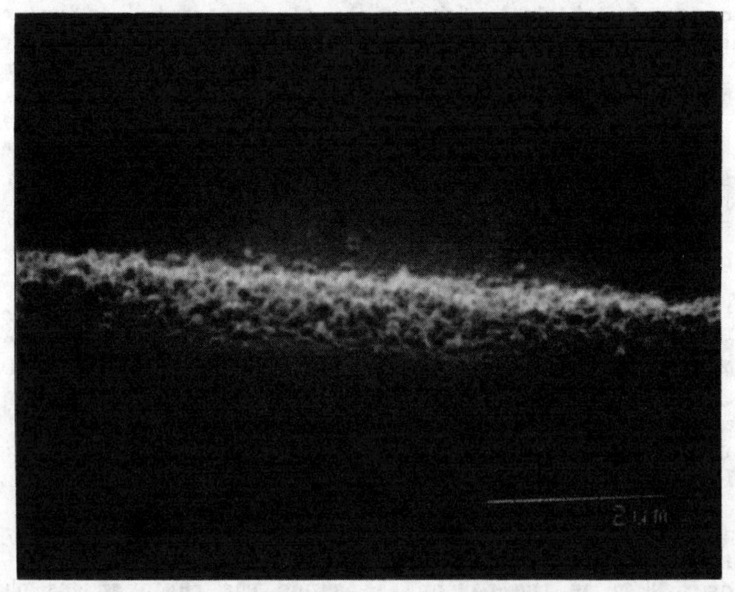

Fig. 1 A zinc line deposited using 1.5 mW of 257-nm radiation focussed to 3 μm, 10 Torr of DMZn and a sample translation speed of 2 μm/sec. The non-localized deposits to either side of the line are due to gas phase deposition.

To investigate the effect of adding a thermal component to the process, lines were also deposited and subsequently irradiated with a 514-nm beam without breaking vacuum. A thermal component to the deposition process is often cited as inducing thermal curing of the film resulting from the desorption of residual organic by-products and, at sufficiently elevated temperatures, more complete dissociation of the organometallic through a hybrid dissociation mechanism.[18]

As laser-deposited Zn desorbs at low temperatures (see below), only relatively low incident powers were investigated. Typical powers of incident 514-nm radiation used were from 15-100 mW focused to a 4 μm spot. The calculated[7] temperature rise above room temperature for this case is 25-210° C, respectively. A reflectivity of 0.37 was assumed for the 514-nm beam.[8] Typical resistivity ratios for thermally treated lines were the same as for untreated lines, within experimental error. This indicates that either the deposition conditions chosen allowed for more or less complete photodissociation of the DMZn (i.e. no thermal curing was evident) or that the limited temperature rise employed was not sufficient to desorb residual impurities. This result holds true for incident powers below 70 mW (temperature rise ≤ 140°C), since greater powers induced changes in microstructure characterized by reduced thickness of the lines, complicating accurate resistivity measurements.

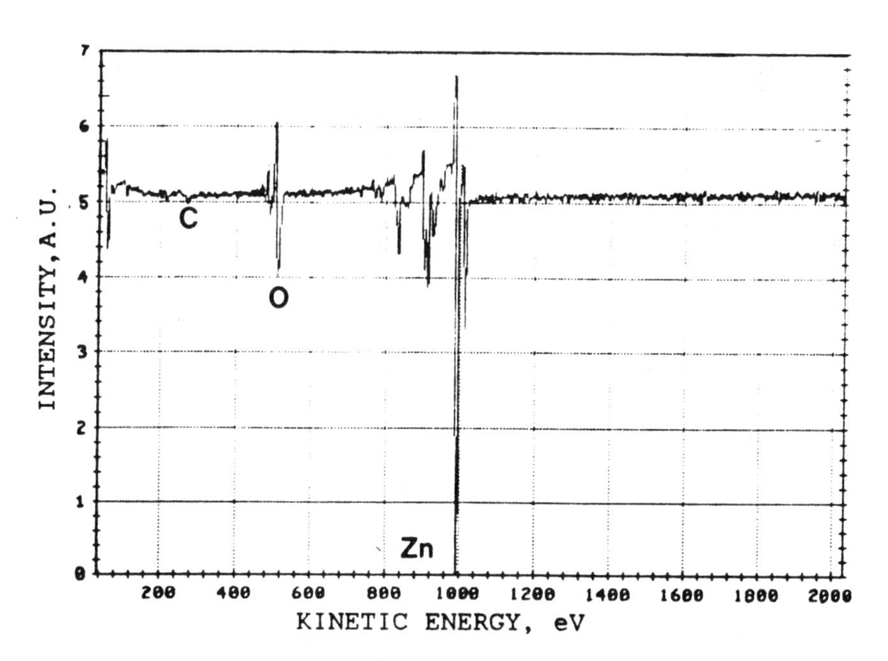

Fig. 2 Scanning Auger electron spectrograph for a Zn line with a 50 Å surface layer removed. The Zn/O/C ratio is 100/10/2.

Along with the monitoring of resistivity ratios, the quality of deposited material was gauged through a variety of direct analytical techniques. Scanning Auger electron microscopy was employed on deposited lines. Figure 2 shows an Auger electron spectrum for a Zn line with a 50 Å surface layer removed. Depth profiling of the lines showed a maximum of 4% and 6% each of carbon and oxygen incorporation in the bulk of the films, respectively. The actual impurity contents in the films can only be less than the values cited here since the shadowing effect during ion milling of granular surfaces leaves persistent contamination from surface layers formed upon exposure to atmosphere. Elemental depth profiles were also generated through Auger electron spectroscopy (AES) and x-ray photoelectron spectroscopy (XPS) for large area excimer-deposited pads. Measurements showed a marked drop from surface levels in both C and O contamination with a maximum in the bulk of the films of about 5% and 10%, respectively. Again, shadowing considerations appertain. To facilitate high temperature treatment, the thermal stability of large area Zn deposits was studied by exposing them to elevated temperatures inside the ultra-high vacuum AES/XPS system. Significant desorption of the laser deposited Zn occurred at about 330°C.

In separate experiments, electrical contacts between excimer-deposited Zn and n-GaAs were characterized. The GaAs samples were cleaned by immersion in trichloroethylene, acetone, methanol and H_2O followed by a H_2O/NH_4OH, 1:1 rinse. A shadow mask with 0.008 cm^2 holes was used to define the exposure to the 248 nm radiation and thus the size of the diodes, as this produced greater reproducibility of results over unmasked samples exposed over a much larger area. Current-voltage (I-V) and internal photoemission barrier height measurements were used to extract ideality factors and barrier heights. Contacts deposited at low incident powers ($\leq$ 20 mJ/cm^2) showed poor ideality factors ($\geq$ 1.2) and poorly defined barrier heights ($\geq$ 1.1 eV). As the incident powers were increased to 35 mJ/cm^2, the diodes looked increasingly better, showing ideality factors of 1.1 and reproducible barrier heights around 0.93 eV. In addition, turn-on voltages became progressively sharper. The estimated temperature rise for 35 mJ/cm^2 incident power is 117°C as calculated using the method of Gallant and van Driel[19] with reflectivity of 0.38[8] and absorption depth of 67 Å.[20] At incident powers greater than about 40 mJ/cm^2, ideality factors again degraded while barrier heights rose. As the power was raised further, gross elevation of both the ideality factor and barrier heights occurred and was associated with a roughening of the GaAs surface and depletion of arsenic at the interface as characterized through AES measurements. Such diodes exhibited a more gradual turn on than the optimal diodes at 35 mJ/cm^2.

Both the control over deposition geometry and the stability of the contacts were evaluated by fabricating gates and interconnects for metal-semiconductor field effect transistor (MESFET) and charge-coupled device (CCD) integrated circuits. In addition, these experiments were used to demonstrate how the flexibility inherent in laser deposition makes it useful during the design process, since gates of different widths and thicknesses can be fabricated either in the center of channels or to one side if desired, as shown in Fig. 3a. The MESFETs were fabricated on $1.0 \times 10^{16}/cm^3$ n-type epitaxial GaAs mesas on semi-insulating GaAs. I-V characteristics of the device are shown in Fig. 3b. Measured transconductances for the laser-formed MESFETs range from 15 to 25 mS/mm. Pinchoff occurs at $\sim$7V and the breakdown voltage is $\sim$9V. These transconductance values are low since the substrates were optimized for CCD rather than MESFET applications. The same transconductances were obtained for MESFETs fabricated with evaporated aluminum gates through conventional photolithographic processing. Finally, the laser deposited lines showed good step coverage over the 45°, 1.3 μm high mesas as shown in Fig. 4. In fact, an advantage to the laser deposition technique is that good step coverage is obtained even for more abrupt steps.

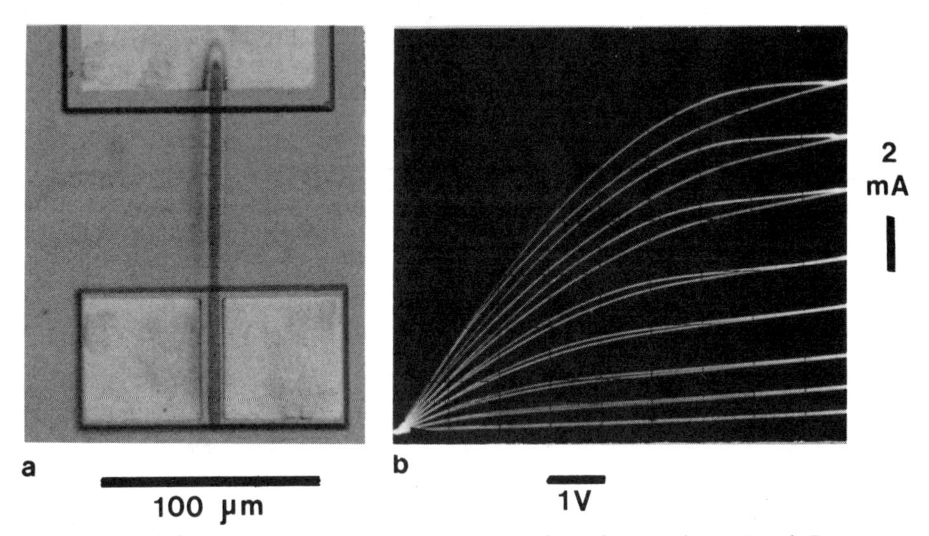

Fig. 3 (a) Photograph of a MESFET with a laser-deposited Zn gate. In order to customize the circuit and define the source and drain electrodes, the gate is positioned to one side of the channel. (b) I-V characteristics for the MESFET shown in (a). The looping is due to the properties of the substrate used.

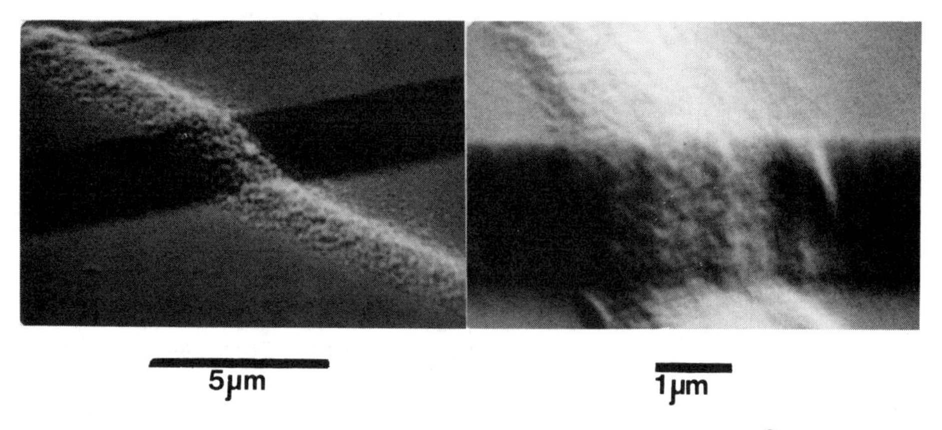

Fig. 4 Step coverage of laser-deposited Zn over a 45° step and a more abrupt step.

CONCLUSIONS

Good quality metallic Zn has been deposited on GaAs at room temperature through photolysis of DMZn using a focused, frequency doubled argon ion laser at 257 nm and an unfocused KrF excimer laser at 248 nm. The focused beam allowed for fabrication of conducting lines suitable both for chip wiring and active circuit elements while the excimer beam allowed for fabrication of large area diodes suitable for metal semiconductor barrier height measurements. The monitoring of resistivity ratios and direct analysis of impurity content indicate that high quality material can be obtained without the addition of a thermal component. This is desirable since compound semiconductors and Zn are unstable at elevated temperatures. The application of moderately elevated temperatures to previously deposited Zn lines resulted in no measurable improvement in material quality. Finally, an initial study of laser-deposited Zn/GaAs contacts is reported.

ACKNOWLEDGEMENTS

The authors would like to thank R. Krchnavek, P. Shaw, C. F. Yu and V. Lieberman for their useful comments and suggestions. In addition we would like to thank D. Rossi and M. Schmidt for their assistance in device characterization. Portions of this research were supported by the Defense Advanced Research Projects Agency/Air Force Office of Scientific Research and the Columbia University Center for Telecommunications Research.

REFERENCES

1. R. M. Osgood, Jr. and H. H. Gilgen, Ann. Rev. Mater. Sci. 15, p. 549 (1985).

2. N. H. Karam, N. A. El-Masry and S. M. Bedair, Appl. Phys. Lett. 49, p. 880 (1986).

3. Y. Rytz-Froidevaux, R. P. Salathe and H. H. Gilgen, Appl. Phys. A37, p. 121 (1985).

4. D. V. Podlesnik, H. H. Gilgen, A. E. Willner and R. M. Osgood, Jr., J. Opt. Soc. Amer. B3, p. 775 (1986).

5. D. M. Dobkin and J. F. Gibbons, Appl. Phys. Lett. 44, p. 884 (1984).

6. W. Ranke and K. Jacobi, Prog. Surf. Sci. 10, p. 1 (1981).

7. M. Lax, Appl. Phys. Lett. 33, p. 786 (1978).

8. Thermophysical Properties of Matter, The TPRC Data Series, Vol. 8, Y. S. Touloukian and D. P. Dewitt, eds., Plenum, New York, (1972), p. 679.

9. J. S. Blakemore, J. Appl. Phys., 53, p. R123 (1982).

10. R. R. Krchnavek, H. H. Gilgen, J. C. Chen, P. S. Shaw, T. J. Licata and R. M. Osgood, Jr., J. Vac. Sci. Technol. B5(1), p. 20 (1987).

11. C. J. Chen and R. M. Osgood, Jr., J. Chem. Phys., 81(1) p. 327, (1984) and P. S. Shaw (unpublished results).

12. C. J. Chen and R. M. Osgood, Jr., Chem. Phys. Lett., 98 No. 4 p. 363 (1983) and P. S. Shaw (unpublished results).

13. D. J. Ehrlich, R. M. Osgood, Jr. and T. F. Deutsch, Appl. Phys. Lett. 38, p. 946 (1981).

14. CRC Handbook of Chemistry and Physics 66, p. F-120 (1985-86).

15. H. Gilgen in Laser Chemical Processing of Semiconductor Devices, F. A. Houle, T. F. Deutsch, R. M. Osgood, Jr. ed., Materials Research Society, Pittsburgh, (1984), p. 55.

16. H. H. Gilgen, C. J. Chen, R. R. Krchnavek and R. M. Osgood, Jr. in Laser Processing and Diagnostics, D. Bauerle ed., Springer, Berlin, (1984), p. 232.

17. H. H. Gilgen, T. Cacouris, P. S. Shaw, R. R. Krchnavek and R. M. Osgood, Jr., Appl. Phys. B42, p. 55 (1987).

18. Ibid., p. 61-62.

19. M. I. Gallant and H. M. van Driel, Phys. Rev. B26 No. 4, p. 2133 (1982).

20. S. M. Sze, Physics of Semiconductor Devices, Wiley, New York, (1981) p. 42.

ELECTRON BEAM INDUCED SURFACE NUCLEATION AND LOW TEMPERATURE THERMAL DECOMPOSITION OF METAL CARBONYLS

R. R. Kunz and T. M. Mayer
University of North Carolina, Chapel Hill, North Carolina, 27514

Abstract

Selective area deposition of iron, chromium, and tungsten thin films via thermal decomposition of their respective carbonyls has been performed on electron beam deposited pre-nucleated layers. The prenucleated layers were deposited either by gas phase or surface electron induced dissociation of the respective carbonyl, depending on the chemical system and operating conditions used. The presence of this layer effectively lowers the activation energy for the decomposition event. The activation energies for the decomposition of the metal carbonyls on their respective prenucleated layers were measured over the temperature range 125 to 350 C and found to be lower than for decomposition on clean silicon. The kinetics of formation of the prenucleated layers were measured over the temperature range 125 to 350 C and found to be lower than for decomposition on clean silicon. The kinetics of formation of the prenucleated layer suggest that the formation of stable nuclei is dependent on the rate of desorption of CO from the surface. Electron stimulated desorption of CO is proposed as the mechanism for production of nuclei at low temperature by electron bombardment. Structure and composition of the nucleated layer and deposited film were examined by TEM, SEM, AES, and EDX.

SEMICONDUCTOR-BASED HETEROSTRUCTURE FORMATION USING LOW ENERGY ION BEAMS: ION BEAM DEPOSITION (IBD) & COMBINED ION AND MOLECULAR BEAM DEPOSITION (CIMD)

N. Herbots[a], O.C. Hellman[b], P.A. Cullen[b] and O. Vancauwenberghe[c]
Department of Materials Science and Engineering

MASSACHUSETTS INSTITUTE OF TECHNOLOGY, 77 MASSACHUSETTS AVE, CAMBRIDGE MA 02139

TO THE MEMORY OF THOMAS S. NOGGLE

ABSTRACT

In our previous work, we investigated the use of ion beam deposition (IBD) to grow epitaxial films at temperatures lower than those used in thermal processing (less than 500°C). Presently, we have applied IBD to the growth of dense (6.4×10^{22} atom/cm^3) silicon dioxide thin films at 400°C. Through these experiments we have found several clues to the microscopic processes leading to the formation of thin film phases by low energy ions. Using Monte-Carlo simulations, we have found that low energy atomic collision cascades in silicon have unique features such as a high probability of relocation events that refill vacancies as they are created. Our results show that the combination of a low defect density in low energy collision cascades with the high mobility of interstitials in covalent materials can be used to athermally generate atomic displacements that can lead to ordering. These displacements can lead to epitaxial ordering at substrate temperatures below the minimum temperature necessary for molecular beam epitaxy (550°C). It can also lead to the formation of high quality silicon dioxide at temperatures well below that of thermal oxidation in silicon (i.e. < 850°C). A growth model which we derived from these observations provides a fundamental understanding of how atomic collisions can be used to induce epitaxy or compound formation at low temperatures.

Applications of low energy ion beams to the fabrication of artificially structured materials at temperatures below which thermal decomposition and diffusivity become negligible have been explored using the insights gained from our model. This has led to the conception of a technique combining ion and molecular beam deposition (CIMD). We show that such a combination resolves the limitations of molecular beam epitaxy (MBE) and ion beam deposition (IBD) when either is used alone.

---

[a]Sponsored by the IBM grant to MIT and the National Science Foundation, under contract 84-18718-DMR.
[b] IBM Doctoral Fellow
[c] Belgian American Educational Foundation Doctoral Fellow
[d]Work sponsored by the division of Materials Science of the US Department of Energy, under contract DE-AC05-840R21400, with Martin Marietta Energy Systems, Inc.

I. INTRODUCTION

This paper reviews recent results on the growth of materials by direct low energy ion beam deposition, especially new results pertaining to the formation of oxides at low temperatures. It also discusses, for the first time, the growth of materials by the simultaneous use of a molecular beam and a low energy ion beam, where the latter contributes directly to the deposition of material in addition to enhancing growth processes. We call this novel technique Combined Ion and Molecular Beam Deposition (CIMD) [1].

1. ION BEAM DEPOSITION

This paper discusses the features of low energy ion beams when used to directly deposit thin films from an ion beam *only* , hence true ion beam deposition, based on the results obtained from independent research efforts at Hitachi [2-4], Philips Laboratories [5], and most recently Oak Ridge National Laboratory (ORNL) [6-10, 16]. Each of these investigations involved both metallic and semiconductor film growth, while ORNL has also undertaken the growth of insulating overlayers [11]. Our intent is to discuss, on the basis of these previous studies and the new results we recently obtained on ion beam oxides, our current understanding of the fundamental physical processes by which low energy ion beams can induce epitaxial growth and chemical reactions at temperatures significantly lower than the ones used in other existing growth techniques. We have derived from these experiments a phenomenological model that quantitatively links computer simulations of low energy atomic collisions with IBD experiments. We have found evidence, both theoretically and experimentally, that IBD provides the capability of manipulating atomic mobilities at solid surfaces and several monolayers below these surfaces. This manipulation occurs through atomic collision processes that require little thermal activation and produce the same effects as conventional thermal processing. It is on the basis of these observations that the unique properties and intrinsic limitations of IBD will be discussed for applications to the growth of semiconductor-based heterostructures of electronic quality.

2. COMBINED ION AND MOLECULAR BEAM DEPOSITION

Overcoming the limitations of IBD is a second point we address. Using our interpretive model and recent results, we have derived the concept of CIMD where an ion beam and a molecular beam simultaneously hit a substrate during deposition. In this process, the ion beam contributes a significant fraction of the deposited species or provides one of the elements of the material to be grown. For instance, epitaxial silicon can be grown by combining a flux of ^{+}Si ions with a Si molecular beam. To form an oxide or a nitride, the ion beam supplies ^{+}O or ^{+}N and the molecular beam provides the material to be oxidized or nitridized (for instance Si), while the substrate is a material from which a stable dielectric film may or may not be formed (Ge, GaAs, InP). The key role played by the species contributed by the ion beam is to modify or enhance the kinetic processes leading to thin film formation. We will show that CIMD can provide a new degree of freedom in the athermal control of atomic mobility by adjusting the rate of atomic displacements during simultaneous growth from molecular and ion beams. CIMD can also take place in the low temperature conditions enabled by IBD, while using the molecular beam to adjust a parameter we have found critical in controlling microstructural evolution during thin film formation with an ion beam: the ratio R_F between the areal density of atomic displacements F_D (displacements/cm^2·sec) produced by the incoming ions, and the total flux F_T of ions F_i and molecules F_m into the surface (species/cm^2·sec). This new parameter R_F (displacements/species), whose value is predetermined by the deposition energy in IBD, can be adjusted independently in CIMD through the relative intensity of each beam. A compromise can thus be established between the enhancement of atomic mobility during growth and the number of residual displacements after growth, making the latter compatible with the epitaxial quality required for electronic and optoelectronic applications.

In addition, the issue of cross-contamination from a volatile species, which is pervasive in the MBE process, can be resolved by an adequate design of the deposition apparatus and by introducing potential contaminants as an ion beam rather than as a vapor phase. For instance, we have been able to successively conduct ion beam oxidation and ion beam epitaxial deposition in the same UHV environment. The capability of confining ionized species through electrostatic or magnetic steering enables the introduction of high vapor pressure materials such as arsenic in a Si MBE system, for instance, and a rapid recovery to uncompromised conditions for the deposition of uncompensated p-type material. The low temperature of the substrate and system during the arsenic ion beam processing also limits arsenic deposition to where the beam impinges. Ion beam induced epitaxy also makes possible the growth of arsenic doped Si from a molecular beam at low temperature without the necessity of post-annealing for electrical activation. In section III, we will describe a model from which growth parameters for CIMD such as temperature, ion energy and dose-rate can be computed, in the same way we have previously suggested for IBD [6].

II. EVOLUTION OF THE IBD CONCEPT AND ITS IMPLEMENTATION

1. THE NECESSITY OF A TRUE ION BEAM

While techniques such as sputter deposition can also be called ion deposition, deposition from ion beams whose energy is both well-defined and well-controlled provides more insight into the fundamental processes involved in such deposition, and results in better epitaxial properties for ion-grown semiconductor materials. From this point of view, only techniques using an energy specific ion beam should be considered IBD techniques.

Two types of implementations have been investigated to obtain a monomodal energy distribution of ions relative to the target surface. The first consists of attaching a low energy (10 to 3000 eV) ion accelerator of the Calutron type to a deposition chamber. The other is more complex and involves the use of a medium energy (20 keV to 200 keV) ion implantation accelerator and a deceleration apparatus to slow the ions from their extraction voltage to the desired energy. Several considerations make the second option preferable.

2. LIMITATIONS OF LOW ENERGY ION SOURCES

Despite its simplicity, the first option is problematic because the theoretical brightness of an ion source is limited by the space charge, an effect which increases with higher current densities. The effect of the space charge becomes more important with decreasing ion extraction energy and with increasing time of flight from ion source to target. Thus, the combination of low energies, high current density and the use of a beamline which lengthens the ions' paths results in extreme defocusing and severe current losses before the ion beam reaches the target. In addition, the possibility of energy analysis and especially mass analysis at low energies is limited, thus making difficult the deposition of a high purity or monoisotopic films. Hence, Calutron sources have enabled worthwhile studies on the growth of IBD films, but their low deposition rates (1 - 10 nm per hour) have led investigators to conclude that such an IBD configuration is impractical. This type of system, however, did allow the first true UHV deposition of high purity epitaxial films of silicon and germanium on silicon at temperatures as low as 150°C [5].

3. HIGHER ENERGY ION SOURCES AND DECELERATION

The use of higher energy ion beams combined with a deceleration system has provided deposition rates higher by as much as an order of magnitude (> 100 nm per hour): as yet impractical for epitaxial deposition, but quite competitive for oxidation. The first attempts with this configuration took place at Hitachi and used a deceleration system that was part of the substrate holder [2]. The extraction voltage of the implanter minus the desired ion deposition energy (200 eV in those experiments) was directly applied to the holder. Pressures reported for epitaxial deposition were quite high (10^{-6} Torr) and no surface preparation was used prior to the growth of epitaxial Si and Ge on silicon at 200°C. It was proposed that the native oxide was sputtered during deposition [3-4]. Experimental evidence for these claims was based on *in-situ* RHEED analysis only; these results could not be reproduced by other independent researchers, who used ion channeling and cross-sectional TEM to establish the microstructure of the deposited films [5-10]. Zalm, who performed IBD at the same energies (150-200 eV) as the Hitachi group but in UHV conditions, demonstrated that the lowering of epitaxial temperature was a direct function of the partial pressures of the residual gases CO, N_2, CO_2 and H_2O. The minimum temperature for epitaxy was found to be a function of the partial pressures of these gases, and was established to be 150°C for a total base pressure of 10^{-10} Torr.

It is likely that the impossibility of maintaining a clean surface in a contaminated system before initiating deposition was the cause of these observations rather than residual gas adsorption during growth. Nevertheless, we have found in our own attempt to reprodúce the results of the Hitachi group that not only are native oxides and surface contamination a problem, but that residual gas incorporation and reactivity with the deposited material is quite significant, due to ion beam induced chemical processes [7,11]. Even at a total pressure of 10^{-9} Torr in a baked system, a massive incorporation of the main residual gas (hydrogen) can take place during silicon deposition, making it mandatory to maintain efficient hydrogen pumping during epitaxial growth [7].

III. ATOMISTIC MECHANISMS DURING ION BEAM DEPOSITION

1. ENERGY RANGE FOR IBD

We speak about an ion beam deposition process when overlayers are grown on a substrate directly from *low energy ions,* and the number of species leaving the substrate does not exceed the number of ions deposited. A simple parameter which determines the ratio of outcoming species to incoming species is the transverse kinetic energy of the ion with respect to the surface. Thus, the IBD criteria are met when the incoming ion energy is less than a certain energy, E_{IBD}, at which the sputtering loss equals the incoming ion flux. In other words, E_{IBD} is the energy at which global loss from the surface is equal to one species per ion. The loss mechanisms include physical and chemical sputtering, as well as particle backscattering and enhanced desorption. E_{IBD} is thus simply the energy below which material from an ion beam can accumulate on a substrate at a stationary rate. This energy E_{IBD} lies below 1 keV for most materials at temperatures below 600°C. When neutral species are simultaneously used for the deposition process, deposition can occur using ion energies significantly higher than E_{IBD}, due to the low sputtering yield of neutrals ($S^m \sim 10^{-3}$) with energies $\approx$ kT. The global sputtering yield is

$$S^T = (S^i F_i + S^m F_m)/F_T \text{ or } \sim S^i F_i / F_T < S^i$$

and can be controlled via adjustment of the intensity of either beam. Hence, CIMD introduces one degree of freedom in fulfilling the deposition criteria, and broadens the range of ion energies allowable.

2. ATHERMAL ENHANCEMENT OF EPITAXIAL ORDERING: PHENOMENOLOGICAL MODEL

Subsurface Mobility

The next consequence of using low energy ions for growth is enhanced solid phase epitaxial ordering [1-14] or enhanced chemical reactivities at low temperature [11,15]. These effects can be described simply in terms of *athermal* enhancement of atomic mobility, i.e. atomic displacements induced by low energy ion collisions in the first few monolayers in the target [16]. Such collision events can only take place at ion energies large enough to displace at least one atom in an elastic process (>10 eV), i.e. at energies well above the thermal energies of particles in a molecular beam (<1 eV). In a low energy (hence, low density) collision cascade, most atomic displacements will result in a net motion of an individual atom from one lattice site to another; it is these displacements which enhance epitaxial ordering near the substrate surface, while preserving the coherency of the lattice, as shown in the calculations below.

The lower limit of the energy range for the enhancement of subsurface atomic mobility is thus the energy required to induce an atomic displacement by a collision process in the substrate or in the growing film, E_D. This threshold for an atomic displacement is of importance for lowering the temperature for epitaxial growth, because ions with sufficient energy can induce atomic jumps that are substitutes for thermally activated atomic jumps.

Surface Mobility

Ion energies lower than E_D can also be used to enhance epitaxial growth. In this case, however, the ions' energy enhances adatom mobility at the surface *only*, because only weakly bound surface atoms can be displaced by collision energies lower than E_D. The surface binding energy for atoms in a far-from-equilibrium processes is of the order of twice the sublimation energy, a few eV for most elements (4.9 eV for Si, 3.9 eV for Ge) [6]. Thus, if the ion energy drops below E_D, a fundamental change occurs in the epitaxial growth mechanism; a subsurface process is replaced by a process limited to the surface only. Functionally, this implies that the occurrence of epitaxial deposition of ions with energies below E_D will be more dependent on surface structure, temperature and composition than for those with energies above. In other words, while the physics of subsurface atomic mobility enhancement by atomic collisions can be described through a fairly simple model based on a single energy, E_D, as shown below, the mechanisms by which the mobility of surface atoms can be enhanced are more complex. The crystallographic structure of the surface, the large number of defects such as steps, terraces and ledges which play an active role in epitaxial growth from the surface multiply the number of possible mechanisms and activation energies associated with them. Hence, the understanding of processes at these lower energies requires a much more detailed knowledge of growth atomistics at surfaces. In summary, while there is an energy threshold for subsurface atomic mobility, there are several lower thresholds for atomic displacements processes at the surface, making the interpretation of experiments involving lower energies more complex.

Practically, the temperature lowering of the global process will be more significant for $E > E_D$ but the epitaxial quality will be poorer (more displacements) than IBD used alone. From a fundamental point of view, a detailed study of the dependence of epitaxy on energy and temperature for ion energies lower than E_D could provide both new insights into atomistic mechanisms taking place during epitaxy and information on their individual activation energies. The difficulty of working at such low energies could be overcome by using a low energy beam impinging the surface at a chosen angle, so that the ion normal velocity with respect to the surface can be selected by geometry. An interesting property of this configuration is that most ions can be made to reflect away from the surface after energy exchange, so that no ion incorporation occurs. In that case, the combination of MBE with the ion beam would allow manipulation of surface processes without having to produce an ion beam of a compatible material. Such a technique would be classified as Ion Assisted Deposition (IAD) rather than IBD or CIMD because the *sole* role of the ion beam is to modify the kinetics of the deposition rather than to participate in the deposition itself.

Microstructural Dependence of E_D

An approximate average energy for atomic displacement E_D is 13 eV for semiconductors and 25 eV for metals. This energy depends on the orientation of the crystal with respect to the direction of displacement. The probability P that a collision will cause atomic displacement is a function of the energy transferred from the ion to the target atom in an elastic collision. $P \Rightarrow 0$ below a minimum threshold of a few eV, and increases slowly towards 1.0 at energies that can be as high as 100 eV or more, depending on the target and its microstructure. For our calculations, we take an average value for E_D, such that $P= 0$ for $E < E_D$, $P =1$ for $E > E_D$.

Differences between Metals and Semiconductors

An important dependence on microstructure is due to the variation of atomic density in different crystal structures. The open crystal structure of a diamond cubic or zinc blende structure of semiconductors allows interstitials to exist in the lattice without strain fields as large as those which would accompany an interstitial in a close-packed metal.

Another difference between close-packed metals and semiconductors is the collective nature of the interactions between the target atoms and the incoming ion. It is well-known that at some low energy threshold, the binary collision approximation (BCA) breaks down. But the threshold itself is not well established. One way to determine whether the BCA is still legitimate is to conduct a molecular dynamics (MD) calculation, in which all atoms potentially participate to the interaction with the moving species, and establish whether one atom at a time still dominates the interaction. An MD calculation in a metal using the isotropic Lennard-Jones potential shows that, on average, about 60 different atoms participate in the interaction when a particle of energy on the order of kT impacts on the substrate [17]. Atoms considered to be participating in the collision are those that gain an energy that can be considered significant . For an impact energy > E_D, three or more atoms in general dominate the interaction by gaining simultaneously energies which are significant with respect to E_D [18]. One can say that in these conditions, the BCA breaks down since it considers that only one atom at a time participates in the interaction.

In semiconductors, the anisotropic interatomic potentials such as the Stillinger-Weber potential, have a shorter range [17]. At energies above E_D, the binary collision approximation is still a valid hypothesis. The BCA continues to be legitimate for semiconductors at lower energies than those for metals. That interstitials are more easily formed in semiconductors than in metals is directly reflected in the lower threshold energy found for the former. Interstitials are also more likely to diffuse within the crystal without disturbing other atoms in their lattice sites. In a collision in a close packed metal, the energy is dissipated by displacing several atoms around the site of the collision, thus resulting in a collective motion of atoms that qualifies as plastic deformation. This is what we consider the most fundamental difference between ion beam processes in metals and semiconductors. As a consequence we expect metals to also exhibit enhanced epitaxial rates at low temperature under ion bombardment, but with many more residual extended defects. In conclusion, the response of a material to ion beams can be qualified as elastic for materials such as silicon under low energy or low dose-rate bombardment that allow for point-defect formation followed by spontaneous elastic recombination (as described below). Metals will tend to offer a plastic response over a wider range of energies and dose-rates, because little free volume is available for point defect motion, leading to a larger number of atomic displacements during cascade formation. This can explain the difficulty of forming an amorphous metal with an ion beam and of achieving high quality ion beam epitaxy of metals.

Classification of Collision Events : Replacements and Relocations

The concept of a minimum energy to create an atomic displacement implies that an atomic collision resulting in one of the species (either target atom or incoming atom) having an energy less than E_D, *cannot* lead to the formation of a stable vacancy. In a collision with an ion of energy E_i, an energy ΔE is transferred from the incoming ion to the target atom; the two end up with energies E_i' and E_{at}, respectively. If E_i' and E_{at} are both larger than E_D, then a defect pair is created, with a fixed vacancy and an atom with kinetic energy $E_{at} - E_D'$, where E_D' is the fraction of E_D the atom loses in escaping the lattice site (The ion loses a corresponding energy $E_D - E_D'$ escaping the lattice site, thus the total energy lost in vacancy creation is E_D). Such defect pair formation occurs with a high probability during medium energy collisions processes such as ion implantation and sputtering with keV ions, as has been demonstrated many times experimentally and through computer simulations. They can leave permanent damage in the target if their density is large enough to lead to a local collapse of lattice ordering over several interatomic distances.

At typical energies used for IBD, however, the collisions often result in one of the particles having an energy less than E_D (we call it a slow particle). If this slow particle is the atom originally residing on that site, one can say that no defect has been formed; the short-lived interstitial recombines with the vacancy within the timescale of the collision, and the incoming particle has merely transferred some fraction of its energy to the lattice. The reason for this spontaneous annihilation is that in this configuration, the elastic strain around the interstitial resulting from the lattice and the close proximity of a vacancy produces an attractive force between interstitial and vacancy leading to *elastic recombination.* The timescale at which elastic recombination takes place is about 10^{-13} sec. On the other hand, if the slow particle after collision is the incoming ion, and the struck atom has received enough energy to leave the collision site, the lattice site will be refilled by the slow ion. Such an event is called a *replacement* (Fig. 1d-e). The net result of this process is that a foreign species, introduced as an ion into the subsurface region of the substrate, now resides on a lattice site after producing only one interstitial, without the creation of a vacancy. If the subsurface is reasonably defect-free to start with, the closest recombination sink for the surviving interstitial is the surface (Fig. 1h). If the incoming atom is a target atom put into motion by a prior collision, this is a *relocation* (Fig. 1f-g).

Elastic recombination only happens if elastic strain is coherent enough to result into a net force leading to the motion of the interstitial towards the vacancy, rather than, for example, lattice collapse and amorphization. This condition is fulfilled if there is a low density of atomic displacements around that site within the timescale of the cascade and if the dose-rate is low enough to avoid cascade overlap. Replacement/relocation events can be calculated independent of whether the target is monocrystalline, polycrystalline or amorphous. Elastic recombinations can be detected in calculations by testing for a slow particle after defect pair formation without reference to a well-defined crystalline arrangement, or a random network.

The number of displacements N_d produced by an ion of energy E_i can be larger than E_i/E_D. This is because not all of E_D is consumed by a relocation event. The energy necessary to activate a relocation is E_D, but some of this energy may be used in activating a second displacement, thus recycling a certain portion of that energy. The energy E_r dissipated in relocations is then found to be equal to the energy of the slow particles E_s (typically a few eV). N_vE_D is consumed in the N_d displacements that lead to N_v vacancies which remain after N_r relocation and replacement events occur (2+1=3 in Fig. 1g). The total number of displacements N_d is then related to the energy of the incoming ion by

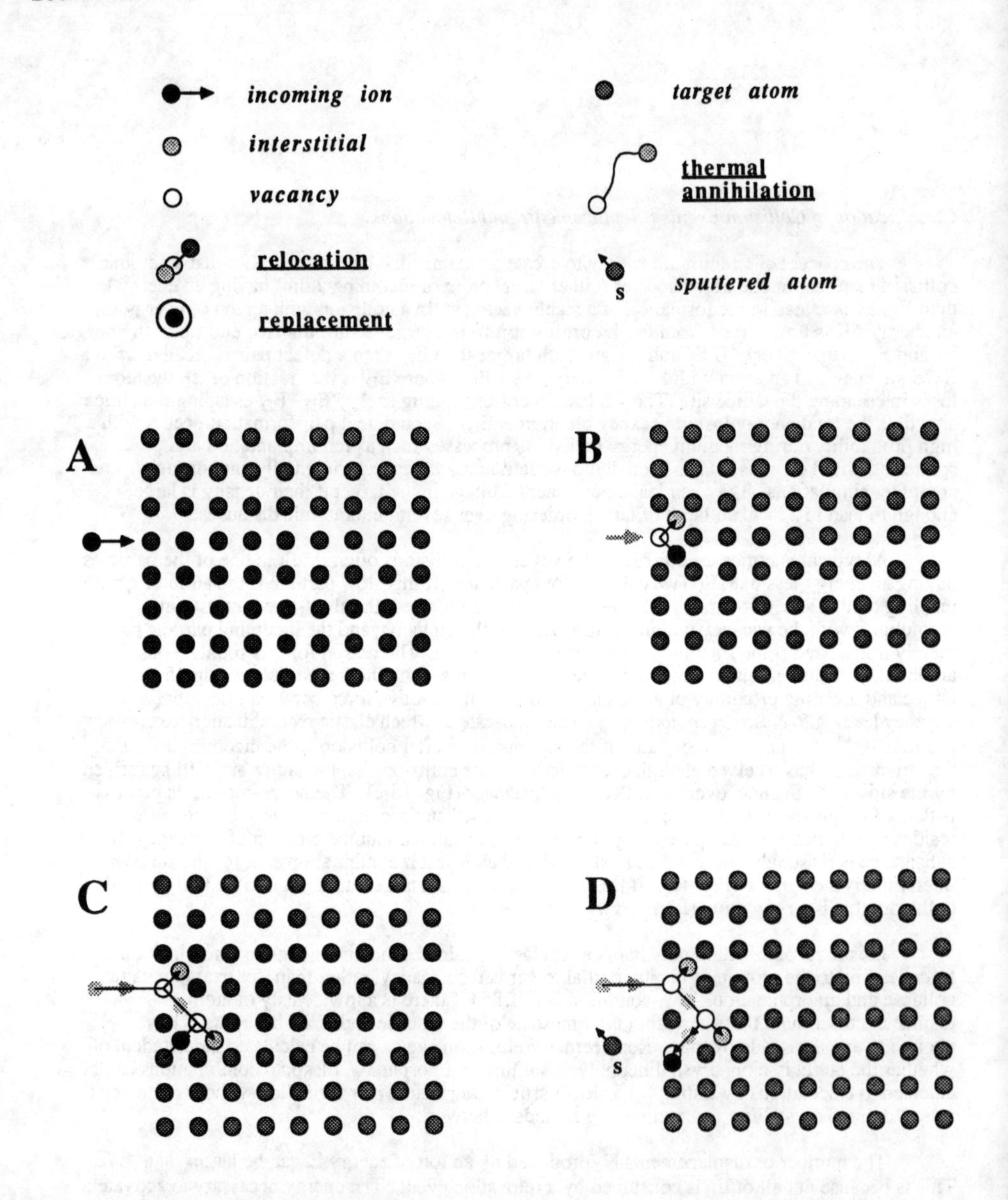
incoming ion
interstitial
vacancy
relocation
replacement
target atom
thermal annihilation
sputtered atom
s
A
B
C
D
s

Figure 1, A-D

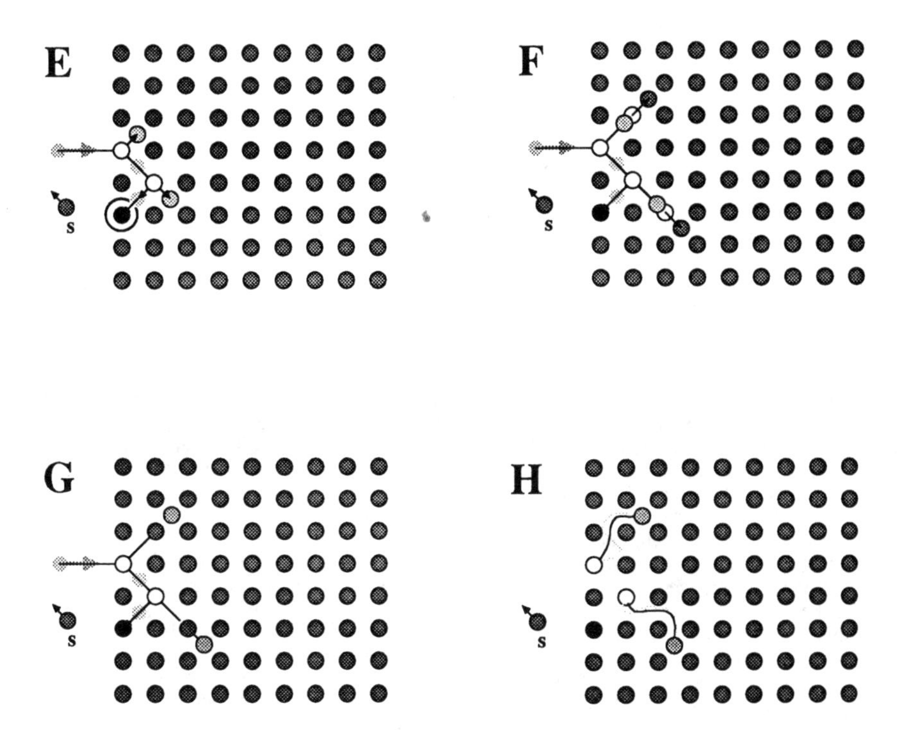

Figure 1, E-H

This sequence of events a-g illustrates the different types of collision events as described in the text. Steps B and C illustrate defect pair creation, steps D and E illustrate a combined replacement/sputtering event, steps F and G illustrate relocations, and step H illustrates thermal defect annihilation. The cascade illustrated here involves the motion of five target atoms. Two of these atoms are relocated, two of these are the interstitials annihilated in H, and the fifth atom is the sputtered atom.

$$E_i \approx E_D N_v + N_r E_s < N_d E_D$$

and

$$N_d = N_v + N_r$$

and the ratio R_o of athermal atomic jumps from one lattice site to another versus the number of athermal atomic jumps which create vacancies is N_r/N_v. When $R_o \geq 1$, we find that epitaxial ordering rather than damage generation can take place.

Fig. 1a-g gives the configuration of a collision cascade for a 40 eV Ge ion in a Si substrate, that was found to have a high statistical incidence (> 50% of 10,000 trajectories), except for the ejection of a sputtered atom, through Monte-Carlo simulations using a modified version of the TRIM code [25]. This version discriminates among each type of event we have defined through appropriate tests on the ion and recoils energies, and hence permits statistical classification of trajectories. The energy of the incoming ion (40 eV) allows for the creation of only two stable defect pairs ($N_v = 2$) but the number of relocations is 2 ($N_r = 3$), and thus the total number of atomic jumps induced by the collision N_d is 5. Hence, $R_o = 1.5$. In Fig. 1h, the resulting configuration is schematized. Two vacancies have been created in the subsurface region, and are immobile below 400°C. The two interstitials are mobile even at room temperature, and are thermally diffusing. Because of the proximity of the vacancies (a few interatomic distances), the probability for capture is high (Fig. 1h): epitaxy is actually taking place. It should also be noted that the role of relocations and ion replacement becomes significant only for low energies where E_i is within an order of magnitude of E_D (15 to 150 eV). Beyond that limit, the contribution of relocations and replacements towards the total number of displacements becomes negligible: R_o is between 0.1 and 0.2 for energies generally used for ion implantation and sputtering [19].

3. ATHERMAL ENHANCEMENT OF EPITAXIAL ORDERING: COMPUTATIONAL RESULTS

Having defined the model, vocabulary and techniques we used to analyze step by step the interaction of a low energy ion with a solid target, we can now quantify IBD and CIMD processes as a function of three primary variables: energy, substrate temperature and ion-to-target mass ratio. The conditions we chose for the simulation data shown as examples are typical for our experimental variables (Ge or oxygen ions, silicon substrate, 25°C < T < 600°C, 20 eV ≤ E ≤ 200 eV, dose-rate ~10^{13} ions/cm^2·sec). After calculating the statistical properties of individual low energy collision cascades, we can derive a global model of thin film growth as a function of the cascade parameters that we are now able to quantify.

Defects and Ion Depth Distributions during IBD at the Timescale of Collision Events

Fig. 2a represents the statistical depth distribution of the total number of displacements generated by a 65 eV Ge ion in a Si substrate. Fig. 2b depicts the depth distribution of the residual number of defects at the end of the collision process. Although a 65 eV Ge ion can produce, at most, only 4 defect pairs, a statistical average of 8 displacements per ion is found, due to the several relocation events. The ion mean projected range R_p is 0.85 nm, which is about 4 monolayers below the surface*. For ion energies above E_D, this is as expected since their energy

* The calculated range is a function of the interatomic potential between the incoming ion and the target atom. The potential function used in the calculations is the universal potential proposed by Ziegler, Biersack, and Littmark, which contains semi-empirical parameters for the five exponential terms giving the potential as a function of the interatomic distance. The validity of this potential at very low energies has been demonstrated recently both experimentally and theoretically by A. Tenner and A. Kleyn [20].

is sufficient to displace atoms in the bulk. This non-zero range indicates that material accumulation during IBD takes place in the subsurface region, not at the surface itself; we call this effect *layer anchoring* because the growing film effectively becomes anchored into the substrate [8].

Defect profiles in Fig. 2a show that a larger number of vacancies are formed between the surface and the mean projected range of the ion, R_p, while interstitials are found in larger numbers beyond R_p, an observation also derived from the analysis of an individual cascade such as in Fig.1. This implies that excess interstitials are pushed deeper into the substrate, an effect we call *snowplowing*. In a film of heavier ions being grown on a substrate of lighter atoms (e.g. Ge on Si), snowplowing effectively depletes the subsurface region of lighter substrate atoms and leads to the formation of a sharper interface than would be expected if only range straggling of ions were considered.

After taking into account the instantaneous annihilation of vacancies by slow particles (Fig. 2b), one can see that about 70% of the vacancies and interstitials disappear (the final distribution of interstitials in this figure also includes the implanted ions). The spatial separation between interstitials and vacancies is further enhanced by this process.

The fact that the ions penetrate to a depth of a few monolayers, as indicated by the value of R_p (0.8 nm in Fig. 2a), leads us to define a growth interface located at that depth, where the density of displacements is close to its maximum, and to take that as the critical depth to determine whether lattice perturbation will affect the occurrence of epitaxy. It also clearly means that IBD and CIMD can be considered *subsurface* growth techniques rather than purely surface processes. This particularity opens the possibility that IBD and CIMD might be less demanding than MBE in terms of surface cleanliness and perfection, which is apparent in our experimental data, where epitaxial films were obtained on very poor surfaces as judged by MBE standards (see below).

Dose Rate Limitation

The evolution into the distribution of Fig. 2b cannot, however, take place if the collision cascades develop in perturbed regions in the lattice, which would lead to a high density of defects, and thus clustering of extended defects or amorphization. This is the basis of high energy ion beam damage, because a single collision cascade causes such a high density of vacancy/interstitial pairs. As outlined above, low energy collisions can lead to ordering provided multiple collisions do not occur, leading to cascade overlap. This theoretically limits the dose rate at which epitaxial IBD can occur. One can define the dissipation time t of the collision cascade and an area per collision A_c, which is the cross-section of the elastic recombination volume with the atomic plane located at the depth of the mean projected range. The fractional surface area A covered by collisions can then be calculated:

$$A = A_c\,(\mathrm{cm^2/ions})\;\; t\,(\mathrm{s})\;\; F_i\,(\mathrm{ions/cm^2{\cdot}sec})$$

Statistically, if A= 0.01, less than 2% of all individual cascades will overlap. We can use this expression to calculate the dose-rate limit for IBD of 65 eV $^{74}Ge^+$ ions on Si, using the results of the calculations shown in Fig. 2b. Using a typical dissipation time of 10^{-12} s, a collision area of 10 nm^2, the dose rate limit is found to be 10^{24} ions/cm^2·sec, which is well above the practical limitations on ion sources and current densities.

4. THE ROLE OF TEMPERATURE

The two "snapshots" of defect distributions in Fig. 2a-b are valid within the short time interval at which the collisions takes place (10^{-13} s). The next step is to calculate thermal diffusion when the collision cascade terminates, the short lifetime of the cascade making it legitimate to

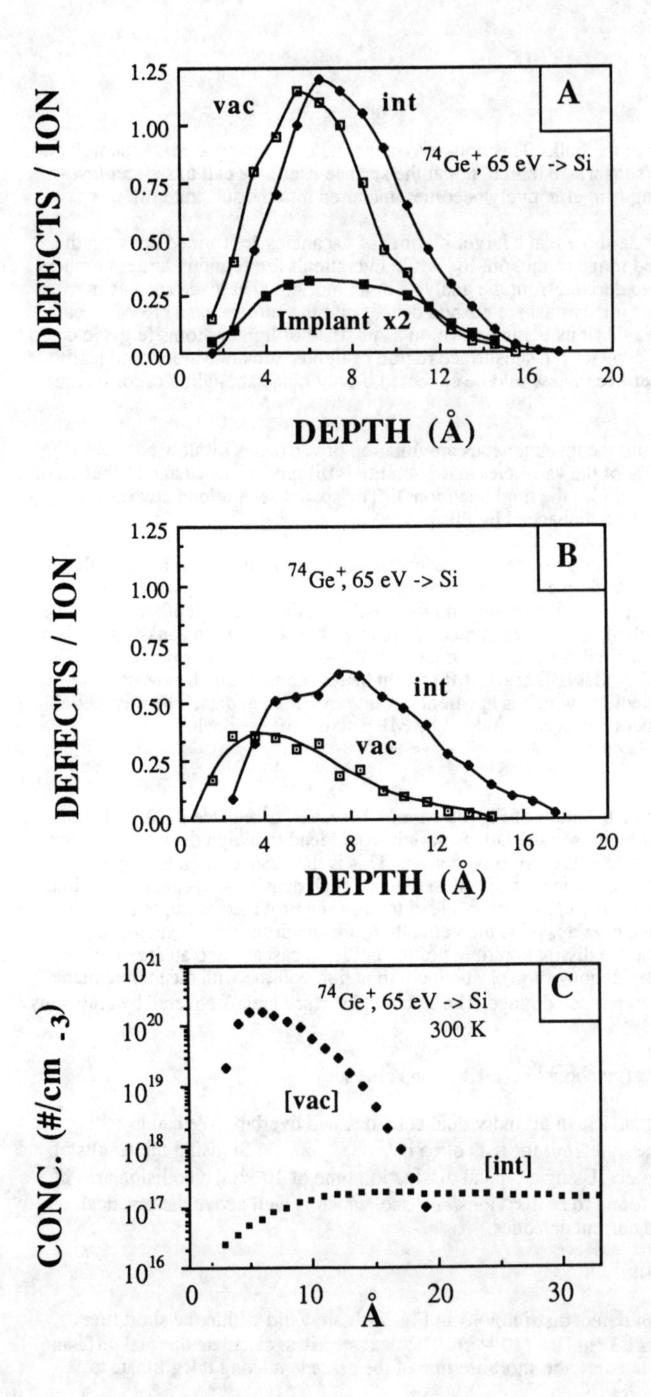

Figure 2

These graphs show the progression of vacancy and interstitial concentrations through calculations of defect creation by low energy ion "implantation" (2a), athermal defect recombination (2b), and thermal defect recombination and diffusion (2c). The first two calculations are made by TRIM, using the parameters shown. The second calculation included a test for athermal recombination. The third calculation is performed using results from TRIM for defect creation, and a finite element method for defect diffusion, and a recombination volume radius of 5 Å.

consider thermal effects separately form collision effects. Thermal mobilities are quite significant for interstitials in Si even at 300°C, and nearly negligible for vacancies below 400°C . Using recent values of these mobilities, we have calculated the steady-state defect distribution near a substrate surface during IBD for a typical dose rate of 10^{13} ions/cm^2·sec. This distribution is shown in Fig. 2c. While vacancies are retained in the amount of 2 x 10^{20}/cm^3 (0.4% of the lattice sites are vacant) at the peak of the distribution, interstitials are annihilated when reaching the surface and are distributed over several 100 nm at an average concentration of less than 3X10^{17} /cm^3. In summary, the residual defects left by athermal displacements are mostly immobile vacancies which are confined to the top 1.5 nm of the growing film. Interstitials introduced by athermal displacements dissipate away by diffusion both to the surface and into the growing film. The high concentration of interstitials thus produced throughout the film also results in the annihilation of vacancies which may exist there, thus making the vacancy concentration in the film negligible. Interstitials which diffuse into the film may coalesce as extended defects. We will show in the experimental section that such an extensive migration of interstitials and their coalescence deep in the substrate take place, and agree qualitatively with our calculations.

5. THIN FILM FORMATION BY IBD

From what we have learned from the previous calculations, we can now describe the process of thin film formation by IBD in three phases, as shown in Fig. 3a-c. The first is an initial implantation during which incoming ions progressively replace substrate atoms in the first few atomic layers. We call it the *ion implantation regime* (Fig. 3a). Upon saturation of the subsurface distribution peak at a concentration close to the atomic density, another regime is initiated in which the whole subsurface region becomes progressively converted into the material deposited, by the combination of snowplowing and also sputtering in the case of an ion-to-target mass ratio ≥1. We call this the *transition regime* (Fig. 3b). Layers of the new film are anchored in the substrate, and the subsurface region is converted into the material being deposited. Eventually, the whole subsurface region becomes converted into the species being deposited, and the third phase is a compositionally homogeneous growth of the deposited ion in its own thin film matrix. It is a stationary growth regime that we call the true *IBD regime* (Fig. 3c). Fig. 4 gives a pictorial representation of regime and highlights the specific characteristics derived from the calculations, and which we also observed in the films we deposited.

6. LOW TEMPERATURE ION BEAM OXIDATION

During ion beam induced chemical reactions, ion kinetic energy is used to overcome or to bypass rate-limiting activation barriers in a reaction process, as illustrated by the ion beam oxidation of silicon. In the thermal oxidation of silicon, molecular dissociation, physisorbtion, chemisorbtion, and solid-state diffusion are all reaction steps which require thermal activation energy (Fig. 5a). However, an oxygen ion of $E > 15$ eV impinging on a silicon target has sufficient energy to bypass all of these steps (Fig. 5b). Even solid-state diffusion can be unnecessary if the ion comes to rest in the target where it can bond to adjacent silicon atoms. At low temperatures (50 °C), where diffusivity of oxygen in silicon dioxide is limited, the ion energy can be used to control the penetration of oxygen into the silicon, thus determining a maximum oxide thickness. At 450 °C, which can be considered low for silicon processing since dopant diffusion and segregation is negligible, oxygen diffusivity is still significant. However, the generation of excess interstitials and their pile-up in the amorphous oxide will rapidly decrease the diffusivity of oxygen into the crystalline substrate. Hence, the oxidation process will be thickness limited beyond a certain point, determined by the number of interstitials generated (hence the oxygen dose and the ion energy), rather than by the range of the ion. Ion beam oxidation is thus an athermal technique for growing a dense oxide whose thickness is determined primarily by the oxygen ion energy.

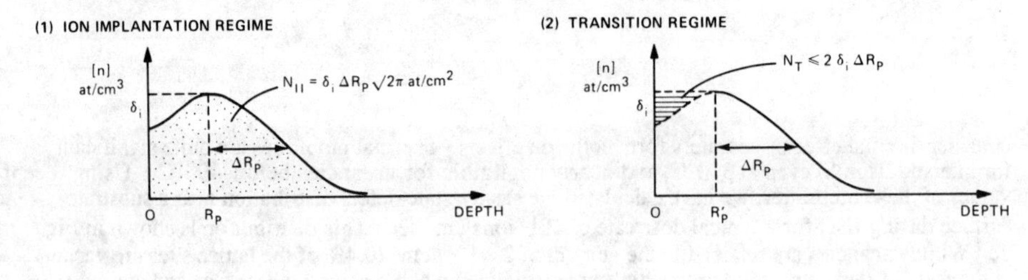

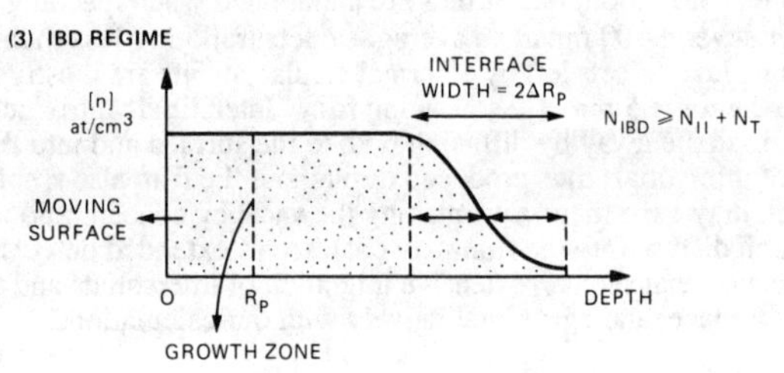

Figure 3

The three IBD regimes: (1) and (2) are the initial stages of IBD growth and remain the dominant regime for ion energies where the true IBD regime cannot be reached.

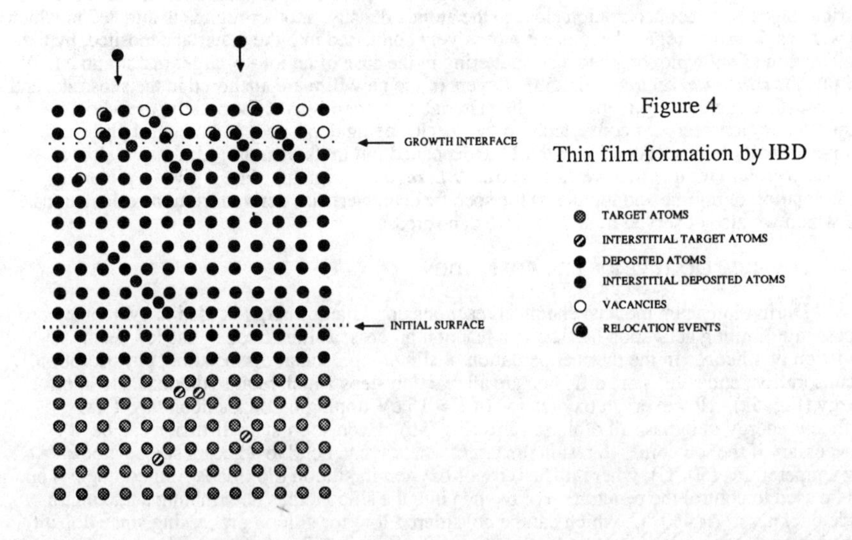

Figure 4

Thin film formation by IBD

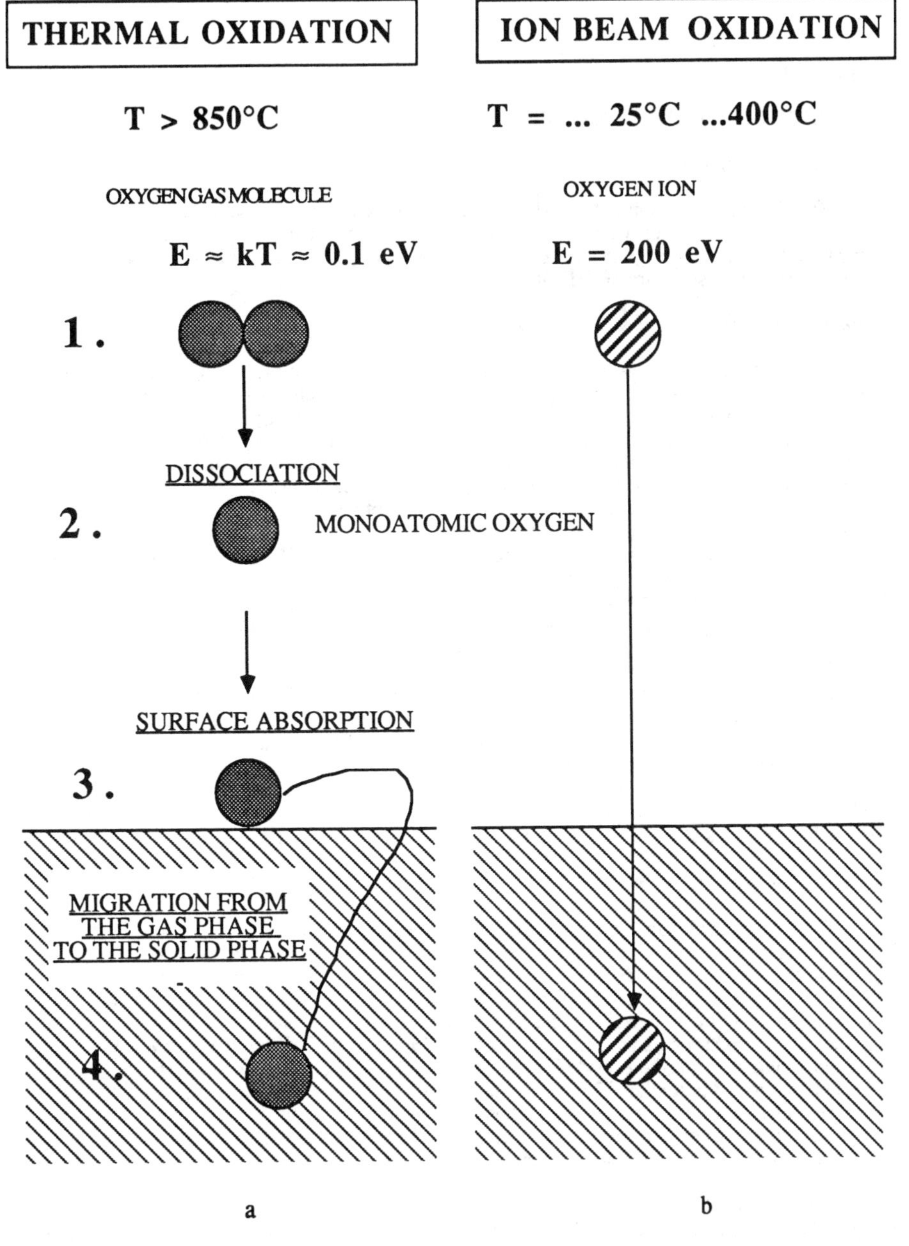

Figure 5
Schematic comparison of oxidation mechanisms during thermal oxidation (a) and ion beam oxidation (b). Several temperature-activated steps are bypassed when an oxygen ion beam is used rather than a molecular gas flow. The primary benefit brought by the use of an ion beam is the capability of introducing the oxygen species directly into the first monolayers below the surface, using the ion kinetic energy rather than the temperature of the substrate.

IV. EXPERIMENTS & DISCUSSION

1. IBD APPARATUS

The IBD apparatus built at ORNL has been the object of several publications [6-12, 16, 22]. We will thus only describe the principle of the experimental set-up and focus on secondary effects of the design that have recently been found to be important for the ultimate purity and the desired microstructure of IBD films.

Ion Beam Production and Transport

The production of a low energy ion beam with a flux comparable in magnitude to that of a molecular beam is a complex problem primarily because of space charge effects. The space charge causes a low energy beam to diverge as it is transported towards the target. This effect becomes more significant at lower energies and at higher current densities, limiting the maximum flux achievable for a given source and beam geometry, independent of the ionization efficiency of the source. In order to ensure good vacuum conditions and ion beam purity, ion beam analysis and differential pumping are necessary to avoid incorporation of desorbed gas and spurious ions into the films. This lengthens the path of the low energy ions before they reach the target and makes the space charge effect unavoidable. This problem is solved in the ORNL apparatus by producing the ions at a fairly high energy (35 keV), at which space charge effects are negligible over distances of the order of 100 mm. The ion beam is mass analyzed and energy analyzed in a high resolution magnet ($m/\Delta m = 1200$) at 35 keV. The beam has a diameter of ~2.2 cm after extraction, analysis and collimation through a first set of slits. It is then transported at 35 keV through a differentially pumped beamline into the deposition chamber. The beamline consists of three differential pumping stages separating a 10^{-7} Torr pressure in the ion source from the UHV chamber's base pressure of 10^{-10} Torr. Each stage has its own pumping unit and a pair of copper collimating apertures of ~2.9 cm diameter. The apertures are liquid nitrogen cooled and a fifth set of apertures defines the beam spot in the UHV chamber before it reaches the target. No copper was ever observed in the films, as measured by Auger Electron Spectroscopy (AES) and Secondary Ion Mass Spectrometry (SIMS), which means that the contamination was less than 10^{-4} a/o. The ions are decelerated only after ion beam mass analysis and transport to the target, in the immediate vicinity of the substrate.

Ion Beam Deceleration

Only when the beam is about to reach the target is it decelerated (Fig. 6). This is done by biasing the sample holder, which is a flat metal disk aligned in the same plane as the sample. The assembly acts as a flat electrode, biased at the extraction voltage (35 kV), minus the desired deposition voltage. Another disk, of the same diameter (10.16 cm) faces the substrate holder at a distance of an 2.54 cm, and has in its center a 2.54 cm aperture to transmit the beam to the target. This second electrode serves as secondary electron suppressor and is biased at -1 kV with respect to ground potential. A third electrode is present upstream of the suppressor at a distance of 2.54 cm and is held at ground potential. This configuration has the strong advantage of simplicity, and minimizes defocusing after deceleration. The design requires good alignment of each electrode with respect to the beam, because misalignment away from normal to the substrate can lead to deflection rather than deceleration, and hence no deposition.

The ion source used is a modified filament source of the Freeman type. The maximum energy spread E_r to be expected from this source can be estimated from the arc voltage. The energy distribution in the decelerated ion beam can be considered monomodal, with an asymmetry in the distribution due the fact that ions with an initial energy lower than 35 keV - E_I , where E_I is the energy selected for deposition, cannot reach the target.

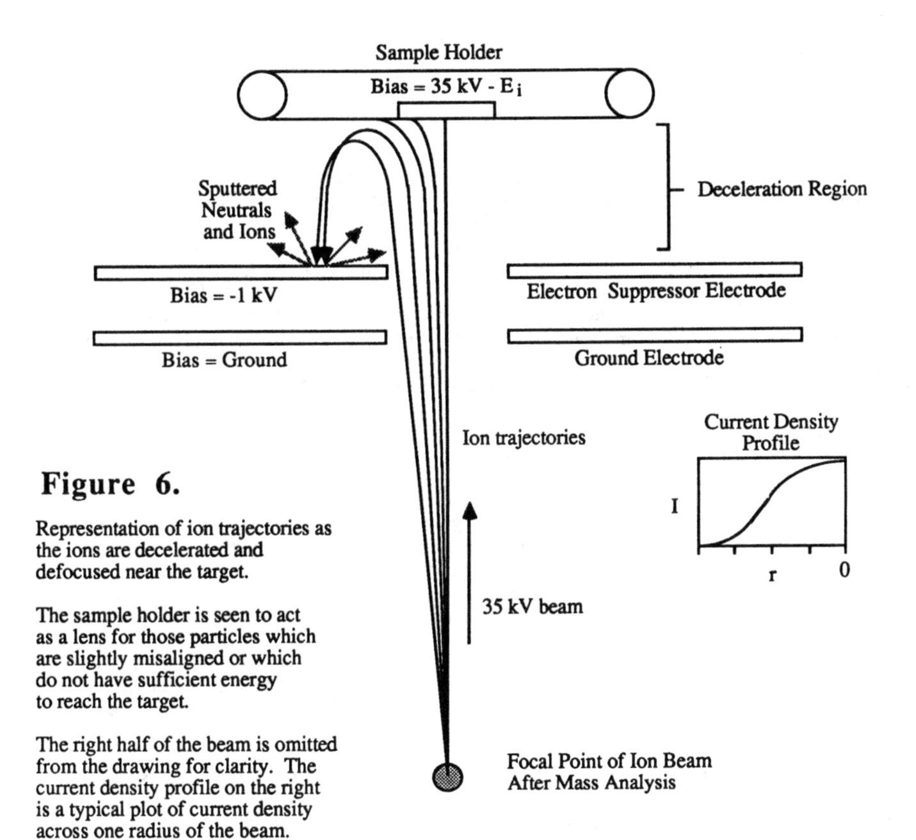

Figure 6.

Representation of ion trajectories as the ions are decelerated and defocused near the target.

The sample holder is seen to act as a lens for those particles which are slightly misaligned or which do not have sufficient energy to reach the target.

The right half of the beam is omitted from the drawing for clarity. The current density profile on the right is a typical plot of current density across one radius of the beam.

A direct observation of ion reflection is the collection of reflected ions on the electron suppressor disk held at -1 kV. Ions are reflected at the end of the deceleration when their energy approaches zero with respect to the substrate holder. They are then reaccelerated to a potential of 36 kV when they turn around and impinge on the electron suppressor. The reflected beam is strongly defocused and a annular deposit of approximately 5.1 cm in diameter is observed around the suppressor aperture. A highly undesirable consequence of this is the sputtering of suppressor material onto the film by reflected ions. Suppressor material incorporation into IBD films was initially very significant. For instance, 7 a/o of stainless steel was found incorporated uniformly into a 3 nm thick germanium film grown by 65 eV Ge^+ ions on a silicon substrate held at room temperature. No evidence of stainless steel was found beyond the film interface within the substrate by SIMS, indicating that high energy neutral incorporation is very unlikely. The problem of film contamination can be drastically reduced by using a suppressor material identical to the material being deposited, which brings the contamination level down to less than 0.035 a/o (determined using Rutherford backscattering analysis).

Further reduction can be achieved by using a substrate holder of the same material. It was observed that the stainless steel sample holder was gradually covered with material from the ion beam over several depositions, resulting in a gradual reduction of contamination. This effect decreases contamination to amounts below the detection limits of ion scattering and SIMS. It should be noted that material from the sample holder itself can migrate to the surface of the sample during heated deposition. This was especially apparent when the part of the holder in direct contact with the sample was made out of molybdenum, which being heavier is easily detected by ion scattering and was the only metallic part of the electrode not made of stainless steel. We found an initial contamination level on the order of 0.01 a/o for samples grown at 600°C, which subsequently decreased as the molybdenum became coated with silicon through successive depositions.

In conclusion, it is essential in such a design that the suppressor as well as the sample holder be made of a material compatible with the deposit, or becomes progressively coated by it though a series of preliminary depositions, which is the case for the data shown here. As far as thin multilayered deposits of monoisotopes are concerned, a purity of 1 to 0.1 a/o is the best that can be expected, since each deposition leaves a trace of contamination for the next. A pure monoisotopic film can be realized after successive depositions of the same isotope, as shown in Fig. 7. This SIMS profile of a single film of ^{30}Si grown up to a thickness of 150 nm demonstrates an isotopic purity better than 10^{-4} a/o, the sensitivity of SIMS to silicon. It is also interesting to note that the contamination level in C and O in the film are equal to that found in the electronic grade Si substrate, which is within the sensitivity of SIMS ($\sim 10^{16}$ atoms/cm^3).

Because the width of the interface can be verified by cross-section TEM to be less than 1 nm, such a sharp, monoisotopic interface is ideal for use as an interface width calibration standard for SIMS. The interface width measured in SIMS can be deconvoluted into a true width and an instrumental broadening width. Because the true width is known and small compared to the instrumental broadening, the instrumental broadening can be measured with great accuracy.

2. GENERAL PROPERTIES OF IBD THIN FILMS

Atomic Density

An interesting characteristic common to all IBD films we have grown (which amounts to about one hundred at the time of this writing) is the fact that their densities are very close to bulk densities when grown at a base pressure of 10^{-10} Torr , whether their microstructure is amorphous, polycrystalline or monocrystalline [7]. We have calculated the atomic density of these films by combining 2 MeV He ion scattering measurements, which provides the absolute areal

Figure 7
SIMS sputter-profile of ^{30}Si on Si(100) grown to a thickness of 150 nm; argon ions were used for sputtering. The width of the Si(100) interface was found to be less than 15 nm by RBS, and less than 2 nm by cross-section TEM. The interface width in this SIMS profile is 25 nm, which can be mostly attributed to instrumental broadening as well as to ion beam mixing during the sputtering process.

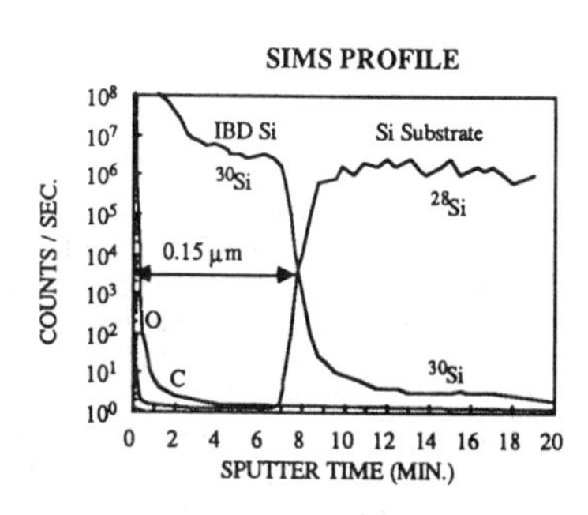

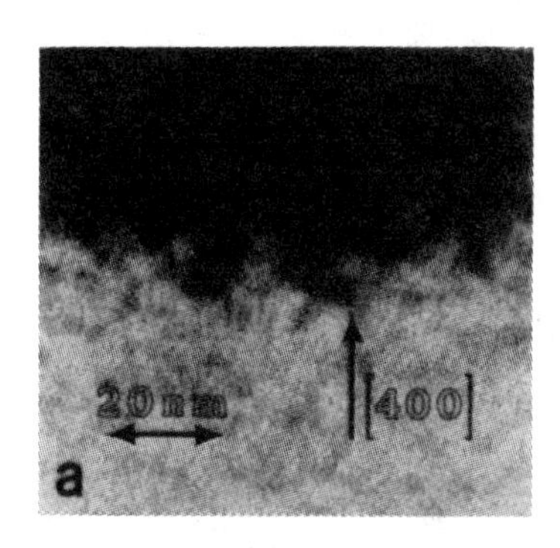

Figure 8a
3 nm thick IBD film grown using ^{74}Ge; 900 eV on Si(100) at room temperature. In this cross-section TEM, the Ge film appears as a dark contrast region 3 nm thick on top of the structure.

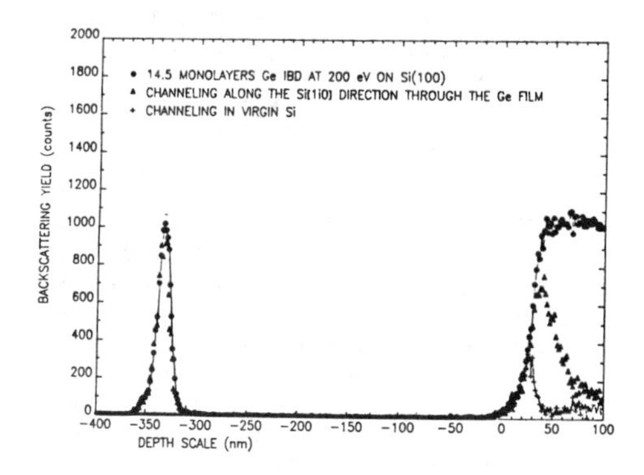

Figure 8b
RBS spectra showing the Ge deposit on top of the substrate, confirming the presence of ion damage 30 nm into the substrate,seen in (8a) as a medium contrast region between the Ge film and lower part of the substrate.

density in the film with an accuracy of a few percent, and cross-section TEM , which provides the absolute geometrical thickness of the film with a resolution of 0.5 nm. Fig. 8 shows a ^{74}Ge thin film grown from 200 eV Ge on silicon at room temperature. The cross-sectional TEM shows that an amorphous film of 3 nm thickness has been grown on a heavily damaged silicon substrate while the ion scattering measurement provides the number of Ge atoms present on the surface (1.35×10^{16} atom/cm^2) and confirms the presence of damage about 30 nm below the surface. The atomic density of the amorphous film is found to be 4.5×10^{22} atom/cm^3 (4.8×10^{22} atom/cm^3 for bulk Ge). Another amorphous Ge film, grown at a lower energy, 40 eV, is shown in Fig. 9. It shows the same quality in atomic density (5×10^{16} atom/cm2 divided by 10 nm yielded a density of 5×10^{22} atom/cm^3), this time with a greater numerical accuracy. The high density consistently found in amorphous and polycrystalline films grown in an uncontaminated ambient [7], can be attributed to the saturation of the film with self-interstitials during IBD growth. This property is attractive for thin film applications such as coatings.

Interface Width and Composition

The width of the interface in the cross-section TEM of Fig. 8 is 1 nm: this is in agreement with our model [6] which predicts a width of twice the range straggling of the ion, $2\times\Delta Rp$, which amounts to 0.9 nm for 200 eV Ge in silicon. A similar interface definition is observed in Fig. 9 in a film grown with 40 eV germanium (less than 1 nm). In addition, no damage is observed in the substrate by cross-section TEM.

A high resolution Auger depth profile by 500 eV Argon sputtering shows that the interface width in this structure scales to less than 9% of the equivalent thickness of the film (10 nm). At the interface location, residual oxygen from the native oxide, in the amount of 1.4×10^{14} atom/cm^2, is still present after deposition. The original amount of oxygen measured by ion scattering is 10^{16} atom/cm^2, corresponding to a thickness of 2.5 nm. The reduction of the native oxide to 1.4% of the original concentration during IBD is consistent with the "anchoring" process we proposed, by which thin film growth is initiated below the surface and depletes the initial surface of substrate atoms. These results illustrate how thin film growth by IBD can reduce the spurious native oxide that usually affects adhesion and contact resistance when evaporation methods are used. It should be noted that despite native oxide reduction, the films do not exhibit epitaxial ordering with respect to the substrate, because the range of the ion (0.8 nm) is well within the original disordered oxide film.

Fig. 10 shows another striking example of the quality of interface formation. In this case, an amorphous Si-Ge superlattice with a 10 nm periodicity has been formed at 600°C with 65 eV ions, by switching the magnetic selection between Si and Ge [7,8,10]. The cross-section clearly shows the highly uniform 2.5 nm native oxide on the last deposited film (silicon), which can be compared to the initial native oxide of the silicon substrate present prior to multilayer deposition. The original oxide has been nearly completely consumed in the deposition process and the Ge/Si interface replacing this original SiO_2/Si interface is 1 nm thick. One can compare it to the following Ge/Si interface formed by sequential depositions in uninterrupted UHV which is about twice as sharp ($2\Delta Rp = 0.4$ nm). A faint contrast at the location of the original surface, 1 nm into the first Ge film, can be seen and is further evidence of the anchoring effect and of the strong reduction of the native oxide film detected by the Auger measurement of Fig. 9.

Thin Film Uniformity, Homogeneous Nucleation and the Effect of Contaminants

The uniformity of the films was established on a microscopic scale by cross-section TEM, which showed that one such as that of Fig. 8 was perfectly continuous over several mm, despite its thickness of only 3 nm. This perfect uniformity is in agreement with our model where we propose that IBD initiates thin film formation by homogeneous nucleation, through the creation of

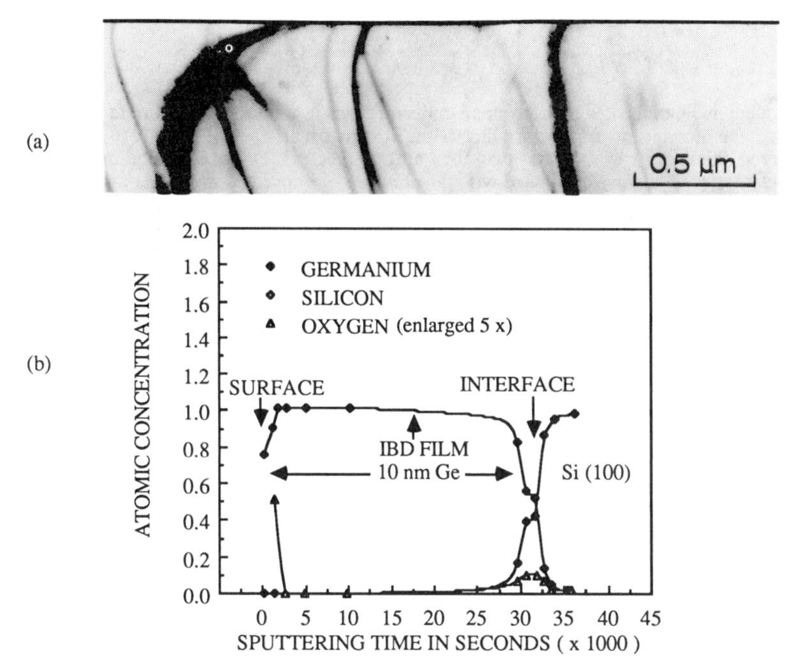

Figure 9 10 nm thick film grown using ^{74}Ge 40 eV ions on Si (100).
The cross-section TEM (a) shows at low magnification the film uniformity.
The Auger profile (b) through the film shows the reduction of the oxide at the interface and the abrupt transition in composition from Ge to Si.

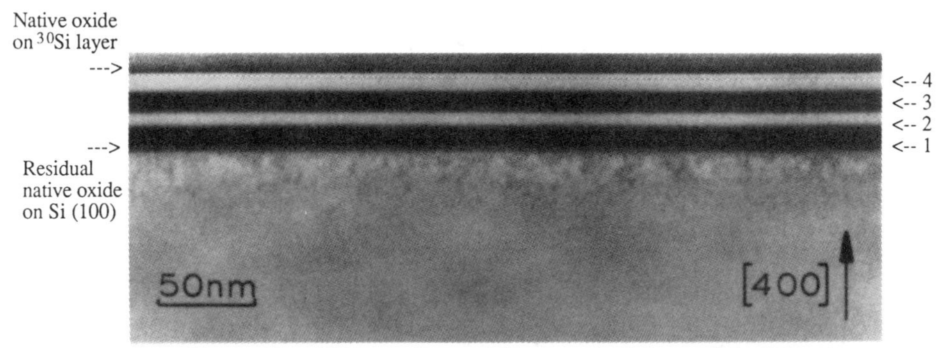

Figure 10 Si/Ge mutilayer of 10 nm periodicity made by IBD of ^{74}Ge (1), ^{30}Si (2), ^{74}Ge (3) and ^{30}Si (4) at 65 eV on Si (100) at Room Temperature.

equivalent nucleation sites by each small collision cascade anchoring the film into the substrate. The same uniformity was found for the 10 nm film grown at 40 eV in Fig. 9 (no coalescence observed, thickness variation less than 0.35 nm). In fact, the few cases of non-uniformity we have encountered were always correlated with the presence of impurities where the material had coalesced. Such a case is illustrated in Fig. 11, where a Ge deposit was made on a contaminated silicon surface. The film exhibited strong islanding, in contrast with a film grown at the same energy on a clean surface. Local stainless steel contamination was established through Auger mapping, and showed an exact correspondence between the location of the islands identified in the SEM mode.

Damage Production

Damage production during IBD is put in evidence by comparing films grown with the same ion energies and dose rates, but at different temperatures. Fig. 12 shows a 40 eV deposit of ^{74}Ge made at 400°C compared to one at room temperature in Fig. 9. Buried damage loops, identified as interstitials [8], were found between 50 and 200 nm below the surface, hence two orders of magnitude deeper than the range of 40 eV Ge ions in silicon (0.7 nm). The total number of displaced atoms can be estimated as 5×10^{14}atom/cm^2 from the cross-section TEM. This number is only 1% of the total dose used to grow that film, and represents thus a small fraction of the total number of displacements and interstitials generated during IBD. This theoretical figure includes all interstitials escaping into the substrate, while the cross-section TEM can only reveal those that coalesced into extended defects, hence the value obtained by TEM should be lower. Post-annealing of this structure at the same temperature for the same duration as that of deposition did not modify the defect distribution or size, as measured by TEM, indicating that residual interstitials have achieved a stable configuration for that temperature and cannot be annealed easily. A higher temperature anneal in an epitaxial film (1100°C) yielded the same results, for the same reasons: interstitials accumulate in the film, with the consequence that very high temperatures are needed in order to thermally generate enough vacancies to annihilate them, as discussed further below (Fig. 13).

3. EPITAXIAL GROWTH BY ION BEAM DEPOSITION

According to our model, epitaxial growth can take place using collisional processes to create defects which participate in that growth. Specifically, some fraction of vacancies are required for epitaxy because they act as sites at which atomic motion and ordering can take place. However, vacancy overpopulation can result in amorphization. IBD takes advantage of ion energy to create a high density of vacancies at the growth interface, while using a surplus of mobile interstitials to annihilate vacancies in the rest of the film, and to limit the concentration of vacancies at the growth interface. Because these defects are created athermally, their rate of creation depends on ion flux and ion energy, not on temperature. Thus, epitaxial growth by IBD exploits ordering mechanisms which are temperature independent, thus allowing for either an increase in epitaxial growth rate, or for a lowering of processing temperature.

Experiments, while still fragmentary because temperature, energy, surface contamination and IBD deposition rate at present constitute four independent parameters each to be explored systematically to quantify the whole process, brought observations that could be consistently interpreted within the frame of our model. In this section we will summarize our results rather than go into extensive experimental detail; for these details see [7-11].

Role of Temperature during Epitaxial Growth

Several depositions of ^{30}Si in Si(100) were made at temperatures ranging between 350°C and 600°C and energies ranging between 20 and 40 eV, at a deposition rate of approximately 0.6

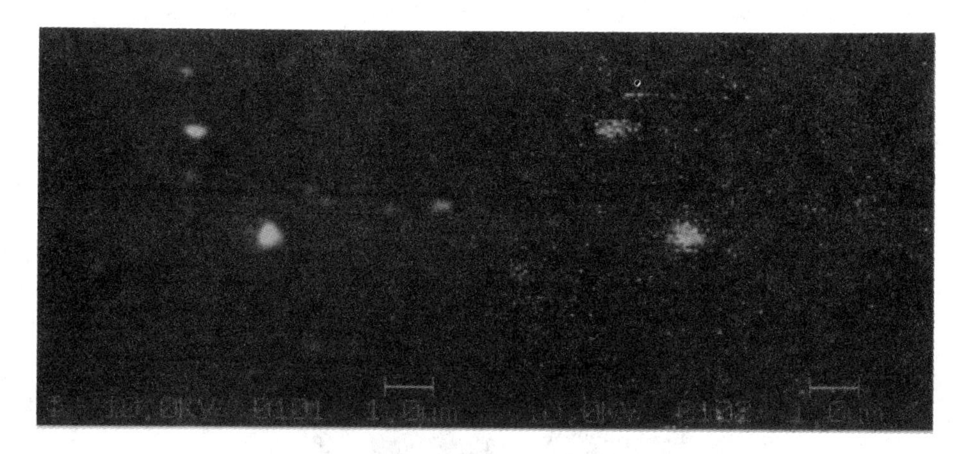

a b

c

Figure 11
Ge film grown by IBD of 65 eV Ge at room temperature on a Si(100) surface contaminated with iron, which is dispersed on the surface as a precipitate. The uneven distribution of the contamination leads to the coalescense of Ge deposit in islands rather than to layer-by-layer growth as seen in Figures 7-10. In (a), a SEM micrograph shows the presence of islands. In (b), an iron composition map obtained by Scanning Auger Microscopy shows the location of iron and demonstrates the association of iron precipitates with the nucleation of Ge islands. In (c), a cross-section TEM of the same sample is shown (note that a 300 nm thick silicon cap is covering the Ge deposit and was grown in-situ after Ge IBD to preserve the structure before it was exposed to atmosphere.

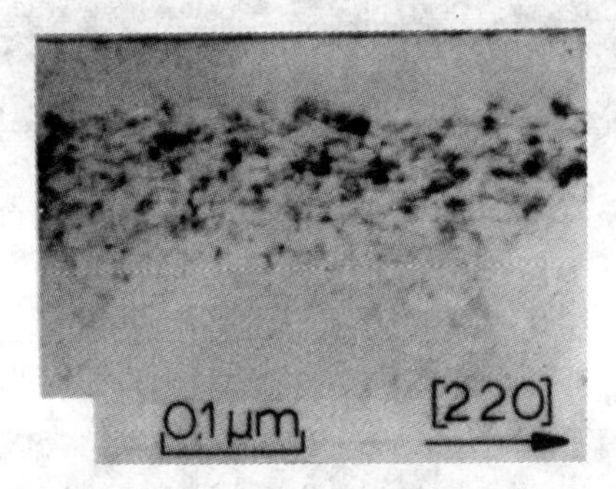

Figure 12
Cross-section TEM micrographs of ^{74}Ge thin films formed on Si(1000) by IBD at 40 eV and 400°C. The buried damage layer is observed only at high temperatures. The number of deposited ^{74}Ge was 10^{16} atoms/cm^2 and the number of displaced atoms in the substrate about $5X10^{14}$ atoms/cm^2.

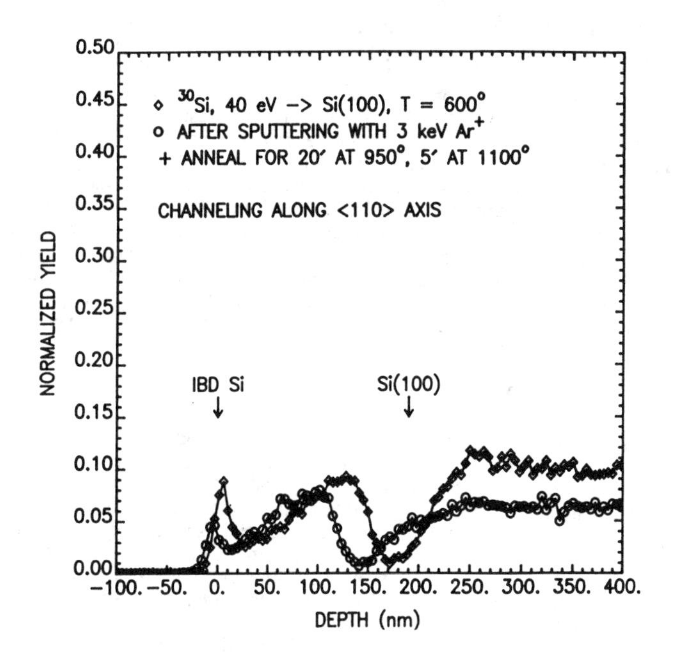

Figure 13
RBS data taken on the IBD film shown in Figures 7 and 14b. The film was annealed after IBD and argon sputter-cleaning. After 10 minutes at 950°C and 5 minutes at 1100°C little damage reduction was observed in the film.

nm/min. Epitaxial growth was observed both for 20 eV and 40 eV ions for three temperatures (350°C, 400°C and 600 °C). These data are shown in Fig. 14a-g, which also shows the maximal solid phase epitaxial regrowth rate of silicon amorphised by ion implantation of a single crystal, for comparison (from reference [23]). It can be seen that for IBD at 600°C, thermal mobility should play a significant role during growth since thermally activated motion of the amorphous/single crystal interface takes place at a rate that is significantly larger that the rate at which silicon atoms are deposited on the substrate. One can argue that the IBD deposition rate (0.6 nm/min) is slow enough that epitaxial ordering occurs through thermal annealing rather than through ion-induced atomic collisions processes. At 400°C and below, however, the contribution of thermal epitaxy is insignificant, as the solid phase epitaxial growth rate becomes infinitesimal (<0.06 nm/min). Epitaxial growth in that range of temperature can thus be attributed to the effect of the ion beam.

Ion Beam Epitaxy at 600°C

These observations are directly confirmed by the comparison of cross-section TEM micrographs of films grown at 600°C and 400°C respectively (Fig. 14b-e). At higher temperatures, extended defects are found in the films, close to the thin film/substrate interface, which is the region submitted to the longest anneal duration during deposition. Dislocation loops or stacking faults are found in the first half of the film that is closest to the interface, while the other half, which is closer to the surface, does not exhibit as many defects (Fig.14b-c). It appears thus that extended defects are nucleated at the interface. Ion channeling measurements confirmed this depth distribution of defects, with minimum yields close to that of single crystals near the surface region (3.5-4%) and highest nearby the interface (10-20%) in 100 nm thick films (Fig. 14e). These observations are also consistent with the fact that the surface acts as a recombination sink for defects while the interface acts as a trap. Thus, migration of ion beam generated defects occurs during growth and can be attributed to thermal diffusivity, and, most probably, if the interface were defect-free, no extended defects would be nucleated. X-ray Bragg profiling has provided evidence that the films are *uniaxially* strained. The substrates where buried damage loops are found, as in Fig. 12, are also uniaxially strained [9].

Ion Beam Epitaxy at 350-400°C

On the other hand, films grown at 350°C and 400°C exhibit *fewer* visible defects in cross-section TEM micrographs, and a much *more uniform* epitaxial quality. The defects nature and distribution is also radically different. A few threading dislocations, instead of the many dislocation loops, are nucleated from the interface and reach the surface. Their number and location is a direct function of interface quality in terms of microstructure: the more numerous dislocations in Fig. 14d directly correlate with the large number of interface defects while the low density of such dislocations in Fig. 14e scales with the smaller number of interface defects observed. In addition, the defect densities of these interfaces are much higher than in Fig. 14b-c, but the total number of defects in the films is significantly lower. This indicates that if defects are nucleated at the interface during growth at 350-400 °C, they do not grow much further and do not extend into the films as in the case of the higher temperature depositions. Ion channeling data confirms part of these TEM observations: the minimum yields are typically 4 % near the surface and 6 % nearby the interface in a 100 nm thick film. But the minimum yields do not drop below 4 %, revealing that at least 1 % of the atoms are still displaced from lattice sites (Fig.14g). Such a large density of defects seems at first to be incompatible with the TEM observations. TEM has a better sensitivity to defects than RBS and the defect density in Fig. 14e-g is lower than those for the higher temperatures (Fig. 14b-c). X-Ray Bragg profiling demonstrated that these films were also strained.

While still pursuing more detailed X-ray investigations on these structures, which have until now mostly confirmed these first measurements, we speculate that the peculiar growth mode by which the films are formed may lead to specific defects configurations leading to present

(a) Epitaxial growth rate for Si (100)

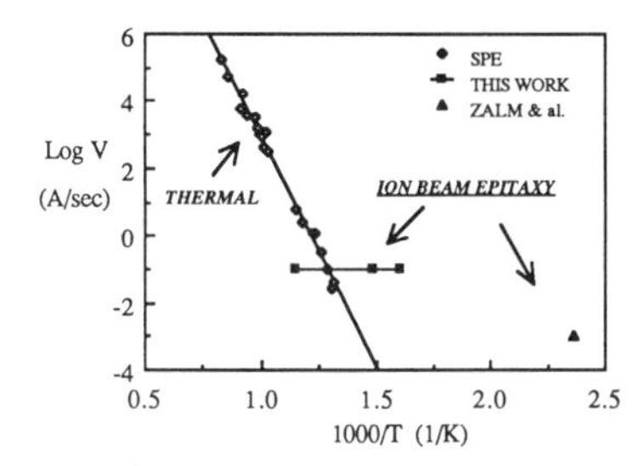

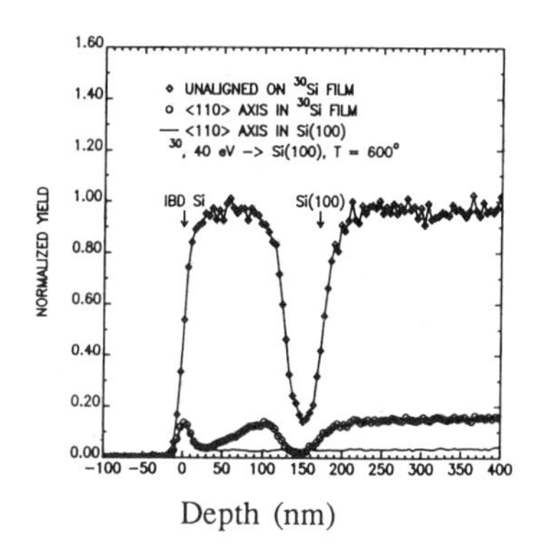

Depth (nm)

RBS spectrum taken on sample of cross-section TEM (b)

(b) Cross-section TEM on film grown with ^{30} Si ions, 40 eV on Si (100) at T = 600 °C, t = 150 nm.

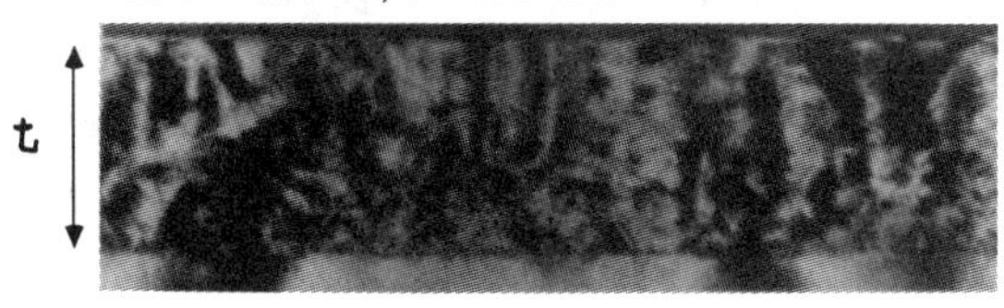

(c) Same as (b) but with 20 eV ions, t = 100 nm

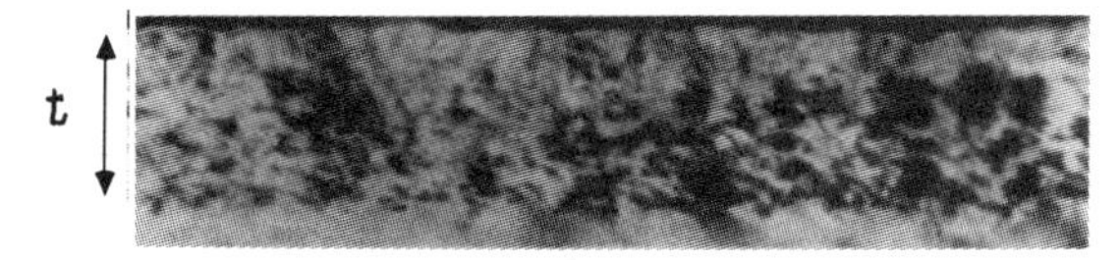

(d) Same as (c) but T = 400 °C, t = 70 nm

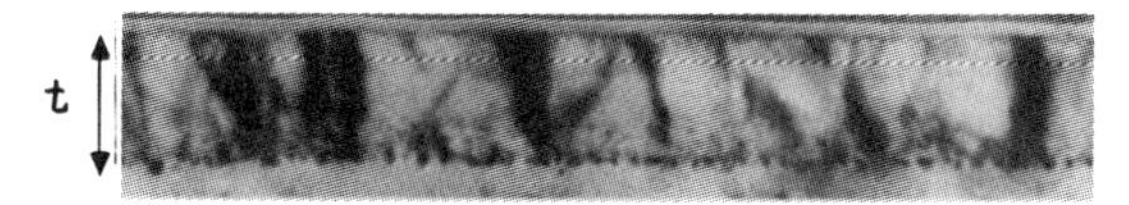

(e) Same as (d) but with a cleaner interface, t = 70 nm

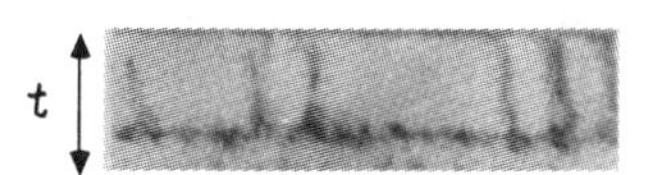

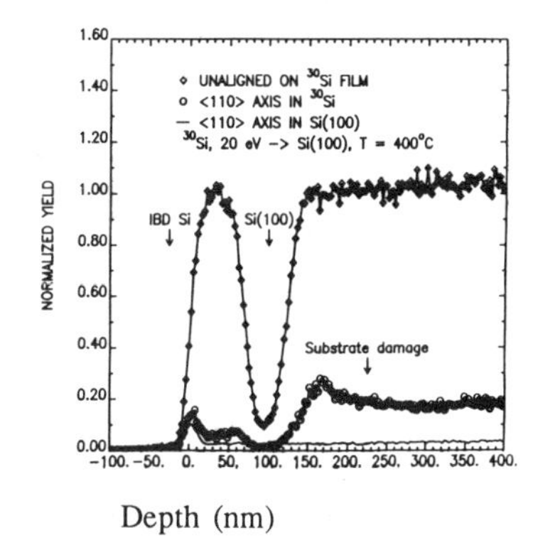

Depth (nm)

(g) Same as (f) but for sample shown in cross-section TEM (e)

Figure 14

observations. Defect formation specific to low temperature IBD processes are also a natural consequence with the more limited thermal mobilities found during IBD growth at 400°C. According to our model, the IBD films are supersaturated in interstitials during deposition as compared to thermally grown films.

If films are grown at 600°C, the mobility of both vacancies and interstitials permits their clustering in extended defects such as interstitial dislocation loops. In all IBD films, interstitial loops are formed 25-40 nm below the surface. Subsequent thermal annealing (Fig. 13) at temperatures as high as 1100°C for as long as 20 minutes does not lead to defect annihilation, while high-energy irradiation damage has been shown to be unstable under annealing at 1000°C for a duration of less than three minutes, even in contaminated samples. In the case of Fig. 13, the sample was also analyzed by SIMS and the resulting profiles (Fig. 11) clearly indicate that no residual impurities are found at the interface, within the sensitivity limit of SIMS. The defects did finally anneal away after a laser recrystallization which melted the sample surface. This is an indication that the defects present in the film were extremely resistant to annealing, whether due to their great numbers, their especially stable configuration, the uniaxial strain observed by x-ray techniques, or some other cause. It should also be noted that this film had a low impurity-content, as may be seen in the SIMS profile taken of that sample (Fig.7).

4. ION BEAM OXIDATION

As described above in section III.6, a low energy ion beam can be used to oxidize a surface, taking advantage of the ions' ability to bypass such reaction steps as dissociation, physisorption, chemisorption, and diffusion. This type of oxidation has been performed and a representative sample is illustrated here. A more complete description of this study will be published elsewhere [24].

TEM Results

The sample shown in Fig. 15a is an Si(100) substrate on which has been deposited a succession of oxygen, silicon and oxygen. The deposition temperature was 400°C and the substrate was subjected only to standard ex-situ cleaning. 10^{17} $^{16}O^{+}/cm^2$ were deposited at 200 eV, followed by 10^{17} $^{30}Si^{+}/cm^2$ at 65 eV, followed again by 10^{17} $^{16}O^{+}/cm^2$ at 200 eV. Fig. 15a is a cross-section TEM micrograph which clearly shows the silicon/silicon dioxide superlattice grown by this process with a 7 nm periodicity. From this micrograph, oxide thicknesses were measured to within 0.35 nm (the resolution of the instrument), and the width of the Si/SiO_2 interfaces was verified to be less than 0.35 nm, which is comparable to that of a high quality thermal oxide.

In contrast to the results from IBD epitaxy, no defects are found in the substrate underneath the oxide. This implies that silicon interstitials created in the collision cascades are not allowed to diffuse into the bulk of the substrate in numbers great enough to form extended defects; rather they remain trapped on the forming oxide layer, because their mobility decreases and because they can bond to stationary oxygen atoms.

RBS Results

The correlation of the TEM results with the RBS results shown in figure 15c allowed the calculation of the atomic density of these films. Whereas TEM provided geometric thickness, RBS provides an absolute measure of areal atomic density. This measurement yields a density of 6.4×10^{22} at/cm^3 for both oxide films. This density is comparable to that of a high quality thermal oxide. The stoichiometry of the oxides was also verified by RBS as being that of SiO_2.

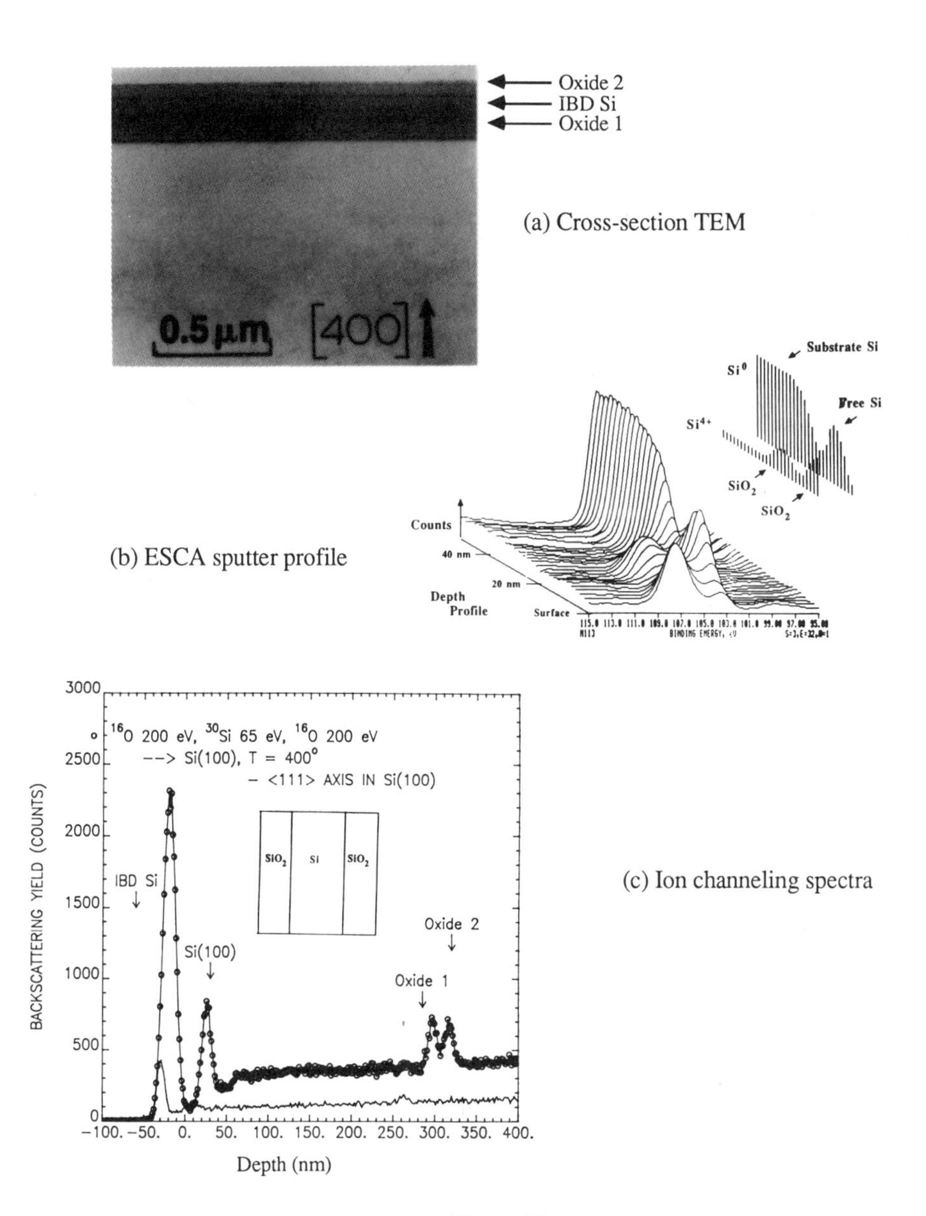

Figure 15
Si/SiO2 superlattice formed by IBD

ESCA Results

Chemical bonding states in the IBD oxide were characterized using angular-resolved ESCA measurements on a PHI Model 5400 at the Physical Electronics Division of Perkin-Elmer. A sputter profile through the sample illustrating two photoemissions measured from the silicon as a function of depth is shown in Fig. 15b. The binding energy peak near 107 eV corresponds to silicon twice bonded to oxygen, while the peak near 103 eV corresponds to silicon bonded only to silicon. The modulation in chemical state through the film is demonstrated by the transition observed from one photoemission to the other. No other photoemission was detected, indicating that no partial oxidation states were present. The angular resolution was used to verify that the overlapping of peaks, for example the appearance of a free silicon signal at the surface, was due to photoemission from subsurface regions and not due to the presence cf free silicon at the surface.

Overcoming limitations of IBD for growing oxides

These observations illustrate the interest of oxides prepared by IBD. The structures of these oxides are of a quality comparable with that of thermal oxides, and the oxide can be grown at surprisingly low temperatures and in an environment compatible with UHV materials processing. However, there are limitations on growing thick oxides by IBD. That is, once the oxide grows thicker than the mean projected range of the ion, film growth is limited by solid phase diffusion, and the time for film growth increases exponentially with film thickness, as observed by independent investigation. This limitation can be overcome by alternating, in our example, deposition of silicon and oxygen using an ion source with great material versatility.

A more elegant solution of this problem in terms of efficiency and technological potential is the simultaneous use of a silicon molecular beam and an oxygen ion beam: CIMD. Besides being able to grow high quality oxides on silicon, CIMD would allow the growth of *heterodielectrics*. For instance, silicon dioxide could be formed on germanium or gallium arsenide using a silicon molecular source and an oxygen ion source, thereby circumventing the problems of germanium oxide instability and gallium and arsenic oxides' immiscibility. This capability of CIMD is an attractive solution to the formation of high quality dielectrics on compound semiconductors.

V. CONCLUSIONS

1. FUNDAMENTAL CONCEPTS OF IBD AND CIMD

A simple collisional model has been developed to describe thin film deposition by low energy ion beams. Through this model, we have been able to study the growth mechanism and interpret certain properties of IBD films, such as compositionally sharp interfaces, high atomic density, low temperature epitaxy and the formation of silicon and germanium oxides. Our model has enabled an interpretational analysis of specific properties of IBD films, such as interstitial loop formation, unique damage distribution and a temperature dependence of epitaxial quality. The fundamental understanding elicited by this modeling has led to the conception of a novel processing technique which can overcome the limitations of IBD both in low temperature epitaxy and low temperature formation of dielectric films: CIMD.

2. FUTURE WORK

The investigations of materials growth by IBD has brought forth important insights into some aspects of epitaxial ordering processes and defect formation by low-energy ion beams. The next step is to use this knowledge to overcome current limitations in materials processing due to the role of temperature in deleterious effects as the dimensions of heterostructures shrink. The control of atomic mobility by collisions processes rather than by temperature offers the possibility of growing silicon-based heterostructures in a temperature range within which dopant diffusion, surface segregation, and thermal decomposition will be reduced to insignificant levels. The inherent limitations due to the IBD process itself, excess interstitial generation, can be overcome by combining simultaneous deposition by IBD and MBE into the CIMD technique which has the potential to bring displacement rates under control and thus hold defect generation to a tolerable level. This is the direction of our future theoretical and experimental work.

ACKNOWLEDGEMENTS

The authors wish to acknowledge J.P. Biersack, L.C. Feldman, M. Grabow, G. Kalonji, D.F. Pedraza, M.T. Robinson, P.C. Zalm and D.M. Zehner for stimulating discussions, helpful comments and advice. We also wish to acknowledge J.S. Moore and D.K. Thomas for their assistance in maintaining, preparing and running the ion implantation accelerator, C.W. Boggs and J.T. Luck for the delicate work of thinning the cross-sections for the TEM measurements, and Lillian Gilbradsen of the Physical Electronics Division of Perkin-Elmer for her technical help and patience in careful and repeated angular-resolved ESCA measurements.

REFERENCES

[1] N. Herbots and O.C. Hellman, US Patent pending (1987).

[2] P. Tsukizoe, T. Nakai and N. Ohmae, *J. Appl. Phys.*, **42** p. 4770 (1976).

[3] K. Yagi, S. Tamura, T. Tokuyama, *Jap. J. Appl. Phys.* **16**, 245-251 (1977).

[4] T. Tokuyama, K. Yagi, K. Miyake, M. Tamura, N. Natsuaki, and S. Tachi, *Nucl. Inst. and Meth.* **182/183** pp.241- 250 (1981).

[5] P.C. Zalm and L.J. Beckers, *Appl. Phys. Lett.*, **41** p. 167 (1982).

[6] N. Herbots, B.R. Appleton, T.S. Noggle, R.A. Zuhr and S.J. Pennycook, *Nucl. Instr. and Meth.*, **B 13**, pp. 250-258 (1986).

[7] N. Herbots, B.R. Appleton, S.J. Pennycook, T.S. Noggle and R.A. Zuhr, in Beam-Solid Interactions and Phase Transformations, Eds. H. Kurtz, G.L. Olson and J.M. Poate, *Mat. Res. Soc. Symp. Proc.*, Vol. 51, pp. 3669-74 (1986).

[8] N. Herbots, B.R. Appleton, T.S. Noggle, S.J. Pennycook, R.A. Zuhr and D.M. Zhener, Semiconductor-based Heterostructures: Interfacial Structure and Stability, Eds. M.L. Green, J.E.E. Baglin, G.Y. Chin, H.W. Deckman, W. Mayo and D. Narashinham, pp. 335-349 (1986).

[9] R.A. Zuhr, B.R. Appleton, N. Herbots, B.C. Larson, T.S. Noggle and S. J. Pennycook, *J. Vac. Sci. Tech.*, **A5** (4), pp. 2135-2139 (1987).

[10] B.R. Appleton, S.J. Pennycook, R.A. Zuhr, N. Herbots and T.S. Noggle, *Nucl. Instr. and Meth.*, **B 19/20**, pp.975-982 (1987).

[11] N. Herbots, D.M. Zhener, B.R. Appleton, S.J. Pennycook, T.S. Noggle, and R.A. Zuhr, to be submitted to *Appl. Phys. Lett.* (1988).

[12] J.E. Greene, A. Rockett and J.-E. Sundgren, in Photon, Beam, and Plasma Stimulated Chemical Processes at Surfaces, Eds. V.M. Donnelly, I.P. Herman and M. Hirose, *Mat. Res. Soc. Symp. Proc.*, Vol. 75, pp. 39-35 (1987).

[13] J.E. Greene, Crit. Rev. in Sol. St. and Mater. Sci., Vol. 11, 1, pp. 47-227 (1983).

[14] J.M.E. Harper, *Solid State Technology*, **30**(4), pp. 129-134 (1987). [See also J.J. Coumo, J.M.E. Harper, and H.T.G. Hentzell, *J. Appl. Phys.*, **58** (1) (1985) for an extensive review and bibliography from these authors.]

[15] S.S. Todorov, S.L. Schillinger, and E.R. Fossum in Photon, Beam, and Plasma Stimulated Chemical Processes at Surfaces, Eds. V.M. Donnelly, I.P. Herman and M. Hirose, *Mat. Res. Soc. Symp. Proc.*, Vol. 75, pp. 349-354 (1987).

[16] B.R. Appleton, R.A. Zuhr, T.S. Noggle, N. Herbots, and S.J. Pennycook in Photon, Beam, and Plasma Stimulated Chemical Processes at Surfaces, Eds. V.M. Donnelly, I.P. Herman and M. Hirose, *Mat. Res. Soc. Symp. Proc.*, Vol. 75, pp. 39-35 (1987).

[17] M.H. Grabow and G.H. Gilmer in Initial Stages of Epitaxial Growth, Eds. R. Hull, J.M. Gibson, and D.A. Smith, *Mat. Res. Soc. Symp. Proc.*, Vol. 94, pp. 15-24 (1987).

[18] These observations were made during discussions on MD simulations on metals, with D.H. Harrison, Jr., who is here gratefully acknowledged for having made the time and the data available for such discussions.

[19] Values established by running the modified TRIM code used in this work for higher energies and confirmed by independent calculations by J.P. Biersack.

[20] A. Tenner, Rainbow Scattering, PhD thesis, FOM-Instituut, Amsterdam, The Netherlands (1986).

[21] U. Gösele in Semiconductor Silicon 1986, Eds. H.R. Huff, T. Abe and B. Kolbesen (Pennington: The Electrochemical Society), p. 541 (1986).

[22] R.A. Zuhr, G.D. Alton, B.R. Appleton, N. Herbots, T.S. Noggle and S.J. Pennycook in Materials Modification and Growth Using Ion Beams, Eds. U.J. Gibson, A.E. White and P.P. Pronko, *Mat. Res. Soc. Symp. Proc.*, Vol. 93, pp. 243-251 (1987).

[23] E.P. Donovan, F. Spaepen, D. Turnbull, J.M. Poate and D.C. Jacobson, *J. Appl. Phys.*, **57** (1985).

[24] N. Herbots, P. Cullen, S.J. Pennycook, B.R. Appleton, T.S. Noggle and R.A. Zuhr, to be published *Nucl. Instr. and Meth.*, **B** (1988).

[25] J.P. Biersack and W. Eckstein *J. Appl. Phys*, **A34**, p. 73 (1984).

CHARACTERISTICS OF THIN TITANIUM LAYERS ON SILICON DEPOSITED BY IONIZED CLUSTER BEAMS

S.E. Huq, V.K. Raman, R.A. McMahon and H. Ahmed
Microelectronics Research Group, Cavendish Laboratory, Department of Physics, University of Cambridge, Cambridge Science Park, Cambridge CB4 4FW, UK.

ABSTRACT

This paper reports investigations of the properties of thin titanium films, around 500Å thick, deposited on p-type silicon by ionized cluster beams (ICB). The titanium layers were grown at low rates, 1-2Å/s, at a pressure of $3x10^{-7}$torr. The effect of varying beam conditions, such as the fraction of ionized clusters and the acceleration potential, on the film properties is explored. Planar and cross-sectional SEM shows the films to be uniform. Electrical measurements on Schottky barriers indicate good interfacial properties. The structural and electrical properties of the thin titanium silicide films formed by precisely controlled electron beam annealing under varying conditions of temperature and time have been compared with conventionally formed layers.

INTRODUCTION

The scaling down of interconnecting track dimensions in VLSI circuits causes their resistance to increase, leading to higher power dissipation and reduction in circuit speed. Thus the main advantage of refractory metal silicides over doped polysilicon as VLSI interconnections is their lower resistivity. Of the various refractory metal silicides, titanium silicide is the most promising material because it has the lowest resistivity. The refractory metal films are also useful for circuit interconnections down to micron dimensions [1]. Different methods of refractory metal deposition have been tried including evaporation, sputtering, and chemical vapour deposition [2]. This paper reports an investigation of titanium layers deposited on single crystal silicon by the Ionized Cluster Beam (ICB) technique [3,4] and their subsequent silicidation by Rapid Isothermal Electron Beam Annealing (RIEBA).

ICB deposition is an ion-assisted film growth technique. The evaporant material is heated in a crucible with a small aperture. Clusters are produced when the gas pressure inside the crucible rises so that the evaporant expands adiabatically through the aperture into a region of low pressure. Initially, Takagi et al.[5] reported that the size of such clusters was anywhere between a few hundred to a few thousand atoms per cluster. But subsequent investigations indicate that the number of atoms per cluster is considerably smaller [6,7]. A fraction of the clusters is ionized by electron impact and any multiply-charged clusters break into smaller fragments as a result of electrostatic repulsion. Finally the charged clusters are accelerated towards the substrate, which may be held at a variable potential. By altering the beam conditions, such as the ionization current and the accelerating potential, the kinetics of film growth can be modified, resulting in improved quality [8]. ICB deposition parameters were investigated and the films were characterized by electrical measurements and microscopy.

SAMPLE PREPARATION

The substrates used throughout this investigation were single crystal p-type <111> silicon chips of resistivity 10-15 Ω-cm. Prior to loading, each sample was cleaned with dilute HF. Evaporation was carried out from a graphite crucible with an aperture diameter of 1mm. Pre-evaporation vacuum pressure was 10^{-6} torr and during evaporation approximately $3x10^{-7}$ torr. The rate of deposition was 1-2Å/s. Radiation from the source and the ionizer caused some increase in the substrate temperature, but it never exceeded 40 deg.C. With all the heating elements turned on, Argon gas (99.99% pure) was introduced into the chamber through a leak valve and the pressure taken up to 10^{-4} torr. The gas was passed for 5 minutes and then turned off. The procedure was repeated several times to reduce the residual gas contamination, especially that from oxygen and nitrogen.

A large number of films, of thickness 500Å, were deposited under different conditions of ionization current and acceleration potential. A number of depositions were also carried out without any ionization or acceleration. Three batches of samples were prepared by depositing titanium on silicon for the study of silicide formation. In all these samples the titanium layer thickness was 500Å. In the first batch, the deposition conditions were 50mA ionization current and 2kV acceleration voltage. The second batch was prepared with 50mA ionization current and 5kV acceleration, and the third batch was a control sample prepared without ionization or acceleration.

To study the as-grown titanium-silicon interface, Schottky barrier diodes were fabricated. The diodes were made by evaporating 1000Å thick aluminium on both the front and back side of the samples in a conventional evaporator. To ensure a good ohmic contact the deposited aluminium was sintered by e-beam for 1s at 450 deg. C. The diode areas were defined by mesa etching.

Silicides were formed on a number of ICB deposited titanium films by the rapid isothermal electron beam annealing (RIEBA) technique [9]. This uses a multiple scan e-beam method to obtain isothermal heating over the whole wafer or chip. In this work chips were heated to temperatures between 800 and 1200 deg.C. In the shortest heating cycle the peak temperature was reached in 0.1s. In longer time cycles the peak temperature was maintained for precisely-defined periods as required. Precise and independent control of temperature was achieved using feedback from a pyrometer. Equilibrium temperatures of 800, 900, 1000 and 1100 deg.C correspond to power densities of 8, 11, 17 and 24 W/cm^2. During annealing, the system pressure was maintained at $6x10^{-6}$ torr or better .

CHARACTERIZATION

The effect of anneal time and temperature on the sheet resistance of the titanium silicide was studied. A standard four point probe measurement was carried out. Results obtained under different conditions are shown in Figs. 1(a), (b) and (c) for the three batches prepared by ICB deposition.The initial sheet resistance of the unannealed titanium layers was around 15 Ω/sq. for all the 500Å thick films. On annealing the sheet resistance always decreases but in a different manner for different deposition conditions. For the titanium films deposited at 50mA ionization current and 2kV acceleration the sheet resistance (Fig.1(a)) drops to <1Ω/sq. at 800 deg.C for a 0.1s anneal time and stabilises at a value of 0.8 Ω/sq for the full range of temperatures from 800 deg.C to 1200 deg.C and times up to 90s. For 50mA ionization current and 5kV acceleration (Fig.1(b)) the behaviour is similar but the sheet resistance stabilises at a higher value around 1.7Ω/sq. and no appreciable change of sheet resistance was observed either due to temperature or time variations. For the unionized condition of deposition (Fig.1(c)) the value of sheet resistance is 2.5 Ω/sq. after 30s annealing at 800deg.C which is relatively high, but falls to an average value of 1.2 Ω/sq. for 900deg.C and 1100 deg.C annealing temperatures. Of the three deposition conditions the best values of sheet resistance were obtained for 50mA ionization current and 2kV

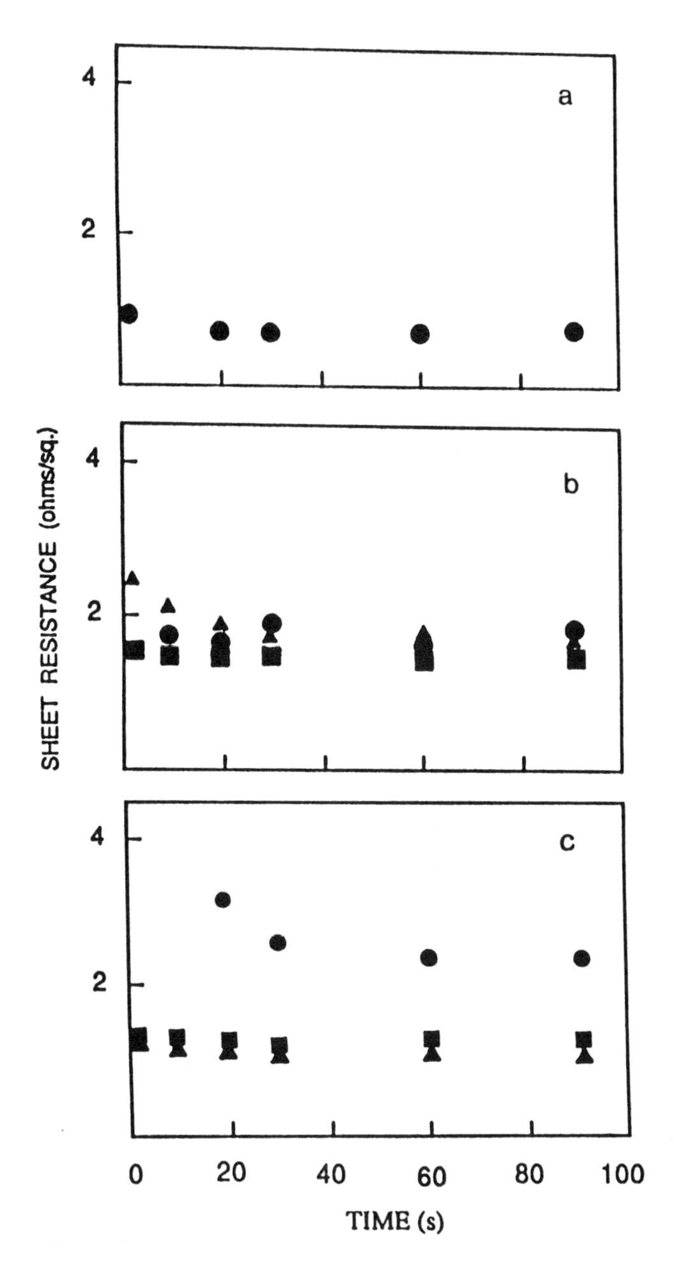

Fig.1. Sheet resistance of $TiSi_2$ under various conditions of metal deposition by ICB.
(a) 50mA ion current at 2kV acceleration, other temperatures not shown since points overlap
(b) 50mA ion current at 5kV acceleration.
(c) without ionization.

● 800 deg.C ▲ 900 deg.C ■ 1100 deg. C

acceleration. Fig.2 is a plot of sheet resistance as a function of time after 800 deg.C anneals for all the three deposition conditions. It has also been observed that for thinner titanium layers deposited on single crystal silicon by ICB and subsequent silicidation by RIEBA at 1000 deg.C the minimum sheet resistance was still almost 1Ω/sq. and this value remained constant up to 90s. annealing time.

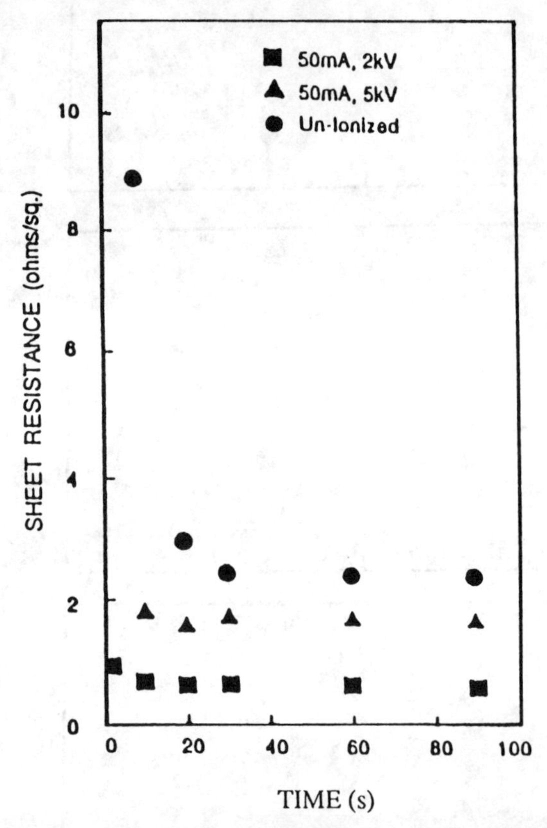

Fig.2. Comparing sheet resistances of titanium disilicides after 800 deg.C. annealing for different deposition conditions.

Scanning electron microscopy was used to study the as-deposited titanium layers and the layers formed after silicide formation. Both planar and cross-sectional studies were carried out to observe the topographical changes due to different deposition conditions and to examine the silicide-silicon interface. Fig.3 shows relative surface features under all three conditions of deposition. Fig.4(a) shows a planar view after silicide formation and Fig.4(b) a cross-section of the silicide-silicon interface.

Current-voltage and capacitance-voltage measurements on Schottky barrier diodes were used to evaluate the ideality factor and barrier height. Fig.5 shows the forward current characteristics. The ideality factor from the forward I-V was found to be 1.2. The barrier height as measured from the C-V characteristics was 0.9 eV, which is in agreement with reported values for films of acceptable quality.

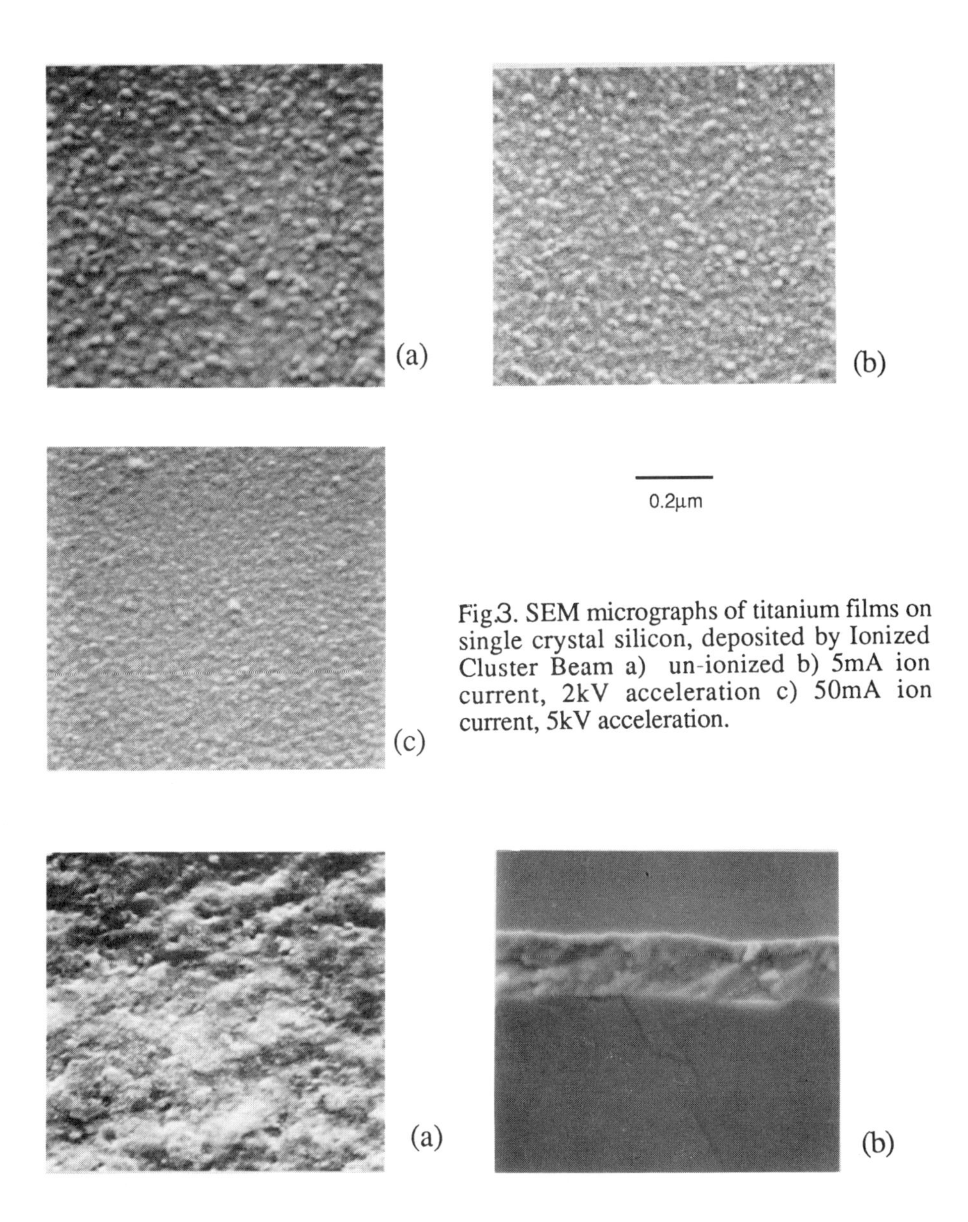

Fig.3. SEM micrographs of titanium films on single crystal silicon, deposited by Ionized Cluster Beam a) un-ionized b) 5mA ion current, 2kV acceleration c) 50mA ion current, 5kV acceleration.

Fig.4 SEM micrographs of titanium silicides formed by annealing at 800 deg.C for 30 seconds a) surface morphology b) cross-section. Deposition conditions were 50mA ion current and 2kV acceleration.

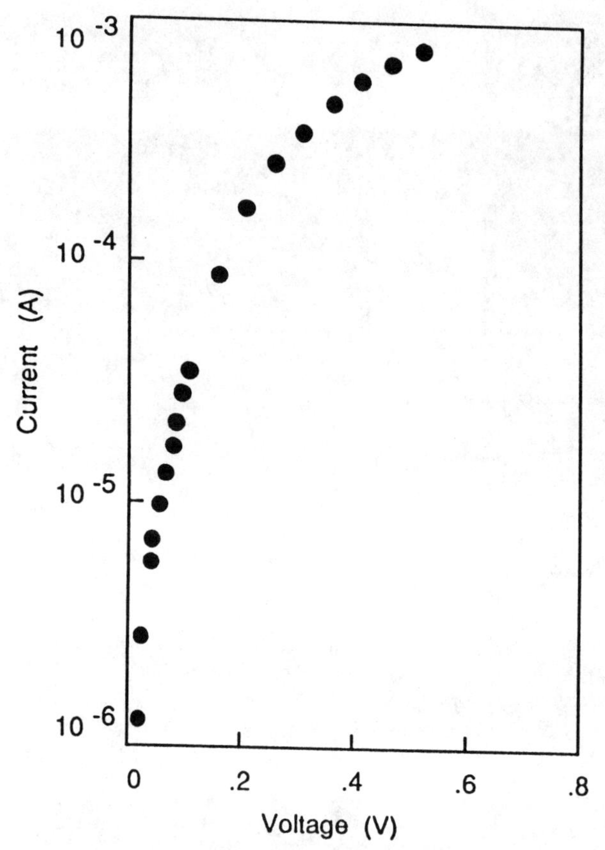

Fig.5. Room temperature I-V characteristics of a titanium-silicon Schottky diode. Titanium deposited by ICB at 50mA ion current and 2kV acceleration.

DISCUSSION

There are several features that distinguish $TiSi_2$ layers formed by depositing titanium films by ICB technique from layers produced with titanium films deposited by conventional techniques. It is interesting to note that the minimum silicide sheet resistance for a 500Å thick titanium layer deposited by ICB is 0.8 Ω/sq. and this value is independent of annealing temperature above 800 deg.C up to 1200 deg.C for annealing times up to 90s. This value of sheet resistance seems to be relatively low as compared with $TiSi_2$ formed by deposition of titanium by conventional methods and silicidation by Rapid Thermal Annealing [10,11,12]. Measurements on films thinner than 500Å indicate that the sheet resistance does not increase markedly; this contrasts with other results, where for a 200Å thick titanium film deposited on single crystal silicon and annealed in nitrogen ambient at 950 deg.C the reported sheet resistance is about 110 Ω/sq. and for a 300Å film the same value is reached at about 1000 deg.C [13] . The values of sheet resistance are relatively higher when the samples are annealed in pure He.

It is believed that, among other factors, the sheet resistance of titanium silicide is influenced by the deposition process, which affects the cleanliness of the titanium-silicon interface, the film thickness, the density and the grain size [14]. These properties are all influenced by the nature of the ICB technique. The main reasons for ionizing the clusters are: firstly, to increase their velocity by applying an electric field and, secondly, to enhance the critical parameters for film formation, such as nucleation, coalescence and chemical activity by utilizing the electric charge of the ions. The kinetic energy of the cluster ions is converted into thermal energy on impact with the surface. This causes the surface temperature to increase without having to heat the bulk. Surface heating in turn affects growth and also increases chemical activity. The high energy flux of the cluster ions also increases the defect concentration which enhances diffusion. It has been demonstrated with Ag deposited onto crystalline silicon that the diffusion characteristics of the ICB technique are different from those of conventional evaporation [15]. The resistivity of the silicon, which is affected by the diffusion of Ag, could be altered by varying the acceleration potential.

In the ICB process the energetic clusters clean the substrate surface by sputtering, improving the film adhesion. Sputtering also helps to remove the native oxide from the silicon surface during the deposition of titanium. It is believed that the native oxide film at the titanium-silicon interface slows down the reaction and causes it to proceed in a non-uniform manner, which results in a rough surface. The distribution of stress in the film is often altered by the surface roughness. SEM micrographs reveal that the silicides formed by ICB deposited titanium films are very uniform. In addition to sputter cleaning the substrate surface, the cluster ions also blend the sputtered substrate material with the evaporant material to form an interfacial layer. The growth morphology of the film is also altered by the fact that the sputtering continues during deposition.

It may be noted that, as the accelerating potential is increased from 0kV to 5kV the $TiSi_2$ resistivity is lower for 2kV but then increases at 5kV. This may be explained by assuming that as the accelerating potential increases, the packing density of the as-deposited metal increases, which contributes to the lowering of resistivity [15]. However, after a certain limiting accelerating potential, although the packing density probably continues to increase (as indicated by the SEM images), it is possible that damage enhanced diffusion of the dopant boron atoms into the silicide and across the silicide-to-silicon interface can result in higher resistivity. The grain size of the as grown titanium film can play an important role in controlling such diffusion mechanisms.

It is concluded that the ICB deposition method offers an alternative technique for depositing Ti films which can give $TiSi_2$ layers with superior properties, when compared with films formed by more conventional methods .

REFERENCES

1. D.S. Gardner, J. D. Meindl and K. C. Saraswat, IEEE Trans. on Electron Devices, **ED-34**, No.3, 633 (1987).

2. S.P. Murarka, *Silicides for VLSI Applications* (Academic Press, N.Y., 1983), p115.

3. T. Takagi, I. Yamada, M. Kunori, and S. Kobiyama, Proc. 2nd Int. Conf. on Ion Sources, 1972, Vienna, p. 790.

4. P. R. Young, J. Vac. Sci. Technol., **A3(3)**, 588 (1985).

5. T. Takagi, I. Yamada, and A. Sasaki, J. Vac. Sci. Technol., **12(12)**, 1128, (1975).

6. S.N. Yang and T.M. Lu, Appl. Phys. Lett. **48** (17), 28 April (1986).

7. A. Kuiper, G. Thomas, and W. Schouten, J. Cryst. Growth **51**, 17 (1981).

8. T. Takagi, I. Yamada and A. Sasaki, Thin Solid Films, **39**, 207 (1976).

9. R.A. McMahon, D.G. Hasko and H. Ahmed, Rev. Sc. Instrum. **56(6)**, 1257 (1985).

10. E.A.M. Ondrusz, R.E. Harper, A. Abid, P.L.F. Hemment and K.G. Stephens, Mat. Res. Soc. Symp. Proc. **25**, 99 (1984).

11. D. Pramanik and A.N. Saxena, J. Vac. Sci. Technol., **B2(4)**, 775, (1984).

12. D. Pramanik, M. Deal, and A.N. Saxena, Semiconductor International, May 1985, p.94.

13. C.Y. Ting, F.M. d'Heurle, S.S. Iyer, and P.M. Fryer, J. Electrochem. Soc., **133(12)**, 2621, (1986).

14. M.E. Alperin, T.C. Hollaway, R.A. Haken, C.D. Gosmeyer, R.V. Karnough and W.D. Parmantie, IEEE J. Solid State Circuits, **SC-20(1)**, 61, (1985).

15. T. Takagi, I. Yamada and A. Sasaki, Thin Solid Films, **45**, 569 (1977).

High-Aspect-Ratio Via Filling with Al Using Partially Ionized Beam Deposition

S.-N. Mei, S.-N. Yang, and T.-M. Lu
Center for Integrated Electronics
Rensselaer Polytechnic Institute, Troy, NY 12181

S. Roberts
IBM Corporation
Essex Junction, VT 05452

ABSTRACT

We have employed a Partially Ionized Beam (PIB) Al deposition technique to fill deep oxide vias with the aspect ratio ranging from one to two. The PIB contained ~ 1.5% of Al self-ions and a few kilovolts bias potential was applied to the beam during the deposition. Deep via filling with Al was achieved at a substrate temperature ≤ 210°C. Unlike conventional evaporation, Al deposited by PIB grew in a layer-by-layer manner from the bottom of the vias and did not develop an overhang structure at the top corner of the vias. This allowed the complete filling of the vias without the creation of cracks at the bottom corner of the vias. It is believed that ion-milling played the major role in achieving this result. The self-ion bombardment also reduced surface roughness of the thick (2 - 4 μ) Al films drastically. The sheet resistivity of the PIB Al films was measured to be 2.8 $\mu\Omega$ cm. The potential applications of the PIB technique in depositing Al "plugs" for multilevel interconnections are discussed.

I. INTRODUCTION

With today's submicron technology for device fabrication, the speed of a VLSI chip is more likely limited by the delay of interconnects rather than the transistor action of the devices. The multilevel metallization is one of the most powerful technologies which are able to tremendously reduce the total length of interconnects. In the multilevel metallization, different layers of patterned conductors are isolated from each other by insulators, and contacts are made between conductors on different layers through vias in the insulator. For the purpose of reducing the capacitance between the conductor layers, a thicker insulator layer is desirable. The high packing density design rule requires the via dimension to be small. These considerations in the multilevel metallization lead to the requirement of filling up high-aspect-ratio vias with low resistivity metals (such as Al) to form reliable contacts. Due to the shadowing effect, step coverage of the conventionally evaporated Al films at step corners and at the sidewall of vias is known to be poor [1]. This causes severe reliability problem in VLSI metallization [2,3]. Several techniques, such as biased sputtering [4,5], pulsed laser irradiation [6,7], selective tungsten chemical vapor deposition [8], and electroless plating methods [9], have been introduced recently to overcome the problem. Aside from higher process cost, these techniques usually yield films with higher resistivity or defect density. In this paper we report the results of high-aspect-ratio via filling with Al using the Partially Ionized Beam (PIB) deposition technique. We also discuss a possible mechanism which gives rise to the unique thin film morphology around the vias.

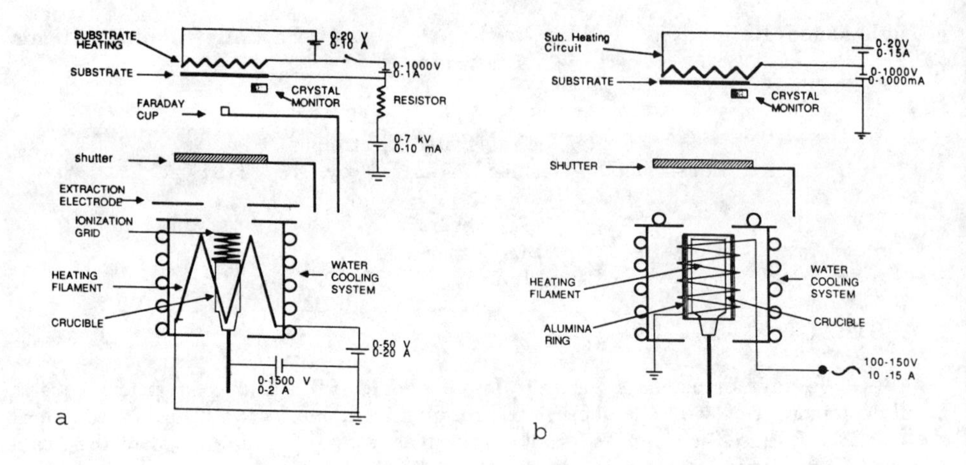

Figure 1: Schematic diagrams of a) the Partially Ionized Beam deposition system and b) the conventional evaporation setup.

II. EXPERIMENTAL SETUP

In the PIB technique, Al is evaporated in a graphite crucible and partially ionized by electron impact ionization. The ionized species are then accelerated toward the substrate by applying a bias potential to the substrate. The bias potential V_{ac} varies from 1 to 4 kV in our experiments to control the ion bombardment energy. Fig.1a is a schematic diagram of the PIB deposition system. The setup consists of five parts: (1) a graphite crucible containing the source material (i.e. Al for the present case), (2) a helix shaped ionization grid sitting on top of the crucible, (3) an electron bombardment heating filament, (4) a water cooling system, and (5) an extraction electrode. The electron bombardment heating filament serves two purposes. It provides the electron bombardment current for heating of the crucible, and at the same time provides the electrons for ionizing the vapor beam. This arrangement enables the vapor beam to be ionized most efficiently in the region where the atomic density is the highest. By the proper combination of bombardment voltage and current, crucible temperature and the ion-to-atom ratio can still be controlled independently within certain limits. This is because the crucible temperature depends on the total input power, while the ion-to-atom ratio is mainly affected by the electron current. The crucible is capable of operating at a temperature up to 1800°C. The deposition rate can be controlled up to 50 Å/sec. The ion-to-atom ratio, α, can be varied in the range of 0 - 5%. The ionization efficiency was determined by measuring the ratio of the total ion current and particle current arriving at the substrate at a given time. (The particle current was measured by a calibrated quartz crystal thickness monitor placed close to the substrate.) A uniform ion current distribution was obtained over an area of 6 cm diameter at a distance of 30 cm (where the substrate was placed) from the crucible. The substrate bias potential used in this experiment varied from 1 to 4 kV. In order to maintain the system pressure at about 3 x 10^{-7} Torr (or 4 x 10^{-5} Pa) during the deposition, the water cooling system was essential. The round shaped extraction electrode, made of stainless steel with a 20 mm diameter aperture, was mounted 10 mm above the crucible.

There is another filament set located about 1 cm behind the substrate, which is able to raise the substrate temperature by radiation heating up to 600°C during the deposition.

For comparison, the via filling experiments have been carried out in the same vacuum chamber with a conventional evaporation source as shown in Fig. 1b. All the geometrical factors, such as crucible dimension, deposition area, and crucible to substrate distance, were kept the same as that in the PIB system. However, instead of using electron-bombardment heating which could introduce ions in the vapor beam, a resistance heating method was used to raise the crucible temperature. No ions were observed in the vapor beam with this setup.

III. RESULTS AND DISCUSSION

Fig.2 shows a SEM cross-section picture of an Al film deposited on oxide vias at room temperature with the conventional evaporation setup. It is obvious that the overhanging structure on the top corner of the vias caused the crack and void at the bottom corner of the vias. As a result, this process cannot form an electrical contact through the via for multilevel metallization. For the vias with higher aspect ratio, the situation is worse. Besides the electrical connection problem, the film is also weak mechanically. It exhibits the typical columnar growth structure which is known to be less dense and brittle. The adhesion of the film on the oxide substrate is very poor. Very often the Al film peels off and leaves a cone-shaped Al drop in the via.

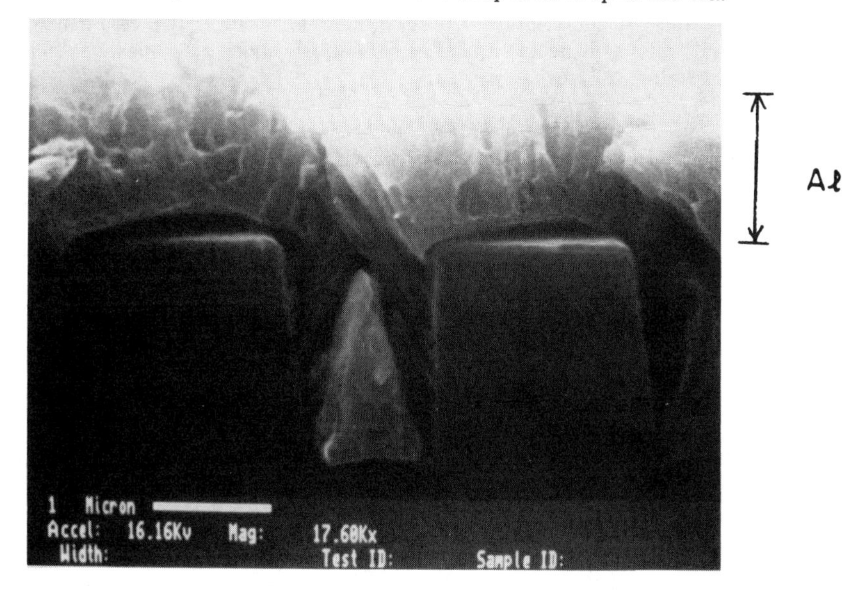

Figure 2: Al via filling by the conventional evaporation technique at room temperature. The Al films do not adhere well to the oxide surface.

Using a partially ionized Al beam we have succeeded in filling high-aspect-ratio (up to 2) vias with an ion to atom ratio of ~ 1 - 2% and a bias potential of several kV at the substrate temperature of ~ 200°C. Fig.3a shows straight wall oxide vias 2 μ deep and 1.5 μ wide, covered with 2 μ PIB Al film. The deposition conditions are:

substrate temperature	T_s = 210°C,
ion/atom ratio	a = 1.5%,
acceleration voltage	V_{ac} = 3.8 kV, and
deposition rate	R = 13 Å/sec.

Notice that the Al layer did not develop an overhang structure over the top corner of the vias during the deposition. This allowed the filling of the vias without creating cracks at the bottom corner of the vias. The morphology of the Al film resembles the original oxide vias, i.e. maintaining the sharp rectangular shape. Fig.3b shows 1 μ wide and 2 μ deep (aspect ratio of 2) vias covered with 2.5 μ PIB Al film. The deposition condition was similar to that of 3a except that the substrate temperature was raised to 260°C during the deposition of the last 0.5 μ Al. The resistivity of the film is ≃ 2.8 μΩ cm which is very close to that of the bulk aluminum (2.65 μΩ cm).

The overhang structure (as shown in Fig. 2) on the top corner of the vias with the Al films deposited by conventional means is initiated by the fact that the deposited film has the tendency to grow in the direction perpendicular to the local substrate surface, rather than along the direction of incident beam. The subsequent shadowing effect leads to the tapering of the Al layer inside the via. Merely changing source dimension without introducing ions into the beam cannot get rid of the void formation in the via.

To achieve a complete via filling, ion bombardment is essential. However, to within the range of ion-to-atom ratio obtainable by our PIB system, moderate substrate heating is still necessary. Figs.4a and 4b show the separate effects of ion bombardment and substrate temperature.

The micrograph shown in Fig.4a was taken from a sample deposited by PIB at room temperature (no intentional substrate heating). By comparing Fig.4a with Fig.2, we found two major structural changes induced by ion bombardment. First, the overhang on the top corners of the via was greatly reduced, which is believed to be a result of simultaneous ion milling during the deposition. The incident Al atoms were constantly able to reach the bottom of the via and the film thickness inside the via is the same as that at the top of the via (except for a very narrow region right next to the via side walls). Second, the film deposited by PIB does not show columnar growth and has a very good adhesion to the oxide substrate. Because of the ion bombardment during deposition, a layer (about a few hundred Å in thickness) near film surface can be heated up substantially [10]. This heating is able to cause interface mixing at the early stage of the film formation and atomic rearrangement in the film during growth. The former improves adhesion and the later can break the columnar structure [10]. However, it is seen that the room temperature PIB process still left very narrow cracks at the edges of the via. Narrow Al contacts can be fabricated using a lift-off metallization technique [11].

The sample showing in Fig.4b was deposited by the conventional evaporation technique at 200°C substrate temperature without the use of ions. It is obvious that substantial surface diffusion took place at that temperature. This diffusion caused more material to flow into the via and, consequently, the film around the via became

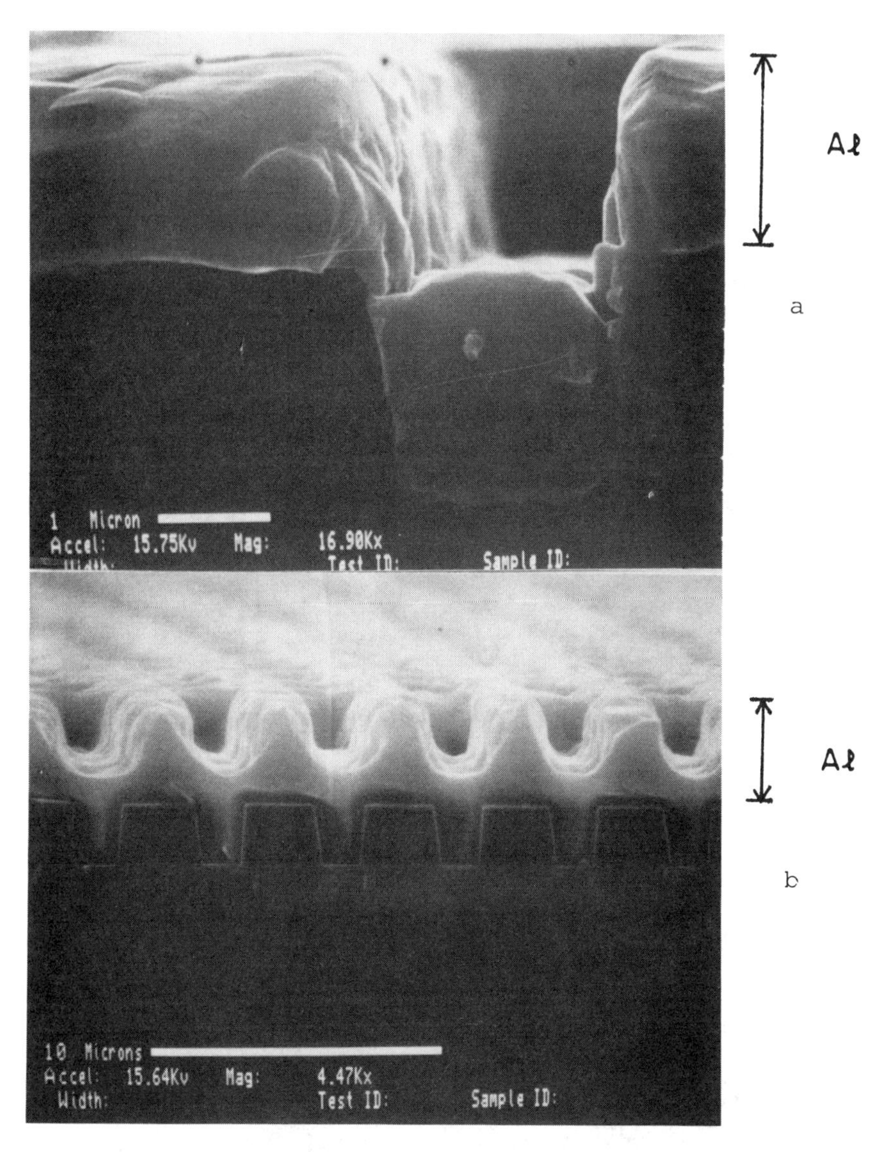

Figure 3: a) A via of 2 μ deep and 1.5 μ wide covered with 2 μ PIB Al film deposited under the following conditions: $T_s \sim 210°C$, $a = 1.5\%$, $V_{ac} = 3.8$ kV, and R = 13 Å/sec.
b) Vias 2 μ deep and 1 μ wide covered with 2.5 μ PIB Al film deposited under similar conditions as for (a) except that T_s was raised to 260°C during the last 0.5 μ Al deposition.

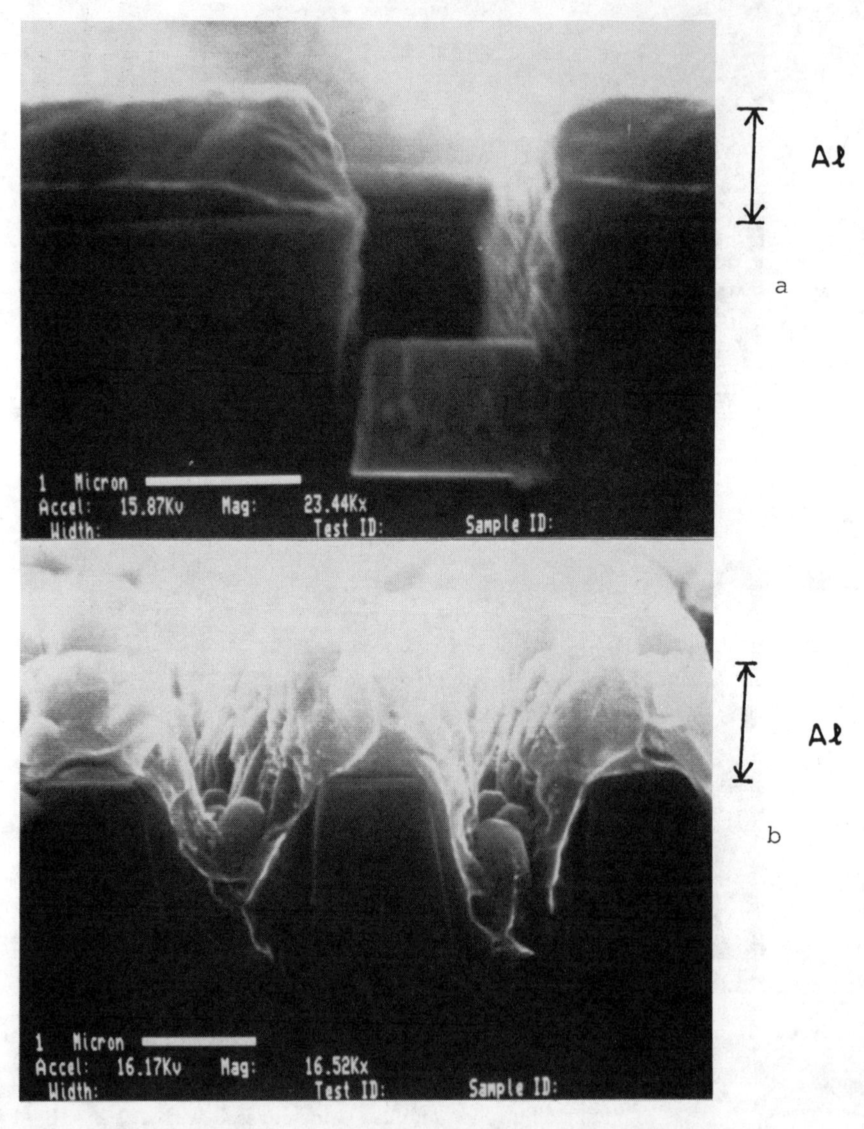

Figure 4: a) Al via filling by partially ionized beam at room temperature.
b) Al via filling by the conventional evaporation at 200°C substrate temperature without the use of ions.

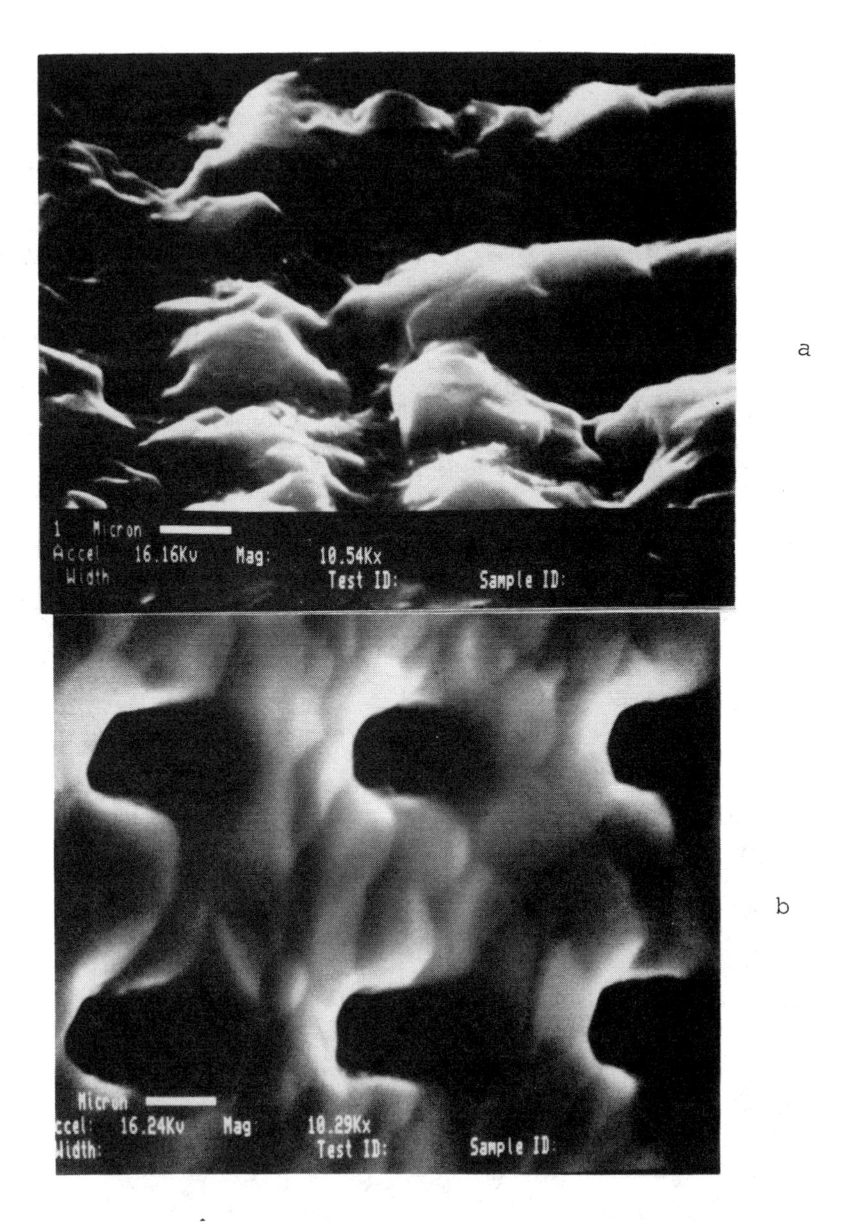

Figure 5: Surface morphology of 2μ thick Al films deposited at ~ 300°C substrate temperatures by a) the conventional method and b) the PIB technique.

semi-conformal. For this kind of semi-conformal growth mode, a void is very likely to be formed near the center of the via. It is also seen that the film has a much rougher surface.

However, with the assistance of this surface diffusion caused by ~ 200°C substrate heating during the PIB deposition, the narrow cracks which appeared in Fig.4a were healed as shown by Fig.3. It is also very important to notice that, unlike what is seen in Fig.4b, the film deposited by PIB at ~ 200°C retained a smooth surface. Figs. 5a and 5b show the surface morphology of 2 μ thick Al films deposited at 300°C substrate temperature by the conventional and PIB techniques respectively. We have also used the PIB technique to deposit thick (4 μ) and smooth Al lines for wafer scale integration which cannot be obtained by conventional means [12].

It should be mentioned that the mechanism of via filling in the PIB deposition is rather different from that of the bias sputtering technique [4,5]. In the bias sputtering technique, a conformal step coverage is formed right at the initial stages of deposition. In the PIB deposition, the initial deposition is completely non-conformal as shown in Fig.6. Very little coating was found on the side walls. Without the shadowing effect caused by the overhang structure, the vias can be filled up layer-by-layer from the bottom.

The ion percentage, ion energy and the substrate temperature we used in this work for obtaining complete via filling may still not be the optimum combination. We feel that higher ion-to-atom ratio may further reduce the minimum temperature required for the complete via filling. The variation of ion energy in the range of 2 keV to 5 keV has not been found to change the via filling result in a significant manner.

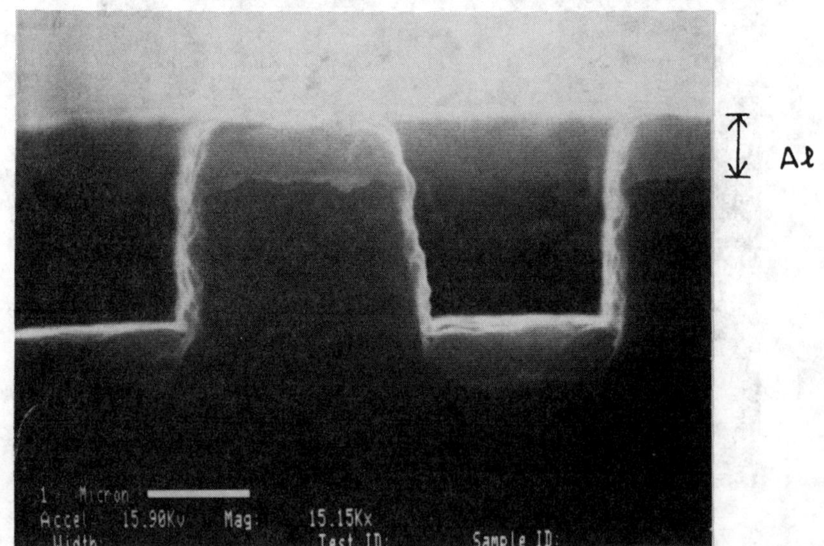

Figure 6: The non-conformal step coverage of PIB Al film at the initial stage of via filling. The deposition condition was the same as in Fig.3. The Al film grew in a layer-by-layer manner without developing an overhang structure.

We have also shown in a brief report [13] that in the PIB deposition of 1 μ deep vias and stepped structure, if the substrate temperature is raised to 320°C at the final stage of the deposition, a partial planarization of the Al film can be achieved.

IV. SUMMARY

We have used a partially ionized Al beam with a very small amount of Al self-ions (~ 1.5%) and a bias potential of several kV to deposit Al film in oxide vias with an aspect ratio up to two. A completely non-conformal via filling can be obtained at $\leq$210°C substrate temperature. The Al films grew in a layer-by-layer manner from the bottom of the vias without developing overhang structures at the top of the vias. Wall-to-wall filling of the vias was achieved. Ion milling effect is believed to be responsible for the elimination of the overhanging structure.

The effect of substrate heating has also been investigated. It was found that high substrate temperature during deposition usually yields films with rougher surface partly due to the increase of grain size. And a small amount of self-ion bombardment during film growth can reduce this surface roughness drastically.

ACKNOWLEDGEMENT

This work is supported in part by the Semiconductor Research Corporation and IBM corporation. We would like to thank C.-K Hu, S. Murarka, and C.-H. Choi for valuable discussions.

REFERENCES

1. I.A. Blech, Solid State Technology (Dec. 1983), p.123.
2. D. B. Fraser, in "VLSI Metallization", edited by S. M. Sze, McGraw-Hill, New York, (1983) p.347
3. A. N. Saxena and D. Pramanik, Solid State Technology, Oct. (1986) p.95
4. Y. Homma and S. Tsunekawa, J. Electrochem. Soc. **132**, 1466 (1985).
5. D. W. Skelly and L. A. Gruerke, J. Vac. Sci. Technol. **A4**, 457 (1986).
6. D. B. Tuckerman and A. H. Weisberg, IEEE Electron Device Lett. vol. EDL-7, 1 (1986).
7. R. Mukai, N. Sasaki and M. Nakano, IEEE Electron Device Lett. vol. EDL-8, 76 (1987).
8. D. M. Brown, B. Gorowitz, P. Piacente, R. Saia, R. Wilson, and D. Woodruff, IEEE Electron Device Lett., vol. EDL-8, 55 (1987).
9. Y. Harada, K. Fushimi, S. Madokoro, H. Sawai, and S. Ushio, J. of Electrochem. Society **133**, 2428 (1986).
10. K.-H. Muller, J. Vac. Sci. Technol. **A4**, 184 (1986)
11. R. Ramanarayanan, K. Polasko, D. Skelly, J. Wong, S.-N. Mei and T.-M. Lu, J. Vac. Sci. Technol. *B5 (1)*, 359 (1987).
12. R. Selvaraj, S.-N. Yang, J. McDonald, and T.-M. Lu, in the Proceedings of ***Fourth International IEEE VLSI Multilevel Interconnection Conference***, IEEE Electron Devices Society, New York (1987), p.440
13. S.-N. Mei, T.-M. Lu, and S. Roberts, IEEE Electron Devices Lett., **EDL-8, 503(1987).**

CHAPTER VI
ALTERNATIVE GROWTH TECHNIQUES AT THE FOREFRONT II

SILICON-ON-INSULATOR: WHY, HOW, AND WHEN

C.-E. Daniel Chen and P. Chatterjee

Semiconductor Process and Design Center
Texas Instruments, Inc.
MS 944, P. O. Box 655012
Dallas, TX 75265

ABSTRACT

SOI technology offers improved circuit performance due to the reduced junction capacitance, latch-up free dielectric isolation, higher packing density and higher radiation tolerance. Some of the SOI approaches can offer an IC fabrication which is compatible with, but simpler than, the bulk CMOS process, especially with scaled geometries; and they can make the Bi-CMOS process integration easier, due to the simplicity in the device isolation process. Some approaches can also be applied to 3-D IC's through multi-layered SOI structures. The realization of these advantages, however, hinges on obtaining a device quality single crystal silicon on top of an insulator. Status, advantage, and disadvantage of various SOI approaches are presented in this paper with particular emphasis placed on the SIMOX approach which in the past couple of years has emerged as the leading SOI technology and offers an opportunity to be a viable cost effective alternative to the sub-half micrometer bulk CMOS scaling.

INTRODUCTION

CMOS/SOI technology offers improved circuit performance due to reduced junction capacitance, latch-up free dielectric isolation, higher packing density and radiation hardness. Figure 1 compares the cross-sectional schematics of a CMOS/SOI with the bulk CMOS. The bulk CMOS structure (1b), even without a trench isolation, is considerably more complex than that of the CMOS/SOI (1a). Due to this simple device structure, some of the SOI approaches can offer a simpler (compared to bulk CMOS/Bi-CMOS) IC fabrication process, and some can be applied to 3-D IC's through multi-layered SOI structures. The realization of these advantages, however, hinges on obtaining a device quality single crystal silicon on top of an insulator with good Si/insulator interface properties.

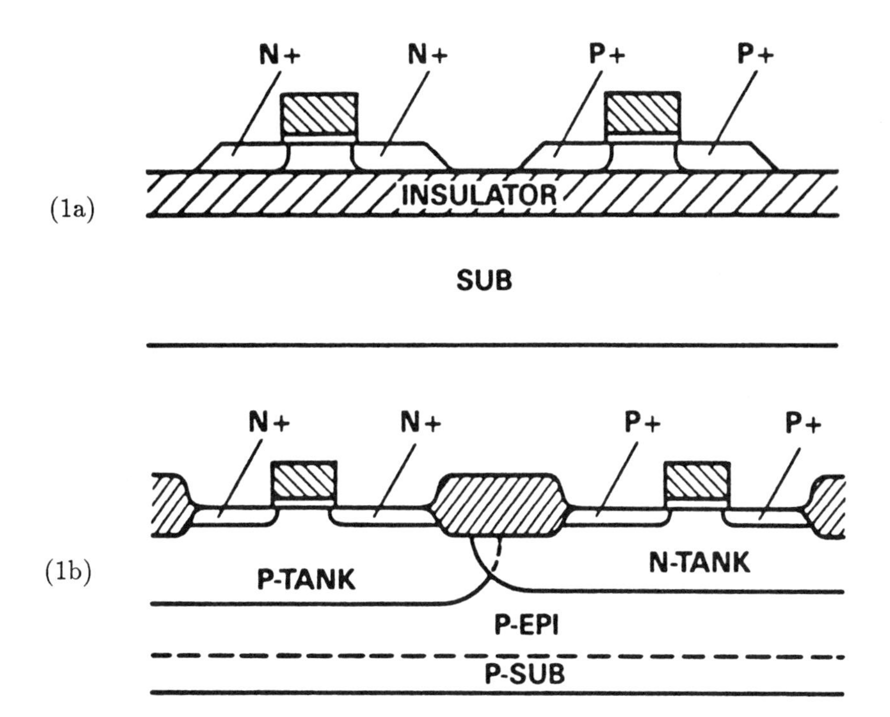

Figure .1: Schematic cross-sections of (1a) CMOS/SOI, (1b) Bulk CMOS

At the present time, the SOI technology thrust is sponsored mainly by the needs in the space and the defense industry for high radiation tolerant integrated circuits. SOI circuits, built on a very thin (typically a few tenths of a micrometer thick) silicon film on the top of an insulator layer, are more tolerant to the Single Event Upset (SEU) and prompt gamma upsets because the junction volume available for the radiation induced electron-hole pair generation is significantly reduced, and the carriers generated in the supporting silicon substrate are not collected by the device nodes because the devices are electrically isolated from the substrate by the insulator layer. The complete dielectric isolation, which eliminates the parasitic SCR latch-up structures in a CMOS circuit, also eliminates the parasitic field devices which in turn can improve the gamma total dose hardness of the circuit. Other advantages that SOI can offer, i.e., better performance due to lower parasitic capacitance, easier scaling due to the simplicity in its device structures, and easier process integration and higher packing density, are not being intensively pursued presently. However, these advantages will be exploited in the next couple of years when the device feature sizes are extended further into the sub-micrometer regime, and quality SOI substrates are available in manufacturing quantities.

For the SOI technology to be widely employed in the semiconductor industry (in the mid-late '90s), SOI researchers need to move quickly from the research/development phase into the prototype/production phase and bring out products in the next couple of years that meet the space and military needs. With good field reliability records demonstrated, more people will accept SOI devices, and more resource will be available to realize the performance advantage and easier IC fabrication process that SOI offers in devices with sub-half micrometer features. With this goal in mind, the chosen SOI technology must be able to provide device quality SOI substrates with at least 6 inch diameter or larger, and it must be able to support devices that integrate more than 1 million sub-micrometer transistors with good yield. To date, SIMOX is the most promising SOI approach with no known limitations in the scaling up of the wafer size and the scaling down of device features, and integrated circuits with more than 100 thousand transistors have been demonstrated with good yield.

MAJOR SOI APPROACHES

The three major SOI approaches, other than the Silicon-On-Sapphire technology, are: Full Isolation by Porous Oxidized Silicon (FIPOS)[1,2,3,4], recrystallization of polycrystalline or amorphous silicon[5,6,7,8], and Buried Oxide SOI [9,10,11], also known as SIMOX (Separation by IMplanted OXygen)[12]. The recrystallization approach can be further split into: zone melting recrystallization (ZMR)[5,6], beam scanning recrystallization (BSR)[7], and solid phase epitaxy (SPE)[8]. The last two approaches, BSR and SPE, can be applied to multi-layered 3-D IC's[7,13] since the underlying structures are less affected by these two processes forming the SOI layers above. Currently, most of the 3-D IC research is done in Japan. The wafer-bonding approach[14], with difficulties in bonding and in etch back control, will have its impact mainly in the bipolar applications (thick SOI film) where conventional poly DI wafers are currently used. The heteroepitaxial SOI [15,16], and other Semiconductor-On-Insulator[17], are still at the early research stage with possible 3-D applications.

FIPOS

Among the one-layer SOI technologies, namely FIPOS, Buried Oxide, and ZMR, the FIPOS approach is less compatible with the bulk CMOS

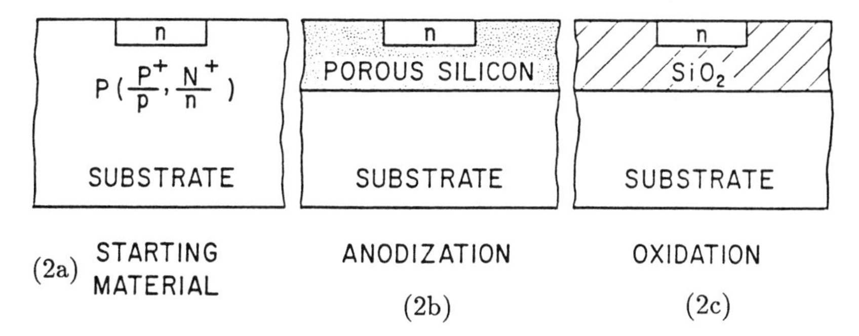

Figure .2: FIPOS process: (2a) Island Definition, (2b) Anodization, (2c) Porous Silicon Oxidation

process because its SOI formation is not done prior to the start of the device fabrication sequence; but rather, the anodization and oxidation are done after the active device area is defined. And because of this, there is a limit on the size of the island that can be dielectrically isolated. Figure 2 shows the SOI island formation in the FIPOS process. In the original FIPOS approach, where p^- silicon was anodized while n^- silicon was retained, the island size was limited to about 10 micrometers, and there was a large anodization tail underneath the island. A 64K SRAM [18] was demonstrated using this process.

Several process modifications have been made to increase the island dimensions and to reduce the anodization tails. The improvements were made by increasing the selectivity of lateral anodization rate to the vertical rate through the use of p^+ on p^- substrate[2] in a current-controlled anodization or the use of n^+ on n^- substrate[3,4] in a voltage-controlled anodization. The single crystal nature of the anodized porous silicon was also utilized to support a silicon Molecular Beam Epitaxy (MBE)[19]. In the voltage-controlled anodization and in the Si-MBE approaches, the silicon tail was eliminated completely. SOI formation was completed when the porous silicon layer underneath the device islands were converted into thermal oxide. High pressure oxidation was shown to reduce the mechanical stress generated in the porous silicon layer oxidation, which had to proceed laterally from the island edge to right underneath the center of the device island.

SOI formation in the middle of the IC fabrication process and the constraint of a maximum allowable size of the active device area make a direct implementation of existing circuits on the FIPOS technology difficult.

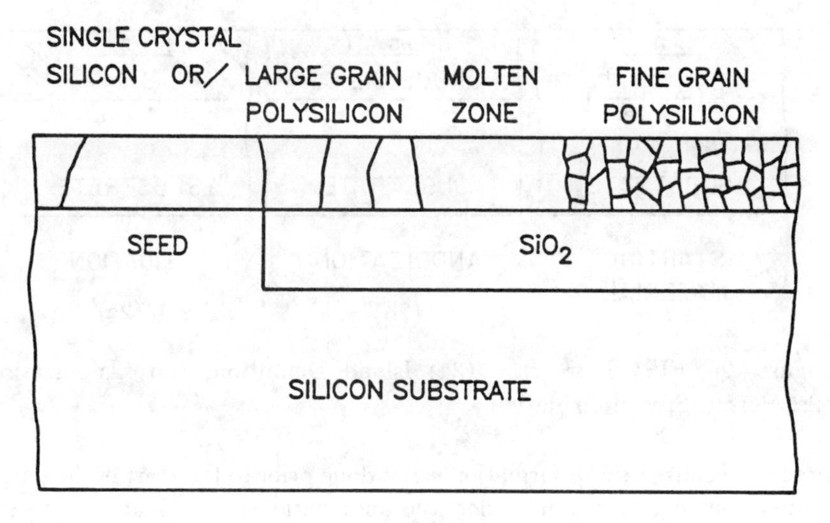

Figure .3: Schematics of the recrystallization process. An external heat source controls the size and the motion of the molten zone.

FIPOS does offer, in principle with a proper anodization and a subsequent oxidation, the potential of obtaining an SOI film with the crystalline quality identical to that of its starting bulk silicon.

ZMR

ZMR approach is an inexpensive way to make SOI substrates. The most commonly used equipment is a graphite strip heater system[5] while some work was done using focused lamps[6] which presumably reduced contamination and increased the control of the heating (radiation). Both seeded and un-seeded approaches were studied with comparable results. Figure 3 depicts the basic principles of the recrystallization process where the molten silicon zone is induced and guided by an external heat source. The predominant defects observed were sub-boundaries[20]. It was shown that these sub-boundaries did not affect the majority carrier transport[21]. However, enhanced doping diffusion along the sub-boundaries was observed. This enhanced doping diffusion makes the fabrication of short channel devices difficult[22]. A few techniques were attempted to reduce or eliminate the sub-boundaries, and it was shown that with thicker top silicon film (thicker than 1 micrometer) and an improved capping technique[23], the sub-boundaries dissociated into individual threading dislocations. Other sporadic defects

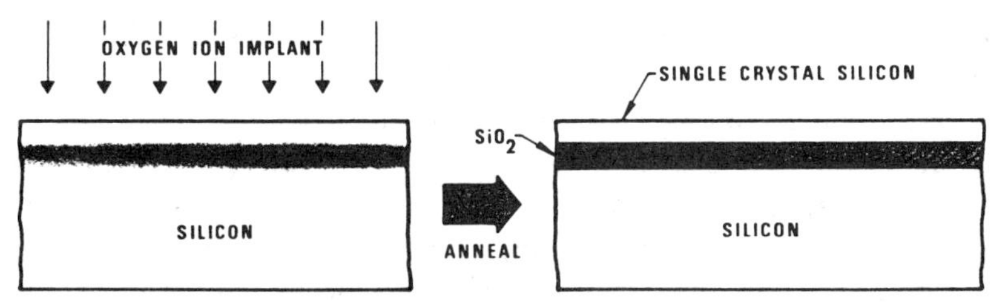

Figure .4: Buried Oxide Formation

observed on the ZMR materials were the protrusions due to local supercooling, film thickness non-uniformity due to the silicon mass transport in the recrystallization process, and wafer distortion due to the thermal stress.

A 1K SRAM (5 μm design rule) and a 1.2K gate arrays (3 μm design rules) were demonstrated using the ZMR SOI technology[22]. However, the scalability of ZMR SOI substrates to larger wafer sizes and the implementation of short-channel VLSI circuits with the presence of the sub-boundary defects remain to be demonstrated.

SIMOX

Buried oxide SOI is the leading one-layer SOI technology because it offers the best quality single crystal silicon material for device fabrication with proven parametric uniformity for VLSI applications[9,10]. Figure 4 shows the buried oxide formation process where a typical oxygen dose is 1.4 to $2.2 \times 10^{18}/cm^2$ and the anneal is done at above 1150°C. Major advances in the understanding of the SIMOX material and buried oxide synthesis have been made in the past two years[11,24,25,26,27]. Many studies were also made in the buried nitride[28], and buried oxynitride (oxygen and nitrogen mixed implant)[29] syntheses.

In the buried oxide formation, because of the high oxygen diffusivity in SiO_2 , the implanted oxygen atoms above the SiO_2 stoichiometric level diffuse through the buried oxide to the implant tails and sharpen the Si/SiO_2 interfaces upon anneal; On contrast, in the buried nitride formation, excessive nitrogen atoms are trapped in gas form in the buried nitride and make it porous while the implanted nitrogen atoms below the Si_3N_4 stoichiometric level diffuse against the concentration gradient and form Si_3N_4 upon high temperature anneals. After a post-implant anneal, the crystallinity of the residual top silicon layer is restored and it can support an epitaxial silicon growth.

Advantages of nitrogen implant are that it requires less dose, it has less crystalline defects after anneal, and that it has thicker superficial top silicon layer which eliminates epi process in some MOS applications. The disadvantages of nitrogen implant, however, are that it has more donor states, higher capacitance to the substrate, poor silicon/insulator interface, and most critically, very small (if any) dose window of achieving simultaneously both good insulating property and mechanical strength of the buried nitride layer[30]. At low doses where the implanted nitrogen peak concentrations were below the Si_3N_4 stoichiometric level, the resulting buried nitride layer did not show good insulating property; while at higher doses, the resulting buried nitride layer had poor mechanical strength because there was a porous region in the middle of the synthesized polycrystalline nitride layer.

For the oxygen implant, significant advances in the material quality has been achieved through the use of higher temperature anneals in the post-implant processing[11,24,25]. Rutherford Backscattering Spectroscopy (RBS) showed that higher post-implant anneal temperature improved the crystallinity of the superficial silicon layer and sharpened the silicon/buried oxide interface. The microstructures determined from cross-sectional TEM studies correlated well with the RBS data showing lower defect densities with higher annealing temperatures. XTEM also identified the defects in the top silicon layer as oxygen precipitates and threading dislocations, and showed that with annealing temperatures at or above 1250°C, the oxygen precipitates could be annihilated, and that the dislocation density reduction leveled off at a value about $10^9/cm^2$. Further increase in the annealing temperature, even up to 1405°C[27,31], did not lower the dislocation density further, but did shorten the time required to annihilate the oxygen precipitates.

Although these dislocations seemed to have little effects on the yield and performance of CMOS circuits (discussed below), one would like to reduce them further especially for bipolar applications. Dramatic reduction of the dislocation density has recently been obtained by sequentially implanting and annealing the SIMOX wafers with oxygen dose below a critical value of $4\times10^{17}/cm^2$ at 150KV between anneals[32]. Dislocation density approaching bulk silicon value (less than $1\times10^3/cm^2$) has been achieved.

Remarkable circuit and device results have been demonstrated on SIMOX substrates with high dislocation densities ($10^9/cm^2$). Surface carrier mobilities identical to bulk CMOS values are constantly achieved[25].

With no abnormal doping diffusion[33], the SIMOX SOI offers a fabrication process compatible with and simpler than the bulk CMOS process, especially in the sub-micrometer regime. In addition, bulk designs can be easily implemented on buried oxide substrates. We have demonstrated a 4K SRAM with 2.5 μm design rules[9] and a 16K SRAM with 1.25 μm design rules[10]. Good radiation results on the circuit[34] and device[29,35] were demonstrated. More significantly, high yields on the 16K SRAMs confirmed the manufacturability of the SIMOX SOI technology. With current materials understanding and the progress achieved so far, it is conceivable to see LSI products available in sampling quantities in two years.

REMAINING ISSUES

For the Silicon-On-Insulator technology to be widely adopted by the semiconductor industry, there are four major issues that need to be addressed. First is the cost and through-put of the starting SOI substrates which may call for equipment developments such as the very high current implanter in the buried oxide SOI technology. Currently, 100mA class very high current oxygen implanters are commercially available, and they can produce 4 or 5 inch SIMOX substrates at a rate of about 1 wafer per 10min. Second is the possible yield loss due to the material defects generated in the SOI substrate preparation process. In the SIMOX approach, high VLSI circuit yield has been demonstrated proving its manufacturability. Third is the floating body effect which generates the "kink" in the transistor I-V curve and may affect the circuit performance if not properly accounted for. On the other hand, the floating body effect may be used to one's advantage. Fourth is the development of a more elaborate electrostatic discharge (ESD) protection circuit since the insulator layer prevents the use of large area diodes clamping to the silicon substrate and also hinders the heat dissipation generated by the ESD pulses.

REFERENCES

[1] K. Imai and H. Unno, IEEE Trans. Elect. Dev., ED-31, 297(1984)

[2] L. A. Nesbit, IEDM Technical Digest, 800(1984)

[3] R. P. Holmstrom and J. Y. Chi, Appl. Phys. Lett., 42, 386(1983)

[4] E. J. Zorinsky, D. B. Spratt, and R. L. Virkus, IEDM Technical Digest, 431(1986)

[5] J. C. C. Fan, M. W. Geis, and B.-Y. Tsaur, Appl. Phys. Lett., 38, 365(1981)

[6] D. P. Vu, M. Haond, D. Bensahel, and M. Dupuy, J. Appl. Phys., 54, 437(1983)

[7] C.-E. Chen, H. W. Lam, S. D. S. Malhi, and R. F. Pinizzotto, IEEE Elect. Dev. Lett., EDL-4, 272(1983)

[8] H. Ishiwara, H. Yamamoto, S. Furukawa, M. Tamura, and T. Tokuyama, Appl. Phys. Lett., 43, 1028(1983)

[9] C.-E. Chen, T.G.W. Blake, L.R. Hite, S.D.S. Malhi, B.-Y. Mao, and H.W. Lam, IEDM Technical Digest, 702(1984)

[10] C.-E. Chen, M. Matloubian, B.-Y. Mao, R. Sundaresan, C. Slawinski, H.W. Lam, T.G.W. Blake, L.R. Hite, and R. K. Hester, IEEE Trans. Elect. Dev., ED-33 1840(1986)

[11] B.-Y. Mao, P.-H. Chang, H. W. Lam, B. W. Shen, and J. A. Keenan, Appl. Phys. Lett., 48, 794(1986)

[12] K. Izumi, M. Doken, and H. Ariyoshi, Electron Lett., 14, 593(1978)

[13] K. Yamazaki, M. Yoneda, S. Ogawa, M. Ueda, S. Akiyama and Y. Terui, IEDM Technical Digest, 435(1986)

[14] J. B. Lasky, S. R. Stiffler, F. R. White, and J. R. Abernathey, IEDM Technical Digest, 684(1985)

[15] T. Asano, and H. Ishiwara, J. Appl. Phys., 55, 3566(1984)

[16] H. Onoda, T. Katoh, N. Hirashita, and M. Sasaki, IEDM Technical Digest, 680(1985)

[17] H. Ishiwara, H. C. Lee, S. Kanemaru, and S. Furukawa, to be published in the Proceedings of the 2nd International Symposium on Silicon Molecular Beam Epitaxy (Electrochem. Soc. Conference Proceedings), and references therein.

[18] E. Ehara, H. Unno, and S. Muramoto, Electrochem. Soc. Extended Abstract, 85-2, 457(1985)

[19] S. Konaka, M. Tabe, and T. Sakai, Appl. Phys. Lett., 41, 86(1982)

[20] R. F. Pinizzotto, H. W. Lam, and B. L. Vaandragger, Appl. Phys. Lett., 40, 388(1982)

[21] B.-Y. Tsaur, J. C. C. Fan, M. W. Geis, D. J. Silversmith, and R. W. Mountain, IEEE Elect. Dev. Lett., EDL-3, 79(1982)

[22] B.-Y. Tsaur, IEEE Circuits and Device Magazine, vol. 3, no. 4, 12(1987)

[23] C. K. Chen, M. W. Geis, M. C. Finn, and B.-Y. Tsaur, Appl. Phys. Lett., 48, 1300(1986)

[24] C. Jaussaud, J. Stoemenos, J. Margail, M. Dupuy, B. Blenchard, and M. Bruel, Appl. Phys. Lett., 46, 1064(1985)

[25] B.-Y. Mao, M. Matloubian, C.-E. Chen, R. Sundaresan, and C. Slawinski, IEEE Elect. Dev. Lett., EDL-8, 306(1987)

[26] B.-Y. Mao, P.-H. Chang, C.-E. Chen, and H.W. Lam, J. Appl. Phys., 62, 2308(1987)

[27] G.K. Celler, P.L.F. Hemment, K.W. West, and J.M. Gibson, Appl. Phys. Lett. 48, 532(1986)

[28] P.-H. Chang, C. Slawinski, B.-Y. Mao, and H. W. Lam, J. Appl. Phys., 61, 166(1987)

[29] B.-Y. Mao, C.-E. Chen, G. Pollack, H. L. Hughes, and G. E. Davis, to be published in the December, 1987 issue of IEEE Trans. Nucl. Sci.

[30] C. Slawinski, B.-Y. Mao, P.-H. Chang, H. W. Lam, and J. A. Keenan, Proc. Mat. Res. Soc., 53, 269(1986)

[31] C.-E. Chen, B.-Y. Mao, and G.K. Celler, un-published results

[32] D. Hill, P. Fraundorf, and G. Fraundorf, to be published in J. Appl. Phys. Also, this work has been repeated, and expanded at Texas Instruments.

[33] K. Hashimoto, T. I. Kamins, K. M. Cham, and S. Y. Chang, IEDM Technical Digest, 672(1985)

[34] G. E. Davis, L. R. Hite, T. G. W. Blake, C.-E. Chen, and H. W. Lam, IEEE Trans. Nucl. Sci., NS-32, 4432(1985)

[35] B.-Y. Mao, C.-E. Chen, M. Matloubian, L. R. Hite, G. Pollack, H. L. Hughes, and K. Maley, IEEE Trans. Nucl. Sci., NS-33, 1702(1986)

A LOW-TEMPERATURE GROWTH PROCESS OF GaAs BY ELECTRON-CYCLOTRON-RESONANCE PLASMA-EXCITED MOLECULAR-BEAM-EPITAXY (ECR-MBE)

Naoto KONDO and Yasushi NANISHI

NTT Opto-electronics Laboratories, Atsugi, Kanagawa 243-01, Japan

ABSTRACT

Taking advantage of plasma excitation, surface cleaning and growth process are realized at low temperatures by electron-cyclotron-resonance (ECR) plasma-excited molecular-beam-epitaxy (MBE). Prior to growth, substrates are cleaned by exposure to hydrogen plasma at temperatures ranging from 300 to 550°C. Arsine gas is introduced and cracked in an ECR plasma generation chamber. Gallium is supplied either as trimethylgallium (TMG) or as metallic Ga. Epitaxial films are successfully grown at substrate temperatures as low as 430°C for the TMG-arsine system and 350°C for the metallic Ga-arsine system. The growth rate for the TMG-arsine system is found to be governed by a balance between TMG decomposition and surface atom desorption. By contrast, the metallic Ga-arsine system is only governed by the desorption process. Exposure to plasma is found to promote desorption of atoms migrating on the substrate surface. The interface between the substrate and the epitaxial layer produced by the ECR-MBE process is found to be clean without piling up of impurity.

INTRODUCTION

With progress in device fabrication processes, low-temperature growth techniques have emerged as a key technology, especially for re-growth processes where it is important that composition and/or dopant profiles should not be transformed. Attempts to reduce growth temperature have been carried out in several ways. For conventional MBE, device-quality GaAs films have been grown at 380°C by simply decreasing the growth rate.[1] Growth rates that are too small, however, say in the vicinity of 0.02 um/h, are not suitable for practical applications. Another strategy for reducing growth temperature has been investigated by Horikoshi et al.[2] They describe a process of alternately feeding Ga and As_4 molecular beams on the substrate surface to enhance their surface migration. Epitaxial layers grown at as low as 200 - 300°C show high photoluminescence (PL) intensities comparable to the layers grown at high temperatures. Both of these methods, however, require high-temperature processing for surface cleaning prior to growth. This is because surface oxides on GaAs substrates cannot be removed at temperatures lower than 500°C. It is thus apparent that low temperature process for both surface cleaning and epitaxial growth are urgently required to realize further advances in device fabrication technology.

The application of an ion or plasma assisted technology to the crystal growth of compound semiconductors is a very promising approach for achieving both growth temperature reduction and substrate surface cleaning at low temperatures. Enhancement of surface atom migration by ion irradiation and its effect on growth temperature reduction was recognized in a study of GaAs lateral epitaxial growth over tungsten grating.[3] It was found in this same study, however, that the occurrence of irradiation damage to the crystal caused by impinging ions was a major drawback, especially for III-V semiconductors. It was found that the grown layers showed high resistivity in spite of a Sn doping of 3×10^{17} cm^{-3}, when irradiated with gallium and hydrogen ions.[4] Although the acceleration voltage in this experiment was as low as 0.1 kV, it was suggested that deep level defects were generated by ion irradiation.

By using ECR plasma, good quality SiO_2 and SiN films are obtained at a fairly low growth temperature.[5,6] Since the ions impinging on the growing surface have energies of a few tens of electron volts in the ECR deposition system, surface migration is effectively enhanced without introducing irradiation damage to the growing film. In order to study the irradiation effects on GaAs crystal, epitaxially grown GaAs layers have been exposed to ECR plasma by the authors using hydrogen or nitrogen discharge gas.[7] Judging from evaluation of 4.2 K PL spectroscopy results, it has been concluded that ECR plasma exposure does not give rise to extensive damage to epitaxially grown GaAs layer.

In this paper, the ECR plasma deposition technique is applied to an MBE growth of GaAs. By using this technique for both cleaning and deposition, epitaxial layers have been successfully obtained at relatively low substrate temperatures all through the growth process. The growth mechanism is also studied to explain growth rate dependence on input microwave power.

EXPERIMENT

Figure 1 shows a schematic diagram of the apparatus used in this study. It consists of three chambers: a plasma generation chamber, a growth chamber, and a load-lock chamber. Discharge gas and

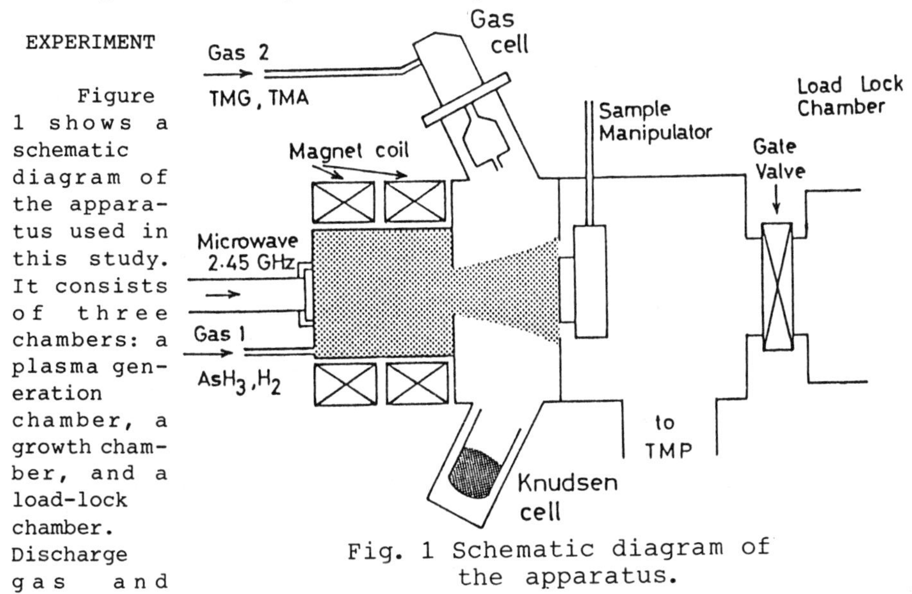

Fig. 1 Schematic diagram of the apparatus.

group V hydride gas are introduced into the plasma generation chamber. In this study, arsine (10 % in hydrogen) gas is used for plasma generation. Microwave power at 2.45 GHz is supplied to the plasma generation chamber through a rectangular waveguide. Magnetic coils are arranged in order to provide the ECR field inside the plasma generation chamber (875 Gauss corresponding to 2.45 GHz). Plasma is transported toward the substrate by a divergent magnetic field, as illustrated by the shaded region in Fig. 1.

Group III elements can be introduced into the growth chamber by two methods. Gas-phase elements, such as trimethyl gallium (TMG), are introduced through a variable leak valve. For liquid-phase elements, like metallic Ga, a conventional Knudsen cell is employed. The growth chamber is also equipped with a reflection high-energy electron diffraction (RHEED) observation system to evaluate the crystalline features of grown layers without breaking the vacuum. The plasma generation chamber and the growth chamber are evacuated by a turbo-molecular pump, to a background pressure in the range of $0.5 - 1.0 \times 10^{-8}$ Torr.

In our experiment, GaAs substrates of (100) orientation are degreased and chemically etched to remove the mechanical polishing damage. Next, they are rinsed in deionized water and blown dry. Then, the substrates are mounted on molybdenum blocks using indium solder.

The TMG flow rate was about 0.5 sccm, which was estimated by the pressure increase (1.15×10^{-5} Torr) and the pumping speed (550 l/s). The Ga Knudsen cell temperature was set constant at 900°C. The arsine flow rate, the microwave power supplied to the plasma generation chamber, and the substrate temperature (T_s) were varied from 10 to 40 sccm, from 100 to 800 W, and from 250 to 630°C, respectively. The pressure in the growth chamber increased up to 3×10^{-4} Torr during growth. Judging from the mean free path at this pressure (a few tens of centimeters), molecules and/or atoms are not expected to collide with each other, just as in conventional MBE.

RESULTS and DISCUSSION

GaAs growth by ECR-MBE was carried out with two systems: TMG-arsine and metallic Ga-arsine. In these experiments, the substrate temperature was raised to 630°C prior to growth in order to thermally evaporate the surface oxides. The prepared surface was confirmed by RHEED observation to be an arsenic-stabilized surface.

Then, the surface cleaning by ECR plasma is described. Finally, the full growth process at low tempera-

Table I RHEED observation at various substrate temperature (TMG-arsine system).

T_s (^{o}C)	RHEED
250	HALO+RING
350	HALO+RING
400	SPOT+RING
430	STREAK
630	STREAK

tures is performed. The grown layers are investigated by means of RHEED observation and secondary ion mass spectroscopy (SIMS) depth profile.

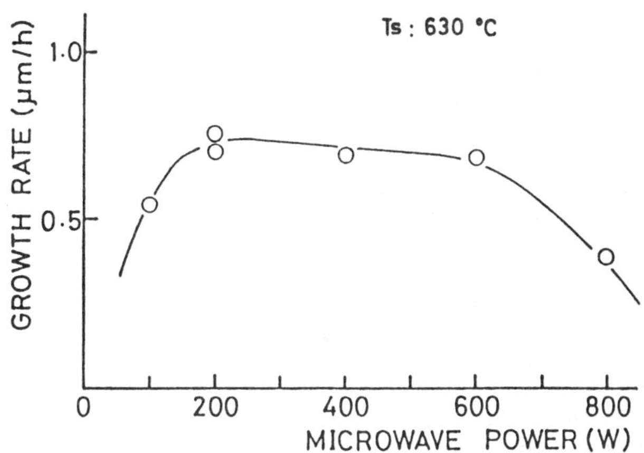

Fig. 2 Growth rate dependence on input microwave power. (TMG-arsine system)

TMG-arsine system

The results of RHEED observations after growth are summarized in Table 1 for various substrate temperatures. The arsine flow rate was 10 sccm and the input microwave power was 400 W. For the layers grown above T_s = 430°C, the RHEED pattern was clearly streaked with integral order diffracted lines. For the layer grown at 400°C, the pattern became spotty with rings. For the layers grown below 300°C, the pattern changed to a halo with rings. The surface morphologies showed generally mirror-like appearances.

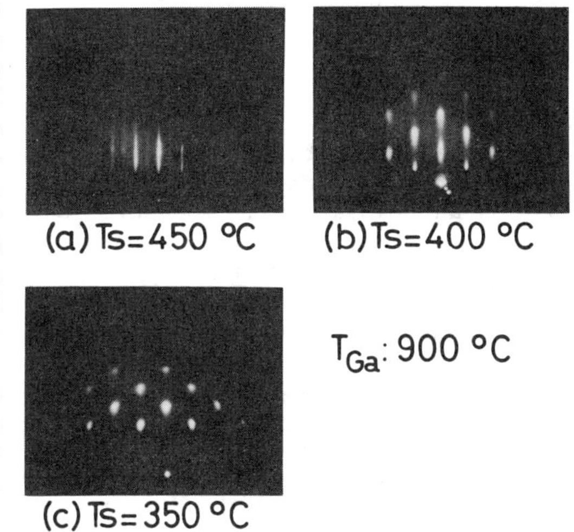

Fig. 3 RHEED patterns of layers grown at substrate temperatures of (a) 450°C, (b) 400°C, and (c) 350°C.

The growth rate was found to be dependent on input microwave power, as seen in Fig. 2. As input microwave power increases, the growth rate increases in the lower microwave power region and then decreases in the higher microwave power region. The declining growth rate over 600 W is considered to be resultant from an enhanced desorption rate of migrating atoms on the substrate surface introduced by excessive plasma exposure, as will be discussed in the next section. The growth rate increase from 0 to 200 W can be explained by an increased decomposition of TMG up to 200 W. This result is consistent with a previous report[8] from the following reasons. Since the number of TMG molecules introduced in the growth chamber is more than one order of magnitude higher than Ga atoms required for the obtained growth rate, TMG which reaches on the substrate surface mostly desorbes from the surface before decomposition takes place under plasma exposure and only Ga atom decomposed from TMG in plasma is incorporated in the epi-layer. Moreover, the surface morphology of the layer grown at 100 W still showed a mirror-like appearance without Ga droplets. Then, it is hard to explain that the growth rate was reduced by insufficient cracking of arsine in the lower power region.

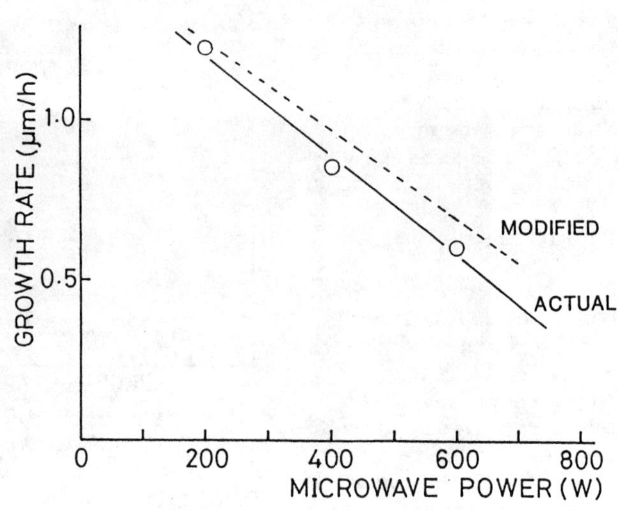

Fig. 4 Growth rate dependence on input microwave power. (metallic Ga-arsine system)

Consequently, the growth rate dependence on microwave power in the TMG-arsine system can be explained by a balance of gas decomposition and migrating atom desorption.

Metallic Ga-arsine system

Figures 3(a), 3(b) and 3(c) show the RHEED patterns of epitaxial layers grown at various substrate temperatures. The arsine flow rate was 10 sccm and the input microwave power was 200 W. For layers grown above T_s = 400°C, the RHEED pattern was streaked with integral diffraction lines, as seen in Figs. 3(a) and 3(b). For the layer grown at 350°C, a spotty pattern characteristic of single crystal was obtained (Fig. 3(c)). For layers grown from 300 to 250°C, the RHEED pattern was composed of rings with twined spots, indicating that the grown layers are poly-crystalline.

The growth rate was also found to be dependent on the input microwave power, as shown in Fig. 4. The growth rate decreases monotonically with the increase of input microwave power, although growth experiment with

Table II The bulk GaAs etching rate due to arsine plasma exposure.

Microwave Power (W)	Etching Rate (Å/h)
200	180
400	330
600	1030

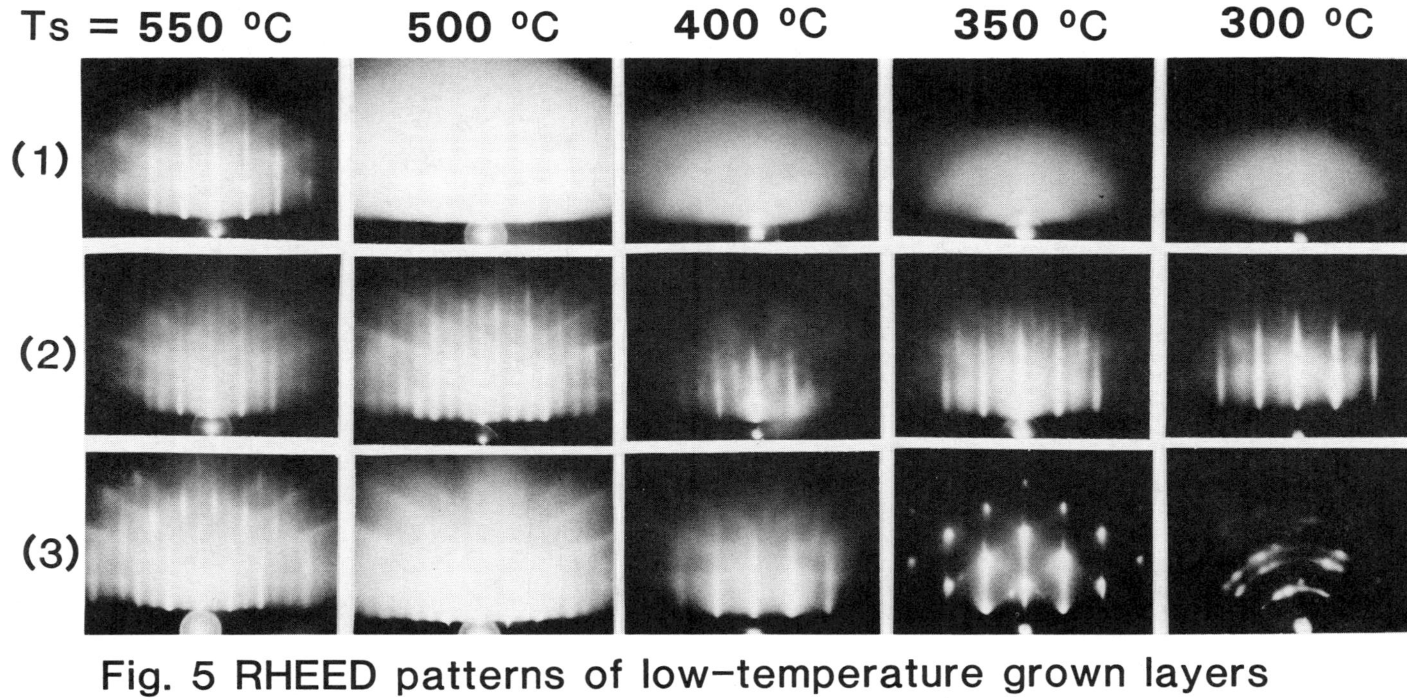

Fig. 5 RHEED patterns of low-temperature grown layers at various stages: (1)before cleaning, (2)after H_2 plasma cleaning, and (3)after ECR-MBE growth.

microwave power less than 200 W was not carried out yet. Since the cell temperature of metallic Ga is kept constant at 900°C, the Ga atom flux impinging onto the growing surface is independent of input microwave power. Therefore, the reduced growth rate due to plasma exposure should be produced either by the etching of the GaAs crystal or by the desorption of Ga atoms migrating on the substrate surface.

In order to identify the main mechanism responsible for growth rate reduction due to plasma exposure, the plasma-induced etching rate was measured by applying only arsine plasma onto a GaAs substrate without any group III element. The etching rate as a function of input microwave power is summarized in Table 2. Here, the arsine flow rate was 10 sccm and the substrate temperature was 630°C. Growth rate dependence on input microwave power was adjusted by these etching rates, as is shown by the broken line in Fig. 4. The adjusted growth rate continues to decrease monotonically with the increase of input microwave power. Therefore, it can be concluded that the plasma-induced desorption of Ga atoms migrating on the substrate surface is primarily responsible for the reduced growth rate.

Surface cleaning by ECR plasma

Surface cleaning of GaAs substrates was examined by ECR plasma exposure. Hydrogen was used for the cleaning gas. The flow rate, input microwave power, and exposure time were fixed at 40 sccm, 100 W, and 30 minutes, respectively. The substrate temperature during plasma cleaning was varied from 300 to 550°C. The results are shown in Fig. 5. In the case of heat treatments in vacuum without plasma exposure (top row in Fig. 5) at temperatures higher than 500°C, substrate surfaces exhibit streaked patterns representing oxide desorption. Below 500°C, however, substrate surfaces exhibit halo patterns indicating that the oxides on the substrate surface were not fully removed. After hydrogen plasma was irradiated onto the substrate, as can be seen in the middle row of Fig. 5, the RHEED patterns appeared as streaked patterns at all substrate temperatures examined. This demonstrates the feasibility of growing high-quality crystalline layers at fairly low temperatures including the surface cleaning process.

Note that these streaked patterns after plasma cleaning are confirmed to be arsenic-stabilized surface, with 1/2-integral orders for electron beam incidence of [110] azimuth and with 1/4-integral orders for [1$\bar{1}$0] azimuth. It is highly advantageous for GaAs growth that the excess arsenic is not removed from the surface by plasma exposure.

Full growth process at low temperatures

By using the ECR-MBE technique for both surface cleaning and epitaxial growth processes, as described in the previous three sections, the full growth process can be realized at low temperatures. In this experiment, a metallic Ga-arsine system is used because this system makes it possible to grow GaAs layers at a lower substrate temperature than the TMG-arsine system. Figure 5 shows the RHEED patterns of stages (before cleaning, after plasma cleaning and after epi-growth). The surface cleaning procedure was the same as described in the previous section. The arsine flow rate and the input microwave power for the growth were kept constant at 40 sccm, and 100 W, respectively. For temperatures higher than 400°C, grown layers show streaked patterns, indicating high-quality crystal characteristic. At 350°C, the grown layer shows a spotty pattern, indicating a rather rough single-crystalline growth. Below 300°C, the RHEED pattern changes to a ring, indicating polycrystalline growth. Although surface cleaning is sufficiently performed by ECR plasma exposure even below 350°C, the single crystalline growth only takes place at temperatures above 350°C. We believe that it is possible to grow single crystalline layers even below 350°C, by optimizing the growth conditions.

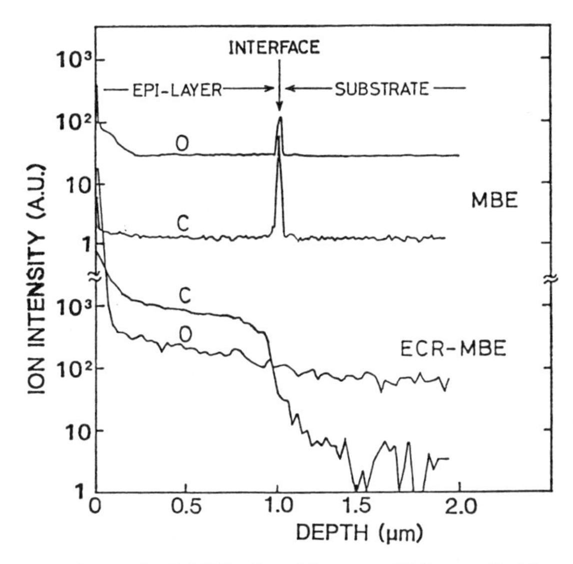

Fig. 6 SIMS depth profile of the layer grown at T_S = 630°C.

Figure 6 shows a SIMS depth profile of an ECR-MBE grown layer compared with that of a conventional MBE-grown layer. Both layers were grown at T_S = 630°C. For the MBE-grown layer, carbon and oxygen are piled up at the layer-substrate interface. Such impurities are not piled up at the interface for the ECR-MBE grown layer, but a higher concentration carbon was found in the epi-layer compared with the MBE-grown layer. It is considered that the surface cleaning induced by plasma exposure can effectively act to reduce impurities at the interface.

CONCLUSION

Both plasma-excited GaAs MBE growth and surface cleaning have been studied with relatively low energy ions of a few tens of electron volts to realize a full, low-temperature growth process. These ions were generated by an ECR ion-source and transported to the growing surface by a divergent magnetic field without causing extensive damage to the growing crystal.

GaAs epitaxial layers have been successfully grown at substrate temperatures as low as 350°C. This implies that ECR excitation is effective for cracking arsine and TMG. Exposure to plasma is found to promote desorption of atoms migrating on the substrate surface. This results in a growth rate reduction with increasing input microwave power. Exposure to plasma is also effective for substrate cleaning even at a temperature as low as 300°C.

Combining of surface cleaning and epitaxial growth using the ECR-MBE technique enables us to grow single crystalline GaAs layers at low temperatures all through the growth process. The interface between substrate and grown layer is found to be clean without piling up of impurity. This technique could prove to be highly significant in the future for re-growth procedures on advanced substrates that are not susceptible of high-temperature processing.

ACKNOWLEDGMENT

The authors are grateful to Seitaro Matsuo for helpful discussions on ECR plasma technologies. They would like to thank Kazuo Hirata, Toshiaki Tamamura and Masatomo Fujimoto for their continuous encouragement.

REFERENCES

1. G.M.Metze and A.R.Calawa, Appl. Phys. Lett., 42, 818 (1983).
2. Y.Horikoshi, M.Kawashima and H.Yamaguchi, Jpn. J. Appl. Phys., 25, L868 (1986).
3. N.Kondo and M.Kawashima, Inst. Phys. Conf. Ser., No.79, 97 (1985).
4. N.Kondo, M.Kawashima and H.Sugiura, Jpn. J. Appl. Phys., 24, L370 (1985).
5. S.Matsuo and M.Kiuchi, Jpn. J. Appl. Phys., 22, L210 (1983).
6. S.Matsuo, Extended Abstracts of the 16th(1984 International) Conf. on Solid State Devices and Materials, Kobe, 459 (1984).
7. unpublished
8. S.Zembutsu and T.Sasaki, Extended Abstracts (The 47th Fall Meeting, 1986); The Jpn. Soc. of Appl. Phys., 27p-H-12.

THE SURFACE STATE OF SI (100) AND (111) WAFERS AFTER TREATMENT WITH HYDROFLUORIC ACID.

M. Grundner, R. Schulz
Wacker-Chemitronic, Research Center, Postfach 1140
D-8263 Burghausen, Fed. Rep. of Germany

ABSTRACT

Si wafers (CZ, boron doped, 3 - 20 Ohm cm) with (100) or (111) oriented surfaces were treated in aqueous HF (0.2%, 1%, 5%, 40%) and measured with X-Ray Photoelectron Spectroscopy (XPS) and High Resolution Electron Energy Loss Spectroscopy (HREELS). The HREELS spectra exhibit vibrational excitations characteristic of a predominant coverage with Si-dihydride on Si (100) and with Si-monohydride on Si (111). After treatment in diluted HF additional OH groups could be observed. The coverage with fluorine is shown to be dependent on the HF concentration and amounts to roughly $1\text{-}1.5 \cdot 10^{14}$ F/cm^2 or 10% of a monolayer after a 40% HF dip. Some hydrocarbon and spurious oxygen contamination could be observed, too.

Water rinsing after the HF dip lowered the fluorine coverage via a substitution reaction Si-F + $H_2O \rightarrow$ Si-OH + HF, as OH groups were detected afterwards. Prolonged water exposure led to the development of a hydrous oxide on part of the surface. A mechanism for the preferential attachment of H to the Si surface is discussed, which is based on the polarisation of the Si-SiX bonds by electronegative groups on the surface.

INTRODUCTION

The treatment of Si with hydrofluoric acid (HF dip) is an important step in microelectronic production, where it serves mainly for stripping oxides. Its usefulness depends on the fact that SiO_2 is dissolved rapidly and that the etching stops precisely at the underlying Si. In wafer cleaning, the HF dip is usually performed to strip the native oxide before applying a wet oxidizing chemical treatment like the RCA clean/1/.

The surface state after the HF dip is empirically characterized as hydrophobic, expressing the fact that the contact angle of a water droplet is rather high on this surface. On the other hand the RCA clean yields a hydrophilic surface. It is known that the oxidation rates in high temperature dry oxidation are different for hydrophobic and hydrophilic surfaces /2/, stressing the importance of a detailed understanding of the nature of these surfaces/3/.

Recently it was shown, that the Si surface immersed in HF has unusual electronic properties/4/. The surface recombination velocity of minority carriers is extremely slow (about 1/4 cm/sec.), which is assumed to be due to a nearly complete saturation of dangling bonds.

Both the etch stop on the Si substrate and the low recombination velocity are manifestations of the unique surface state Si attains during contact with HF. Despite the ubiquitous use of HF treatments in the electronic industry publications on this surface state are rather scarce and moreover contradictory. Weinberger et al /5/ concluded from their XPS and ISS measurements that the passivation is due to a monolayer of Si-F, whereas Yablonowitch et al/4/ deduced from ATR infrared measurements that the Si surface in HF is mainly saturated with hydrogen.

Our measurements with High Resolution Electron Energy Loss Spectroscopy (HREELS) essentially confirm the latter finding. The Si surface is prevalently Si-H covered under the usual conditions of a HF dip. In addition X-Ray-Photoelectron (XPS) measurements were performed to assess the coverage with fluorosilyl species, which could be shown to be a function of concentration. As the HF dip is usually followed by a final water rinse, our investigations include the influence of water on the surface conditions of Si, too, showing that it has a pronounced impact on fluorine coverage and the oxidation state of the surface.

EXPERIMENTAL

The experiments were performed in a combined HREELS/XPS apparatus with a load lock for fast insertion of the samples. The XPS measurements were carried out with MgK_α radiation at a constant pass energy of 50 eV. To enhance the surface sensitivity the electron take-off angle was chosen to be 75° to normal incidence. The HREELS spectrometer was adjusted to a resolution of about 7 meV (56 cm^{-1}). Primary energy of the electrons was 6.94 eV and only specularly reflected electrons were measured.

The samples were polished Si wafers, boron doped in the range of 5 - 20 Ohmcm, with (100) or (111) oriented surfaces. The sample preparation was performed by dipping the wafers in electronic grade HF. If a H_2O rinse was applied, water of a resistivity of 18 MOhmcm was used. The time necessary to insert the sample into the vacuum chamber amounted to less than two minutes. Control measurements of a sample which was stored for five minutes in air revealed no detectable changes in the spectra. The pressure during the measurements was in the low 10^{-9} mbar region or less.

INFLUENCE OF HF

1. XPS Measurements

A survey spectrum of Si after an HF dip (40% HF, 1 min) is depicted in fig. 1. It shows the Si 2p 3/2,1/2 and Si 2s lines, and only minor quantities of C, O, and F as surface contaminants. As a quantification of these contaminants is of vital interest for the interpretation of the overall surface state, an assessment of the various contributions to the surface coverage was performed. This assessment is based on the expression for the XPS intensity ratio adlayer/silicon/6/, which yields the surface density of the measured species in the case of a monolayer coverage or less.

The procedure for the evaluation of the small contribution of chemically shifted Si atoms is depicted in fig. 2 for the case of a

1 min, 40% HF dipped Si (100) sample. After subtraction of a smoothly varying background from the measured Si 2p peak (full line) the difference to a reference Si peak (dashed line) was determined. The residual intensity is depicted separately in fig. 2 and corresponds to chemically shifted (oxidized) Si. The use of a reference Si peak turned out to be undispensable because elemental Si showed a slight tail on the high binding energy side, which could not be fitted by simulating the Si 2p 3/2,1/2 components with symmetrical functions. The reference peak was generated by heating Si to about 900°C in UHV. No carbon and only a minor oxygen contamination (O/Si <0.02) could be detected.

The same procedure was applied in the case of the 5% and 1% HF dipped samples, the results are shown in fig. 2. Diluted HF obviously has the effect of allowing some Si species to exist in a higher oxidation state, as the difference curves extend to higher binding energies than in the case of the 40% HF dip. The latter exhibits only a single peak with a chemical shift of slightly more than 1 eV. According to Himpsel et al this value corresponds to a Si^{1+} oxidation state/7,8/.

The intensity of this state can well be explained by fluorine bonding to Si. The Si^{1+} intensity amounted to $8 \cdot 10^{13}$ at/cm^2 and the F concentration to $9.3 \cdot 10^{13}$ at/cm^2, whereas the oxygen signal is compatible with the intensity of chemically shifted components of the carbon peak. The preferred configuration obviously is only one F atom per involved Si atom, although two bonds are to be saturated on the (100) surface.

In the case of the 1% or 5% HF dipped samples the picture is not so clear, as the concentration of fluorosilyl species is lower (see fig. 3), but the total amount of oxidized Si atoms rises to values between 1 - 1.5 $\cdot$ 10^{14} at/cm^2. So part of the oxidized Si intensity may originate from Si-O-Si bridges or alkoxy species Si-O-C_xH_y in this case.

The fluorine coverage turned out to be a function of the HF concentration as shown in fig. 3 for Si (100) and (111) and 1 min exposure time. There is a distinct difference in fluorine coverage for the very low concentrations of HF.

The main conclusion however is that in any case the total surface coverage from XPS measurements is well below a monolayer equivalent. This means in terms of bond sites that it is far below 1.36 $\cdot 10^{15}/cm^2$ for the Si (100) and 7.8 $\cdot$ $10^{14}/cm^2$ for the Si (111) surface. Hence we have to conclude that these bonds cannot be saturated by elements detectable by XPS. The succeeding HREELS measurements will reveal a clear picture of these surfaces.

2. Vibrational investigations

Typical vibrational spectra obtained on Si (100) and (111) surfaces after a 1 min dip in 40 % HF are shown in fig. 4. The assignment of the principal vibronic excitations is easily performed by a comparison with published spectra of UHV cleaned and hydrogen dosed Si samples, e.g. /9,10/. The spectrum of the Si (100) surface corresponds closely to that of a hydrogen treated (1 x 1) surface. Which develops on a Si (2 x 1) reconstructed surface after exposing it to atomic hydrogen at a higher dosage. This state of the surface is referred to as the dihydride phase in contrast to the monohydride phase which develops under conditions of low H exposure or temperatures above 650 K.

The dihydride structure exhibits four losses corresponding to a stretching mode at 2080 cm^{-1}, a scissor mode at 900 cm^{-1}, a wagging mode at 640 cm^{-1} and a fourth mode at 480 cm^{-1}, which has either been assigned to a rocking mode or a hydride mode coupled to Si phonons /9/. With the exception of the latter vibration all modes can be identified unambiguously in the spectrum of the HF dipped Si (100) sample in fig. 4. Note that no asymmetric Si-O-Si stretching vibration in the frequency region 1050 - 1150 cm^{-1} /11/ can be observed. Furthermore the position of the hydride stretching band at 2080 cm^{-1} corroborates that no Si-O-Si units exist, otherwise this frequency would be shifted to a somewhat higher value /12/.

The relative intensities (referred to the no-loss peak) for the various hydride vibrations agree with those of the hydrogen dosed surfaces to within a factor of two. In most cases the agreement was even better, especially for those samples dipped in diluted HF. This is a rather astonishing result as the hydride vibrational intensities are reported to be sensitive to surface conditions and slight temperature variations during atomic H dosing /10/. The similarity of H dosed and HF dipped samples is based on two conditions. Firstly, only a minor part of the surface reacts with atoms other than H, as was proved by the XPS measurements, and secondly the polished surface must be of high structural quality, i.e. atomically well ordered, which was confirmed recently by LEED measurements /13/.

The remaining lines in the spectra at 2900 cm^{-1} and 1350 cm^{-1} are hydrocarbon stretching and deformation vibrations, respectively. Around 400 cm^{-1} a more or less pronounced hump could be observed, whose origin is as yet unclear.

In the case of the Si (111) surface similar statements hold with respect to the comparison of the spectra of atomic H dosed and HF dipped surfaces. The spectra essentially show a monohydride coverage with a stretching vibration at 2080 cm^{-1} and a bending vibration at 640 cm^{-1}. The occurrance of monohydride is understandable from the structure of the (111) surface with one dangling bond per surface atom. The spectra exhibit some additional intensity at about 900 cm^{-1}, which is not expected to occur on an ideal (1 x 1) termination of the Si lattice. This may be explained by the edge atoms on the steps of the well ordered surface, which result from the misorientation between wafer surface and crystal lattice. These edge atoms can form SiH_2 groups. As was the case with the (100) oriented samples, hydrocarbon vibrations at 2900 cm^{-1} and 1350 cm^{-1} were present.

At lower HF concentrations the 900 cm^{-1} scissor vibration appeared more pronounced relative to other Si-H vibrations. The Si (100) samples dipped in 0.2% HF additionally carried OH groups, which was observed with Si (111) even up to concentrations of 1% HF, as can be seen in fig. 5.

INFLUENCE OF H_2O ON A HF TREATED SURFACE

The influence of a H_2O rinse after the HF dip on the surface state is of practical interest as this is the usual way this procedure is completed. The fluorine and oxygen coverage as a function of the rinsing time of a sample which was dipped for 1 minute in 40% HF is shown in fig. 6 . The initial rapid decrease of the fluorine surface concentration is accompanied by an increase in the oxygen coverage.

What happens during the H_2O rinse is revealed by the corresponding HREELS spectra in fig. 7. The spectrum at time zero is essentially that of fig. 4. Already after 30 sec a vibrational band of singular OH groups has developed. The explanation is simply that a substitution reaction of F with OH occurs.

Further attack of the water leads to the appearance of a vibrational band at 1060 cm^{-1} after 18 min. This band can be unequivocally identified as the asymmetric Si-O-Si stretching vibration /11/. In the course of the further slow oxidation, this band shifts to higher frequencies and the Si-H stretching vibration attains a pronounced shoulder at the higher energy side, which indicates that Si-SiX bonds are oxidized beneath the surface/12/.

DISCUSSION

The surface state of Si after an HF dip is determined by a prevalent coverage with hydrogen, which exists on Si (100) as dihydride and on Si (111) as monohydride. The hydrophobic properties of these surfaces must be attributed to the character of the Si-H bond. The formation of these relatively stable Si-H bonds is explained by fig. 8. The starting condition on the surface with its native oxide can be represented by Si-OH groups covering the surface. An OH group with its high electronegativity lowers the electron density of the adjoining Si-Si bond, resulting in strong partial charge (δ++) at the next (α) and a lesser positive charge (δ+) at the second next (β) silicon atom. These charges direct the attack of H^+F^- in a way that

H–Si< and OH–Si(–F)< groups

are formed. The same holds for the attack of the next HF molecules, giving configuarations of

H–Si(–H)< and HO–Si(–F)(–F)–

the latter being quickly dissolved forming finally SiF_6^{2-} after further reaction. This is why no SiF_2 states could be detected on the surface.

The selectivity caused by the above mentioned polarization is very strong (90% in 40% HF), but a small amount of F, especially at high HF concentrations, may attack the β -Si, leaving

$$\text{Si-F groups on (111) and } Si{<}^{F}_{H} \text{ groups}$$

on (100) surfaces. These Si-F residues may be hydrolyzed by water rinsing, exchanging OH for F.

The electronegativity of hydrogen is about that of silicon and does not induce much of a polarization. Therefore, Si-Si bonds adjacent to Si-H remain stable under these conditions of low pH values.

During a water rinse, however, the OH^- activity is high enough to attack the Si-Si bond in a corrosive way, forming primarily Si-OH and Si-H. With rising pH-values the substitutional reaction

$$Si\text{-}H + H_2O \rightarrow Si\text{-}OH + H_2$$

also takes place.

Depending on the OH^- concentration in the solution, F bonded to silicon can be substituted by OH, and vice versa. This is well known in the chemistry of H_2SiF_6. At low HF concentrations the removal of the oxide is not completely performed on (111) Si surfaces. Therefore Si-F and Si-OH groups can be observed (fig. 3, 5) to a higher extent than on the (100) surface. Between neighbouring OH groups Si-O-Si bridges are formed, which is evident from the increase in intensity of the asymmetric Si-O-Si vibration. Additionally higher Si oxidation states in the XPS spectra and the evidence from HREELS for the shifted SiH_2 vibrations are explainable by species like

$$(-O)_2\ SiH_2$$

groups. Details of these reactions will be the subject of further investigations.

REFERENCES

1. W. Kern, D.A. Puotinen, RCA Rev. 31, 187 (1970)
2. F.J. Grunthaner, J. Maserjian, IEEE Trans. NS-24, 2108 (1977)
3. M. Grundner, H. Jacob, Appl. Phys. A39, 73 (1986)
4. E. Yablonovitch, D.L. Allara, C.C. Chang, T. Gmitter, T.B. Bright, Phys. Rev. Lett. 57 (2), 249 (1986)
5. B.R. Weinberger, G.G. Peterson, T.C. Eschrich, H.A. Krasinski, J. Appl. Phys. 60 (9), 3232 (1986)
6. D. Briggs, M.P. Seah, Practical Surface Analysis (John Wiley and Sons, N.Y. 1983), p. 211
7. G. Hollinger, F.J. Himpsel, Appl. Phys. Lett. 44, 93 (1984)
8. F. R. McFeely, J. F. Morar, N.D. Shinn, G. Landgren, F.J. Himpsel, Phys. Rev. B 30 (2), 764 (1984)
9. H. Froitzheim, U. Köhler, H. Lammering, Surf. Sci. 149, 537 (1985)
10. J.A. Schäfer, F. Stucki, J.A. Anderson, G.J. Lapeyre, W. Göpel, Surf. Sci. 140, 207 (1984)
11. H. Ibach, H.D. Bruchmann, H. Wagner, Appl. Phys. A 29, 113 (1982)
12. J.A. Schäfer, D. Frankel, F. Stucki, W. Göpel, G.J. Lapeyre, Surf. Sci. 139, L 209 (1984)
13. P.O. Hahn, Mat. Res. Soc. Symp. Proc. 54, 645 (1986)
14. H. Ibach, H. Wagner, D. Bruchmann, Solid State Commun. 42, 457 (1982)

FIGURES

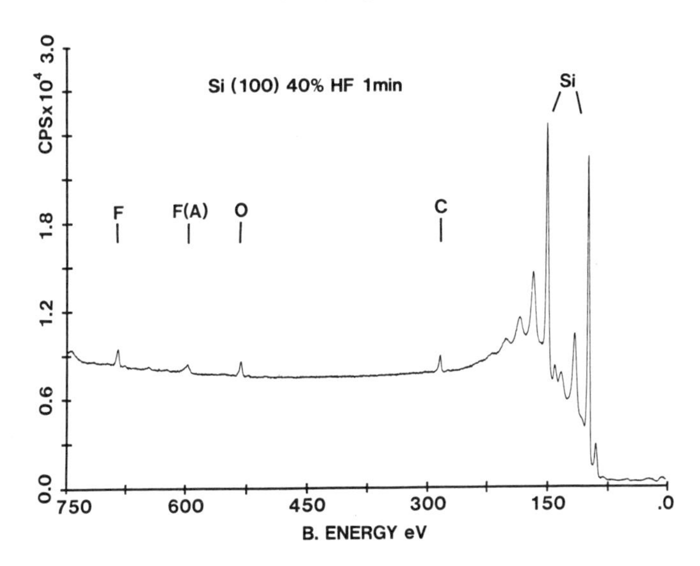

Fig. 1: Wide scan spectrum of Si (100) after an HF dip (40% HF, 1 min), showing Si2s,2p, O1s, F1s,F(Auger) and C1s lines.

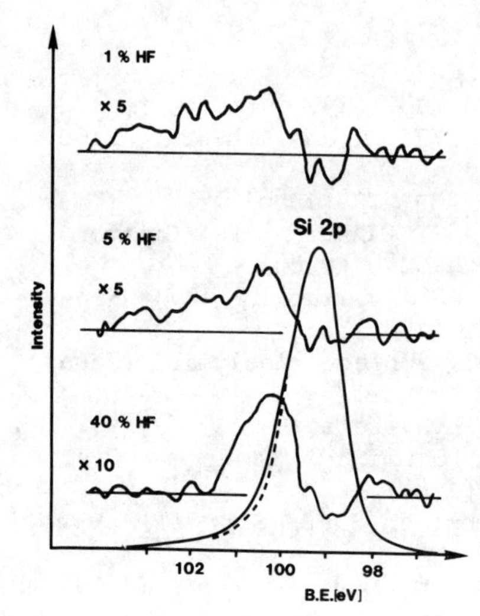

Fig. 3: Fluorine coverage for Si (100) and (111) surfaces as a function of HF concentration.

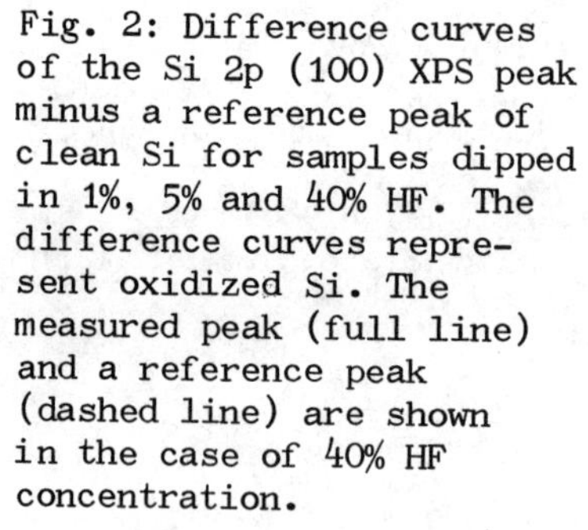

Fig. 2: Difference curves of the Si 2p (100) XPS peak minus a reference peak of clean Si for samples dipped in 1%, 5% and 40% HF. The difference curves represent oxidized Si. The measured peak (full line) and a reference peak (dashed line) are shown in the case of 40% HF concentration.

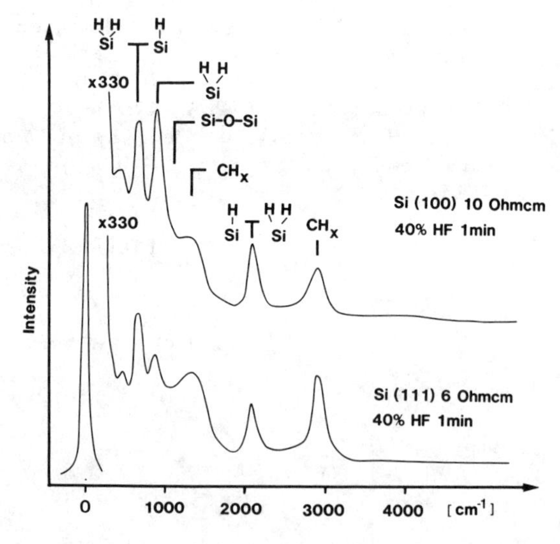

Fig. 4: Vibrational spectra of Si (100) and (111) surfaces after 1 min exposure to 40% HF. Mainly hydride lines can be identified.

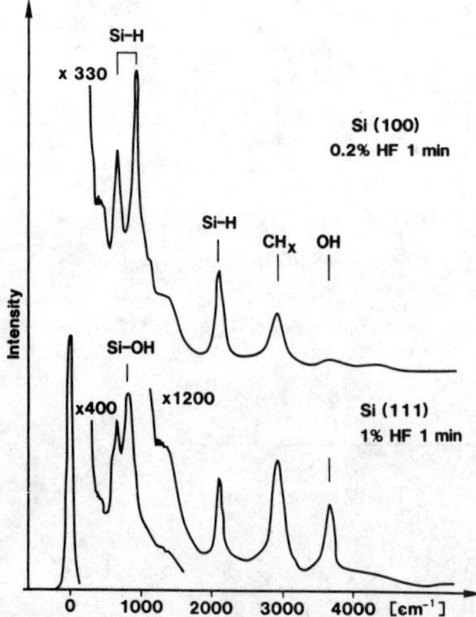

Fig. 5: At low HF concentrations OH groups at 3680 cm^{-1} appear on Si (100) up to 0.2% and on Si (111) up to 1% HF concentration.

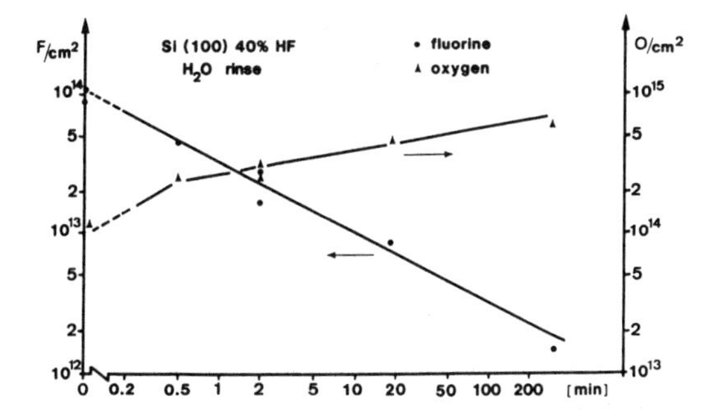

Fig. 6: Fluorine and oxygen coverage of a 1 min, 40% HF dipped sample as a function of rinsing time.

Fig. 7: Vibrational spectra corresponding to the data points in fig. 6. Note the appearance of OH groups after 30 sec. Invasion of Si backbonds by oxygen is indicated by the asymmetric Si-O-Si vibration and a chemically shifted compoent of the Si-H stretching vibration.

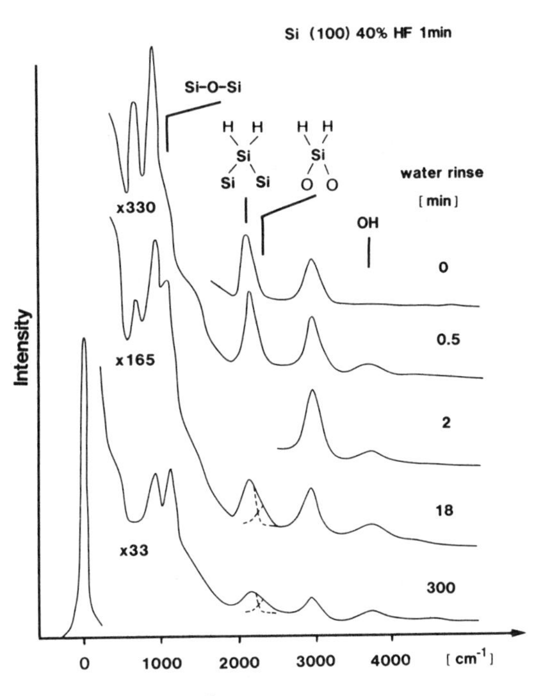

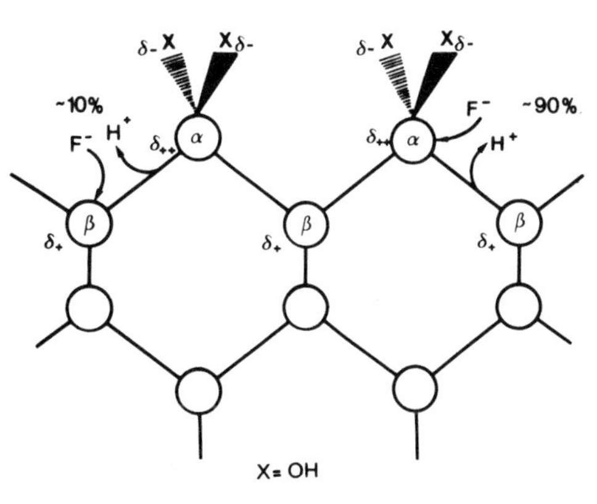

Fig. 8: The electronegative substituents induce a polarization of the Si-SiX bond, enabling the HF molecule to attack such a Si-SiX bond.

Low Temperature Si Processing Integrating Surface Preparation, Homoepitaxial Growth, and SiO_2 Deposition into an Ultrahigh Vacuum Compatible Chamber

G. G. Fountain, R. A. Rudder, S. V. Hattangady, D. J. Vitkavage
and R. J. Markunas
Research Triangle Institute, Research Triangle Park, NC 27709

Integration of low temperature Si processing steps using interconnected ultra-high (UHV) systems addresses two concerns of the semiconductor industry, low temperature processing and control of wafer environment between processing steps. We report results from a single remote plasma enhanced chemical vapor deposition (RPECVD) reactor with UHV capability. In situ surface preparations using a 300 °C hydrogen plasma treatment have been successful in reconstructing Si(100) surfaces. SiO_2 layers deposited on these surfaces at 250 °C have resulted in MOS capacitors with minimum interface state densities of 1.8×10^{10} $cm^{-2}eV^{-1}$. Homoepitaxial Si epitaxy originally nucleated at 520 °C renucleated for growth temperatures as low as 235 °C. This work clearly demonstrates the versatility and potential for the RPECVD process to become a member of a low temperature, integrated silicon processing facility.

I. INTRODUCTION

Integration of low temperature Si processing steps using interconnected ultra-high vacuum (UHV) systems addresses two concerns of the semiconductor industry, low temperature processing and control of wafer environment. Research focusing on low temperature silicon processing has been motivated by a variety of technical problems.[1,2] Fabrication of hyperabrupt doping profiles and maintenance of those profiles throughout subsequent steps demands low temperature processing. Increasingly complex, multilayer metalization and other upper story processes demand low temperature processing. Wafer warpage associated with processing at high temperature has become a problem as wafer diameters have increased beyond 7.5 cm. As device structures have shrunk in physical dimensions, they have become more sensitive to particulates and surface contaminants. This sensitivity, when coupled with higher levels of integration and larger die sizes, demands strict attention to the environment these devices are exposed to for acceptable yields. Control of the wafer environment using isolated processing steps with wafer transfers under clean room conditions is becoming increasingly costly and may have fundamental limitations. An integrated UHV processing system may offer the environmental control necessary. The integration of Si processing steps into an UHV environment is an alternative approach for future generations of fabrication lines.

We report here on results using a remote plasma enhanced chemical vapor deposition (RPECVD) reactor with UHV capabilities. A discussion of the remote plasma process as applied to the deposition of silicon and SiO_2 will be included. In one reactor, we demonstrate the ability to perform *in situ* surface cleaning and analysis, SiO_2 deposition, and homoepitaxial deposition of Si. Hydrogen plasma treatments at substrate temperatures of 300 °C have been successful in producing reconstructed Si(100) surfaces. SiO_2 depositions at 250 °C on Si(100) have yielded MOS capacitors with a minimum interface state density of 1.8×10^{10} cm^{-2} eV^{-1}. Epitaxial Si layers have been nucleated on Si(100) at 520 °C and continued to renucleate for growth temperatures as low as 250 °C. From these results the RPECVD process emerges as a viable technique which may become a prominent member of a silicon, low temperature processing line.

II. THE REMOTE PLASMA PROCESS

Plasma enhanced processes have figured prominently in research efforts to lower process temperatures. In conventional plasma enhanced chemical vapor deposition (PECVD), the parent gas molecules are dissociated into precursor atoms and radicals which can deposit on substrates at lower temperatures than in thermal chemical vapor deposition. The deposition occurs at lower temperatures than purely pyrolytic processes because the plasma supplies energy to break chemical bonds in the parent molecules that would only be broken by thermal decomposition if the plasma were not present. Parent molecule dissociation is accomplished in the plasma through various processes involving collisions with electrons, ions, photons, and excited neutral species. Unfortunately, the precursor species are now subject to the same active environment which dissociated the parent molecules. This can lead to further dissociation or reaction of gas phase species to form more complicated radicals before the radicals can condense on the substrate. Thus, a wide spectrum of deposition precursor species are incident on the substrate.[3] A further complication is that in conventional PECVD the substrate is immersed in the plasma region. This results in a large flux of charged species incident on the substrate during film deposition. The incident energies of these ions may be as high as 160 eV in some immersion systems. This can lead to ion implantation, energetic neutral embedment, sputtering, and associated damage. Thus, there are two major problems associated with conventional PECVD, *one* adequate control over incident gas phase species and *two* ion damage as result of the substrate being immersed in the plasma region.

Remote plasma enhanced chemical vapor deposition, RPECVD, is an innovative variant of conventional PECVD which seeks to avoid these problems. A block schematic of the RPECVD process is given in Fig. 1. There are two primary differences between RPECVD and PECVD. *First,* the parent gas molecules are not excited in the plasma region but instead react with excited, metastable gas species that flow from the plasma region. These metastable species typically are at 4 - 20 eV above their ground state. The coupling of this energy into the parent molecules during collisional events determines the gas phase species. The plug velocity of gas through the plasma tube prohibits back diffusion of parent molecules into the plasma region. *Second,* unlike conventional PECVD, the substrates are well removed from the plasma region, minimizing the plasma densities near the substrate. This should result in virtually no sheath fields between the substrate and the plasma in contrast to immersion systems. Ions created by Penning processes in the vicinity of the substrate see no large sheath fields to accelerate them. Furthermore, with typical deposition pressures between 0.100 and 0.300 Torr, the ions are thermalized, reducing their incident energy on the substrates. Considering the damage and embedment that has been observed in silicon from even moderately low energy ions (< 50 eV), reduction of ion flux and energy is certainly an advantage offered by RPECVD.

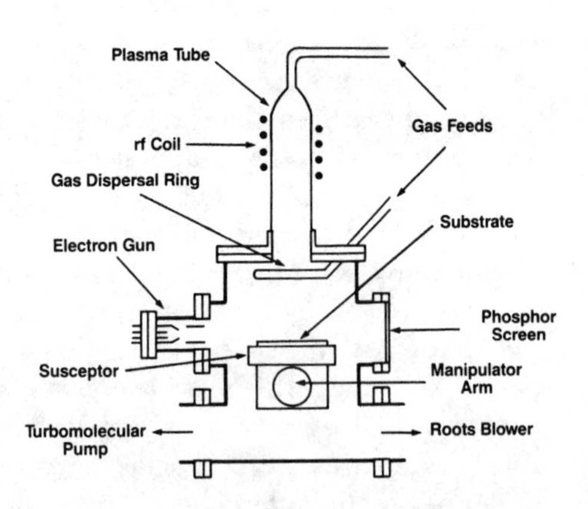

Fig. 1 shows a schematic of the RPECVD reactor. Details of that reactor have been reported previously.[4] Emphasized here are attributes of this reactor pertinent to the Si processing results. The chamber is an UHV compatible, stainless steel chamber which following a 250 °C bakeout reaches pressures below 10^{-9} Torr. The chamber is equipped with a 10 keV reflectance high energy electron diffraction (RHEED) system for *in situ* surface studies prior to and after vacuum processing. The gasses used are 99.999% H_2, 99.9999% He, 18.9% O_2 in 99.9998 Ar, and 20.2% SiH_4 in 99.999% He. Substrates are heated by a quartz halogen lamp that resides in the vacuum chamber.

III. RESULTS AND DISCUSSIONS

A. Wet chemical preparation of substrates

All silicon substrates used for these experiments were (100) p-type with a resistivity of 10 - 20 ohm-cm. Samples are first ultrasonically cleaned for 5 min in baths of low mobile ion grade trichloroethylene, acetone, and methanol followed by a rinse in flowing deionized water. The samples are then etched in a modified RCA clean[5] followed by $H_2O(50)\cdot HF(1)$ etch and deionized water rinse. At this point samples are placed in a dry nitrogen environment for transfer to the reactor. This cleaning procedure leaves the surface with a minimal of surface oxide or contamination, less than 5 Å as determined by X-ray Photoelectron Spectroscopy.

B. In situ surface treatments

Workers have reported that the removal of surface contamination is critical to the success of any low temperature epitaxial process.[6,7] As will be discussed later, the cleanliness and order of the surface is paramount in low temperature SiO_2 depositions. In an effort to remove contamination remaining on the Si surfaces after a thermal bake at 300 °C, we used hydrogen plasma treatments on the Si(100) surface.

During hydrogen plasma treatments, 80-100 sccm of H_2 gas flows through the plasma region. The chamber pressure is maintained between 4 and 5 mTorr. With the substrate at 300 °C, a hydrogen plasma is initiated. The afterglow from the plasma region is observed to engulf the entire chamber due to the high plug velocity and long mean free path at 5 mTorr. Fig. 2 shows the before and after RHEED patterns for a 80 sccm, 5 mTorr, 35 Watt, 20 s hydrogen plasma treatment. Exposure of the Si(100) surface to the hydrogen plasma resulted in the surface displaying *more* prominent 1/2-order diffraction streaks after the exposure. Some samples do not show as much improvement as others. Small variations in gas quality or system leak integrity can adversely effect the results by reoxidizing the silicon surface during the plasma treatment. A quadrapole mass spectrometer has been recently installed on the system. Experiments are underway to quantify both the purity of the as delivered H_2 to the plasma and the partial pressure of water vapor present in the system following the overnight bakeout. We would like to note that similar hydrogen plasmas treatments have been used to effectively reorder Ge(111) surfaces at 300 °C. There, the plasma treatments *always* yield reconstructed surfaces.[8]

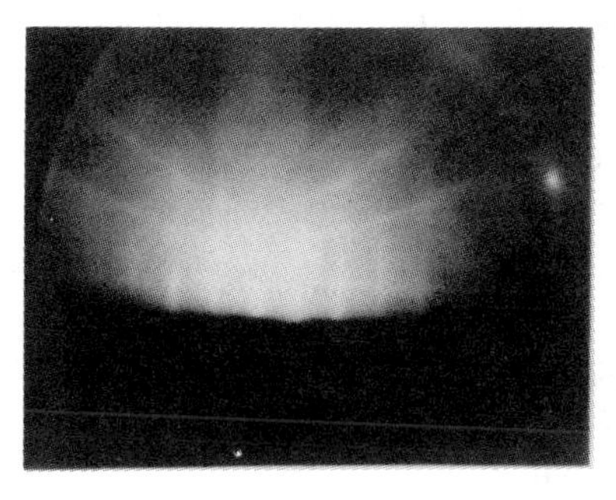

2a

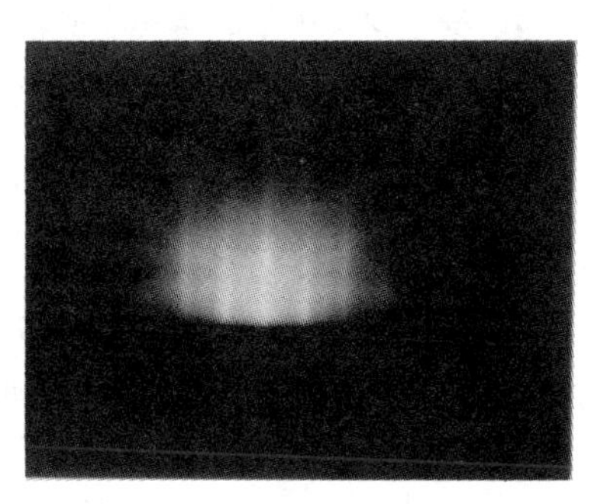

2b

While RHEED is a very sensitive probe of surface quality, it should be noted that the presence of a sharp RHEED pattern with half-order streaks does not guarantee that the surface is atomically clean. Using Auger electron spectroscopy (AES) or X-ray photoelectron spectroscopy (XPS) on silicon surfaces, several workers have detected appreciable amounts of carbon and metallic contamination on Si surfaces which had sharp RHEED or LEED patterns. The inability to perform in situ AES or XPS analysis in our present apparatus makes it difficult to estimate the quantity of contamination present on the Si(100) surfaces after hydrogen plasma treatments. Certainly, the presence of carbon, oxygen, and metals on the surface in amounts less than 2×10^{14}atoms cm^{-2} is entirely possible. The presence of hydrogen is also likely. Schaefer et al.[9] have studied hydrogen coverage on Si(100) surfaces in the 100-300 °C temperature range. By using low energy electron loss spectroscopy in combination with low energy electron diffraction, they determined that surface reconstruction on Si(100) surfaces persisted even with partial hydrogenation under some conditions. Specifically, monohydride (SiH) termination of the (100) surface still permits surface reconstruction while dihydride (SiH_2) termination of the (100) surface resulted in bulk-like structure. To obtain dihydride termination substrate temperatures below 200 °C were used. Above that temperature the hydrogen adsorption was mainly in monohydride groups.

C. SiO_2 deposition

The fundamental difference between thermal oxidation of silicon and dielectric deposition of oxides is that few surface Si atoms are consumed during a dielectric deposition. The number consumed will depend on the number of oxygen atoms diffusing to the interface and the rate at which those atoms can react with the silicon. Lower deposition temperatures used during deposition reduce the silicon substrate oxidation rate. Contaminants and imperfections on the surface prior to SiO_2 deposition are likely to cause defects at the SiO_2/Si interface. In the deposition process, silane molecules react with metastable oxygen species flowing from the plasma region. Oxygen bearing precursor species such as siloxanes are formed which can deposit at very low substrate temperatures.[10] Once these species have condensed on the substrate, they must cross-link themselves and liberate excess hydrogen. Siloxanes, however, do not have the stoichiometry of SiO_2. To overcome this deficiency, oxygen to silane ratios of typically 20 to 1 are used to provide excess oxygen at the growth surface.

Dielectric deposition in the RPECVD reactor follows approaches by other workers.[10,11] However, all samples are subjected to an *in situ* hydrogen plasma treatment prior to dielectric deposition. The details of SiO_2 deposition on Si(100) have been more fully reported elsewhere.[12] Briefly, SiO_2 is deposited while flowing 180 sccm of 20% O_2 in Ar through the plasma tube and by flowing 7 sccm of 20% SiH_4 in He through the ring. A rf discharge of 30 W is sustained during deposition. The deposition occurs at a total pressure of 0.300 Torr. The deposition rate is 10 nm/min.

Dielectric layers have been deposited on 300 °C Si(100) surfaces with varying degrees of reconstruction. Following the deposition MOS capacitors were fabricated. Electrical interface state densities were calculated from 1 MHz and quasi-static capacitance-voltage data. As previously reported,[12] the dielectrics showing the lowest density of interface traps D_{it} were the ones deposited on reconstructed Si(100) surfaces. Correspondingly, samples deposited on unreconstructed Si(100) surfaces displayed higher D_{it}. Under otherwise identical conditions, an order of magnitude variation in interface state density has been correlated with the quality of the RHEED patterns. This correlation establishes the importance of clean, well ordered surfaces prior to SiO_2 depposition. A minimum in interface state density $3.5 \times 10^{10} cm^{-2} eV^{-1}$ and a high breakdown field of 9 - 10 MV/cm indicate that the oxide and interface are relatively defect free.

Furthermore, we have found a pronounced dependence of D_{it} on growth temperature. Figure 3 shows the minimum interface state density as a function of deposition temperature from 150 °C to 400 °C. The graph shows a minimum at 250 °C with the D_{it} at $1.8 \times 10^{10} cm^{-2} eV^{-1}$ Breakdown field and mobile ion density are 9 MV/cm and $1.5 \times 10^{10} cm^{-2}$, respectively. Figure 4 shows the quasi-static and high frequency capacitance voltage curves for that sample. The exact reasons for pronounced dependence on deposition temperature are not established yet. We postulate that the increase in D_{it} below 250 °C is the result of a poorer quality dielectric layer. Remember that the model presented for SiO_2 deposition depends on the cross-linking of siloxanes on the surface and the liberation of excess hydrogen. Both events would not be as efficient at lower temperatures. We speculate that the increase in interface state densities above 250 °C is related to either surface oxidation or silane pyrolysis

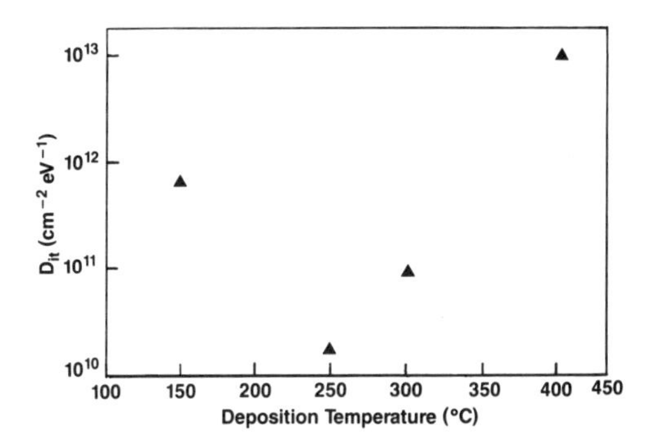

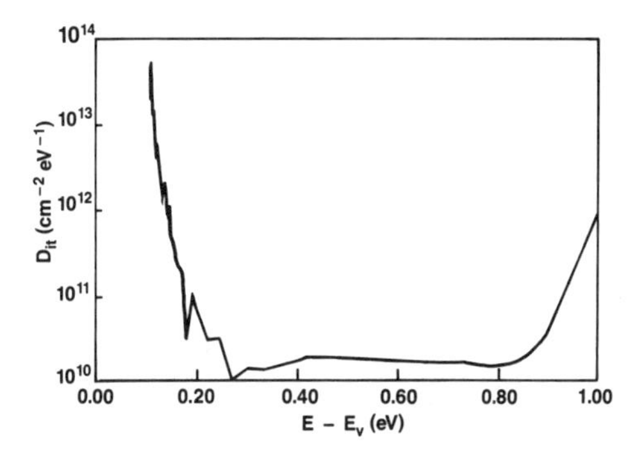

on the substrate during the initiation of the deposition process.

While there remain many questions to be answered concerning SiO_2 deposition, it is clear that RPECVD offers a low temperature alternative to the fabrication of SiO_2 MOS capacitors. At 250 °C the interface state density, breakdown field, and mobile ion density are state of the art for SiO_2.

D. Epitaxial deposition

Epitaxial depositions of Si are driven by collisions between excited, metastable He atoms and silane molecules. The collisions occur in a region outside the intense plasma area. System design and operating parameters prohibit silane from diffusing into the plasma region. Collisions between metastables and silane dissociate the silane forming radicals such as SiH_x which can deposit at very low temperatures. Once these radicals condense on the substrate they must cross-link to form a silicon network. If this process is to form epitaxial layers of silicon, excess hydrogen carried by the free radicals must be liberated and the reactant species must have sufficient surface mobility to form an ordered solid. Other workers[3, 13] have suggested mechanisms for hydrogen removal that involves the formation of H_2 and Si-Si at the expense of two SiH bonds. These workers have been concerned with chemical vapor deposition in general and not specifically RPECVD. We suggest that in RPECVD hydrogen removal can occur when He metastables collide with the growth surface. The same energy which can break Si-H bonds during gas phase collisions seems likely to break Si-H bonds at the nucleating surface.

Epitaxial silicon layers have been deposited in the temperature range of 235 - 520 °C using the current RPECVD reactor. Higher substrate temperatures have not been attempted due to heating limitations in this reactor. Details of the deposition process are similar to those we previously reported for epitaxial Ge deposition.[4] Briefly, epitaxial silicon is deposited while flowing 200 sccm of He through the plasma tube and by flowing 100 sccm He and 1 sccm 20% SiH_4 in He through the gas dispersal ring. A rf discharge plasma of 30 W is sustained during deposition. The deposition process occurs at a total pressure of 0.200 Torr. Under these conditions, a helium afterglow extends into the ring area but does not engulf the sample.

Nucleations were initiated on Si(100) 1×1 surfaces at 520 °C. (This work has not been repeated since the wet chemistry and hydrogen plasma treatments have yielded reconstructed surfaces.) After a 30 min deposition, the RPECVD reactor was evacuated of process gas, and RHEED were taken of the epitaxial layer. Extensive streaking on both integral and 1/2-order lines were observed. Fig. 5a shows the RHEED patterns observed along the [011] direction. Only integral order lines were observed along the [010] or the [001] directions.

To determine at what temperature the above deposition conditions would no longer result in single crystal Si, successive layers of Si were deposited at lower and lower temperatures. While this method results in successive layers being deposited on surfaces not necessarily as ordered as the original surface, it does offer each subsequent layer a clean surface on which nucleation can occur. Silicon depositions were run for 30 min at each temperature. After each layer deposition, RHEED was used to examine the quality of epitaxy. Fig. 5 shows layers deposited at 520, 375, 235, and 40 ° C. Notice that even at 235 ° C the epilayer is single crystal with no polycrystalline rings apparent. One sees the expected transitions on going to lower and lower temperatures. First, the integral order line become broader(520 - 375 ° C). The integral lines continue to broaden and the 1/2-order lines disappear (375 - 235 ° C). And finally, the polycrystalline rings appear at 40 ° C.

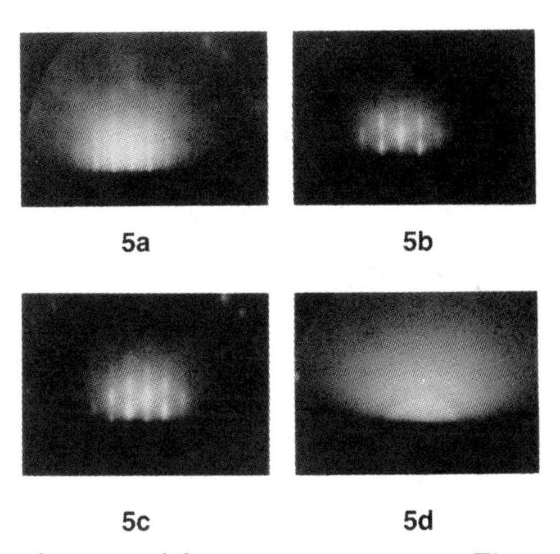

Upon removing from the reactor, a mechanical stylus trace across a masked edge revealed the total deposit to be about 700 Å. The deposition rate of 0.01 nm/s assuming unity sticking would be equivalent to an impingement rate of $1.0 \times 10^{14} cm^{-2}s$. While we have not measured our metastable density, this number is consistent with metastable densities of $10^{10} cm^{-3}$ reported in flowing afterglow systems.[14] This is the first report of Si epitaxy at temperatures below 300 ° C using silane. Nishida et al[6] have reported PECVD of silicon epitaxy at 250 ° C using difluorosilane. There the authors reported the use of difluorosilane necessary to rid the surface of native oxides. The use of a fluoro-compound process does not seem compatible with the integration of silicon processing steps into an UHV environment.

IV. CONCLUSIONS

In conclusion, we have demonstrated a low temperature silicon processing technology using the RPECVD technique. The RPECVD technique has been used to successfully reconstruct Si(100) surfaces using a hydrogen plasma treatment at 300 ° C. Epitaxial silicon layers have been deposited on Si(100) surfaces at substrate temperatures less than 300 ° C. SiO_2 depositions at 250 ° C have resulted in MOS capacitors with minimum interface state densities of $1.8 \times 10^{10} cm^{-2}eV^{-1}$. These accomplishments have occurred all in the same reactor at moderate temperatures. The reactor is one of unique design combining ultra-high vacuum capability with low pressure processing for surface preparation, dielectric and epitaxial depositions. This report clearly demonstrates the versatility and potential for integration of the RPECVD process into a low temperature, silicon processing facility.

The authors would like to thank George Hudson for his technical support in operating the reactor and Robert Hendry for his labor in designing and fabricating the RPECVD reactor. We acknowledge the helpful discussions and ongoing collaboration with Prof. Gerald Lucovsky at North Carolina State University in developing the insulator technology using RPECVD. This work has been supported by the Office of Naval Research, SDIO-IST, and the Semiconductor Research Corporation.

[1]B. S. Myerson, Appl. Phys. Lett. **48**, 797(1986).

[2]R. G. Frieser, J. Electrochem. Soc. **115**, 401(1968).

[3]Akihisa Matsuda and Kazunobu Tanaka, Thin Solid Films, **92**, 171(1982).

[4]R. A. Rudder, G. G. Fountain, and R. J. Markunas, J. Appl. Phys. **60** (10), 3519, 1986.

[5]Werner Kern, "Purifying Si and SiO_2 Surfaces with Hydrogen Peroxide." Semiconductor International, April 1984, pp. 94-99.

[6]Shoji Nishida, Tsunenori Shiimoto, Akira Yamada, Shiro Karasawa, Makoto Konagai, and Kiyoshi Takahashi, Appl. Phys. Lett. **49** 79(1986).

[7]S. Suzuki and T. Itoh, J. Appl. Phys **54** (3), 1466, 1983.

[8]S. V. Hattangady, R. A. Rudder, G. G. Fountain, D. J. Vitkavage, and R. J. Markunas, to be published in proceedings of Fall 1987 MRS meeting, Boston, Mass.

[9]J. A. Schaefer, F. Stucki, D. J. Frankel, W. Gopel, and G. J. Lapeyre, J. Vac. Sci. Technol. A **2**, 359(1984).

[10]G. Lucovsky, P. D. Richard, D. V. Tsu, S. Y. Lin, and R. J. Markunas, J. Vac. Sci. Technol. A **4**, 681(1986).

[11]L. G. Meiners, J. Vac. Sci. Technol. **21**, 655(1982).

[12]G. G. Fountain, R. A. Rudder, R. J. Markunas, P. S. Lindorme, submitted to Appl. Phys. Lett.

[13]T. Motooka and J. E. Greene, J. Appl. Phys. **59**, 2015(1986).

[14]John Balamuta, Michael F. Golde, and Yueh-Se Ho, J. Chem. Phys. **79**, 2822(1983).

Se PASSIVATION AND RE-GROWTH ON ZnSe AND (Zn,Mn)Se (001) EPILAYER SURFACES

B.T. Jonker, J.J. Krebs and G.A. Prinz
Naval Research Laboratory, Washington, D.C. 20375-5000

ABSTRACT

We report the use of a thin Se coating to passivate and protect the surface of ZnSe and $Zn_{1-x}Mn_xSe$ (001) epilayers on GaAs(001) prior to removal from the molecular beam epitaxy (MBE) system and exposure to atmosphere. After prolonged exposure to the laboratory ambient, the samples are returned to the MBE system and the Se is thermally desorbed. The resultant surface is clean and well-ordered as demonstrated by Auger electron spectroscopy and reflection high energy electron diffraction (RHEED), and either homo- or hetero-epitaxial growth is continued. For ZnSe epilayers grown in two equal stages to a total thickness of 1 μm with an intervening passivation and 24 hour air exposure, the (004) x-ray double crystal rocking curve exhibits no increase in linewidth relative to continuously grown reference samples of the same thickness. In the case of heteroepitaxial growth of Fe films on $Zn_{0.8}Mn_{0.2}Se$, RHEED shows well-ordered single crystal growth of α-Fe(001) even at very low coverages. Ferromagnetic resonance and vibrating sample magnetometry data show these films to be of excellent magnetic quality. This Se passivation technique thus offers a ready means of transferring these II-VI epilayers from the growth system to other facilities.

INTRODUCTION

The II-VI semiconductor ZnSe and its diluted magnetic derivatives, e.g. (Zn,Mn)Se, are emerging as promising candidates for a variety of potential wide-gap device and magneto-optic applications.[1-6] These materials have been grown as epilayers on GaAs substrates by molecular beam epitaxy (MBE), and may thus be incorporated into current GaAs-based fabrication technology. A preliminary to actual device fabrication is a thorough characterization of the clean surface properties and the behavior exhibited during the initial stages of interface formation with either metals or other semiconductors. Since a clear picture of a process as complicated as heteroepitaxial growth or Schottky barrier formation requires the use of several complementary analysis techniques not available on the growth system, it is often necessary to transfer the sample to other facilities without contaminating the surface. In addition, device fabrication generally involves several processing steps in which the reliability or reproducibility would benefit if done on a clean surface. Thus a simple means of protecting the sample surface to allow transfer through atmosphere to other facilities is very desirable.

Miller, Kowalczyk and co-workers[7,8] have reported the use of an As overcoat to prevent degradation of (Al,Ga)As samples during exposure to atmosphere. Arsenic passivation of MBE-grown epilayers of GaAs and its related ternaries is now commonly employed, and permits such samples to be simply mailed from one location to another for further growth or processing.

We describe here a similar procedure which utilizes a Se layer to passivate the surface of ZnSe and $Zn_{1-x}Mn_xSe$ epilayers grown on GaAs(001) bulk substrates by MBE. After growth of the desired epilayer, the sample is cooled and a Se overcoat deposited to passivate and protect the surface from contamination. A sample thus protected may be exposed to atmosphere for a prolonged period of time until needed for further growth or transferred to another facility. When returned to a vacuum system, the Se overlayer is easily thermally desorbed at 115-150°C, leaving a clean, stoichiometric and well-ordered single crystal surface for further epitaxial growth or surface studies.

We have investigated the use of this technique in two cases: i) the continued growth of ZnSe on a Se-passivated and air-exposed buffer layer of ZnSe previously grown on GaAs(001),[9] and ii) the heteroepitaxial growth of ferromagnetic α-Fe(001) films on a Se-passivated and air-exposed epilayer of the diluted magnetic semiconductor $Zn_{1-x}Mn_xSe$ on GaAs(001). Auger electron spectroscopy (AES) and reflection high energy electron diffraction (RHEED) were used to confirm the cleanliness and single crystal order of all sample surfaces at each stage of the growth process. X-ray double crystal rocking curves were obtained to determine the crystalline quality of the ZnSe epilayers, and ferromagnetic resonance (FMR) and vibrating sample magnetometry used to evaluate the magnetic quality of the α-Fe films. We compare results with those obtained for reference samples produced under nearly identical conditions without an intervening passivation/exposure step, and find no significant differences. Thus the Se passivation, air exposure, desorption and regrowth procedure does not degrade the quality of the final growth layer to the extent measured by the analysis techniques employed.

EXPERIMENTAL

The samples were grown in a PHI Model 400 MBE system equipped with AES and RHEED. GaAs(001) device grade wafers were used as substrates, and prepared in a conventional fashion using a 0.2%Br-methanol polish and etch, with final cleaning accomplished by heating to 585°C in ultra-high vacuum just prior to growth. The ZnSe and (Zn,Mn)Se epilayers were deposited using elemental Knudsen cell sources at a substrate temperature of 330°C and growth rates of ~ 0.33 μm/hr. The fluxes were monitored with a quadrupole mass spectrometer, and bulk stoichiometry verified with energy dispersive x-ray fluorescence following growth.

After growth of 0.1-0.5 μm of ZnSe or (Zn,Mn)Se, the source ovens were shuttered and the sample cooled to 30°C or less in approximately 20 minutes by bringing the edge of the sample holder into contact with one of the liquid nitrogen cooled cryopanels in the growth chamber. The sample was then returned to the growth position and 200-1000 Å of Se deposited. Varying the thickness of the protective Se layer within this range had no detectable effect on the results.

The sample thus protected was then removed from the MBE system and stored in a plastic container. The period of time during which the sample was exposed to atmosphere ranged from several hours to several days to simulate transfer of the sample to another vacuum system within the building or mailing it to another geographical location.

After some period of time had elapsed, the sample was returned to the MBE system and the protective Se coating thermally desorbed in ultra-high vacuum while monitoring the surface order with RHEED and the desorbed flux with the quadrupole mass spectrometer. Initial Se desorption occurs at approximately 115°C as indicated by the reappearance of streaks in the RHEED pattern, and is largely completed by 150°C. No Se flux is detected after heating to 250-300°C. The condition of the surface was then evaluated with AES and RHEED, followed by either homoepitaxial growth of ZnSe on the initial ZnSe buffer layer, or heteroepitaxial growth of single crystal α-Fe(001) films on the initial $Zn_{1-x}Mn_xSe(001)$ layer.

RESULTS AND DISCUSSION

After thermal desorption of the Se overlayer, the resultant single crystal surfaces were found to be clean and well-ordered as demonstrated by AES and RHEED. Figure 1 shows the RHEED patterns obtained during various stages of the ZnSe growth. Figures 1(c) and (d) are the patterns produced by the initial ZnSe buffer layer prior to Se passivation, and figures 1(e) and (f) show the patterns obtained following air exposure and thermal desorption of the protective Se coating. These final patterns show well-defined streaks and Kikuchi lines comparable to those produced by the as-grown surface, and indicate that no degradation of single crystal surface order has occurred. The AES spectrum obtained from this surface is shown in Figure 2 -- no contamination is evident above the signal-to-noise level (detection sensitivity < 1% monolayer).

To evaluate the effect of the Se passivation, air exposure and thermal desorption process on further epitaxial growth, ZnSe(001) epilayers were grown in two equal stages to a total thickness of 1 µm with an intervening passivation and 24 hour air exposure. X-ray θ-2θ measurements of the (00ℓ) $\ell = 2,4,6$ and (444) reflections were used to determine both the perpendicular lattice parameter $a_\perp$ and the lattice parameter parallel to the interface $a_\parallel$, respectively. These measurements confirmed that no distortion of the cubic zincblende lattice had occurred, and showed $a_\perp = a_\parallel = 5.6676 + 0.0005$ Å, the bulk ZnSe lattice parameter. X-ray double crystal rocking curves obtained with Cu $K\alpha_1$ radiation for the (004) reflection were used to further evaluate the crystalline quality of

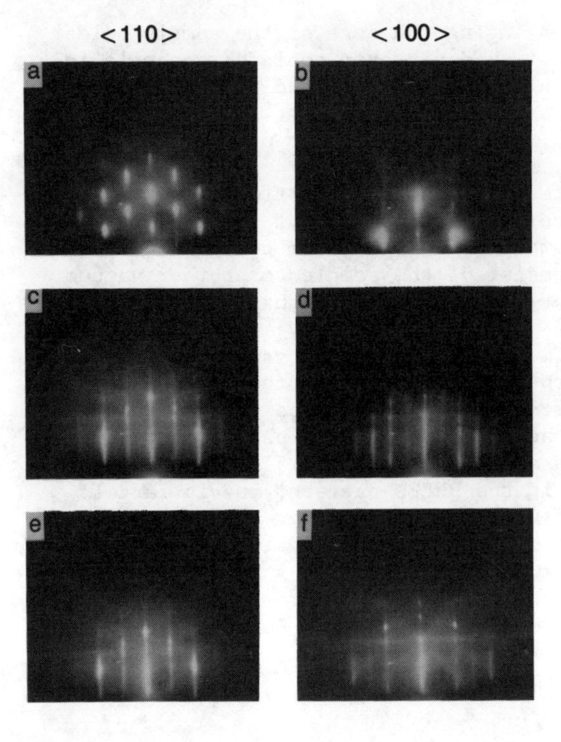

Figure 1. RHEED patterns obtained at 10 keV with the electron beam incident along the <110> and <100> azimuths for: a,b) the clean GaAs(001) surface just prior to ZnSe deposition; c,d) a 0.2 µm ZnSe(001) epilayer before Se passivation; and e,f) the ZnSe epilayer following passivation, air exposure and thermal desorption of the Se layer in ultra-high vacuum.

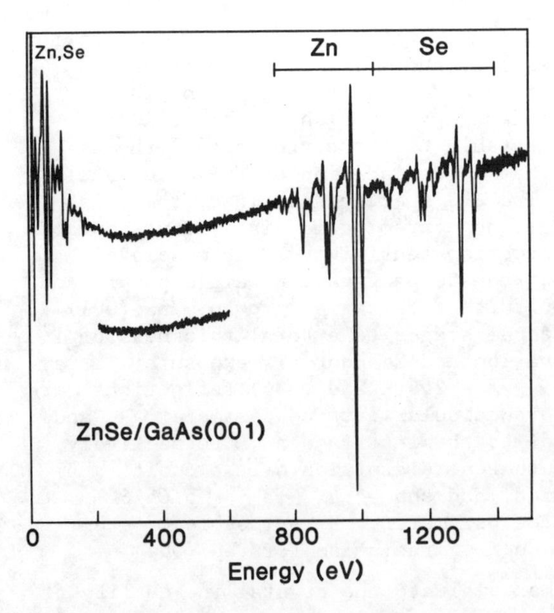

Figure 2. AES spectrum (3 keV) of the ZnSe epilayer surface following Se passivation, 24 hour air exposure, and thermal desorption of the Se overcoat. No impurities are detected.

these 1 μm thick epilayers. The rocking curves showed no increase in linewidth (FWHM = 260 arc sec), i.e. no degradation of crystalline order, due to the passivation/exposure process relative to continuously grown reference samples of the same thickness.

The second system studied was the heteroepitaxial growth of α-Fe(001) films on Se-passivated and air-exposed epilayers of the diluted magnetic semiconductor $Zn_{1-x}Mn_xSe$ on GaAs(001). The RHEED patterns produced by a $Zn_{0.8}Mn_{0.2}Se$ epilayer just prior to Se passivation are shown in Figure 3(a), and after air exposure and thermal desorption of the passivating layer in Figure 3(b). These patterns again show that the final surface is as well-ordered as the intial growth surface.

The Fe was deposited from a high-temperature Knudsen cell source at a substrate temperature of 175°C and growth rates of ~ 5 Å/min. RHEED shows well-ordered single crystal Fe growth even for relatively thin films, as indicated by the patterns produced at 45 Å Fe coverage in Figure 3(c). The films grow as (001) planes, with Fe[100] // $Zn_{1-x}Mn_xSe$[100] and Fe[110] // $Zn_{1-x}Mn_xSe$[110].

In ferromagnetic metals, certain magnetic parameters are very useful indicators of overall film quality. Ferromagnetic resonance (FMR) and vibrating sample magnetometry were therefore used to evaluate the magnetic character of the resultant Fe films. Two of the most important parameters are the FMR linewidth ΔH and the coercive field H_c -- narrow linewidths and small values for H_c indicate high quality material.

The FMR spectrum of a 573 Å Fe film grown on a Se-passivated and air-exposed $Zn_{0.8}Mn_{0.2}Se$(001) epilayer is shown in Figure 4 for H // [100] at 34.8 GHz and room temperature. The spectrum exhibits a symmetric Lorentzian lineshape with a linewidth of only 62 Oe. The effective magnetization $4\pi M'$ obtained from these measurements is 20.9 kG, over 97% of the bulk Fe value. These data are summarized in Table I and compared with similar data obtained from a 530 Å Fe reference film grown on a ZnSe(001) epilayer with no intervening passivation and air exposure.[10] Although this thinner film exhibits a slightly narrower linewidth (51 Oe), the difference is within the variation observed from sample to sample. These narrow linewidths indicate low losses at microwave frequencies and demonstrate excellent magnetic quality.

Vibrating sample magnetometry plots of the magnetization versus applied field (M vs H loops) for H applied along an easy in-plane <100> axis exhibit abrupt rectangular loops with small values for the coercive field H_c. These values are also shown in Table I -- for the 573 Å Fe film grown on the passivated $Zn_{0.8}Mn_{0.2}Se$ surface we obtain H_c = 9 Oe, compared with a value of 10.3 Oe for the 530 Å continuously grown reference sample. Thus the Se passivation, air exposure and thermal desorption process introduces no defects at the Fe/$Zn_{0.8}Mn_{0.2}Se$ interface which act as domain wall nucleation or

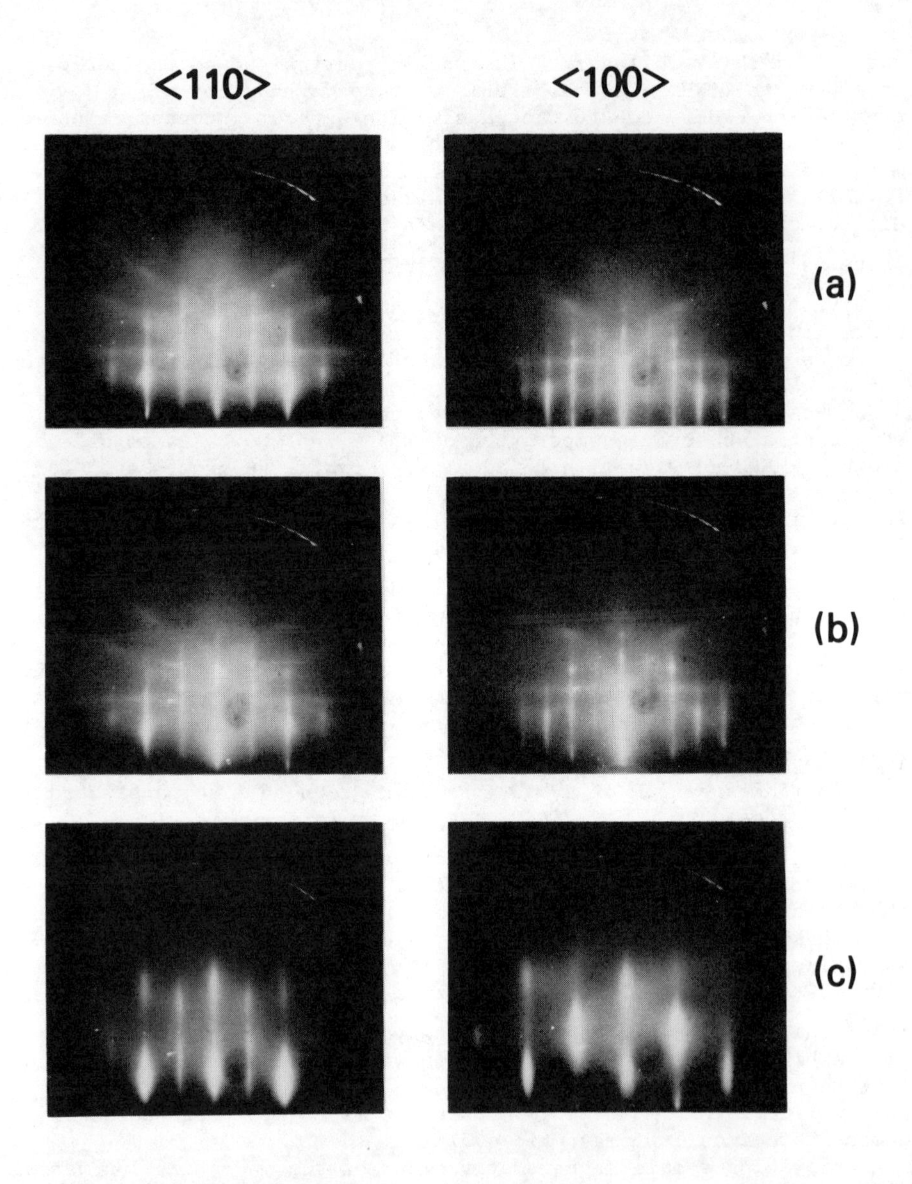

Figure 3. RHEED patterns obtained at 10 keV with the electron beam incident along the <110> and <100> azimuths for: a) a $Zn_{0.8}Mn_{0.2}Se$ (001) epilayer just prior to Se passivation; b) the epilayer following passivation, air exposure and thermal desorption of the Se overlayer; and c) a 45 Å Fe(001) film grown on this surface.

pinning sites to prevent the Fe films from magnetically switching. These low values for H_c indicate that these Fe films would be useful in applications requiring easy switching.

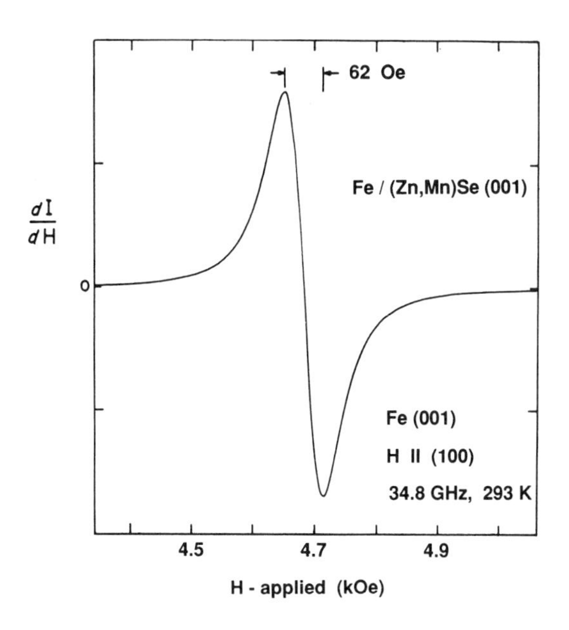

Figure 4. FMR spectrum of a 573 Å α-Fe(001) film grown on a $Zn_{0.8}Mn_{0.2}Se$ epilayer surface following Se passivation, air exposure and thermal desorption of the Se overcoat.

Table I. Magnetic parameters for an α-Fe(001) film grown on a Se passivated, air-exposed (Zn,Mn)Se epilayer compared with reference sample.

Sample	ΔH(Oe)	$4\pi M'$(kG)	H_c(Oe)
573 Å Fe/pass. $Zn_{0.8}Mn_{0.2}Se$	62	20.9	9
530 Å Fe/ZnSe -- reference	51	21.0	10.3
bulk Fe		21.5	

SUMMARY AND CONCLUSIONS

Se passivation of ZnSe and $Zn_{1-x}Mn_xSe$ (001) epilayer surfaces is very effective in preventing contamination during prolonged exposure to atmosphere. The Se overlayer is readily desorbed at temperatures of 115-150°C in ultra-high vacuum, yielding clean, well-ordered and stoichiometric single crystal surfaces. This procedure permits simple transfer of such II-VI epilayers from the growth system to other facilities for surface studies, processing or additional epitaxial growth. The crystalline quality of ZnSe films grown with an intermediate passivation and air expsoure is not measureably effected. Ferromagnetic α-Fe(001) films grown on passivated, air-exposed $Zn_{1-x}Mn_xSe$ epilayers exhibit excellent magnetic properties comparable to those of continuously grown reference samples, with low losses at microwave frequencies and low switching fields. This passivation, transfer and re-growth process may thus be used to facilitate multiple overgrowths incompatible in a single growth chamber system.

ACKNOWLEDGMENTS

This work was supported by the Office of Naval Research. The authors gratefully acknowledge the technical assistance of D. King and F. Kovanic.

REFERENCES

1. J.E. Potts, T.L. Smith and H. Cheng, Appl. Phys. Lett. 50, 7 (1987).
2. T. Mitsuyu, K. Ohkawa and O. Yamazaki, Appl. Phys. Lett. 49, 1348 (1986).
3. D.J. Olego, Appl. Phys. Lett. 51, 1422 (1987).
4. R.B. Bylsma, W.M. Becker, T.C. Bonsett, L.A. Kolodziejski, R.L. Gunshor, M. Yamanishi and S. Datta, Appl. Phys. Lett. 47, 1039 (1985).
5. T. Yasuda, I. Mitsuishi and H. Kukimoto, Appl. Phys. Lett. 52, 57 (1988).
6. H. Cheng, J.M. DePuydt, J.E. Potts and T.L. Smith, Appl. Phys. Lett. 52, 147 (1988).
7. D.L. Miller, R.T. Chen, K. Elliot and S.P. Kowalczyk, J. Vac. Sci. Technol. B3, 560 (1985); J. Appl. Phys. 57, 1922 (1985) and references therein.
8. S.P. Kowalczyk and D.L. Miller, J. Vac. Sci. Technol. B4, 625 (1986).
9. B.T. Jonker, J.J. Krebs and G.A. Prinz, to be published, J. Appl. Phys., June 1 (1988).
10. J.J. Krebs, B.T. Jonker and G.A. Prinz, J. Appl. Phys. 61, 3744 (1987).

VAPOR PHASE DEPOSITION AND GROWTH OF POLYIMIDE FILMS ON COPPER

M. Grunze, J.P. Baxter, C.W. Kong, R.N. Lamb and W.N. Unertl
Laboratory for Surface Science and Technology and Dept. of Physics,
University of Maine, Orono, ME 04469

C.R. Brundle
IBM Almaden Research Center, San Jose, CA. 95120

ABSTRACT

The formation of thin polyimide films from vapor phase deposited 4,4 Oxydianiline (ODA) and Pyromellitic dianhydride (PMDA) was studied by X-ray photoelectron spectroscopy and vibrational spectroscopies. Codeposition of ODA and PMDA onto polycrystalline copper substrates at room temperature, followed by heating in vacuum, led to polymerization and the formation of thermally stable ($T \leq 723$ K) polyimide films. Films with thicknesses ranging from ultra-thin (12-30 Å) to several hundred nanometers thick were prepared by this method. Adhesion of the polymer to the surface involve chemical bonding to fragments of PMDA and ODA initially chemisorbed on the clean metal surface.

INTRODUCTION

The formation of thin dielectric polymer films by chemical vapor deposition techniques rather than by spin coating procedures opens the possibility of simplified or alternate manufacturing steps in the microelectronic industry. It is necessary, however, to produce these films with the required electrical and mechanical properties and adhesive strength. Whereas films produced by conventional wet chemical methods are well characterized[1], it would be presumptuous to assume that films formed by vapor deposition techniques, and therefore in the absence of solvents, have the same properties. The absence of the solvent can effect the polymerization process itself and influence the interaction of the film with the substrate leading to an adhesion behavior different from that of spun-on films.

This article describes the results of our experiments on the production of thin polyimide films by codeposition of its respective constituents 4,4' oxydianiline (ODA) and 1,2,4,5 benzenetetracarboxylic anhydride (pyromellitic dianhydride, PMDA) from the vapor phase in a vacuum apparatus. The overall reaction scheme is illustrated in Fig. 1. The principles of solventless preparation of polyimide films were first described in the production of very thick (> 1μm) films[2]. Since the thermochemical characteristics of polyimide preclude its vapor deposition, the appropriate constituents which make up the polyimide polymer are co-deposited on the substrate. Under conditions of carefully controlled temperature the coadsorbed molecules will react to form the polymer. The interaction of the constituents with the metal surface and the polymerization process in the codeposited layer was followed by X-ray photoelectron spectroscopy (XPS), infrared reflection absorption spectroscopy (IRRAS) and Raman spectroscopy to determine the interaction with the substrate and the chemical processes during imidization.

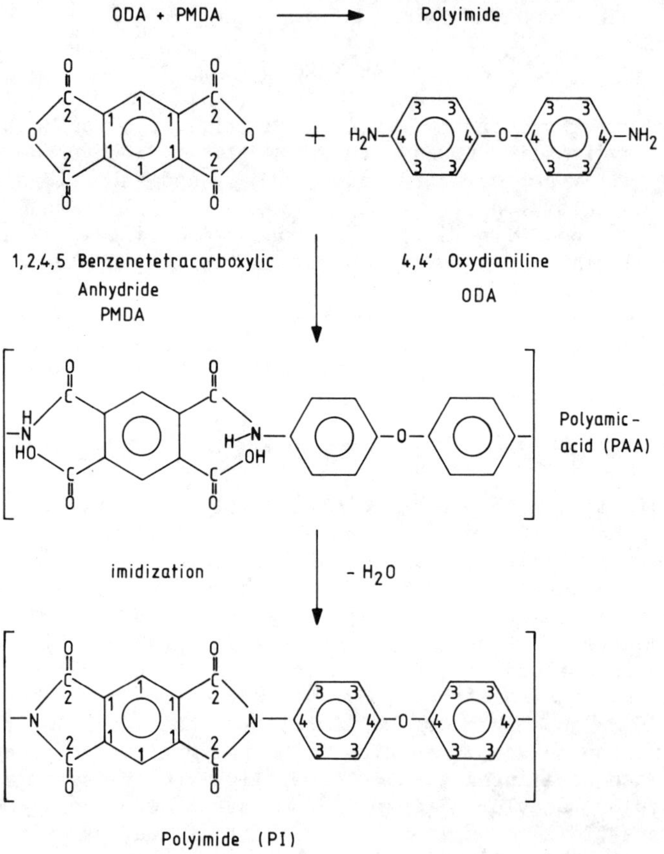

ODA + PMDA → Polyimide

1,2,4,5 Benzenetetracarboxylic Anhydride PMDA

4,4' Oxydianiline ODA

Polyamic-acid (PAA)

imidization $- H_2O$

Polyimide (PI)

Fig. 1. Schematic representation of the reaction of ODA and PMDA to form polyimide. Labelling of the carbon and oxygen atoms is used to aid the discussion of the XPS spectra.

We first briefly review the pertinent literature on polyimide/ metal interfaces. Then we describe our experiments and discuss the chemical characteristics of the films produced by vapor deposition. Finally, we address the nature of the adhesive bond to the copper substrates used in our studies.

1. BACKGROUND

Polyimides (PI) are a class of high temperature polymers that exhibit a unique combination of thermal stability, high softening point and easy processability into coatings or films. In microelectronic device applications, polyimides are used in packaging (e.g. alpha particle barriers, protective overcoats in passivation layers) and as insulating interlevel dielectrics (e.g. pattern delineating material). Polyimide films have usually been applied to the substrate by spin coating. The initial spin-coated layer consists of a solution of the polyimide precursor polyamic acid (PAA) dissolved in a polar solvent. This layer is given a thermal treatment to evaporate the solvent and to imidize the polyamic acid to polyimide.

Reliable adhesion between PI and the substrate is a crucial aspect in all applications. In the absence of extrafacial inhomogeneity (e.g. stress free films) the strength of the adhesive bond is dependent directly on the physics and chemistry of the polymer/metal interface.[3] This has prompted a number of investigations to probe the microscopic origins of the bonding which have involved a variety of interfaces and techniques.

Studies involving metallized plastics, where a thin metal film is in contact with a much thicker (usually bulk) fully-cured polyimide phase, have provided the main source of chemical information about the polymer/substrate interface.[4-9] For example, room temperature deposition of chromium leads initially to bonding to the PI substrate, possibly via the carbonyl groups, and subsequently with increasing chromium coverage to formation of a carbide like carbon species.[10] Similarly, other electropositive metals such as aluminum,[7] titanium[8] and nickel[4] also appear to react through this carbonyl entity. Copper[4,6,7] and silver[4], however, show only a weak interaction with the oxygen in the ether part of the chain.

An alternative method for producing the metal/polyimide interface is spin coating the polymer precursor (PAA) onto a supported metal film, prior to curing and the formation of polyimide. Bulk polyimide/metal interfaces formed in this way[9] have been shown to produce a marked increase in, for instance, the peel strength of a PI/copper oxide interface compared to conventional metal deposition. The precursor/metal interfacial reaction is apparently much stronger compared to that of the metal/PI where the polymer is fully cured prior to metal deposition. Spin coating has not, until recently,[11] been used successfully to generate films thin enough for use with electron spectroscopic techniques because electron propagation away from the interface and out to the vacuum for analysis requires that the insulating films be relatively thin (<100 Å).

It is clear that the way in which the interface is formed is a main factor in determining its chemical and physical properties. While spin-coating and vapor deposition rely upon initial interaction of the substrate with the precursor prior to imidization, formation of a polymer/metal interface by evaporation of the metal onto the fully cured polyimide represents a class by itself. The effect of the solvent must also be considered in a comparison of spin coated and vapor deposited films. As demonstrated recently by Kim et al.[9] and Kowalczyk et al.[12] for polyimide/copper oxide interfaces, the presence of the solvent (N-methylpyrrolidone) leads to formation of cuprous oxide particles in the polymer film, whereas interfaces prepared by solventless methods, i.e. vapor deposition of the organic constituents or copper evaporation onto fully cured polyimide, showed no such precipitates.

2. EXPERIMENTAL

The vapor deposition and XPS experiments were carried out in an arrangement consisting of three interconnected vacuum chambers. These were capable of (relatively) high pressure (10^{-6} - 16 bar), high vacuum (to 10^{-9} mbar) and ultra-high vacuum (10^{-11} mbar) conditions, respectively. They were connected via a sample transfer rod which supported the copper samples (1-2 cm^2) and was also equipped with both cooling and resistive heating facilities. The temperature was monitored by a chromel-alumel thermocouple. Sample cleaning prior to film deposition was carried out by heating the sample to 800 K in 13 mbar O_2 followed by 3 mbar of H_2 to remove surface oxygen and/or argon ion sputtering. The organic vapor sources consisted of two small quartz tubes (50 mm length, 5 mm diameter) containing the crystalline PMDA and ODA (Aldrich Gold Label), respectively. Thin tungsten wire was coiled around each tube which was subsequently encased in a ceramic block that supported both the wire (heated resistively) and also a chromel alumel thermocouple which was inserted through the mouth of each tube. Prior to deposition, the materials were degassed for thirty minutes at ~393 K. Deposition was carried out in the high vacuum chamber with the sample in the temperature range 200 K-300 K. Optimum deposition conditions were obtained for sublimation temperatures between 373 K and 423 K with concomitant pressures of $2x10^{-6}$ to $8x10^{-6}$ mbar. Ultra thin films could be deposited in less than 2 minutes. An arbitrary exposure scale, L = pt [$x10^{-6}$ mbar · s] where t is exposure time and p the measured background pressure, is used to indicate the extent of deposition.

In comparison to a similar study with very thick (> 1μm) films,[2] this present arrangement for co-deposition of organic vapor was significantly simpler in design. In particular, it avoided the necessity of a mixing chamber prior to deposition. No effort was made to maintain a stoichiometric mixture of vapor fluxes during the experiments.

The XPS experiments were carried out in the UHV chamber. The spectrometer contained a Leybold-Heraeus EA11 hemispherical electrostatic electron analyzer and a MgK_α x-ray source operated at 100 W. An experimental resolution of 0.92 eV was measured using the Ag 3d emission. The electron binding energies were calibrated against the Au $4f_{7/2}$ emission at E_B = 84 eV.

Analysis of XPS spectra arising from the organic depositions on the metallic substrate involves the determination of peak binding energies arising for the individual constituents plus determination of the mean stoichiometry from the measured peak areas. The ratio of these is calculated as:

$$\frac{N_1}{N_2} = \frac{I_1}{I_2}\frac{\sigma_2}{\sigma_1}\left[\frac{E_2}{E_1}\right]^{m-0.73} \quad . \qquad (1)$$

In eq. 1, N is the number of atoms/unit area with core level intensity I (proportional to the area under the peak) for species 1 and 2. The relative photoelectron excitation cross sections, σ, are 100, 285 and 177 for C 1s, O 1s and N 1s respectively.[14] The electron mean free pathlengths for organic materials and the transmission of the electron spectrometer ($E^{0.73}$) are functions of electron kinetic energy E. A combination of these leads to the form shown in eq. 1. A value for m of zero refers to the limit for ultra thin films. In the case of thick films the correct value of m is not accurately known but probably lies in the range 0.5 to perhaps 0.71;[16] this uncertainty in m results in up to 15 percent uncertainty in relative compositions calculated using eq. 1.

The contribution that final state effects such as shake-ups will have on the integration were discussed in a recent publication with respect to thick vapor deposited films.[13] Since shake-ups effectively borrow intensity from the primary peaks, uncertainties arise in apportioning these additional intensities. Therefore, calibration spectra of pure PMDA and ODA are used to assign the actual shake-up region in the polyimide spectra.[19]

The determination of absolute binding energies is complicated by charging within the film. The general trend observed was shifting of the peaks towards higher binding energies with increasing film thickness and can be explained by a decrease in the final state screening of the photoionized molecule by metal electrons as film thickness increases. Because of the indeterminate contribution of charging to this shift as a function of the film thickness, no corrections have been made to the data presented here.

Film thickness (d) was calculated from the attenuation of the Cu $2p_{3/2}$ intensity as

$$d = -\Gamma \ln (I/I_o) \qquad (2)$$

where Γ is the mean free path length of the Cu 2p photoelectrons in the overlayer and I_o is the intensity measured on a clean surface. The main assumptions inherent in eq. 2 are the continuity and homogeneity of the film. We have no information to what extent these conditions are satisfied in the following experiments.

Uncertainties in the choice of Γ have been addressed previously.[13] The important consideration is, however, its correct order of magnitude rather than any absolute value. The choice of 8 Å for the Cu 2p photoelectrons[15] is considered reasonable in light of previously reported compilations of Γ.[16] The calculated thicknesses reported below are therefore regarded as being suitable approximations. Note that with this value for Γ, films thicker than 37 Å will attenuate the Cu 2p signal to less than one percent of its clean surface value.

The Raman experiments, presented here, were conducted on cured polyimide films which were prepared in the high vacuum chamber and then left exposed to the environment, prior to measurement with the Raman instrument.

3. RESULTS AND DISCUSSION

The overall reaction[17] in the formation of polyimide from ODA and PMDA is shown schematically in Fig. 1. The numbers in the structural formulae are given to facilitate the discussion of the XPS spectra. Initial interaction leads to the formation of polyamic acid which converts upon heating to polyimide.

The following section is concerned with reviewing characteristic features of thin/ultra-thin film spectra for each of the primary constituents, PMDA and ODA. The data for the intermediate (PAA) and the product (PI) are subsequently discussed in light of the analysis of the pure component deposition. This is followed by the examination of a sequence of experiments for determination of the minimum thickness of the codeposited layer necessary to produce a polyimide film.

3.1 ODA AND PMDA

ODA and PMDA were deposited separately on polycrystalline copper substrates to determine the interaction of the two polyimide constituents with the metal surface. The reactivity of the substrate with these will to some extent determine the processes occurring within the codeposited layers and therefore effect the interfacial chemistry between the polyamic acid (and consequently polyimide) and the metal surface.

Figures 2, 3 and 4 show the C 1s, O 1s and N1s spectra of ODA deposited on polycrystalline copper as a function of exposure. These spectra provide clear evidence for a chemical reaction between ODA and the metal at room temperature.

The adsorption of a carbon containing species is evident for the smallest exposure in Fig. 2, but only at the larger exposures (L > 240) is there sufficient oxygen and nitrogen for the adsorbed layer to be detected above the noise level. Spectrum 2c already begins to show the characteristic splitting of the C 1s band for ODA,[13] whereas the O 1s spectra, Fig. 3c, indicate by the weak emission around 530 eV that besides "organic" oxygen (O 1s emission at $E_B \sim 534$ eV) an oxidic-like species is formed. The N 1s emission corresponding to L = 240, however, is too weak to be clearly detected. At L = 1140, the C 1s signal increased and a N 1s signal is detected, and there is a small increase in the O 1s spectrum (note the change in intensity scale for the O 1s data between spectra c and d).

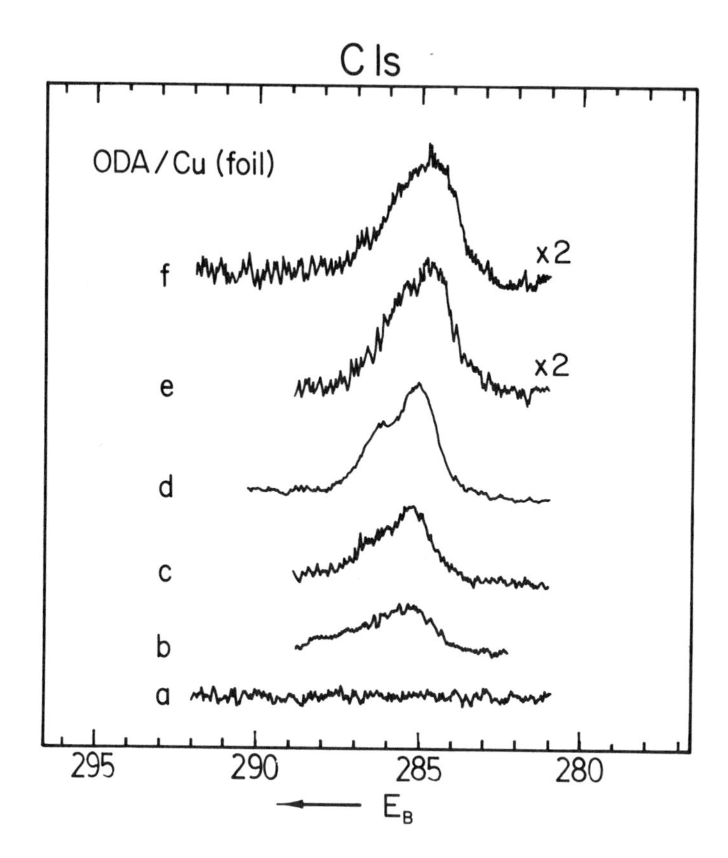

Fig. 2. XPS C(1s) spectra for adsorption and decomposition of ODA on polycrystalline copper. **a:** clean substrate (298 K), **b:** L = 60 (298 K), **c:** L = 240 (298 K) **d:** L = 1140 (298 K), **e:** Substrate heated to 346 K, **f:** substrate heated to 383 K. In trace d the scan speed was decreased from 4 eV/minute to 1 eV/minute and the time constant was increased from 1 s to 5 s.

Heating the sample to 346 K and 383 K (spectra e, f) results in a broadening and decrease in intensity of the C 1s and a decrease in the N 1s peaks and in a reduction of the "organic" oxygen peak located near 534 eV. Some ODA may also sublime.

An evaluation of the stoichiometry of the overlayer after L = 1140 results in a total C:O:N ratio of 12:0.9:0.8, if the oxidic oxygen at E_B = 530 eV is excluded, the ratio is 12:0.7: 0.8. Clearly the layer does not consist of molecular ODA, since the nitrogen content is reduced to about 40% of the expected stoichiometry value of 12:1:2. We therefore conclude that ODA partially dissociates upon adsorption on clean polycrystalline copper. The fact that the carbon to oxygen ratio is close to the value expected (in particular where the oxidic species are included) but that nitrogen is lost in the initial adsorption process, suggests that the molecule dissociates

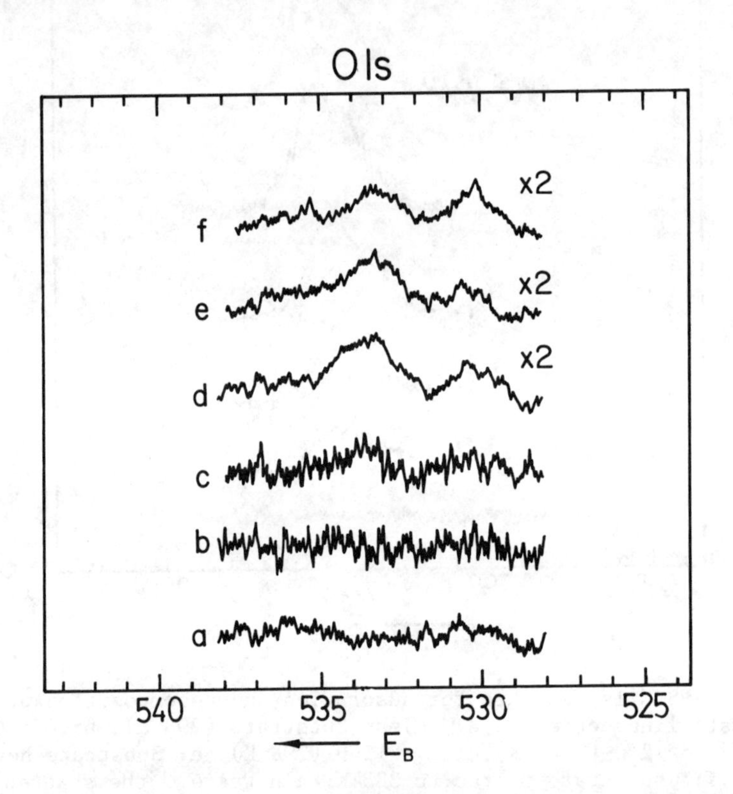

Fig. 3. XPS O 1s spectra for the adsorption and decomposition of 4, 4' oxydianiline on polycrystalline copper. Traces a-f are defined in Fig. 2. In traces d, e, and f the scan speed was decreased from 4 eV/min to 1 eV/min and the time constant was increased from 1 s to 5 s.

via the release of nitrogen or nitrogen carbon entities. A similar reaction scheme has been postulated for the interaction of ODA with silver surfaces[19]. As will be discussed elsewhere, heating an ODA layer to $T \geq 380$ K leads to further decomposition leaving primarily amorphous carbon residues on the surface.[18]

PMDA deposited at room temperature on a clean Cu {111} surface also decomposes as can be seen from Figs. 5 and 6. Data for PMDA on polycrystalline copper is not available, but it is safe to assume that the chemically more active polycrystalline substrate would show a similar behavior.

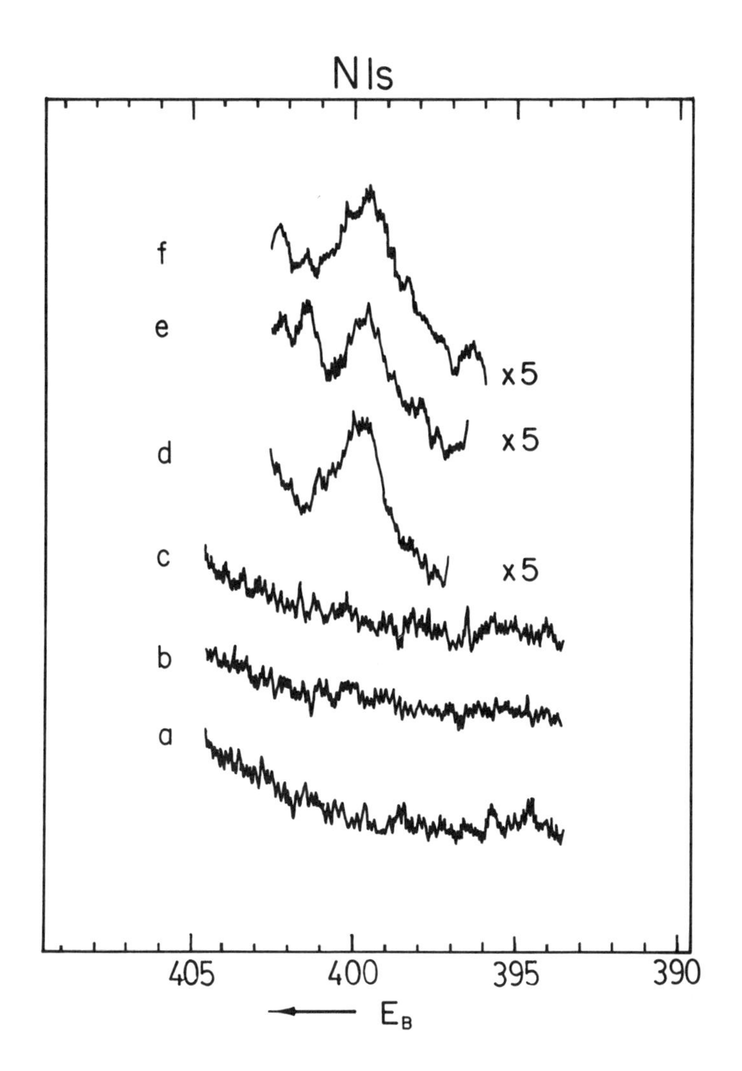

Fig. 4. XPS N 1s spectra for the adsorption and decomposition of 4, 4' oxydianiline on polycrystalline copper. Traces a-f are defined in Fig. 2. In traces d, e, and f the scan speed was decreased from 4 eV/min to 1 eV/min and the time constant was increased from 1 s to 5 s.

Figure 5 shows the C 1s spectra for L = 30 and L = 2040 at room temperature (Figs. 5a and 5b respectively) and, for comparison, a spectrum recorded after L = 360 with the sample held at 200 K (Fig. 5c). The low temperature spectra (Fig. 5c) is for a thick (d > 40 Å film) and exhibits the expected[13] C 1s doublet from the phenyl carbons (C 1 in Fig. 1) and the carbonyl carbons (C 2 in Fig. 1) with a peak area ratio of 6:4. The O 1s spectra (Fig. 6c) corresponding to the C 1s spectra in Fig. 5c also is indicative of condensed molecular PMDA. The unresolved doublet originates from the anhydride oxygen atoms (E_B ~ 534.3 eV) and the carbonyl oxygen atoms (E_B = 533.0 eV).

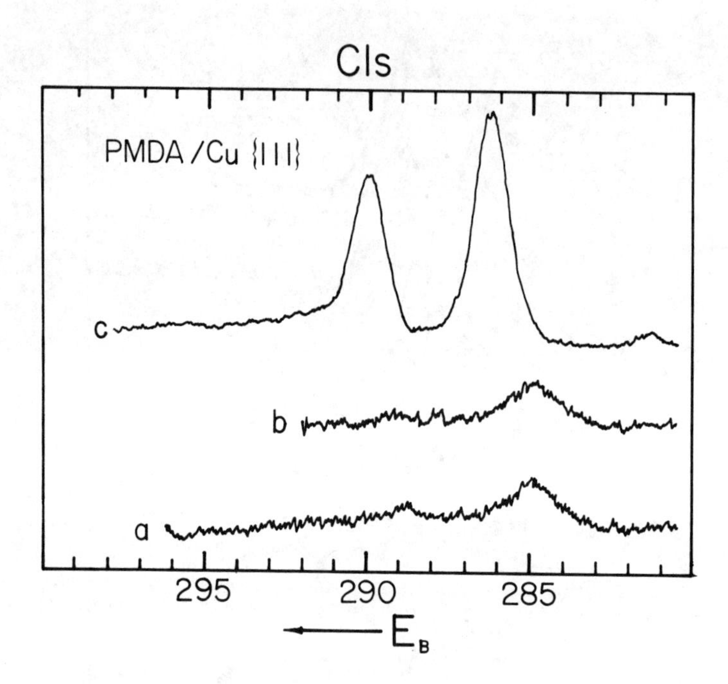

Fig. 5. XPS C1s spectra of pyromellitic dianhydride (PMDA) on Cu{111}. **a:** L = 30 (298 K), **b:** L = 2040 (298 K), **c:** L = 360, exposure at 200 K. In trace c the scan speed was decreased from 4 eV/min to 2 eV/min and the time constant was increased from 1 s to 2 s.

Thus, the carbon 1s and oxygen 1s spectra Figs. 5a, b and Figs. 6a, b respectively are clearly <u>not</u> consistent with molecular PMDA. The presence of a low binding energy O 1s emission indicates that PMDA oxygen reacts with the surface although the presence of "oxidic" oxygen at E_B = 530.5 eV is less clear than in the case of ODA (Fig. 3). The C 1s spectra show a broad low binding energy C 1s emission (E_B ~ 284.9 eV) and some residual carbonyl carbon bands (E_B ~ 289.0 eV). Clearly, the ratio between phenyl and carbonyl carbon emission of 6:1.7 (in Fig. 5a) is not consistent with adsorbed molecular PMDA, but indicates that PMDA dissociates upon adsorption on the surface. The total C 1s:O 1s ratio of 10:4.6, together with the apparent deficit of carbonyl carbon suggests that PMDA decomposes on the surface with release of one or two CO molecules. The oxygen 1s emission in Fig. 6a and b centered at E_B = 531.5 eV is consistent with hydroxyl groups adsorbed on the surface. Formation of hydroxyl species could result from contamination during our deposition experiments, or by reaction of molecular or atomic hydrogen produced on the hot filaments in the preparation chamber with the PMDA fragments on the surface. From the C 1s spectra it is indicated, however, that PMDA in contact with the clean metal will dissociate.

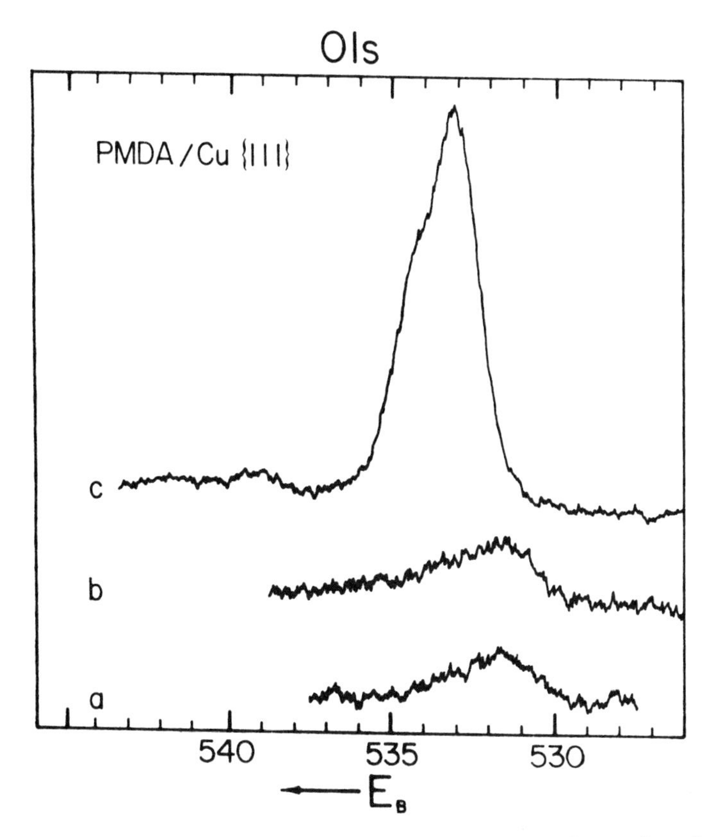

Fig. 6. XPS O 1s spectra of pyromellitic anhydride (PMDA) on Cu{111}. **a:** L = 30 (298 K), **b:** L = 2040 (298 K), **c:** L = 360, substrate temperature 200 K.

3.2 CODEPOSITION OF ODA AND PMDA

The XPS spectra for thin ($d > \Gamma$) and ultra-thin ($d \leq \Gamma$) films of PMDA and ODA on polycrystalline silver substrates have been presented and discussed in detail previously.[13,19] Deconvolution of the C 1s, O 1s and N 1s bands into the spectral features representative for the different functional groups in the polyamic acid and polyimide chains demonstrated that the resulting polymer films after curing do not consist of pure polyimide exclusively, but must contain a substantial fraction of terminal groups of polyimide chains and/or unreacted ODA molecules. The details of these experiments and discussion are given in Refs. 13 and 19. Here we only show the general features of the codeposition and imidization process again for a thin film (d > 40 Å) of ODA and PMDA codeposited at room temperature onto a polycrystalline copper substrate.

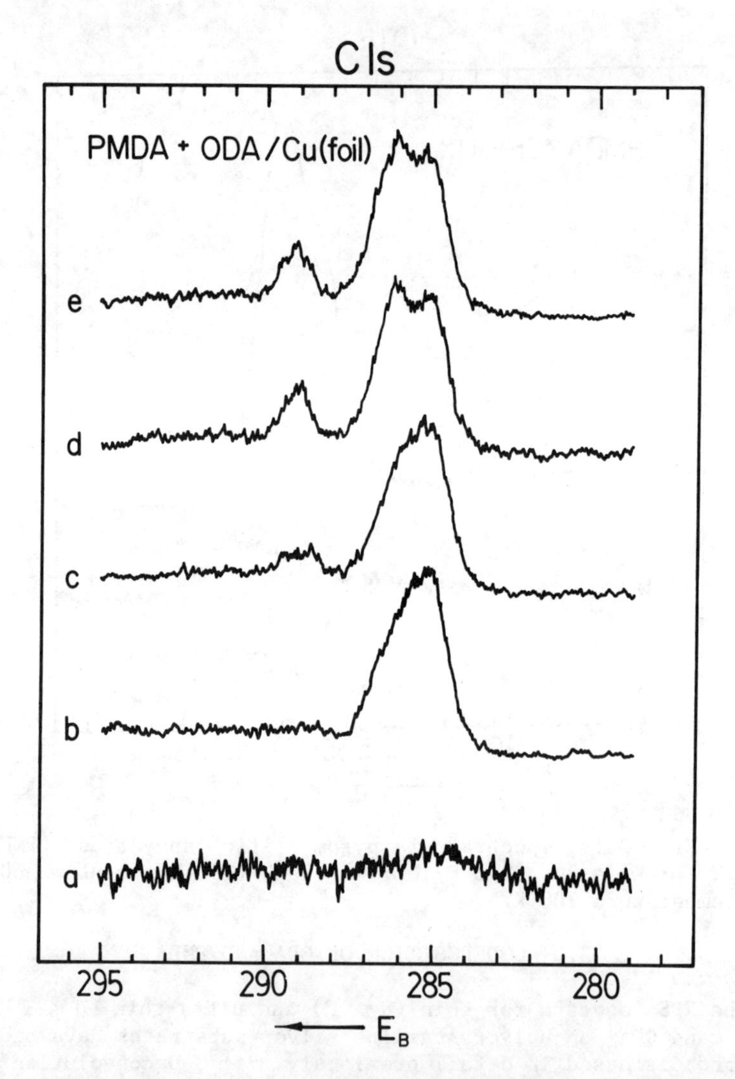

Fig. 7. XPS C 1s spectra for the coadsorption of ODA and PMDA on polycrystalline copper and subsequent heating to form polyimide. **a:** clean copper foil, **b:** L = 120 exposure of PMDA and ODA at 298 K, **c:** after heating the substrate for 5 min at 423 K, **d:** heating for 10 min at 573 K, **e:** heating for 10 hours at 573 K. Trace a is magnified by a factor of 2.

The C 1s, O 1s and N 1s spectra Figs. 7b, 8b and 9b are representative of polyamic acid formed by reaction of PMDA and ODA in the adsorbed film. The particulars have been discussed elsewhere,[13] here we only point out again (i) the absence of a defined carbonyl (C 2) emission, near 290 eV, due to the fact that carbonyl carbons are in a variety of chemical environments, (ii) the pronounced high binding energy shoulder on the O 1s peak attributed to oxygen atoms in

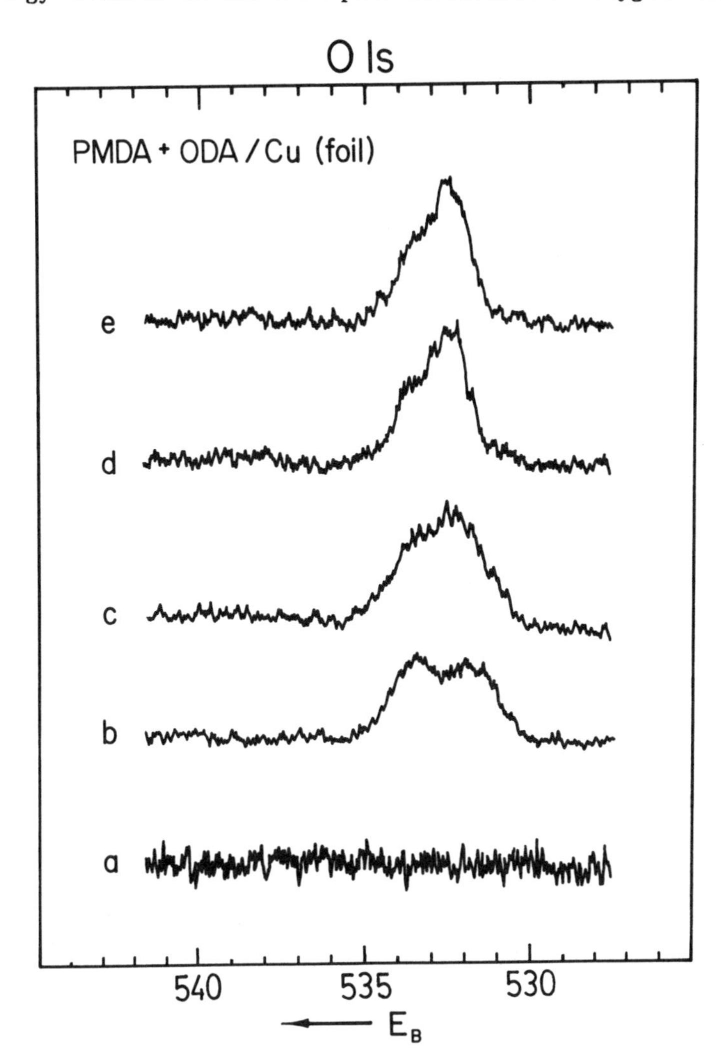

Fig. 8. XPS O 1s spectra for the coadsorption of ODA and PMDA on polycrystalline copper and subsequent heating to form polyimide. Traces a-e are defined in Fig. 7. Trace a is magnified by a factor of 2.

hydroxyl groups in polyamic acid and (iii) the broad N 1s emission indicating a chemical reaction of the ODA amino groups. The stoichiometry of the codeposited layer is C:O:N = 22:4.7:3.2 showing an excess of nitrogen and an apparent deficit of oxygen. In polyamic acid, the stoichiometry is expected to be C:O:N = 22:7:2. An excess of ODA is consistent with the experimental data, since the higher carbon content of ODA with respect to PMDA will give an apparent lower value to the oxygen stoichiometry.

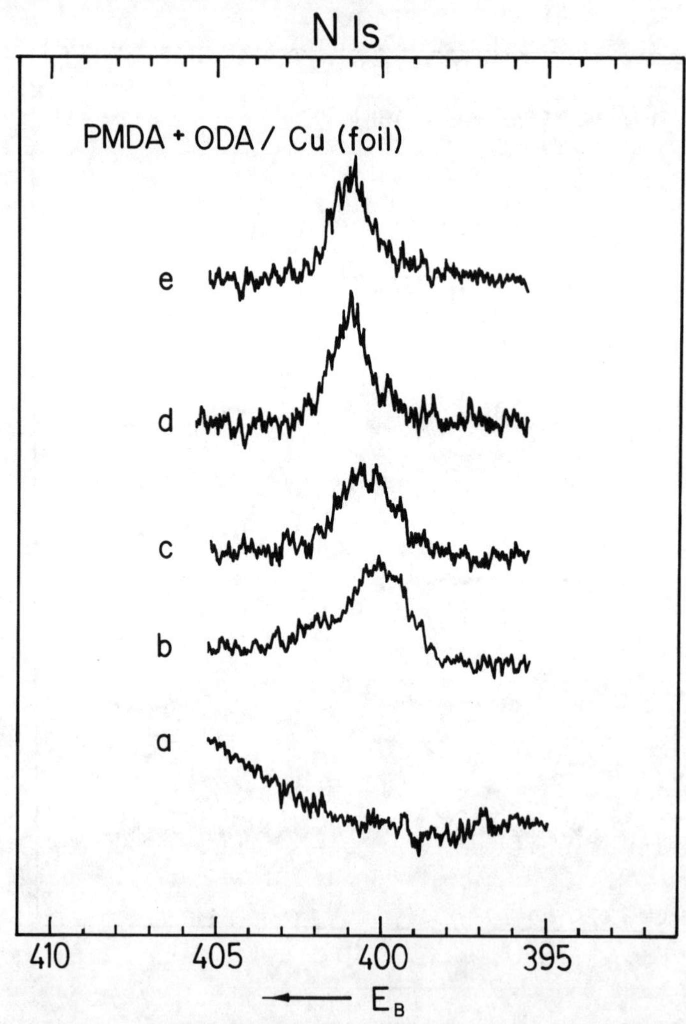

Fig. 9. XPS N 1s spectra for the adsorption of ODA and PMDA on copper and subsequent heating to form polyimide. Traces a-e are defined in Fig. 7.

Heating the film to 423 K for 5 minutes leads to changes in the XPS spectra similar to those observed on silver[13,19]. In particular, the hydroxyl O 1s emission at E_B = 534 eV decreases relative to O 1s emission at the lower binding energies and a carbonyl C 1s emission is resolved again above the background. In the N 1s emission, a shift to higher binding energies is observed indicating the onset of the imidization reaction.[19] Heating to 573 K for 10 minutes then completes the imidization reaction and results in spectra identical to those for a thick polyimide film formed on silver.[13] In analogy with silver, we conclude that this film even after curing for 10 hours at 573 K does not consist entirely of polyimide. This is supported by the slight deviation of the measured stoichiometry (C:O:N = 22:4.1:1.75) after curing from that of polyimide (22:5:2). In addition, the persistence of a low binding energy tail on the N 1s band indicates the presence of amino terminal groups or perhaps unreacted ODA in the polymer film.

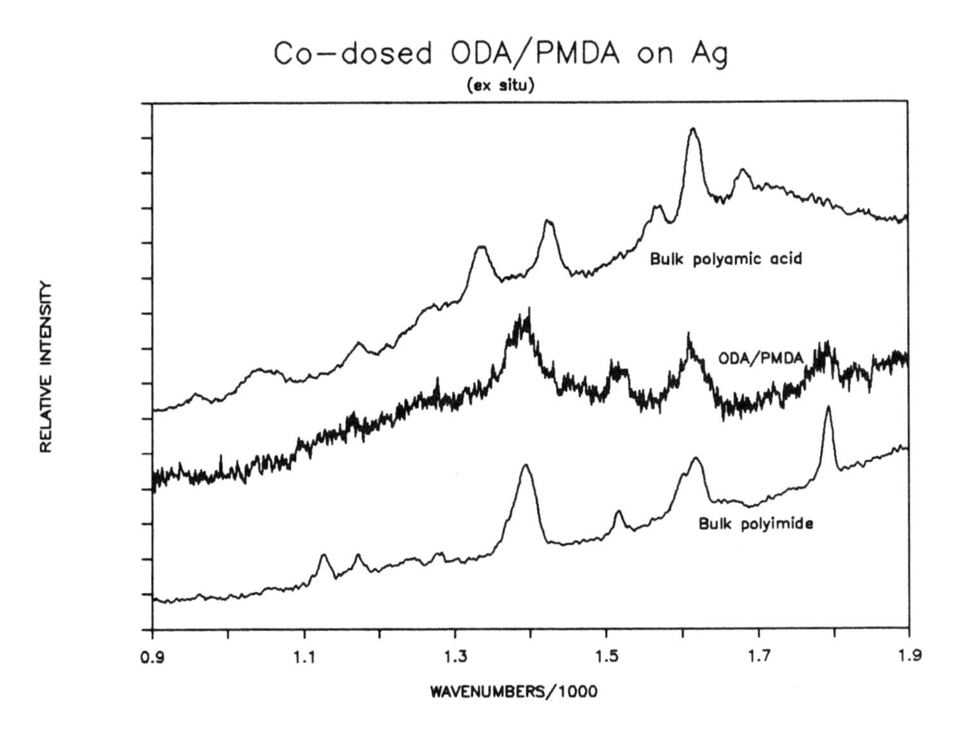

Fig. 10. Raman spectra of polymeric species. The top trace shows a spectra of bulk polyamic acid. The center trace shows a vapor deposited and cured film of about 200 Å thickness on polycrystalline silver. The bottom trace shows the Raman spectrum of a thick polyimide film prepared by spin-coating techniques.

The XPS data shown demonstrate that the core-level photoemission data can be used to follow the deposition and imidization process and can be used to estimate the composition of the final cured films. That these films prepared by chemical vapor deposition indeed are primarily polyimide, follows also from the Raman spectra shown in Fig. 10. Figure 10 compares the Raman spectra of bulk polyamic acid and a spun-on polyimide film with a spectrum obtained from a vapor deposited polyimide film on a polycrystalline silver substrate.[19] This film was cured at 673 K and estimated to be about 200 Å thick based on relative exposure values. The detailed description of these experiments will be given elsewhere, but it is obvious that the vibrational spectrum of the vapor deposited film resembles closely the one for the film prepared by spin coating techniques. However, the vibrational bands in the vapor deposited film (in particular the carbonyl stretching band at 1800 cm^{-1}) are broader. We attribute this to inhomogeneity in the film due to both the uncontrolled stoichiometry of the vapor fluxes and the contribution of the metal/polymer interfacial region.

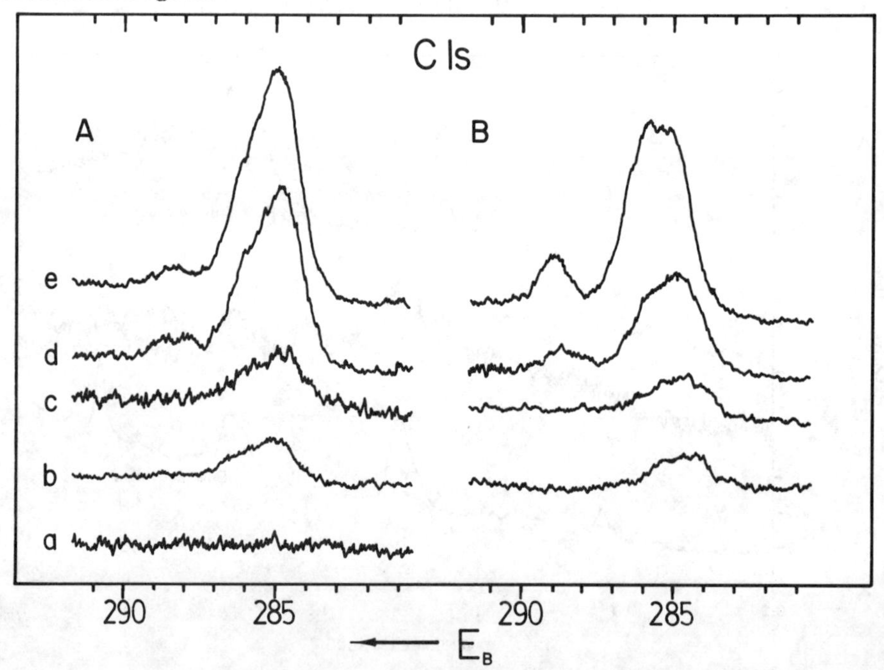

Fig. 11. XPS C 1s spectra of PMDA and ODA codeposited on polycrystalline copper. The spectra on the left (section A) are after deposition at 298 K and the spectra on the right (section B) are after curing the deposit for two hours at 473 K. **a:** clean surface, **b:** after L = 20 exposure of the two monomers, **c:** L = 60, **d:** L = 300, **e:** L = 600. Traces b and c part B were taken at a reduced scan speed (2 eV/min) and with an increased time constant (2 s).

We finally want to address the minimum possible thickness of polyimide films prepared by vapor deposition, and the chemical composition of the polymer/metal interface. To study these questions, we performed a series of experiments where polyamic acid films of different initial thicknesses were prepared at room temperature (Figs. 11A, 12A, 13A) and subsequently heated for 2 hours to 473 K to form the polyimide (Figs. 11B, 12B, 13B). Representative C 1s, O 1s and N 1s spectra for four different coexposures and thus different initial film thicknesses and the respective spectra after heating are shown

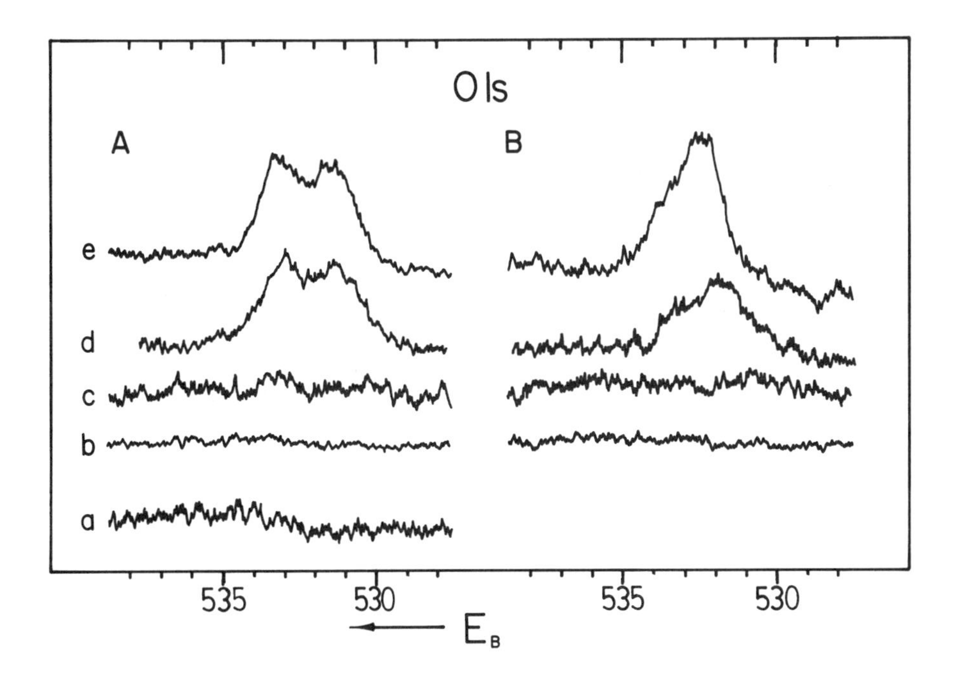

Fig. 12. XPS O 1s spectra of PMDA and ODA codeposited on polycrystalline copper. The traces are defined in Figure 11. Traces b were taken at a reduced scan speed (2 eV/minute) and an increased time constant (2 s).

as curves b, c, d and e in Figs. 11-13. A compilation of the stoichiometries of the different films before and after heating is given in Fig. 14. Spectra b in Figs. 11-13 refer to a very brief coexposure (film thickness ~ 3 Å). Only carbon and some nitrogen can be detected after the initial room temperature exposure, indicating that the layer must consist primarily of ODA. Since our fluxes are not calibrated, this result might reflect an excess of ODA in the flux and/or preferential adsorption of ODA on the polycrystalline substrate. Heating leads to a shift of the C 1s emission to lower binding energies, indicating some decomposition of the film. The somewhat thicker codeposit (spectra c) gives a similar result. A codeposit of 31 Å thickness (d), however, results in C 1s, O 1s and N 1s spectra similar to those expected for polyamic acid.[20]

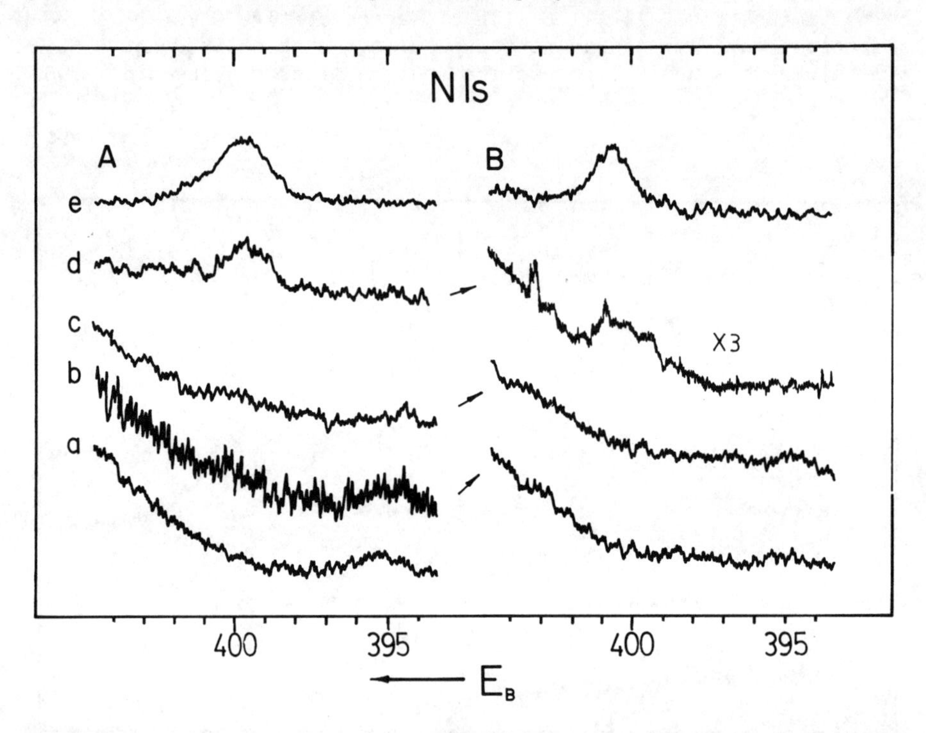

Fig. 13. XPS N1s spectra of PMDA and ODA codeposited on polycrystalline copper. The traces are defined in Fig. 11. In trace b part A the time constant was reduced from 2 s to 1 s.

A broad C 1s carbonyl emission is observed in Fig. 11d as compared to the spectrum in Fig. 7b. We did notice in all of our experiments, that the appearance of the carbonyl C 1s emission in the initial deposits on copper varies strongly, possibly due to different local chemical environments. This variation was not observed for initial deposits on silver. Heating this 31 Å deposit leads to the loss of material (the film thickness decreases to ~11 Å) and to changes in the XPS spectra indicating the presence of organic fragments. It is difficult to identify any spectral features which would unambiguously be defined as due to polyimide. The O 1s emission shows a low binding energy shoulder indicative of oxidic oxygen and, at higher energy, an unresolved doublet which resembles the O 1s from polyimide. Also, the N 1s emission is shifted to slightly higher binding energies and is indicative of imide nitrogen.

The series of spectra obtained for a deposit of initial thickness d > 70 Å (Figs. 11e, 12e, 13e) heated to 473 K for 1 hour clearly resembles the XPS data for thick polyimide films. The resulting average film thickness for these spectra is about 42 Å and the stoichiometry is consistent with that of the thick polyimide film shown in Figs. 7e, 8e and 9e. However, a closer inspection of the C 1s and O 1s lineshapes reveals that both have additional contributions at lower binding energies which are similar to those shown in Fig. 2-7 for initial depositions of pure ODA or PMDA leading to fragmentation.

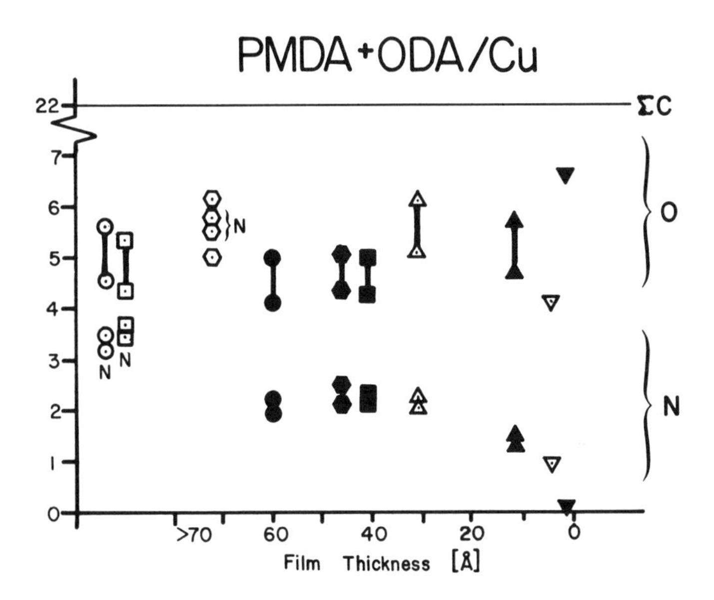

Fig. 14. The stoichiometry of codeposited layers of PMDA and ODA on polycrystalline copper as a function of film thickness reference to a total of 22 carbon atoms. The open symbols are before curing and the closed symbols are after curing. For d > 10 Å the upper and lower symbols connected by the bars refer to the thick film and thin film limit, respectively (see eq. 1). For d < 10 Å, only the stoichio-metry evaluated for the thin film limit is displayed.

We therefore conclude that the minimum thickness of a polyimide film which can be formed on polycrystalline copper is about 42 Å and that there is an intermediate layer at the polymer/metal interface which contains species similar to those observed for fragmented ODA and PMDA. As has been discussed in more detail elsewhere, this intermediate layer is stable at temperatures up to 673 K, whereas pure ODA or PMDA layers decompose into an amorphous carbon layer at these temperatures.[18] This suggests that the interfacial species are stabilized by the polyimide film, which in turn means that they must chemically interact with the polymer, and that adhesion of the polyimide to the metal substrate occurs through fragmented PMDA and/or ODA.

4. SUMMARY AND OUTLOOK

Polyimide films as thin as ~40 Å can be produced by chemical vapor deposition of PMDA and ODA on clean polycrystalline copper substrates. Even thinner films (d ~ 11 Å) were obtained on silver substrates as reported elsewhere.[19] Separate adsorption experiments for the pure constituents show that both PMDA and ODA fragment on clean copper surfaces at room temperature; only at temperatures less than 250 K is condensation to form thick films observed. Decomposition of the constituents is also observed in codeposition experiments. Only for sufficiently thick codeposited layers does imidization of the polyamic acid to polyimide occur concurrently with fragmentation at the interface. Adhesion of the polyimide film to the substrate must thus occur via chemical interaction with the fragmented layer.

Our experiments show that chemical bonding is involved in adhesion at the polyimide-metal interface for polyimide films prepared by vapor deposition. The experiments also demonstrate that these films can be made thin enough so that surface sensitive analytical techniques can be used to study their interfaces. This opens possibilities for systematic study of other aspects of microscopic interface chemistry. From a practical viewpoint, we can now begin to study the behavior of interface specific materials such as adhesion promoters.

ACKNOWLEDGEMENTS

This work was supported in part by the Office of Naval Research and the National Science Foundation through grant No. DMR-8403831. We thank R. Mack, C. Carlin and M. Cook for obtaining the Raman spectra.

REFERENCES

1. P.O. Hahn, G.W. Rubloff and P.S. Ho, J. Vac. Sci. Technol. A, 2, 756 (1984).
2. J.R. Salem, F.O. Sequeda, J. Duran, W.Y. Lee and R.M. Yang, J. Vac. Sci. Technol. A, 4, 369 (1986).
3. K.L. Mittal in "Microscopic Aspects of Adhesion and Lubrication," Tribology Series Vol. 7, Ed. J.M. Georges, Pub. Elsevier, Amsterdam (1982).
4. N.J. Chou and C.H. Tang, J. Vac. Sci. Technol. A, 2, 751 (1984).
5. J.L. Jordan, P.N. Sanda, J.F. Morar, C.A. Kovac, F.J. Himpsel and R.A. Pollak, J. Vac. Sci. Technol. A, 4, 1046 (1986).
6. P.N. Sanda, J.W. Bartha, J.G. Clabes, J.L. Jordan, C. Feger,

B.D. Silverman and P.S. Ho, J. Vac. Sci. Technol. A, 4, 1035 (1986).
7. P.S. Ho, P.O. Hahn, J.W. Bartha, G.W. Rubloff, F.K. LeGoues and B.D. Silverman, J. Vac. Sci. Technol. A, 3, 739 (1985).
8. F.S. Ohuchi and S.O. Freilich, J. Vac. Sci. Technol. A, 4, 1039 (1986).
9. Y.-H. Kim, G.F. Walker, J. Kim and J. Park, International Conference on Metallurgical Coatings (San Diego) p. 23-27, (1987).
10. J.L. Jordan, C.A. Kovac, J.F. Morar and R.A. Pollak Phys. Rev. B, July 15 (1987).
11. N.J. DiNardo, J.E. Demuth and T.C. Clarke, J. Chem Phys., 85, 6739 (1986).
12. S.P. Kowalczyk, Y.H. Kim, G.F. Walker, J. Kim, preprint.
13. R.N. Lamb, J. Baxter, M. Grunze, C.W. Kong and W.N. Unertl, Langmuir (in press) (1987).
14. J.H. Schofield, J. Electron. Spectros., 8, 129 (1976).
15. D.T. Clark in "Chemistry and Physics of Solid Surfaces," Vol. 11, Ed. R. Vanselow, CRC Press, Boca Raton (1979).
16. M.P. Seah and W.A. Dench, Surf. Interface Analy., 1, 2 (1979).
17. P.M. Cotts, "Polyimide - Synthesis, Characterization and Application," Vol. 1, Ed. K.L. Mittal, Plenum Press (1984).
18. M. Grunze, J. Baxter, C.W. Kong, W.N. Unertl and C.R. Brundle, in preparation.
19. R.N. Lamb and M. Grunze, J. Chem. Phys., submitted.
20. H.J. Leary and D.S. Campbell in "Photo, Electron and Ion Probes of Polymer Structure and Properties," ACS Symp. Ser., 162, 419 (1981).

SURFACE CHEMISTRIES AND ELECTRONIC PROPERTIES OF MOLECULAR SEMICONDUCTOR THIN FILMS GROWN BY EFFUSION BEAMS

P. Lee, J. Pankow, J. Danziger, K.W. Nebesny, and N.R. Armstrong
Department of Chemistry, University of Arizona, Tucson, Az. 85721

ABSTRACT

Ultra-high vacuum deposition methods are described for the production of molecular semiconductor (MS) thin films produced from dyes known as phthalocyanines (Pc's). Effusion sources allow for "Pc beams" to be directed to a combination of interdigitated microelectrode arrays to measure the conductivity and photoconductivity during and after the growth of the MS thin film, and quartz-crystal microbalances (QCM's) to monitor film growth and reactivities toward various dopants. Molecular dopants can be introduced by either leak valves or by other effusion beams to tailor the properties of the MS thin film.

INTRODUCTION

There is interest in the semiconductor properties of molecular thin films such as those which can be produced from a class of molecules known as the phthalocyanines (Figure 1). These molecules have good chemical durability and low vapor pressure, which makes them amenable to deposition with technologies already in place for inorganic semiconductors.

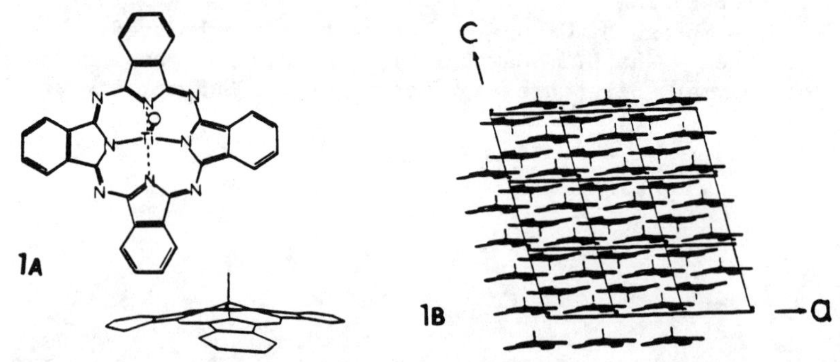

Figure 1 - TiOPc

In the solid state they pack in extended molecular stacks (Figure 1b) which leads to charge delocalization and to broadening of their visible absorbance spectra into the near infra-red region.[1-8] The rules which govern semiconductor properties of molecular systems are still debated.[1,2,4] Ordering of the solid state and extensive delocalization of charge seem to be essential for reasonable semiconductor properties. High contrast between photoconductivity and dark conductivity is a key requirement as well for solar cell applications. Pc's have also proven to be extremely sensitive to the uptake of adventitious dopants.[1,2,5,7] In photovoltaic and photoelectrochemical cells, rectification effects and photovoltages arise from doped interfacial regions of the Pc thin film which may be 1000 Å or less in thickness. Charge carriers generated in thicker films (necessary for high light absorption) are wasted in the energy conversion process. To optimize these materials for energy conversion applications, as well as for their use as MS-chemical sensors, molecular transistors, or NLO materials, etc.,[1-8] it is necessary to assume complete control over their synthesis and deposition processes as thin films, and to optimize the dark and photoconductivities and I/V behavior during deposition.

This paper focuses on vacuum deposition of Pc's and related molecules, the sensitivity of these materials to uptake of dopants such as I_2 or TCNQ (tetracyanoquinodimethane) during or after the deposition process, and the technologies necessary to controllably introduce other molecular materials in codeposition or sequential deposition formats. Our focus is on Pc's which contain a $(Ga\text{-}Cl^{+2})$ ion or a (TiO^{+2}) ion in the center of the ring (GaPc-Cl or TiOPc (Figure 1)), and stack in configurations where the separation between molecular units is ca. 3.25-3.8Å.

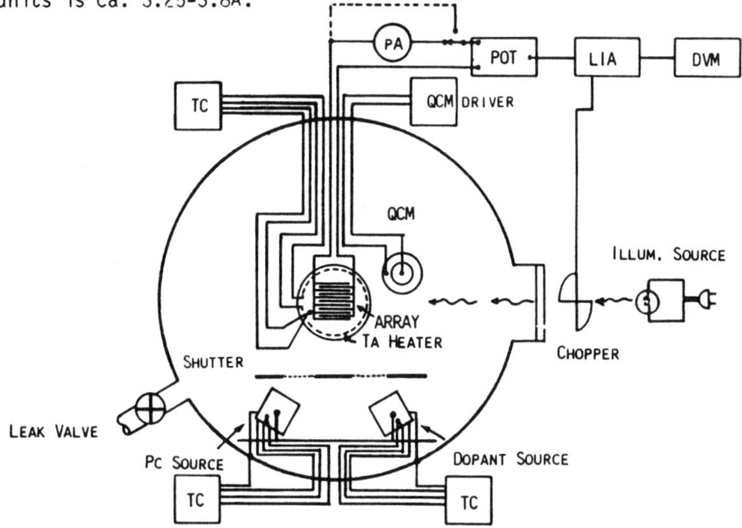

Figure 2 - Deposition chamber and associated measurement hardware

EFFUSION SOURCE AND CHARACTERIZATION METHODOLOGIES

Figure 2 is a schematic of the vacuum system to deposit Pc-MS thin films. Similar systems have been recently described with different end goals.[10] Two types of effusion sources were used in these studies, both of which allowed for the deposition of sublimable molecular systems at rates of less than one monolayer per minute (or faster if desired). The first source used, which produced the TiOPc films for this study, consisted of an aluminum cylinder, ca. 2.5 cm diameter and 3 cm high which was machined to give a small reservoir in the base and loaded with a few milligrams of the Pc solid powder. Above this reservoir was a tortuous pathway which forced the heated Pc vapor to rise, then fall and then effuse through a 3 millimeter orifice toward the substrate held ca. 5 cm away. Resistive heating of the entire aluminum cylinder was used to cause effusion of the Pc. This assembly was mounted at the base of the chamber. During the pump down cycles, this source was heated to a temperature just below the sublimation temperature of the Pc, so as to further remove lower molecular weight volatile impurities that might remain. Deposition temperatures were measured by means of a thermocouple placed in direct contact with the Pc reservoir. Deposition generally occurs in the range from 365 C - 420C at the effusion source. More recent designs for the experiments described later with GaPc-Cl use a simpler source, where the Pc powder was placed in a small (ca. $1cm^3$) boron nitride crucible set into a stainless steel cylinder. The cylinder was also resistively heated. The cylinder was closed off with a cap which allowed for the use of variable apertures to be placed between the crucible and the lid, which then defined the diameter of the Pc beam which emerged from the source. A similar source is used for all of the dopants which can be sublimed, such as TCNQ. Low vapor pressure Pc's require an aperture of ca. 1 millimeter diameter. More volatile dopants require apertures with smaller diameters. The assembly is mounted on a single vacuum flange at the base of the chamber, and three sources can be run independently or in combination. Substrates are suspended ca. 5-10 cm above the effusion sources and the divergence of each source is such that it is possible to evenly coat (less than 10% nonuniformity) substrates with diameters of ca. 5 cm. The effusion sources are separated from each other, such that no detectable "cross-talk" is observed when two or more sources are running simultaneously (as determined by the response of the QCM). This is facilitated by erecting cryoshrouded walls in the vacuum chamber in the region of the effusion sources. Base pressures in this chamber of ca. 5×10^{-10} torr are possible after bakeout, with operating pressures of ca. 1×10^{-9} torr (depending upon the molecules being deposited).

Microcircuits (supplied by either Motorola or Burr-Brown), consist of 59 Au fingers, interdigitated to be 3 microns apart, with a 3 micron finger width, are mounted in the center of the deposition chamber. Accompanying this circuit are a 10 MHz quartz crystal microbalance, and appropriate substrates for producing MS-Pc films on thin Au foils for subsequent spectroscopic photoelectrochemical, or photovoltaic characterization. The microcircuit is illuminated during and after the deposition. Photoconductivities and dark conductivities are measured with the microcircuit under bias (0.5 to 3 volts) with a picoammeter in series with the microcircuit. Occasionally the light source is modulated, and the current through the microcircuit converted to a voltage and demodulated through a lock-in-amplifier.

PHOTOCURRENT YIELD SPECTRA AND ENERGETICS OF CONDUCTION AND PHOTOCONDUCTION

Molecular thin films can exist in a number of different microcrystallographic orientations which are revealed by differences in visible or infrared absorbance spectra, and by differences in photoactivity at different wavelengths. Figure 3 shows the photocurrent yield spectrum for a TiOPc thin film ($x < 50$ nm) on the microcircuit, as deposited. Heat of condensation caused the microcircuit temperature to rise to about 33°C during the deposition. Also shown is the photocurrent yield spectrum after the first activation energy study during which the substrate temperature was raised to about 95°C via resistive heating. The absorbance spectrum of an annealed thin film is also included. The photocurrent action spectrum before annealing on the microcircuit is different than for the annealed films, with strong red-shifting of both the absorbance and photocurrent yield spectra after annealing, and the appearance of a new band in the region of 900 nm. The process of annealing increases the photocurrent yield at all wavelengths. In the 900 nm region, the yield of charge carriers increases by a factor of 100. The substrate temperature during and after the growth of the thin film is extremely important in determining the orientation and extent of long-range ordering of the Pc molecules --and strongly influences the overall photoactivity.[7] The change in the spectral response corresponds to a change in crystallographic stacking of the molecules into at least one form which is more photoconductive than the as-deposited form. Results of studies on trivalent metal Pc's indicate that this transition may be from a perfectly cofacial molecular stack[7c,11] to one where the molecules are staggered, as shown in Figure 1.

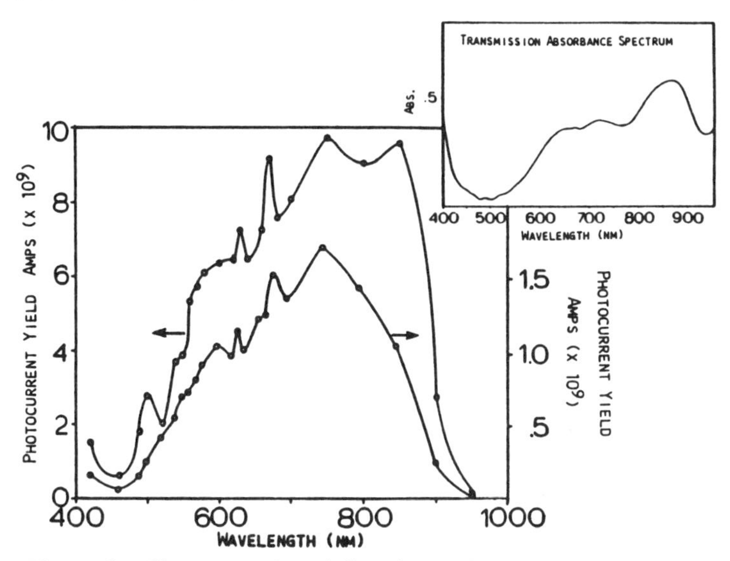

Figure 3 - Photocurrent and Absorbance Spectra

The dark conductivities of UHV-deposited TiOPc (Table I) place it in the range of insulators. Its photoconductivity is at least two orders of magnitude higher (depending upon illumination wavelength and photon flux) and can be as high as 10^{-6} ohm^{-1} cm^{-1} for photoelectrochemically examined thin films.[7b] The dark conductivity and photoconductivity are both sensitive to the presence of adsorbed impurities from the atmosphere, and can be several orders of magnitude higher in atmosphere than in the UHV grown and characterized materials. The activation energy for dark conduction in UHV grown thin films was ca. 1.9 eV before annealing and 1.4 eV after annealing. Activation energies for photoconduction were lower before annealing at most wavelengths, but were lowered to ca. 0.2 eV after annealing. Plots of $\ell n\ (i_{ph})$ vs. 1/T had three distinct slopes between 25° and 95°C before the films were annealed. Such breaks in slope are typical where the photoconduction process is controlled by innergap traps or defect levels.[2] Increases in temperature populate these traps or exhaust defect levels and change the mechanism for photoconduction. After restructuring of the Pc film, there is a single trap level which dominates the photoconduction mechanism about 0.2 eV below the conduction band edge[2] for the entire temperature regime examined.

TABLE I - DARK AND PHOTOCONDUCTIVITIES OF TiOPc AND GaPc-Cl

	$\sigma_d(\Omega^{-1}cm^{-1})$	$\sigma^*_{ph}(\Omega^{-1}cm^{-1}$	$E_{a(d)}$(eV)	$E_{a(ph)}$ a	(eV) b
TiOPc	7.1×10^{-11}	1.9×10^{-9} (25m Watts, 632.8 nm)	1.9[a] 1.4[b]	0.8[c] 0.8[d] 1.0[e] 3.0[f]	0.17 0.16
GaPc-Cl[1]	7.0×10^{-8}	3.7×10^{-8} (630 nm)	0.6	0.1	0.15
GaPc-Cl[2]	1.3×10^{-11}	4.4×10^{-10} (5mW 632.6 nm)	~3	0.56 (5mW 632.8 nm)	

(a) As-deposited film
(b) After 95°C anneal.
(c) 660 nm (d) 750 nm (e) 850 nm (f) 950 nm

1) No circuit treatment (film grown at room temperature)
2) Circuit annealed during pumpdown and bakeout (film grown at 70°C)

Figure 4 shows the change in room temperature photocurrent yield of another TiOPc/ microcircuit thin film (grown at a circuit temperature of about 33°C) as a function of exposure to an electron acceptor, I_2, which is known to dope Pc's to make them good dark conductors.[12] A 25 mW HeNe laser was used as the illumination source. (Exposures carried out by bringing I_2 vapor into the entire vacuum system through a leak valve.) There is a substantial change in the photocurrent response with exposures to I_2 which correspond to submonolayer coverages. The photocurrent yield actually drops with I_2 exposure, in contrast to the expected increase. The absolute change in photocurrent response is 88% of the total photoresponse, and demonstrates how significant a role is played by the photogeneration of charge in the near surface region. Saturation of the effect occurs with a 400 L exposure. (400 seconds at $P_{I_2}=10^{-6}$ torr). Replotting of the response as $L_{I_2}/\Delta i_{ph}$ demonstrated equilibrium adsorption and led to computation of the chemisorption equilibrium constant, $K = 0.149L^{-1}$. The influence of I_2 vapor on the photoconductive properties of the TiOPc thin film indicates how sensitive these materials can be to adsorbates which interact with the organic film in a charge transfer process in the near surface region. Although I_2 is not normally found as an impurity in vacuum systems, other weak oxidants such as O_2, when introduced into this system cause a similar decrease in the photoconductive response of the TiOPc system. Much higher doses of O_2 were required to achieve the same decrease in photoconductivity as the I_2 caused. Interestingly, when the Pc thin films are exposed to atmosphere after their preparation in UHV, they actually show an increase in both dark and photoconductivity. Thus, low concentrations of atmospheric impurities in a vacuum system can have a profound effect on the electrical properties of Pc thin films, and atmospheric pressure dopants can produce quite a different response. The mechanisms for these effects are under exploration. Deposition of molecules at background pressures of ca. 10^{-6} torr may not be low enough to produce semiconductor grade materials. Weak oxidants may act to dissociate charges from the Pc aggregates, making good dark conductors. Others may serve to assist exciton dissociation, enhancing photoconductivity (see below).

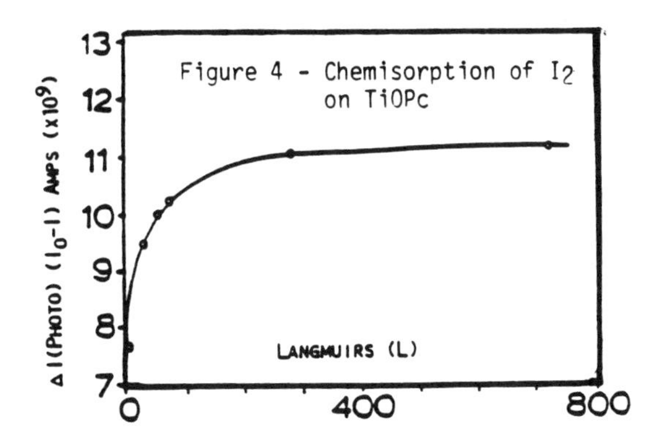

Figure 5 is the photocurrent yield spectrum for a UHV-deposited thin GaPc-Cl film. The dashed line represents the most probable form of the complete spectrum, which mimics thin film absorbance spectra of this material. At certain wavelengths there was enhancement of the photocurrent, which we believe to be due to internal reflection artifacts within the microcircuit channels, coated with the Pc film. Unlike the TiOPc films, GaPc-Cl deposits in a form which is difficult to interconvert with annealing at temperatures below 100°C, but which we know exists in at least two phases with different photoactivity.[7,11] The most red-absorbing phase of GaPc-Cl is more photoactive (as in TiOPc) and is the form which we attempt to maximize during deposition processes. True epitaxy of this material appears possible, based upon previous deposition results for similar Pc's.[13] GaPc-Cl films are also sensitive to adsorbed dopants (see below). The dark conductivity of GaPc-Cl (1) is higher than the TiOPc thin films (Table I), and high contrast between dark and photoconductivities have been observed in our laboratory $\sigma^*_{ph}/\sigma_{dark} > 500$. XPS studies have demonstrated a decreased affinity of GaPc-Cl for O_2 as opposed to TiOPc and other transition metal Pc's.[7]

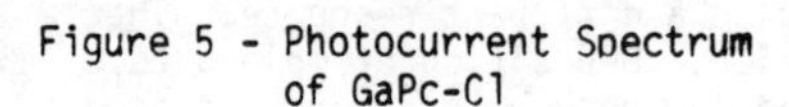

Figure 5 - Photocurrent Spectrum of GaPc-Cl

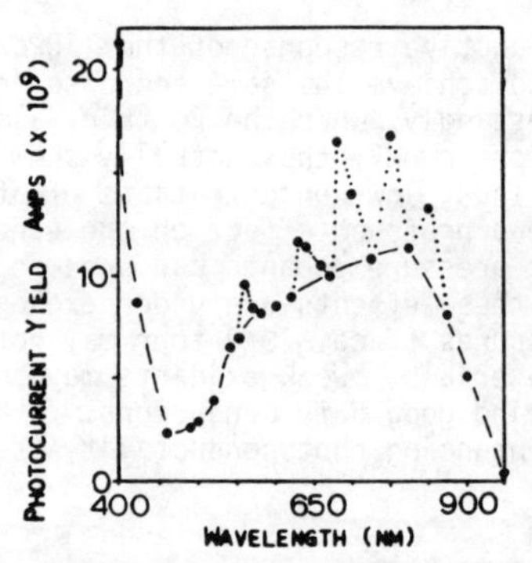

The dark and photoconductivities of GaPc-Cl thin films have been found to be sensitive to the pretreatment of the microcircuit substrate. For films deposited on a microcircuit simply placed in vacuum at room temperatue, the dark and photoconductivities were several orders of magnitude higher (ClGaPC (1)) than when the microcircuit was annealed during pumpdown and bakeout and held at ca. 50°C during Pc deposition (Table 1). The unusually high dark current in GaPC-Cl (1) was probably due to a small short across two or more circuit fingers. The variability of conductivity values obtained in GaPc-Cl (1) and (2) demonstrate the need for careful circuit cleaning prior to vacuum use and annealing during bakeout cycles. High circuit temperatures during deposition (as in GaPc-Cl (2)) also appears to rid the circuit of dopant impurities.

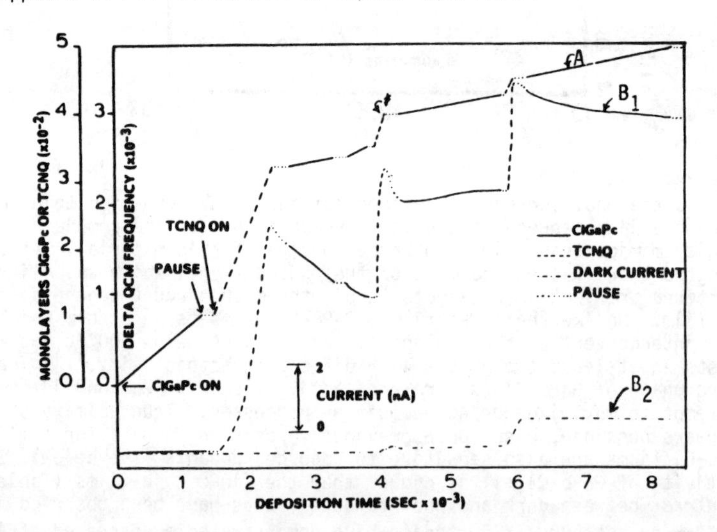

Figure 6 - Sequential Deposition of GaPc-Cl and TCNQ on a room temperature and quartz-crystal microbalance microcircuit.

A) QCM Mass uptake plots showing the equivalent monolayer thickness of GaPC-Cl and TCNQ condensed on the microcircuit.

B1) Photocurrent yield of microcircuit during GaPc-Cl and TCNQ deposition.

B2) Dark current yield during the same sequence.

SEQUENTIAL OR CODEPOSITION OF PHTHALOCYANINES WITH OTHER MOLECULAR DOPANTS

TCNQ is known to act as a dopant for many molecular semiconductor materials.[1,12,14] Its electron withdrawing properties have also made it of interest for the charge transfer salts that it forms with other molecules such as tetrathiofulvalene (TTF) and related molecules.[14] It is a good candidate for use in a sequential or codeposition technology with Pc's. Figure 6a shows recent experiments where GaPc-Cl and TCNQ were deposited sequentially on both the microcircuit, QCM and thin Au foils all at or near room temperature. These plots show the excellent discrimination between the two sources during the deposition process, and show the likelihood of producing spatially well-defined molecular layers. With the initial results reported here each layer is disordered and there is some mixing of the interfaces. Because of the extremely thin films produced, the entire assembly approaches the composition of a codeposited film.

The initial deposition of the Pc layer proceeds until a reasonable photocurrent (at 632.8 nm) is achieved with a contrast between light and dark of at least 10:1 (Figure 6b). Following this, the TCNQ beam is opened to the circuit and the photocurrent activity increases dramatically during the initial stages of TCNQ deposition (σ^*_{ph}/σ_d > 6000:1 through the addition of a few equivalent monolayers of TCNQ). After the TCNQ deposition is halted, the photocurrent yield appears to decrease, but is not due to evaporative loss of TCNQ, as shown by the stable QCM frequency. Subsequent Pc cap layers also decrease the photocurrent yield, which may be due to interdiffusion of the two layers, and/or displacement of TCNQ monomers or clusters from critical charge generation sites. New TCNQ deposition reenhances photoconductivity. Initial experiments with low coverages of TCNQ showed little change on the photocurrent yield spectrum of Figure 5, but it should be pointed out that the near-IR region was not explored. Subsequent experiments with higher dosages of TNCQ have shown that considerable changes have occurred to both the thin film visible absorbance spectra, and to the photocurrent yield spectra, extending the photoactivity over a much broader spectral region (greater activity in the 400 - 700 nm region). The results of the photocurrent spectra for low coverages of TCNQ argue for the fact that the increased photoactivity is due to a true dopant effect and not to the presence of a new photoactive material (the TCNQ layers), which absorb light in the same wavelength region as the Pc film. The major enhancement of photoconductivity occurred at the lowest coverages of TCNQ, which is consistent with the idea that the near surface region is predominantly responsible for photoconduction processes. Electron microscopic characterization shows that the surfaces of the as-deposited Pc film are rough (average crystallite diameter < 0.1μ) with a corresponding high surface area to volume ratio. Chemisorption of TCNQ takes place beyond one geometric equivalent monolayer before saturation of all available surface sites occurs. Beyond a critical coverage of TCNQ (# in Figure 6), the dark conductivity increases dramatically. If TCNQ deposition is continued much beyond that point (not shown), the dark conductivity increases beyond the photoconductivity, as is expected if a charge transfer salt of Pc-TCNQ were formed. The fact that the photoconductivity can be enhanced without influencing the dark conductivity (at low TCNQ coverages) is encouraging and argues for the formation of TCNQ-Pc clusters at the Pc surface which may act as exciton dissociation sites. In the dark, these sites are not in electrical contact with each other (regions of insulating Pc spearate these TCNQ doped regions). Upon illumination, the photoconductive regions of the Pc are able to connect these clustered areas, which enhance the overall photoconductivity. Recent photoelectrochemical studies of these TCNQ-doped GaPc-Cl films have shown that the TCNQ layers are accessible to electrolyte and are redox active. The TCNQ-doped material also shows an enhanced open-circuit photopotential vs. the undoped film (about 40% increase to 0.7 volts) but does not show an improved short circuit photo current.

The challenge for future deposition schemes which strive to optimize photoconductivity is clear: The dopant must be continuously incorporated into the film as it is grown, probably in a gradient fashion. Photoconductivity (and perhaps dark conductivity) must be maximized in the regions which are farthest away from the interface which is illuminated, and the appropriate Schotky barriers must be created. More of the absorbed light may then produce harvestable charge and allow for enhanced efficiencies of solar cells. Such deposition schemes are also useful for the production of new MS-chemical sensors and non-linear optical materials (studies underway).

Achnowledgements: This research was supported by the National Science Foundation and by the Materials Characterization Program - State of Arizona.

REFERENCES

1. J. Simon and J.-J. Andre, "Molecular Semiconductors," Springer-Verlag, NY, 1985 pp. 73-148.

2. H. Meier, "Organic Semiconductors," Verlag Chemie, Berlin, 1974.

3. A. W. Snow, W. R. Barger, M. Klusty, H. Wohltjen and N. L. Jarvis, Langmuir, 2, 513 (1986).

4. M. Martin, J.-J. Andre and J. Simon, J. Appl. Phys. 54, 2792 (1983).

5. a) R. O. Loutfy, J. H. Sharp, C. K. Hsiao and R. Ho, J. Appl. Phys. 52, 5219 (1981).

   b) R. O. Loutfy, A. M. Hor and A. Rucklidge, J. Imaging Sci., 31, 31 (1987).

6. P. Leempoel, F.-R. F. Fan and A. J. Bard, J. Phys. Chem., 87, 2948 (1983).

7. a) T. J. Klofta, P. C. Rieke, C. A. Linkous, W. J. Buttner, A. Nanthakumar and T. Mewborn, J. Electrochem. Soc. 132, 2134 (1985).

   b) T. J. Klofta, J. Danziger, P. Lee, J. Pankow, K. W. Nebesny and N. R. Armstrong, J. Phys. Chem. 91, 000 (1987).

   c) T. J. Klofta, T. D. Sims, J. W. Pankow, J. Danziger, K. W. Nebesney and N. R. Armstrong, J. Phys. Chem 91,000 (1987).

8. Z. Z. Ho, C. Y. Ju, W. M. Hetherington III, J. Appl. Phys., 62(2), 716-18 (1987).

9. J. Simon, F. Tournilhac and J.-J. Andre, Nouv. J. Chem. 11, 383 (1987).

10. a) J. L.-Moigne and R. Even, J. Chem. Phys., **83**, 6472 (1985).

b) M. Maitrot, G. Guillaud, B. Boudjema, J. J. Andre, and J. Simon, J. Appl. Phys., **60**, 2396 (1986).

11. T.D. Sims, P. Lee, J. E. Pemberton and N. R. Armstrong, manuscript in preparation.

12. a) C. W. Dick, E. A. Mintz, K. F. Schoch, Jr., T. J. Marks, J. Macromol. Sci. Chem., **A16**, 275 (1981).

b) R. S. Nohr, and K. J. Wynne, J.C.S. Chem. Comm. **1981**, 1210.

13. a) E. Suito, N. Uyeda, M. Ashida and K. Yamamoto, Proc. Japan. Acad. **42**, 54 (1966).

b) N. Uyeda, M. Ashida, and E. Suito, J. Appl. Phys. **36**, 1453 (1965).

c) J. C. Buchholz and G. A. Somorjai, J. Chem. Phys. **66**, 573 (1977).

d) J. P. Biberian and G. A. Somorjai, J. Vac. Sci. Technol., **16**, 2073 (1979).

14. a) J. B. Torrance, Account. Chem. Res. **12**, 79 (1979).

b) A. F. Garito and A. J. Heeger, Account. Chem. Res., 7, 232 (1974).

9. [illegible] Johnson [illegible], [illegible] Chem., [illegible]

10. a) [illegible] L. McMillan, [illegible] Phys. [illegible] 33, [illegible]

b) [illegible], [illegible], B. [illegible], [illegible] and [illegible], Appl. Phys. [illegible]

11. [illegible], [illegible] Roberson and [illegible], [illegible] communication.

12. a) [illegible], [illegible], [illegible] Phys. [illegible] 175 [illegible]

b) [illegible] 1967, 10.

13. a) [illegible], M. [illegible] and [illegible]

b) [illegible], 1969.

c) [illegible] 26, [illegible]

d) [illegible]

e) [illegible]

14. [illegible] A. F. [illegible], [illegible] Chem. Res., 1, 239 [illegible]

Author Index

AIP Conference Proceedings

		L.C. Number	ISBN
No. 1	Feedback and Dynamic Control of Plasmas – 1970	70-141596	0-88318-100-2
No. 2	Particles and Fields – 1971 (Rochester)	71-184662	0-88318-101-0
No. 3	Thermal Expansion – 1971 (Corning)	72-76970	0-88318-102-9
No. 4	Superconductivity in *d*- and *f*-Band Metals (Rochester, 1971)	74-18879	0-88318-103-7
No. 5	Magnetism and Magnetic Materials – 1971 (2 parts) (Chicago)	59-2468	0-88318-104-5
No. 6	Particle Physics (Irvine, 1971)	72-81239	0-88318-105-3
No. 7	Exploring the History of Nuclear Physics – 1972	72-81883	0-88318-106-1
No. 8	Experimental Meson Spectroscopy –1972	72-88226	0-88318-107-X
No. 9	Cyclotrons – 1972 (Vancouver)	72-92798	0-88318-108-8
No. 10	Magnetism and Magnetic Materials – 1972	72-623469	0-88318-109-6
No. 11	Transport Phenomena – 1973 (Brown University Conference)	73-80682	0-88318-110-X
No. 12	Experiments on High Energy Particle Collisions – 1973 (Vanderbilt Conference)	73-81705	0-88318-111–8
No. 13	π-π Scattering – 1973 (Tallahassee Conference)	73-81704	0-88318-112-6
No. 14	Particles and Fields – 1973 (APS/DPF Berkeley)	73-91923	0-88318-113-4
No. 15	High Energy Collisions – 1973 (Stony Brook)	73-92324	0-88318-114-2
No. 16	Causality and Physical Theories (Wayne State University, 1973)	73-93420	0-88318-115-0
No. 17	Thermal Expansion – 1973 (Lake of the Ozarks)	73-94415	0-88318-116-9
No. 18	Magnetism and Magnetic Materials – 1973 (2 parts) (Boston)	59-2468	0-88318-117-7
No. 19	Physics and the Energy Problem – 1974 (APS Chicago)	73-94416	0-88318-118-5
No. 20	Tetrahedrally Bonded Amorphous Semiconductors (Yorktown Heights, 1974)	74-80145	0-88318-119-3
No. 21	Experimental Meson Spectroscopy – 1974 (Boston)	74-82628	0-88318-120-7
No. 22	Neutrinos – 1974 (Philadelphia)	74-82413	0-88318-121-5
No. 23	Particles and Fields – 1974 (APS/DPF Williamsburg)	74-27575	0-88318-122-3
No. 24	Magnetism and Magnetic Materials – 1974 (20th Annual Conference, San Francisco)	75-2647	0-88318-123-1
No. 25	Efficient Use of Energy (The APS Studies on the Technical Aspects of the More Efficient Use of Energy)	75-18227	0-88318-124-X

No.	Title	L.C. Number	ISBN
No. 26	High-Energy Physics and Nuclear Structure – 1975 (Santa Fe and Los Alamos)	75-26411	0-88318-125-8
No. 27	Topics in Statistical Mechanics and Biophysics: A Memorial to Julius L. Jackson (Wayne State University, 1975)	75-36309	0-88318-126-6
No. 28	Physics and Our World: A Symposium in Honor of Victor F. Weisskopf (M.I.T., 1974)	76-7207	0-88318-127-4
No. 29	Magnetism and Magnetic Materials – 1975 (21st Annual Conference, Philadelphia)	76-10931	0-88318-128-2
No. 30	Particle Searches and Discoveries – 1976 (Vanderbilt Conference)	76-19949	0-88318-129-0
No. 31	Structure and Excitations of Amorphous Solids (Williamsburg, VA, 1976)	76-22279	0-88318-130-4
No. 32	Materials Technology – 1976 (APS New York Meeting)	76-27967	0-88318-131-2
No. 33	Meson-Nuclear Physics – 1976 (Carnegie-Mellon Conference)	76-26811	0-88318-132-0
No. 34	Magnetism and Magnetic Materials – 1976 (Joint MMM-Intermag Conference, Pittsburgh)	76-47106	0-88318-133-9
No. 35	High Energy Physics with Polarized Beams and Targets (Argonne, 1976)	76-50181	0-88318-134-7
No. 36	Momentum Wave Functions – 1976 (Indiana University)	77-82145	0-88318-135-5
No. 37	Weak Interaction Physics – 1977 (Indiana University)	77-83344	0-88318-136-3
No. 38	Workshop on New Directions in Mossbauer Spectroscopy (Argonne, 1977)	77-90635	0-88318-137-1
No. 39	Physics Careers, Employment and Education (Penn State, 1977)	77-94053	0-88318-138-X
No. 40	Electrical Transport and Optical Properties of Inhomogeneous Media (Ohio State University, 1977)	78-54319	0-88318-139-8
No. 41	Nucleon-Nucleon Interactions – 1977 (Vancouver)	78-54249	0-88318-140-1
No. 42	Higher Energy Polarized Proton Beams (Ann Arbor, 1977)	78-55682	0-88318-141-X
No. 43	Particles and Fields – 1977 (APS/DPF, Argonne)	78-55683	0-88318-142-8
No. 44	Future Trends in Superconductive Electronics (Charlottesville, 1978)	77-9240	0-88318-143-6
No. 45	New Results in High Energy Physics – 1978 (Vanderbilt Conference)	78-67196	0-88318-144-4
No. 46	Topics in Nonlinear Dynamics (La Jolla Institute)	78-57870	0-88318-145-2
No. 47	Clustering Aspects of Nuclear Structure and Nuclear Reactions (Winnepeg, 1978)	78-64942	0-88318-146-0
No. 48	Current Trends in the Theory of Fields (Tallahassee, 1978)	78-72948	0-88318-147-9

No. 49	Cosmic Rays and Particle Physics – 1978 (Bartol Conference)	79-50489	0-88318-148-7
No. 50	Laser-Solid Interactions and Laser Processing – 1978 (Boston)	79-51564	0-88318-149-5
No. 51	High Energy Physics with Polarized Beams and Polarized Targets (Argonne, 1978)	79-64565	0-88318-150-9
No. 52	Long-Distance Neutrino Detection – 1978 (C.L. Cowan Memorial Symposium)	79-52078	0-88318-151-7
No. 53	Modulated Structures – 1979 (Kailua Kona, Hawaii)	79-53846	0-88318-152-5
No. 54	Meson-Nuclear Physics – 1979 (Houston)	79-53978	0-88318-153-3
No. 55	Quantum Chromodynamics (La Jolla, 1978)	79-54969	0-88318-154-1
No. 56	Particle Acceleration Mechanisms in Astrophysics (La Jolla, 1979)	79-55844	0-88318-155-X
No. 57	Nonlinear Dynamics and the Beam-Beam Interaction (Brookhaven, 1979)	79-57341	0-88318-156-8
No. 58	Inhomogeneous Superconductors – 1979 (Berkeley Springs, W.V.)	79-57620	0-88318-157-6
No. 59	Particles and Fields – 1979 (APS/DPF Montreal)	80-66631	0-88318-158-4
No. 60	History of the ZGS (Argonne, 1979)	80-67694	0-88318-159-2
No. 61	Aspects of the Kinetics and Dynamics of Surface Reactions (La Jolla Institute, 1979)	80-68004	0-88318-160-6
No. 62	High Energy e^+e^- Interactions (Vanderbilt, 1980)	80-53377	0-88318-161-4
No. 63	Supernovae Spectra (La Jolla, 1980)	80-70019	0-88318-162-2
No. 64	Laboratory EXAFS Facilities – 1980 (Univ. of Washington)	80-70579	0-88318-163-0
No. 65	Optics in Four Dimensions – 1980 (ICO, Ensenada)	80-70771	0-88318-164-9
No. 66	Physics in the Automotive Industry – 1980 (APS/AAPT Topical Conference)	80-70987	0-88318-165-7
No. 67	Experimental Meson Spectroscopy – 1980 (Sixth International Conference, Brookhaven)	80-71123	0-88318-166-5
No. 68	High Energy Physics – 1980 (XX International Conference, Madison)	81-65032	0-88318-167-3
No. 69	Polarization Phenomena in Nuclear Physics – 1980 (Fifth International Symposium, Santa Fe)	81-65107	0-88318-168-1
No. 70	Chemistry and Physics of Coal Utilization – 1980 (APS, Morgantown)	81-65106	0-88318-169-X
No. 71	Group Theory and its Applications in Physics – 1980 (Latin American School of Physics, Mexico City)	81-66132	0-88318-170-3
No. 72	Weak Interactions as a Probe of Unification (Virginia Polytechnic Institute – 1980)	81-67184	0-88318-171-1
No. 73	Tetrahedrally Bonded Amorphous Semiconductors (Carefree, Arizona, 1981)	81-67419	0-88318-172-X

No. 74	Perturbative Quantum Chromodynamics (Tallahassee, 1981)	81-70372	0-88318-173-8
No. 75	Low Energy X-Ray Diagnostics – 1981 (Monterey)	81-69841	0-88318-174-6
No. 76	Nonlinear Properties of Internal Waves (La Jolla Institute, 1981)	81-71062	0-88318-175-4
No. 77	Gamma Ray Transients and Related Astrophysical Phenomena (La Jolla Institute, 1981)	81-71543	0-88318-176-2
No. 78	Shock Waves in Condensed Matter – 1981 (Menlo Park)	82-70014	0-88318-177-0
No. 79	Pion Production and Absorption in Nuclei – 1981 (Indiana University Cyclotron Facility)	82-70678	0-88318-178-9
No. 80	Polarized Proton Ion Sources (Ann Arbor, 1981)	82-71025	0-88318-179-7
No. 81	Particles and Fields –1981: Testing the Standard Model (APS/DPF, Santa Cruz)	82-71156	0-88318-180-0
No. 82	Interpretation of Climate and Photochemical Models, Ozone and Temperature Measurements (La Jolla Institute, 1981)	82-71345	0-88318-181-9
No. 83	The Galactic Center (Cal. Inst. of Tech., 1982)	82-71635	0-88318-182-7
No. 84	Physics in the Steel Industry (APS/AISI, Lehigh University, 1981)	82-72033	0-88318-183-5
No. 85	Proton-Antiproton Collider Physics –1981 (Madison, Wisconsin)	82-72141	0-88318-184-3
No. 86	Momentum Wave Functions – 1982 (Adelaide, Australia)	82-72375	0-88318-185-1
No. 87	Physics of High Energy Particle Accelerators (Fermilab Summer School, 1981)	82-72421	0-88318-186-X
No. 88	Mathematical Methods in Hydrodynamics and Integrability in Dynamical Systems (La Jolla Institute, 1981)	82-72462	0-88318-187-8
No. 89	Neutron Scattering – 1981 (Argonne National Laboratory)	82-73094	0-88318-188-6
No. 90	Laser Techniques for Extreme Ultraviolt Spectroscopy (Boulder, 1982)	82-73205	0-88318-189-4
No. 91	Laser Acceleration of Particles (Los Alamos, 1982)	82-73361	0-88318-190-8
No. 92	The State of Particle Accelerators and High Energy Physics (Fermilab, 1981)	82-73861	0-88318-191-6
No. 93	Novel Results in Particle Physics (Vanderbilt, 1982)	82-73954	0-88318-192-4
No. 94	X-Ray and Atomic Inner-Shell Physics – 1982 (International Conference, U. of Oregon)	82-74075	0-88318-193-2
No. 95	High Energy Spin Physics – 1982 (Brookhaven National Laboratory)	83-70154	0-88318-194-0
No. 96	Science Underground (Los Alamos, 1982)	83-70377	0-88318-195-9

No. 97	The Interaction Between Medium Energy Nucleons in Nuclei – 1982 (Indiana University)	83-70649	0-88318-196-7
No. 98	Particles and Fields – 1982 (APS/DPF University of Maryland)	83-70807	0-88318-197-5
No. 99	Neutrino Mass and Gauge Structure of Weak Interactions (Telemark, 1982)	83-71072	0-88318-198-3
No. 100	Excimer Lasers – 1983 (OSA, Lake Tahoe, Nevada)	83-71437	0-88318-199-1
No. 101	Positron-Electron Pairs in Astrophysics (Goddard Space Flight Center, 1983)	83-71926	0-88318-200-9
No. 102	Intense Medium Energy Sources of Strangeness (UC-Sant Cruz, 1983)	83-72261	0-88318-201-7
No. 103	Quantum Fluids and Solids – 1983 (Sanibel Island, Florida)	83-72440	0-88318-202-5
No. 104	Physics, Technology and the Nuclear Arms Race (APS Baltimore –1983)	83-72533	0-88318-203-3
No. 105	Physics of High Energy Particle Accelerators (SLAC Summer School, 1982)	83-72986	0-88318-304-8
No. 106	Predictability of Fluid Motions (La Jolla Institute, 1983)	83-73641	0-88318-305-6
No. 107	Physics and Chemistry of Porous Media (Schlumberger-Doll Research, 1983)	83-73640	0-88318-306-4
No. 108	The Time Projection Chamber (TRIUMF, Vancouver, 1983)	83-83445	0-88318-307-2
No. 109	Random Walks and Their Applications in the Physical and Biological Sciences (NBS/La Jolla Institute, 1982)	84-70208	0-88318-308-0
No. 110	Hadron Substructure in Nuclear Physics (Indiana University, 1983)	84-70165	0-88318-309-9
No. 111	Production and Neutralization of Negative Ions and Beams (3rd Int'l Symposium, Brookhaven, 1983)	84-70379	0-88318-310-2
No. 112	Particles and Fields – 1983 (APS/DPF, Blacksburg, VA)	84-70378	0-88318-311-0
No. 113	Experimental Meson Spectroscopy – 1983 (Seventh International Conference, Brookhaven)	84-70910	0-88318-312-9
No. 114	Low Energy Tests of Conservation Laws in Particle Physics (Blacksburg, VA, 1983)	84-71157	0-88318-313-7
No. 115	High Energy Transients in Astrophysics (Santa Cruz, CA, 1983)	84-71205	0-88318-314-5
No. 116	Problems in Unification and Supergravity (La Jolla Institute, 1983)	84-71246	0-88318-315-3
No. 117	Polarized Proton Ion Sources (TRIUMF, Vancouver, 1983)	84-71235	0-88318-316-1

No.	Title	LC Number	ISBN
No. 118	Free Electron Generation of Extreme Ultraviolet Coherent Radiation (Brookhaven/OSA, 1983)	84-71539	0-88318-317-X
No. 119	Laser Techniques in the Extreme Ultraviolet (OSA, Boulder, Colorado, 1984)	84-72128	0-88318-318-8
No. 120	Optical Effects in Amorphous Semiconductors (Snowbird, Utah, 1984)	84-72419	0-88318-319-6
No. 121	High Energy e^+e^- Interactions (Vanderbilt, 1984)	84-72632	0-88318-320-X
No. 122	The Physics of VLSI (Xerox, Palo Alto, 1984)	84-72729	0-88318-321-8
No. 123	Intersections Between Particle and Nuclear Physics (Steamboat Springs, 1984)	84-72790	0-88318-322-6
No. 124	Neutron-Nucleus Collisions – A Probe of Nuclear Structure (Burr Oak State Park - 1984)	84-73216	0-88318-323-4
No. 125	Capture Gamma-Ray Spectroscopy and Related Topics – 1984 (Internat. Symposium, Knoxville)	84-73303	0-88318-324-2
No. 126	Solar Neutrinos and Neutrino Astronomy (Homestake, 1984)	84-63143	0-88318-325-0
No. 127	Physics of High Energy Particle Accelerators (BNL/SUNY Summer School, 1983)	85-70057	0-88318-326-9
No. 128	Nuclear Physics with Stored, Cooled Beams (McCormick's Creek State Park, Indiana, 1984)	85-71167	0-88318-327-7
No. 129	Radiofrequency Plasma Heating (Sixth Topical Conference, Callaway Gardens, GA, 1985)	85-48027	0-88318-328-5
No. 130	Laser Acceleration of Particles (Malibu, California, 1985)	85-48028	0-88318-329-3
No. 131	Workshop on Polarized ^{3}He Beams and Targets (Princeton, New Jersey, 1984)	85-48026	0-88318-330-7
No. 132	Hadron Spectroscopy–1985 (International Conference, Univ. of Maryland)	85-72537	0-88318-331-5
No. 133	Hadronic Probes and Nuclear Interactions (Arizona State University, 1985)	85-72638	0-88318-332-3
No. 134	The State of High Energy Physics (BNL/SUNY Summer School, 1983)	85-73170	0-88318-333-1
No. 135	Energy Sources: Conservation and Renewables (APS, Washington, DC, 1985)	85-73019	0-88318-334-X
No. 136	Atomic Theory Workshop on Relativistic and QED Effects in Heavy Atoms	85-73790	0-88318-335-8
No. 137	Polymer-Flow Interaction (La Jolla Institute, 1985)	85-73915	0-88318-336-6
No. 138	Frontiers in Electronic Materials and Processing (Houston, TX, 1985)	86-70108	0-88318-337-4
No. 139	High-Current, High-Brightness, and High-Duty Factor Ion Injectors (La Jolla Institute, 1985)	86-70245	0-88318-338-2

No. 140	Boron-Rich Solids (Albuquerque, NM, 1985)	86-70246	0-88318-339-0
No. 141	Gamma-Ray Bursts (Stanford, CA, 1984)	86-70761	0-88318-340-4
No. 142	Nuclear Structure at High Spin, Excitation, and Momentum Transfer (Indiana University, 1985)	86-70837	0-88318-341-2
No. 143	Mexican School of Particles and Fields (Oaxtepec, México, 1984)	86-81187	0-88318-342-0
No. 144	Magnetospheric Phenomena in Astrophysics (Los Alamos, 1984)	86-71149	0-88318-343-9
No. 145	Polarized Beams at SSC & Polarized Antiprotons (Ann Arbor, MI & Bodega Bay, CA, 1985)	86-71343	0-88318-344-7
No. 146	Advances in Laser Science–I (Dallas, TX, 1985)	86-71536	0-88318-345-5
No. 147	Short Wavelength Coherent Radiation: Generation and Applications (Monterey, CA, 1986)	86-71674	0-88318-346-3
No. 148	Space Colonization: Technology and The Liberal Arts (Geneva, NY, 1985)	86-71675	0-88318-347-1
No. 149	Physics and Chemistry of Protective Coatings (Universal City, CA, 1985)	86-72019	0-88318-348-X
No. 150	Intersections Between Particle and Nuclear Physics (Lake Louise, Canada, 1986)	86-72018	0-88318-349-8
No. 151	Neural Networks for Computing (Snowbird, UT, 1986)	86-72481	0-88318-351-X
No. 152	Heavy Ion Inertial Fusion (Washington, DC, 1986)	86-73185	0-88318-352-8
No. 153	Physics of Particle Accelerators (SLAC Summer School, 1985) (Fermilab Summer School, 1984)	87-70103	0-88318-353-6
No. 154	Physics and Chemistry of Porous Media—II (Ridge Field, CT, 1986)	83-73640	0-88318-354-4
No. 155	The Galactic Center: Proceedings of the Symposium Honoring C. H. Townes (Berkeley, CA, 1986)	86-73186	0-88318-355-2
No. 156	Advanced Accelerator Concepts (Madison, WI, 1986)	87-70635	0-88318-358-0
No. 157	Stability of Amorphous Silicon Alloy Materials and Devices (Palo Alto, CA, 1987)	87-70990	0-88318-359-9
No. 158	Production and Neutralization of Negative Ions and Beams (Brookhaven, NY, 1986)	87-71695	0-88318-358-7

No. 159	Applications of Radio-Frequency Power to Plasma: Seventh Topical Conference (Kissimmee, FL, 1987)	87-71812	0-88318-359-5
No. 160	Advances in Laser Science–II (Seattle, WA, 1986)	87-71962	0-88318-360-9
No. 161	Electron Scattering in Nuclear and Particle Science: In Commemoration of the 35th Anniversary of the Lyman-Hanson-Scott Experiment (Urbana, IL, 1986)	87-72403	0-88318-361-7
No. 162	Few-Body Systems and Multiparticle Dynamics	87-72594	0-88318-362-5
No. 163	Pion–Nucleus Physics: Future Directions and New Facilities at LAMPF (Los Alamos, NM, 1987)	87-72961	0-88318-363-3
No. 164	Nuclei Far from Stability: Fifth International Conference (Rosseau Lake, ON, 1987)	87-73214	0-88318-364-1
No. 165	Thin Film Processing and Characterization of High-Temperature Superconductors	87-73420	0-88318-365-X
No. 166	Photovoltaic Safety (Denver, CO, 1988)	88-42854	0-88318-366-8